GENERAL GEOLOGY

Fourth Edition

Robert J. Foster

San Jose State University

CHARLES E. MERRILL PUBLISHING COMPANY
A Bell & Howell Company
Columbus   Toronto   London   Sydney

Published by Charles E. Merrill Publishing Company
A Bell & Howell Company
Columbus, Ohio 43216

This book was set in Univers
Production Editor: Cynthia Brunk
Cover Designer: Tony Faiola
Cover Photograph: David Muench

Library of Congress Catalog Card Number: 82–62186
International Standard Book Number: 0–675–20020–2
Printed in the United States of America
1 2 3 4 5 6 7 8—87 86 85 84 83

PREFACE

FOCUS

This book is about the solid rock earth, especially the processes that formed its surface features. In this sense it covers classical geology, but in order to understand these features, it will be necessary to consider evidence from a wide range of fields. Thus, the viewpoint is wide, and many of the topics would not have appeared in a geology text only a few years ago. This broadness, however, makes modern geology a more valuable course for the nonscience major.

Most of the changes in this revision are in response to a survey, and it is hoped that they make the book more interesting, more timely, and more useful to both the student and the instructor. The introductory chapter has been revised and retitled, and now introduces plate tectonics. This enables the rock chapters to relate the various rock types to plate tectonics. The classification of pyroclastics has been revised, the discussion of carbonate rocks has been expanded, and a discussion of metamorphic facies has been added. In the mineral resource chapter, metal deposits can now be related to plate tectonics, and the discussion of energy resources has been revised.

The erosion chapters have also been revised, and some have new titles. The environmental aspects of river erosion have been placed at the end of that chapter, so that they do not interrupt the discussion of geologic factors. The discussion of sand dunes has been revised and expanded. The discussion of the causes of glaciation has also been expanded. Shoreline development has been revised and rewritten. The ground water chapter has been revised, and the discussion of water resources has been shortened.

Major changes were made in the structural geology chapters. The original four chapters are now three chapters. The earthquake chapter has been revised and slightly expanded. A new chapter on plate tectonics contains much new material. The continental structures chapter has been tied more closely to plate tectonics.

The geologic time chapter has been retitled and revised, with unconformities and relative dating expanded. The chapter on astronomical aspects of geology has been revised so that even with the coverage of recent discoveries about the moons of Jupiter and Saturn, it is still about the same length. The space program has made it possible to compare the earth with other planets and, perhaps more important, to see more clearly that the earth is finite.

To make the book more interesting and more timely, geological aspects of environment are included in every chapter to help the student see better the many ways in which geology affects our lives. If today's students understand the environment, they may be able to make our earth a better place. Brief historical summaries at the beginnings of the chapters will help the general education student view the development of science as part of our overall intellectual development, as well as help the geology major better understand the development of geologic concepts. Hopefully, the addition of environment and history will help to humanize geology.

Concern for our environment and realization that our resources are limited have made geology courses more popular and relevant today. Unfortunately, many students can take only one geology course, which generally emphasizes physical geology, and so miss historical geology. Historical geology reinforces some of geology's most important concepts, such as the immensity of geologic time, the constantly changing surface of the earth, and how this changing environment has affected life. These ideas are very important in understanding our present environment and in keeping the earth a nice place on which to live. In this text, the essentials of both physical and historical geology are presented in manageable length.

ORGANIZATION—FLEXIBLE USAGE

As in previous editions, this revision begins with minerals and rocks so that they can be introduced in the early laboratory periods, if the course has a laboratory. Then erosion is described, and next, structural geology is covered. Our survey revealed that this is the most favored sequence of topics, but other sequences are used in almost as many courses.

The chapters of this book are designed so that they can be used in almost any sequence. Mineral resources and geologic time can be used at almost any point in the course, and weathering, the first part of Chapter 4, can be studied with either sedimentary rocks or with erosion.

As in the earlier editions, the historical geology chapters are organized around the concept of sedimentary cycles. Recent research has refined these cycles and has extended their recognition world wide.

LEARNING AIDS

In this revision, every effort has been made to make this book more useful and more appealing to the student. Terms are now emphasized in boldface color type where they are introduced, and their definitions are nearby in italics. At the end of each chapter, as in earlier editions, is a summary, which includes both terms and principles. The questions at the end of each chapter are usually simple and factual to help the student review and test retention, although a few go much further. Complete chapter outlines and a detailed table of contents are included. A light color tint on the glossary pages makes its use easier, thus encouraging its use. These features will make study easier for the student.

The Supplementary Reading section at the end of each chapter stresses articles in *Scientific American* and other current periodicals, as well as books available in paperback editions. Such readily available, inexpensive materials may encourage more outside reading.

ACKNOWLEDGMENTS

The preparation of this fourth edition was made easier by the help of many people. My wife, Joan, edited, proofread, typed, and aided in countless other ways. Pamela Cooper, Margaret Malde, and Cynthia Brunk at Charles E. Merrill worked hard on every stage of planning, writing, editing, and publication. I am also indebted to the many individuals and organizations that contributed photographs and other illustrations. They are acknowledged throughout the book. In addition to those who read the earlier editions, the physical geology portion of this fourth edition was reviewed by Kenneth D. Hopkins, University of Northern Colorado; W. Calvin James, Vanderbilt University; Raymond K. Moore, Radford University; Donal M. Ragan, Arizona State University; and Richard E. Uhte, University of Minnesota. The astronomy chapter was read by Paul D. Loman, Jr. of NASA. The historical geology chapters were read by Raymond Sullivan, San Francisco State University, and James W. Trow, Michigan State University. Any inaccuracies that remain can only be attributed to my not heeding their comments.

CONTENTS

6 MINERAL RESOURCES

7 DOWNSLOPE MOVEMENT OF SURFACE MATERIAL

12 GROUND WATER AND WATER RESOURCES

13 EARTHQUAKES AND THE EARTH'S INTERIOR

14 PLATE TECTONICS—OCEANS AND CONTINENTAL DRIFT

15 THE CONTINENTS

16 GEOLOGIC TIME— DATING GEOLOGIC EVENTS

17 EVOLUTION AND FOSSILS

18 EARLY HISTORY OF THE EARTH— ASTRONOMICAL ASPECTS OF GEOLOGY

19 THE PRECAMBRIAN—THE OLDEST ROCKS AND THE FIRST LIFE

20 EARLY PALEOZOIC— CAMBRIAN, ORDOVICIAN, SILURIAN

APPENDICES

1 PLANET 3 – AN OVERVIEW

A generation comes, a generation goes, but the Earth remains forever.

Ecclesiastes 1:4

WHY STUDY THE EARTH?

Geology is the study of the earth. That we live on the earth is reason enough to study it. The more that we know about our planet, especially its environment and resources, the better we can understand, use, and appreciate it. To us the earth is the most important body in the universe.

In the broader view, however, the importance of the earth shrinks. At least a half dozen stars are known to have faint, unseen companions, but it is not known whether these companions are very faint stars or massive planets revolving about the observed star. Thus the earth appears to be an average planet. However, at least in the solar system, *the earth is unique in having abundant water, exemplified by the oceans, and an atmosphere that can support life.* The earth's surface temperature, controlled largely by its distance from the sun, makes these features possible, and these features, in turn, make life possible on the earth. The development and history of life are important aspects of historical geology. The space program has also revealed that the earth is unique among the earthlike planets so far studied in having a magnetic field. The earth's magnetic field, for reasons that will be discussed later, is believed to be caused by a liquid iron core that may also store the energy which causes the formation of such surface features as mountain ranges. These features, which, as far as we know, are unique to the earth, are central to the processes of erosion and deformation of the earth's surface that are the main aspects of physical geology. Thus geology is in large part the study of the consequences of the earth's unique features.

Geology has contributed a great deal to civilization both intellectually and economically. Among the great concepts gained from geological studies are an understanding of the great age of the earth and the development of an absolute time scale. *Geology differs from most other sciences in that it is concerned with absolute time.* Time appears in the equations of physics and chemistry, but these sciences are generally concerned with rates of change, and the time is relative, not absolute. Geologic time extends back almost 5000 million years to when the earth formed. Thus geology is concerned with immense lengths of time when measured against human experience. It is difficult to comprehend the lengths of time involved in geologic processes, but this must be done to appreciate geology fully. (See Fig. 1–1.)

The fundamental principle that underlies most of geology is simply that the present processes occurring on the earth have occurred throughout geologic time. Thus ancient rocks can be interpreted in terms of present processes. This is called the principle of *uniformitarianism* (also called *actualism).* The other basic principles of geology are equally simple and self-evident. Layers of sediments deposited in a lake or the ocean are originally deposited in more or less horizontal layers. This is termed *original horizontality;* and if the beds or layers are not horizontal, later deformation is implied. The third principle is called *superposition* and states that if a series of beds or layers of sedimentary rocks have not been disturbed, the oldest beds are at the bottom and the youngest at the top.

Another important point learned from geology is that constant change, both biological and physical, has been and is occurring on the earth.

The origin and development of life is part of geology. It is closely related to the history and development of the earth's surface and thus cannot be separated from the physical history of the earth. Geology shows that, in the broadest sense, all life is related. Biologists share this concern with life, but much of the evidence is geologic in nature. Thus geology and biology overlap in part.

The economic contributions of geology to civilization show that in many ways, too, geology is a very practical science. *Geologic knowledge is used to locate and to exploit our mineral resources.* Except for water and soil, all mineral resources, such as sand and gravel, petroleum, coal, and metals, are nonrenewable. Once mined they are gone, and new deposits must be found. Geologists have discovered the deposits of metal and energy-producing minerals on which our civilization is based. We take these things for granted now; but a hundred years ago, when the American West was opening up and the industrial revolution was occurring, these mineral deposits were being discovered at a rapid rate and geologists were the most influential scientists of the day. At this time, too, the principles of geology were being formulated.

Today the study of mineral resources is of utmost importance, but geologists are also concerned

A group photo showing members of Hayden Survey taken at Red Buttes, Wyoming, August 24, 1870. In about a ten-year period, the Hayden Survey explored most of the Rocky Mountains geologically. The early geologic surveys of the west resulted in many concepts that helped the development of the science of geology. This scene contrasts sharply with today's space program, which is one of today's geologic frontiers. Photo by W. H. Jackson, U. S. Geological and Geographical Survey of the Territories.

1

ERA	APPROX. AGE IN MILLIONS OF YEARS (RADIOACTIVITY)	PERIOD OR SYSTEM *Period refers to a time measure; system refers to the rocks deposited during a period.*
CENOZOIC	7 26 37–38 53–54	Recent (Holocene) Pleistocene } Neogene Tertiary { Pliocene / Miocene / Oligocene / Eocene / Paleocene } Paleogene
MESOZOIC	65 136 190–195	Cretaceous Jurassic Triassic
PALEOZOIC	225 280 310 345 395 430–440 500 570	Permian Pennsylvanian } Mississippian } Carboniferous Devonian Silurian Ordovician Cambrian
PRECAMBRIAN	700 3,500 4,000 4,500	*First multicelled organisms* *First one-celled organisms* *Approximate age of oldest rocks discovered* *Approximate age of meteorites*

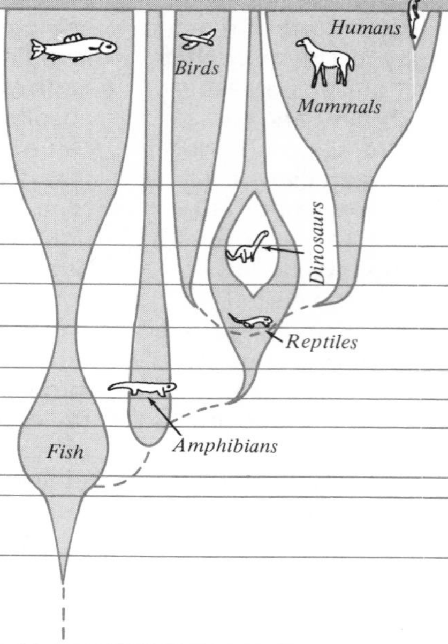

FIGURE 1–1 The geologic time scale. Shown to the right above is a very simplified diagram showing the development of life. Not included on the diagram are many types of invertebrate fossils such as clams, brachiopods, corals, sponges, snails, and so forth, which first appeared in the Cambrian or Ordovician and have continued to the present. This figure is discussed in Chapter 16.

with other economic problems, such as urbanization. The development of large cities has resulted in the building of large structures, such as tall buildings and dams. Geology helps in designing foundations for these structures. Examples of both large and small structures that have failed through neglect of simple geologic principles, easily understood by elementary students, are common. (See Fig. 1–2.) Dams fail because they are built near active faults or on porous foundations. During their first rainy season, new highways are washed out or blocked by landslides. Homes built on hillsides are destroyed by landslides and mudflows. Geologists have also recognized the need for earthquake-resistant structures in some areas and have helped in

FIGURE 1–2 Landsliding at Point Firmin, California. Geologic studies can prevent losses such as this. Spence air photo.

their design. These and many other geological aspects of our environment will be described. (See Fig. 1–3.)

SCOPE OF GEOLOGY

Geology is concerned both with the processes operating in and on the earth, and with the history of the earth including the history of life. In the broadest sense, geology includes the study of the continents, the oceans, the atmosphere, and the earth's magnetic and radiation fields. Clearly, this scope is too broad for any one scientist, so geologists generally, but not exclusively, limit themselves to the solid earth that can be studied directly. Other specialties have developed to study the other aspects of the earth. Geophysicists study the deep parts of the earth and its fields, mainly by indirect methods; oceanographers study the hydrosphere; and meteorologists study the atmosphere.

Even with this restriction, geology is a very broad field. Most geologists specialize in one or more facets of geology, much as engineers specialize in various fields of physical science such as electronics or construction. However, geology is even broader than engineering because it encompasses both physical and biological science.

Mention of a few of the specialties in geology will illustrate. Those who study minerals and rocks need specialized training in chemistry and physics, as does the geochemist who is concerned with chemical processes in the earth. Those who study fossils must be trained in biology of plants and animals, both vertebrate and invertebrate. Those who study deformed rocks must know mechanics. Ground water and petroleum geologists must be familiar with hydrodynamics. A complete listing would be very long, but these examples will illustrate the point. All of these specialties overlap somewhat, and, no matter

what the specialty, a geologist must be familiar with all facets of geology.

THE MOBILE EARTH

The earth is a dynamic planet whose surface is constantly undergoing change. Given the length of geologic time, one can comprehend that the hills and mountains are not eternal features, but that eventually they will be worn flat by erosion. Earthquakes and volcanoes reveal a restless interior. Deformed rocks that have been bent into folds and broken also bear mute testimony to the power of the earth's internal energy.

CONTINENTAL DRIFT

The idea that the continents have moved around on the earth's surface is a very old one. Almost every school child notices that the continents fit together like the pieces of a jigsaw puzzle. The fit is not perfect (see Fig. 1–4), but ero-

FIGURE 1–3   Urban pressures in the area south of San Francisco resulted in homes being built between 1956 and 1966 in the San Andreas fault zone and too near landslides. A number of the homes shown in the 1966 photo had to be removed because of damage. The Chapter 13 opening photo shows an oblique aerial view of this area. Photos from U.S. Geological Survey.

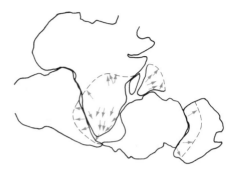

FIGURE 1–4 Reconstruction of the ancient continent of the southern hemisphere. The glaciers and their directions of movement are indicated.

sion and deformation could have distorted the pieces. The fit of the continents involves more than just shape; the rock layers and types also match. Again, the fit is not perfect because of erosion and deformation. The ages of the rocks and their structures also are similar.

In the nineteenth century, geologic studies in the southern continents revealed other kinds of evidence. Large areas of glacial deposits and glacially eroded bedrock were discovered in India, South America, Africa, and Aus-

tralia. The glaciers were continental glaciers such as those in the northern continents during the much more recent Ice Age. The glacial deposits on the southern continents are all about 250 million years old. These glacial rocks are now found on both sides of the equator in areas of warm climates. (See Fig. 1–5.) Even if the south pole is moved to the most favorable location among these ancient glaciers, some would still have been near the equator. However, if the continents are joined as in Fig. 1–4, the glacial areas and the directions of ice movement recorded in the rocks all match well. The rocks associated with the glacial deposits are also similar on all the southern continents. On each of these continents, the rocks contain similar fossils of both plants and reptiles. These organisms could not have migrated across the seas that presently separate these continents.

Therefore, the evidence is strong that at least the southern continents were once a single continent, and that this supercontinent broke up and the pieces drifted apart. Although this concept was widely accepted by geologists in the southern continents, it was not believed by most American and European geologists. Their lack of acceptance was, in part, due to the fact that rocks in the northern continents do not match so clearly as those in the south, but mostly because no mechanism that could cause continental drift was known.

SEA-FLOOR SPREADING

Sea-floor spreading is a much newer idea than continental drift. Although it had been proposed much earlier, it was finally accepted in the 1960s when the evidence for it became overwhelming. All of the ocean bottoms have ridges or rises in them, called *mid-ocean ridges.* These are areas of volcanic activity and shallow earthquakes. *New ocean floor is created at the ridges by the volcanic activity.* (See Fig. 1–6.) This is shown by the age and thickness of the rocks on the ocean floors, which have been sampled by drill-

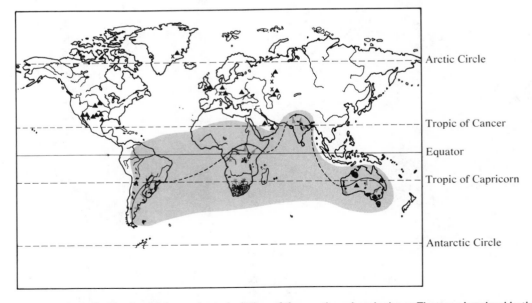

FIGURE 1–5 Map showing the present distribution of the ancient glaciation of the southern hemisphere. The area involved in the hypothetical southern continent existing before drifting is stippled. The areas of glacial deposits are indicated in black and extend from north of the equator to near the south pole. Reefs (crosses) and evaporites (triangles) far to the north indicate much warmer climates. Reefs and evaporites from Opdyke, 1962.

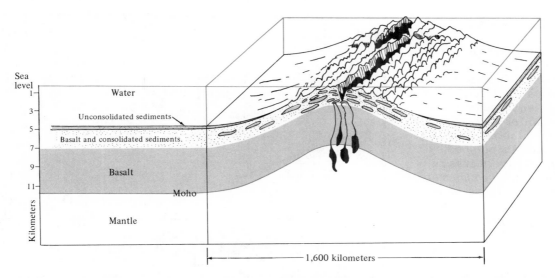

FIGURE 1–6   Mid-ocean ridge. A cross-section of typical oceanic platform is shown on the left. The ocean crust thins at the ridge. Note the vertical exaggeration.

ing. The age of the volcanic bedrock of the oceans is progressively older as one moves away from the ridges in both directions. In the same way, the sediments above the bedrock become thicker, and the deepest sediments older, with distance from the mid-ocean ridges. The conclusion seems inescapable that new ocean floors are being created at the mid-ocean ridges, and that the ocean floors are moving away from the ridges. The rate of movement is only a few centimeters per year, but given the earth's great age, the total movement is very large.

If new ocean floor is created at the mid-ocean ridges, then the ocean floors must be consumed somewhere; otherwise, the earth must expand. *The volcanic arcs are places where the ocean floor apparently returns to the interior of the earth.* (See Fig. 1–7.) At many places the volcanic arcs form volcanic island arcs, islands formed by volcanoes that appear as curves or arcs on a map. The volcanic arcs are geologically active areas with deep earthquakes as well as the volcanic activity.

It is believed that the ocean floor returns to the interior of the

earth at the volcanic arcs because the earthquakes occur on a plane sloping under the volcanic arc. (See Fig. 1–7.) Movement of a slab of ocean floor would cause the earthquakes. The volcanoes could be the result of partial melting of the slab of ocean floor. Downward movement of the ocean floor is also suggested by the deep submarine trenches in front of volcanic arcs. (See Fig. 1–7.) Thus, *in sea-floor spreading new ocean floor is created at the mid-ocean ridges and moves slowly toward the volcanic arcs where it is consumed.*

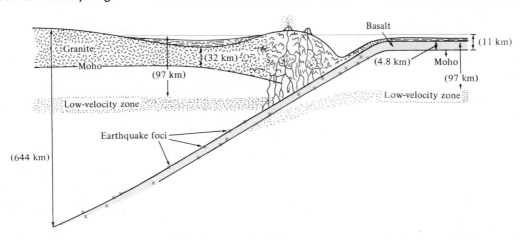

FIGURE 1–7   Volcanic island arc, showing location of earthquake foci. Because of the distances involved, this figure is not drawn to scale.

PLATE TECTONICS

Continental drift and sea-floor spreading come together as parts of *plate tectonics. In plate tectonics, the surface of the earth is believed to be composed of a number of thin slabs or plates that move across the surface.* (See Fig. 1–8.) The plates are outlined by the locations of earthquakes as shown in Fig. 1–8. *The plates are formed at the mid-ocean ridges, and they return to the earth's interior at convergent plate boundaries such as the volcanic island arcs. Where the plates pass each other, earthquakes are also caused by the motion, and these plate boundaries are called transform faults.*

The continents are made of granite, a rock that is less dense than the ocean-floor rocks. *The continents ride atop the ocean-floor basalt, and so as the plates move, they carry the continents with them.* (See Fig. 1–9.) In this way, the same mechanism that causes sea-floor spreading also is the cause of continental drift. Movements of the continents through geologic time are shown in Fig. 1–10.

As the plates move across the earth's surface, they collide from time to time. Apparently, at times new mid-ocean ridges form. The type of collision is determined by the composition of the plates at the point of collision. *If ocean floor collides with continent, the ocean floor is consumed at the resulting convergent plate boundary.* (See Fig. 1–11.) *If ocean floor meets ocean floor, one plate is consumed and a volcanic island arc forms.* (See Fig. 1–12.) *If continent collides with continent, a mountain range is formed.* (See Fig. 1–13.) In all of these collisions, the rocks are deformed by the collision. The rocks, especially the sedimentary rocks that form in layers, are bent and folded and at places they are broken and faulted as well. Igneous rocks are also formed where plates are con-

sumed, and the rising igneous rocks cause metamorphism and deformation. This, then, is the origin of the folded and faulted rocks that form the earth's mountain ranges. In our exploration of other planets, no such structures have been yet found.

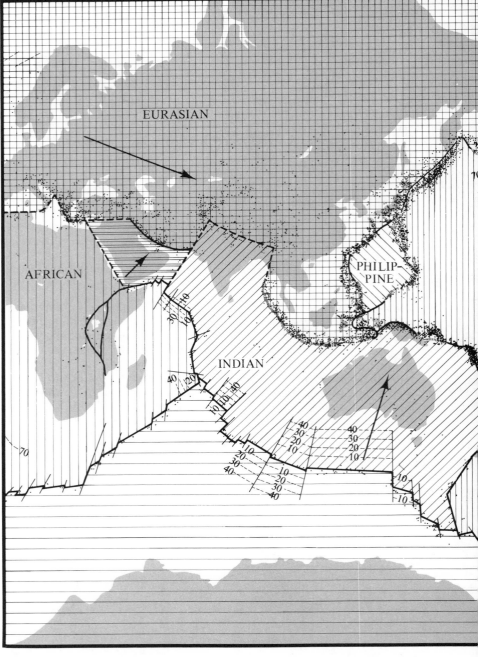

FIGURE 1–8 The crustal plates. In general, each plate begins at a ridge (heavy line) and moves toward a trench (hatching). Earthquake epicenters (dots) tend to outline the plates. Transform faults are shown by thin lines. The dashed lines show the ages (in

DISCOVERING THE EARTH'S PLACE IN THE UNIVERSE

The science of geology developed around the end of the eighteenth century but long before then, the

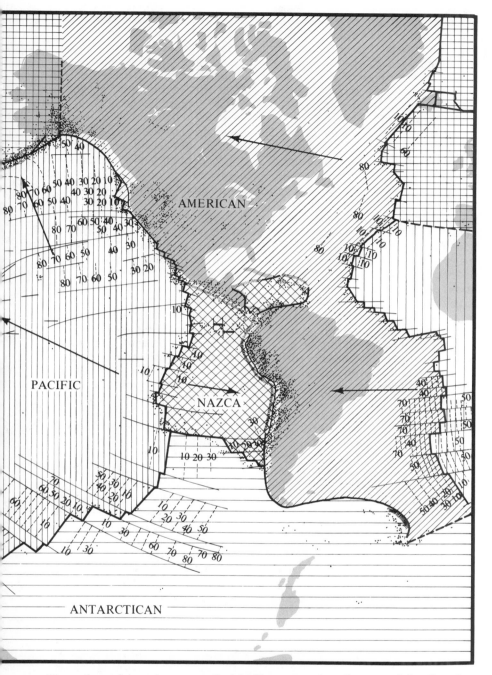

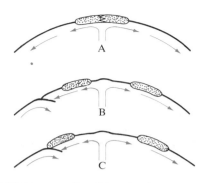

FIGURE 1–9 Possible role of sea-floor spreading or convection currents in continental drift. Formation of ridge beneath continent could result in splitting of the continent and movement to a trench.

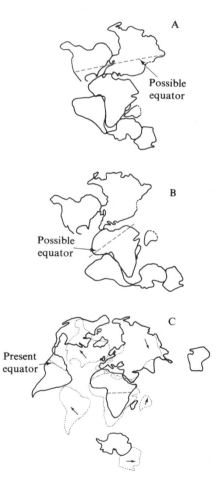

FIGURE 1–10 Movement of continents. A. Late Paleozoic time, 250 million years ago. B. End of Jurassic Period, 136 million years ago. C. End of Cretaceous Period, 65 million years ago, is shown dotted and present positions are shown by the solid lines. Based on R. S. Deitz and J. C. Holden, 1970.

millions of years), based on magnetic data. The arrows show the general direction of movement of the plates. The information and interpretation are from many sources, mainly workers at Lamont-Doherty Geological Observatory.

earth's place in the universe was explored. In many early civilizations, the priests kept track of the movements of the sun, moon, and planets relative to the stars. This was of practical value, as well as religious, because it led to the development of calendars and pre-

diction of such events as eclipses. The Greeks went further and developed a system to explain the motions of the heavens. This was the first big step in finding the earth's place in the universe.

The work of the Greeks was summarized in 140 A.D. by Ptol-

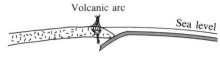

FIGURE 1–11   Continent-ocean plate collision.

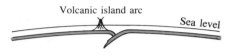

FIGURE 1–12   Ocean-ocean plate collision.

FIGURE 1–13   Continent-continent plate collision.

emy. In the Ptolemaic system, the earth was at the center of the universe and all of the other objects in the solar system—the sun, moon, and planets—revolved around the earth. Although not all of the earlier Greeks would have agreed with this system, the great advantage of Ptolemy's work was that it could be used to predict future movements of the sky. Thus, it was of great use in astrology and navigation. Columbus and the early explorers used the Ptolemaic system.

The first important challenge to the idea that the earth and humanity's place was at the center of the universe came at the Renaissance about 1400 years after Ptolemy. Putting the sun at the center and having the planets revolve around the sun made a much simpler system in the sense that fewer motions are necessary. This system was published posthumously in 1543 by Nicholas Copernicus. Moving humans away from the center created a controversy that raged for a hundred years because proof of the earth's daily rotation and yearly revolution came much later. The main advantage of the Copernican system over the Ptolemaic system was that it was simpler. Both had many inaccuracies in prediction. In

the early part of the seventeenth century, Johannes Kepler, using data collected by Tycho Brahe, was able to describe the motions of planets and showed that Copernicus was basically correct.

Galileo Galilei, an Italian, was a contemporary of Brahe and Kepler. He worked in many fields, but only parts of his studies of astronomy and mechanics will be mentioned. In 1609, Galileo built one of the first telescopes and used it to observe the heavens. He saw things that no one had ever seen before, and many of his observations could not be explained by the Ptolemaic system. He announced his discoveries in two books but was later forced by the Inquisition to recant. As a result, he had to spend the last years of his life in virtual house arrest. It was during this period that he carried out many of his experiments in mechanics, experiments which later helped Newton.

Isaac Newton, an English physicist, using the work of Kepler and Galileo, discovered the laws of mechanics and of gravity. He worked out these laws about 1664 and published them in 1686. It was then possible to understand why the planets orbit the sun. This discovery profoundly influenced eighteenth century thought, and it was widely believed that almost everything could be explained by Newton's laws. It was not until the early part of the twentieth century that Einstein showed that Newton's laws do not apply in all situations.

Thus by Newton's time, the earth's place in the sun's family was established. The distance to the sun, 150 million kilometers, was measured in 1672. But where is the sun in the universe? New instruments, especially big telescopes, had to be built before this question could be answered and our place in the universe known, if indeed it can ever be fully known.

The first step out of the solar system was to measure the dis-

tance to some nearby stars. This was not done until 1838 when Bessel measured the very small change in angle to a nearby star, as the earth moves from one side of its orbit to the other. This geometric method can only be used to an angle of 0.008 second of arc which corresponds to a distance of about 400 light-years. Beyond this distance the angle is too small to measure accurately. The closest star is about 4 light-years away.

It was not until the early twentieth century that greater distances could be measured. During much of the nineteenth century, spectroscopic analysis was developed and applied to starlight. In 1911, Ejner Hertzsprung in Holland and H. N. Russell in the United States noted that there is a relationship between spectral type and absolute brightness of a star. The absolute brightness can be calculated from the apparent or measured brightness for those stars whose distance is known. It then became possible to calculate the distance from the apparent brightness of any star whose spectral type is known, because from the spectral type, one knows the absolute brightness.

The size of the known universe was growing rapidly, but for the most part, the distances measured were to stars in our galaxy, the Milky Way. (All of the stars seen with the naked eye are within our galaxy, although some more distant objects such as the Andromeda galaxy can be seen without a telescope.) During the nineteenth century, astronomers noted many objects whose light was soft and fuzzy. By 1850, several spiral galaxies had been noted, but their nature was in doubt. In 1908, some stars were resolved in some of these objects, suggesting that they might be vast groups of stars like the Milky Way. It was not until 1925, when the new 100 inch (254-centimeter) telescope at Mount Wilson became available, that Edwin Hubble was able to report that these objects were composed of many stars. Objects the size of the Milky Way, so distant that they appear to be stars except in the largest telescopes, reveal a universe whose size is beyond comprehension. The diameter of the Milky Way itself is 80,000 light-years.

In 1912, Henrietta Leavitt made a discovery that enabled measurement of the distance to a galaxy. She observed that there is a relationship between the brightness and the period of a type of variable star called a Cepheid. Using this, as well as other methods, the size of the known universe quickly expanded to over a 1000 million light-years. The total number of galaxies may be as great as 100,000 million.

Perhaps the most astonishing thing revealed was Hubble's discovery in 1929 that all of these galaxies are moving away from us—that is, the universe is expanding. The further away a galaxy, the faster it is receding from us. This motion is measured from the *red shift,* a movement of the lines in a galaxy's spectrum toward the red. This change in frequency is similar to the change in pitch of a train whistle as the train moves away from the listener.

Thus, in about 500 years, our perception of the earth has changed from its being the center

of the universe to its being a small planet orbiting an average star in one of perhaps 100,000 million galaxies. (See Fig. 1–14.) Research continues in astronomy, and new, strange objects (quasistellar objects—QSO's) have been discovered. Some of these QSO's emit radiowaves, and some have very large red shifts, suggesting very high velocities and perhaps great distances. These objects, or others yet to be discovered, may again change our ideas on the earth's place in the universe.

EXPLORING THE EARTH

Concurrent with finding our place in the universe, we also explored our own planet. The Crusades, which lasted from 1095 to 1291, marked the end of the Dark Ages. In the Crusades, the Europeans discovered that the people of the Near East were in many ways more advanced than the Europeans. Marco Polo also described much more advanced peoples in the Far East. He returned to Venice about 1300 after a trip, mainly overland, to China and India and wrote a book describing the things he had seen. Although he attracted much interest, he was not widely believed because few peo-

ple thought that the Orient could be more advanced than Europe. At this time very little was known of the world outside of Europe, although missionaries had ventured into Africa and Asia. Marco Polo's journey and the Crusades were the first step in awakening Europeans to the world around them.

Some attempts at systematic exploration began during the first half of the fifteenth century. Portuguese captains under the direction of their ruler Henry the Navigator (1394-1460) sailed further and further along the African coast. The age of exploration, however, began with the fall of Constantinople to the Turks in 1453. This severed the trade route to Asia and left Europe with no source of silks or spices. They could do without the silk but they had to have spices. The price of spices rose sharply, adding a commercial incentive to the existing religious and political reasons for finding a route to Asia.

Spices were very important at that time because in the winter Europeans had almost nothing to eat except pickled meat. They had no fodder crops, so each fall most of the cattle were slaughtered and pickled. Without spices this was a very dreary diet.

The great voyages of exploration began in 1492 with Columbus.

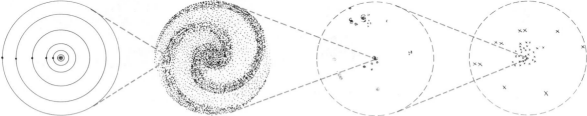

A. Solar System

Innermost circle represents Earth's orbit around sun. Other circles represent orbits of Mars, Jupiter, Saturn, Uranus, Neptune, and Pluto. One light-year is approximately 1600 times the distance from the sun to Pluto.

B. Milky Way

Distance from the center of the galaxy to the sun is 30,000 light-years, 48 million times the sun – Pluto distance. The center to the edge of the galaxy is 40,000 light-years, 64 million times the sun – Pluto distance.

C. Local Group

The Local Group of galaxies extends about 3 million light-years away, 75 times the distance from the center of the Milky Way to its edge. The sizes of the galaxy symbols are greatly exaggerated.

D. Universe of galaxy groups

Large X's represent large groups of galaxies; smaller x's, small groups. Dashed circle approximately 500 million light-years away from center.

FIGURE 1–14   The earth's place in the universe.

He sailed west, hoping to reach the Indies. This was the first time anyone had ventured into the Atlantic far away from the coast. One reason for this reluctance to leave the sight of land was the lack of knowledge of navigation. Columbus used a plumb bob and a quadrant to sight stars and could find his latitude from sighting the North Star, but with his equipment even this was difficult in a wave-tossed ship. He had only an hourglass to measure time, so he was unable to find his longitude with any accuracy. He thought that he had reached the Indies because he believed that the earth was much smaller than it is.

After Columbus came the other great explorers. In 1497 Da Gama sailed around Africa to India. Balboa discovered the Pacific in 1513 and was the first of the Spanish Conquistadors to explore the Americas. In 1519 Magellan sailed around South America to India. There were many other voyages, but interestingly, it was almost one hundred years later that Australia was discovered by the Dutch. The last of the great voyages was that of Captain Cook who in 1768 was sent to find the mythical southern continent. It was not until the eighteenth century that the chronometer and other navigational methods and instruments became available.

Exploration of the continents lagged somewhat behind that of the oceans. Exploration of the Americas was started almost as soon as they were discovered, but Africa was explored much later. As late as 1725, guidebooks to the Alps described dragons who reportedly lived there. The opening of the Alps is especially interesting because many important geological discoveries were made in those mountains. Until well into the last half of the eighteenth century, very few of the Alps had been climbed, although some climbing began around 1500. In 1769 Rousseau wrote *La Nouvelle Héloïse,* a very popular and influential book which glorified nature and encouraged exploration. In 1786 Mont Blanc, the highest peak in the Alps, was climbed, an important milestone. By 1850 about one hundred mountains had been climbed. Between 1854 and 1865 almost all of the peaks in the Alps were conquered.

Although Africa is close to Europe, exploration of its interior by Europeans did not really begin until the late eighteenth century. North Africa is defended by great deserts, and much of the rest is shielded by tropical jungles. Much of the exploration was done by the British, with some by Germans and other nationalities. The objective of most expeditions was to determine the course of one of the major rivers. These expeditions generally lasted longer than one year, and their stories are gripping examples of privation and heroic personal accomplishment. The culmination was the search for the source of the Nile. Expeditions by Burton and Speke, 1857-1859, Speke and Grant, 1860-1863, and others had narrowed the problem.

David Livingstone, a Scottish doctor and missionary, had spent many years in Africa and was well known throughout the world. No word from him had reached the outside world for five years when, in 1871, the *New York Herald* sent Henry Stanley to look for him. Stanley returned in 1872 after spending several months exploring with Livingstone. Stanley led other expeditions in the 1870s and 80s and was responsible for exploring much of central Africa. He also helped in the colonization and opening up of Africa by building roads.

Thus the physical geography of Africa was determined after the time of the American Civil War. This was the same time that the American West was being explored and opened up. It was in 1869 that Major John Wesley Powell's expedition went down the Colorado River in small boats to find its course. (See Fig. 1–15.) Such geographical exploration must precede or accompany geologic study of an area.

BEGINNINGS OF GEOLOGY

At first, people were only interested in rocks and minerals that

A

B

FIGURE 1–15   A. Major Powell's famous armchair boat. This photo was taken in 1872 on his second trip down the Colorado River. B. Photo of the same place taken in 1968. Note the small amount of erosion in nearly one hundred years. Photo A by J. K. Hillers, Photo B by H. G. Stephens, U.S. Geological Survey.

were useful. Later interest expanded to ornamental materials and much later to attempting to understand the origins of the earth and the materials of which it is made.

The first observers of whom we have any record were the early Greeks. They drew logical conclusions from their observations, but many of their discoveries were lost during the Middle Ages. One of the first of these observers was Herodotus (500 B.C.), who discussed the Nile River and saw that the Nile deposited silt during the floods. Thus he recognized some of the slow, continuous processes that modify the surface of the earth; but like his contemporaries, he believed that most of the surface features were caused by rapid, violent actions. This is not unreasonable because he lived in an area of many earthquakes, which were then thought to be caused by winds inside the earth.

Aristotle (384-322 B.C.) was the most important of the Greeks. He was Plato's student. He recognized river-deposited material and realized that fossil sea shells from rock strata were similar to those found along the beach, indicating the fossils were once living animals. From this he concluded that the position of land and sea had changed, and so he was the first to recognize the great length of time necessary for these changes. Some of his ideas that persisted through the Middle Ages were that the earth is a sphere because that is the most perfect shape, that the orbits of planets are circular for the same reason, and that a heavy object falls faster than an lighter one.

Theophrastus (374-287 B.C.) was Aristotle's pupil, and he wrote the first book on mineralogy, called *Concerning Stones.* This was a very important book and was the basis of most mineralogies through the Middle Ages and even after. Unfortunately he also suggested that fossils might not have been once living animals and

started a controversy that lasted through the Middle Ages.

After the fall of Rome (500 A.D.) the Dark Ages began, and little more is heard about science until the Renaissance. During this time the work of the earlier philosophers and observers was preserved in two ways—by the Church and by the Arabs. The Muslim Avicenna or Ibn Sina (980-1037) wrote *On the Conglutination of Stones.* This work was in part original and in part probably copied from Theophrastus. He also translated much of Aristotle's work. He noted that mountains could result from either of two processes—either from uplift by earthquakes or from erosion. In recognizing that mountains can be uneroded remnants, he was very advanced. The Dark Ages ended about 1100 with the establishment of the Feudal System and the First Crusade.

The Dark Ages were followed by the Middle Ages which lasted until the Renaissance. During the Middle Ages the concept of the Ptolemaic universe with the earth at the center of the universe was firmly entrenched. The belief that the position of the sun and moon and planets had a great effect on human lives was widely held. This idea has persisted to modern times as what we now call astrology. Also most believed that everything in nature had some special meaning or use and certain minerals were believed to have special powers such as curing or protecting from some sickness. It was also believed that either fire or the sun's rays at the center of the earth produced metals. The alchemists believed, following Aristotle, that base metals could be transformed into precious metals. Ores that contained several metals, such as the common association of lead and silver with a little gold, were thought to be the proof of this. Stones were believed to grow in the soil, and even today, anyone who farms or gardens will know how this idea

started. The source of water was thought to be underground as the Bible seems to say this.

A rebirth of learning did occur around the middle of the thirteenth century. About this time the great universities of Europe were formed. Probably an important factor in this rebirth was the final failure of the Crusades, which shook the foundation of European thought. Translation of Arabic books into Latin, which at that time was the universal language of scholars, helped bring about the resurgence of learning. Typical of this period were Dante, St. Francis of Assisi, and Marco Polo. The most influential was probably St. Thomas Aquinas (1225–1274), who attempted to weave Aristotle's theories and the Bible into a single system covering all phases of life. As a result of this work, future people were to regard Aristotle's writings as having almost the same stature and authority as the Bible. With the invention of the printing press about 1450, the Renaissance began.

With the Renaissance, people began to observe again as the early Greeks had many years before. The medieval scholars, following Aristotle's methods, had used only logic and philosophy as their tools. These observers now began to rediscover the world around them. This was the age of exploration which culminated with Columbus and Magellan.

GEOLOGIC METHODS

Geology is based mainly on observations and seeks to determine the history of the earth by explaining these observations logically, using other sciences such as physics, chemistry, and biology. Only a small part of geology can be approached experimentally. For example, although the important use of fossils to date or establish contemporaneity of rock strata is based on the simple, basic principle that life has changed during the history of the earth, this principle could not be established experimentally; it was the result of careful observations and analyses over a long period of time by many people of varied backgrounds.

Geologic problems are many, diverse, and complex; almost all must be approached indirectly, and in some cases, different approaches to the same problem lead to conflicting theories. It is generally difficult to test a theory rigorously for several reasons. The scale of most problems prevents laboratory study; that is, one cannot bring a volcano into the laboratory, although some facets of volcanoes can be studied indoors. It is also difficult to simulate geologic time in an experiment. All of this means that geology lacks exactness and that our ideas change as new data become available. This is not a basic weakness of geology as a science, but means only that much more remains to be discovered; this is a measure of the challenge of geology.

Reasoning ability and a broad background in all branches of science are the main tools of geologists. Geologists use the method of *multiple working hypotheses* to test their theories and to attempt to arrive at the best-reasoned theory. This thought process requires as many hypotheses as possible and the ability to devise ways to test each one. Not always is it possible to arrive at a unique solution—but this is the goal. In the sense that geologists use observation, attention to details, and reasoning, their methods are similar to those of fictional detectives.

The most important method used by the geologist is to plot on maps the locations of the rock types exposed at the earth's surface. The rocks are plotted according to their type and age on most maps. (See Fig. 1–16.) From such maps it is possible to interpret the history of the area. The early geologists had to make their own maps, and work in a remote area

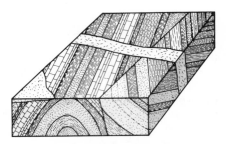

FIGURE 1–16 Block diagram showing a geologic map with cross-sections on the sides. A geologic map shows the distribution of rock types on the surface. From such maps, the history of an area is interpreted.

was very difficult in many cases. Now, excellent maps produced from aerial photographs are available for most areas. Much geologic mapping is done directly on either black and white or color aerial photographs, which have proved to be unexcelled for accurate location of rock units. In addition, the outcrop pattern of the rock units generally shows well on air photos. (See Fig. 1–17.) Radar images that show the surface beneath thick forest cover now extend the use of aerial photographs into such areas. (See Fig. 1–18.) Helicopters and jeeps have largely, but not entirely, replaced pack animals and backpacking. (See Fig. 1–19.) However, in spite of all these advances that have accelerated geologic mapping, much of the earth is not mapped geologically or at best has been mapped only in reconnaissance fashion.

Recent advances have extended classical geologic mapping by providing new things to map. As an example, it is possible to obtain, with some complex instrumentation, aerial views of areas in infrared or heat wavelengths rather than visible light. These are used in the study of active volcanic

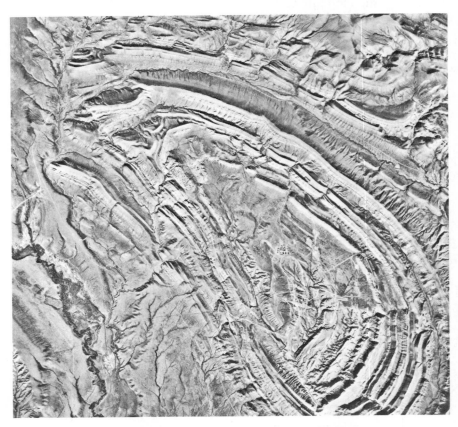

FIGURE 1–17 Vertical aerial photograph of folded rocks in Wyoming. Erosion has etched out the differences in resistance of the various rock layers. Such a photo is essentially a geologic map of the area. Photo from U.S. Geological Survey.

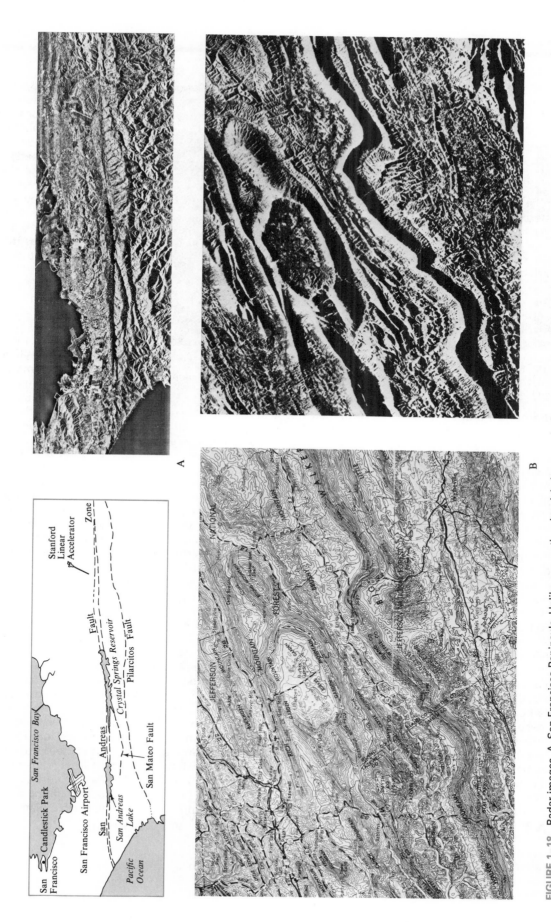

FIGURE 1–18 Radar images. A. San Francisco Peninsula. Unlike conventional aerial photos, radar images can be obtained in cloudy weather or even at night. The radar penetrates the vegetation and reveals the actual surface. The bottom (west) part of this area has thick redwood forests. From U.S. Geological Survey in cooperation with NASA and Westinghouse Electric Co. B. Appalachian Mountains in western Virginia. The structures of these folded rocks are well shown. Courtesy of Raytheon Company, Autometric Facility.

13

FIGURE 1-19 Mapping on the northwest coast of Alaska. The helicopter is used to transport geologists to this remote area and for movement about this trackless region. Photo by Standard Oil Co. of California.

areas and ground water studies as well as other types of studies. (See Fig. 1–20.) Other examples of the types of data now mapped include the earth's magnetic field, radioactivity, heat flow, seismic properties, and many more. Mapping this new data is giving geologists new insights in understanding the earth.

The other more obvious method used extensively by the geologist is the detailed study of rocks themselves. (See Figs. 1–21 and 1–22.) This can be done at several levels, and the first chapters of this book are largely concerned with these topics. The first methods applied were optical microscopy and chemical analysis, and these are both still in wide use. Newer methods such as the electron microscope; X-ray analysis; optical, infrared, and X-ray spectroscopy; and the electron microprobe now permit very detailed studies of rocks.

RECENT TRENDS

Recent discoveries have expanded the interest of geologists far beyond their classical realm—the rocks of the earth's surface. In addition, geology is becoming more quantitative, although it is still qualitative when compared to a science such as physics. In the early part of the twentieth century, radioactive dating enabled the geologist to work with the very old rocks, and the problem of the origin and early history of the earth began to take form. Since World War II oceanography has developed rapidly, and the features of the ocean bottoms have been mapped. (See Figs. 1–23 and 1–24.) These oceanographic discoveries, together with the study of magnetism in ancient rocks, resulted in the theory of plate tectonics.

At present, the space programs are giving us information about other parts of the solar system. (See Fig. 1–25.) Most astronomers believe that all of the solar system formed at the same time, so this new information gives much help to the geologist. This has given rise to a new science of ''geology'' of planets, termed *astrogeology*, or perhaps better, *planetology*. The origin of the solar system is,

of course, a classic problem of astronomy, but is also part of geology. These advances, plus the sharpening by geophysicists of the tools that probe the depths of the earth, have greatly expanded the viewpoint of the geologist. At the present time geology, astronomy, oceanography, geophysics, and meteorology are coming closer together; and it is this combination of sciences that must be used together to study the whole earth.

THIS BOOK

This book is about the solid rock earth, especially its history and the processes that formed its surface features. In this sense it covers classical geology, but in order to understand these features, it will be necessary to consider evidence from a wide range of fields. Thus, the viewpoint will be wide, and many of the topics would not have appeared in a survey of geology only a few years ago. This broadness, however, makes modern geology a more valuable course for the nonscience major.

We begin our study by considering the crust—the outer few miles—of the earth. We study, first, the minerals and rocks of the crust, because they tell us much about the composition of the crust and the processes that take place within the crust as well as on its surface. Then we consider erosion—the dominant surface process. Next, we consider the interactions of the processes in the crust and in the deeper parts of the earth and learn how the present surface features of the earth formed. This leads us to a model of the entire earth. Finally, the methods of dating and deciphering the earth's history are described.

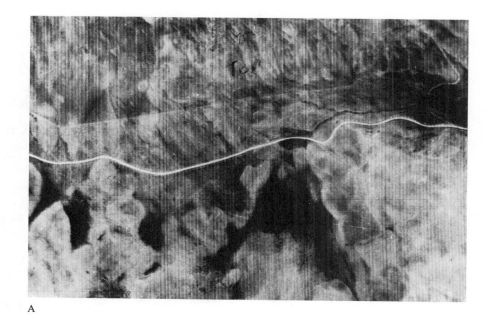

A

B

FIGURE 1–20 Infrared image and aerial photograph of the same area. A. Infrared image shows warm areas in light tones and cool areas in dark tones. The curving light line is a blacktop road, and the straight lines are fences separating fields with different kinds of plant cover. The light pattern near the bottom is landslides, and the dark areas are poorly drained areas between the landslides. B. Aerial photograph shows landslides at the bottom and hills near the top. The landslides moved from the bottom toward the top. From an unpublished report to NASA by U.S. Geological Survey, by R. E. Wallace and R. M. Moxham with J. Vedder.

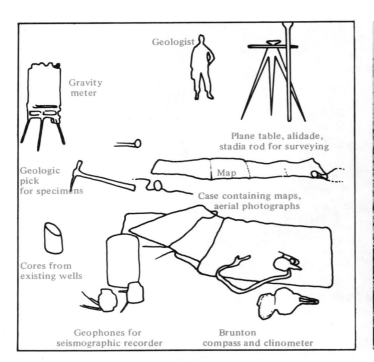

Geologist

Gravity meter

Plane table, alidade, stadia rod for surveying

Geologic pick for specimens

Map

Case containing maps, aerial photographs

Cores from existing wells

Geophones for seismographic recorder

Brunton compass and clinometer

FIGURE 1–21 A geologist and some of his tools. The gravity meter measures the earth's gravity. The plane table, alidade, and stadia rod are used in surveying. Cores from wells, if available, reveal the rocks at depth. The geophones are used in seismic studies of the buried rocks. The Brunton compass and clinometer is used to measure the attitude of layered rocks. Photo by Standard Oil Co. of California.

FIGURE 1–22 This geologist is examining a sample of a sedimentary rock that has abundant fossil fragments. The hand lens is used in the field and is commonly followed by microscopic, X-ray, or many other more detailed methods in the laboratory. Photo by Standard Oil Co. of California.

16

FIGURE 1–23 U.S. Geological Survey research vessel *Don J. Miller* at Prince William Sound, Alaska. Photo by W. R. Hansen, U.S. Geological Survey.

FIGURE 1–24 A two-man submersible research vessel. Note the extended sampling arm. Photo courtesy of Noel P. James, University of Miami.

FIGURE 1–25 Scientist-astronaut Harrison Schmitt beside a huge boulder on the moon. Photo from NASA.

SUMMARY

Geology is the study of the earth.

The earth is unique in having abundant water as well as an atmosphere and temperature that can support life.

The fundamental principle of geology is *uniformitarianism,* which is simply that the present processes occurring on the earth have always gone on.

Geologic studies have revealed that constant change, both physical and biological, has been and is occurring on the earth.

Geology differs from most sciences in that it is concerned with absolute time.

Continental drift is suggested by the shapes of the continents and the match of their rocks and fossils.

Sea-floor spreading is the theory that suggests that new sea floor is created at the mid-ocean ridges, and that the sea floors move slowly away from the ridges and are eventually consumed at volcanic arcs.

The *plate tectonics theory* assumes that the earth's surface is composed of a number of slabs, or plates, that move across the surface. Some of the plates contain continents. The plates form at mid-ocean ridges and are consumed at convergent boundaries, such as volcanic island arcs. Where one plate slides past another, the boundary is called a *transform fault.*

When plates collide, they form convergent boundaries and the surface rocks are deformed. If a plate composed of ocean floor collides with a similar plate, one of the plates is consumed at the resulting convergent boundary and a volcanic island arc forms. If ocean floor collides with a continental boundary, the ocean floor is consumed at the resulting convergent plate boundary. When continent collides with continent, a mountain range is formed.

At first the earth was thought to be at the center of the universe, but by the Renaissance a few recognized that the earth orbits the sun. Newton explained the earth's motions; during his time, the distance to the sun, 150 million kilometers (93 million miles), was measured. In the nineteenth and twentieth centuries, distances to stars and galaxies were measured, and the nature of the galaxies was determined.

Systematic exploration of the earth began around the Renaissance with the great voyages of exploration. In the eighteenth and nineteenth centuries, mountain ranges were conquered by climbers and exploration of Africa began.

Modern geology began in the eighteenth century.

Geology is based mainly on observation and reasoning, together with a knowledge of the other sciences.

The most important method used by geologists is to plot on maps the locations of rock types. Aerial photography and radar images can aid in mapping.

Study of rocks has progressed from the use of a hand lens to a microscope to many modern instruments.

At present, oceanography and the space program are contributing to our knowledge of geology.

QUESTIONS

1 In what ways is the earth unique?
2 What are some of the economic uses of geology?
3 What are some of the ways that geology can aid in planning the growth of an area?
4 What is meant by the term *uniformitarianism?*
5 What is the evidence for continental drift?
6 Outline the theory of sea-floor spreading.
7 What are the three features that are the boundaries of plates?
8 Describe the three types of convergent plate boundaries.
9 Outline the steps in discovering the distance to other galaxies.
10 What is the geologist's most important method of study of an area?
11 Why is not geology based on experimentation?

SUPPLEMENTARY READINGS

Albritton, C. C. (ed.), *The Fabric of Geology.* Reading, Mass.: Addison-Wesley Publishing Co., Inc., 1963, 372 pp.

American Geological Institute, *Geology: Science and Profession.* Washington, D.C.: American Geological Institute, 1965, 28 pp.

Booth, Basil, and Frank Fitch, *Earthshock,* New York: Walker & Co., 1979, 256 pp.

Geikie, Archibald, *The Founders of Geology.* New York: Dover Publications, Inc., 1962, 486 pp. (paperback). Originally published in 1905 by Macmillan and Co.

Gillispie, C. C., *Genesis and Geology.* New York: Harper and Row, 1951, 306 pp. (paperback).

Howard, A. D., and Irwin Remson, *Geology in Environmental Planning.* New York: McGraw-Hill Book Co., 1978, 478 pp.

Murray, R. C., and J. C. F. Tedrow, *Forensic Geology: Earth Sciences and Criminal Investigations.* New Brunswick, N.J.: Rutgers University Press, 1975, 217 pp.

Rabbitt, M. C., *Minerals, Lands, and Geology for the Common Defence and General Welfare,* Vol. I, Before 1879. (USGS) Washington, D.C.: Superintendent of Documents, 1979, 332 pp. (paperback).

Short, N. M., and others, *Mission to Earth: Landsat Views the World.* NASA SP-360. Washington, D.C.: U.S. Government Printing Office, 1977, 459 pp.

U.S. Geological Survey, *Earth Science in the Public Service.* Professional Paper 921, Washington, D.C.: U.S. Government Printing Office, 1974, 73 pp.

Mobile Earth

Dewey, J.F., "Plate Tectonics," *Scientific American* (May 1972), Vol. 226, No. 5, pp. 56–68.

Dickinson, W. R., "Plate Tectonics in Geologic History," *Science* (October 8, 1971), Vol. 174, No. 4005, pp. 107–113.

Hallam, A., "Continental Drift and the Fossil Record," *Scientific American* (November 1972), Vol. 227, No. 5, pp. 56–66.

Heirtzler, J. R., "Sea-Floor Spreading," *Scientific American* (December 1968), Vol. 219, No. 6, pp. 60–70.

Kurtén, Björn, "Continental Drift and Evolution," *Scientific American* (March 1969), Vol. 220, No. 3, pp. 54–64.

Vine, F. J., "Sea-floor Spreading—New Evidence," *Journal of Geological Education* (February 1969), Vol. 17, No. 1, pp. 6–16.

Wilson, J. T., "Some Aspects of the Current Revolution in the Earth Sciences," *Journal of Geological Education* (October, 1969), Vol. 17, No. 4, pp. 145–150.

Wyllie, P.J., *The Way the Earth Works: An Introduction to the New Global Geology and Its Revolutionary Development.* New York: John Wiley & Sons, Inc., 1976, 296 pp. (paperback).

2 ATOMS, ELEMENTS, AND MINERALS

Which was which he could never make out
    Despite his best endeavour.
Of that there is no manner of doubt—
No probable, possible shadow of doubt—
    No possible doubt whatever.

Sir W. S. Gilbert and Sir Arthur Sullivan
The Gondoliers

INTRODUCTION

Mineralogy, which is now only a branch of geology, was probably the parent of geology. The first geological books were mineralogies, and they bear very little resemblance to modern mineralogy books. They made no distinction between minerals and fossils and included myths and folklore associated with the minerals.

Mineralogy came out of the Middle Ages with the writings of Georg Bauer (1494–1555) who, following the custom of the time, wrote in Latin under the Latinized name of Agricola. Like many of the early people in geology, he was a physician, as this was the only scientific training available. He was a Saxon German and lived in Joachimstal, near the famous silver mines. His hobby was the study of minerals and fossils, and he wrote several books on these subjects. His most important book was *De Re Metallica* which was published in 1556. This was the first book on mining and metallurgy. It was very popular and was one of the first books translated into local languages that could be read by the common people. This book was published at the right time, about 100 years after the invention of printing, enabling it to be widely circulated; and it was on a practical subject that was well received. It is interesting to note that this took place during the age of such men as Da Vinci, Copernicus, and Luther.

Another of Agricola's books, *De Natura Fossilum,* was on minerals and fossils. It was published in 1546 and classified minerals on the basis of physical properties.

Nicolaus Steno (1638–1687) was about one hundred years ahead of his time in understanding geology. He will be introduced here for his work in mineralogy, and we will come back to him later when stratigraphy is considered. He was born a Dane, Niels Steensen, but he also Latinized his name because he wrote in Latin. In 1668 he published a book in which he recognized the true nature of fossils. He also noted from his study of quartz crystals that the angles between the faces are always the same. This was the beginning of the science of crystallography, one of the offshoots of mineralogy.

Abraham Gottlob Werner (1749–1817) did much for mineralogy but probably held geology back for 100 years. He studied at Freiberg from 1769 to 1771 and then went on to the University of Leipzig where he wrote a mineralogy book, *Von der äusserlichen Kennzeichen der Fossilien,* in 1774. On the basis of this book, he was invited back to Freiberg the same year and became a Professor at the age of 25. He was the chief proponent of the so-called mixed system of mineralogy as the most practical at the time. He knew that this was not the best system and strived for the ideal system based on composition. He constantly improved his system during the 40 years that he stayed at the Academy at Freiberg. In his last revision in 1817, he recognized 317 minerals in classes such as earthy minerals, saline minerals, combustible minerals, and metallic minerals.

Werner also founded the Neptunist theory of the origin of the earth. This latter work was not at all in the same class with his mineralogy. The theory was based on his idea that all rocks were precipitates from a universal ocean. Although Werner wrote very little, even when his theory of the earth was under attack, he was extremely popular, and his fame, based in part on his mineralogy, was spread far and wide by his students.

During Werner's lifetime, further advances were made in crystallography. Abbé René-Just Haüy (1743–1822), a French botanist, became interested in mineralogy after hearing a lecture at the Royal Gardens in Paris. The consistency of form of plants and minerals intrigued him. After accidentally dropping a calcite crystal, he noted that no matter what the shape of the original crystal, the cleavage fragments all have the same (rhombohedral) shape. From this he deduced that all minerals were made of tiny building blocks, thus beginning our modern ideas of unit cells. He also developed the six crystal systems.

Mineralogy developed very rapidly in the nineteenth century. Crystallography advanced with the development of instruments to measure accurately the angles be-

Calcite crystals from Iron County, Missouri. Photograph by and courtesy of Nathaniel Ludlum. Department of Mineralogy, The Ohio State University.

tween crystal faces. The most important advance, however, was the development of chemical analysis so that the composition of minerals could be determined.

The new mineralogy based on chemistry came from the great Swedish chemist Johann Jacob Berzelius (1779–1848). He discovered selenium and thorium, and was the first to isolate calcium, barium, strontium, silicon, and zirconium. He also devised the system of writing chemical formulas and equations that we still use. In 1814 he published in Stockholm *A new system of mineralogy.* In this book he recognized silicates, bringing order out of chaos.

With these chemical and crystallographic methods available, mineralogists studied many minerals and much new data were produced. This, in turn, led to the publication of books that de-

scribed and classified minerals. Two outstanding books of this era that are still in use (in much more modern editions) are a German text by Groth published in Germany and Dana's *System of Mineralogy* published in the United States. James Dwight Dana (1813–1895) was a professor at Yale who made important advances in structural geology.

In the twentieth century, the discovery of X rays enabled probing the internal structure of minerals. This was first done in 1912 by Max von Laue and his students W. Friedrich and P. Knipping. Their discovery proved that the atoms in a crystal have an orderly arrangement. Very soon after this, Sir William Bragg and his son, Laurence, used X rays to determine the internal structure of halite (rock salt), and a new era in mineralogy began. Since then X-

ray methods have been refined, and the structures of many minerals have been determined.

ABUNDANCE OF ELEMENTS

On casual examination, the surface of the earth appears to be quite complex. Many different-appearing rocks and soils are to be seen. The complexity is much more apparent than real, and close examination reveals a very simple composition. The chemical composition of the crust of the earth will be considered first.

Over 100 elements are known, and, of these, 80 are stable. The crust of the earth is composed almost entirely of only eight elements, and all of the other elements make up less than one percent, as shown in Table 2–1. Thus,

TABLE 2–1 The average composition of continental crustal rocks. (A few very rare elements and shortlived radioactive elements are omitted.)

Symbol	Name	Weight (percent)	Symbol	Name	Weight (percent)	Symbol	Name	Weight (percent)
O	Oxygen	46.4	Nd	Neodymium	0.0028	Mo	Molybdenum	0.00015
Si	Silicon	28.15	La	Lanthanum	0.0025	Ho	Holmium	0.00015
Al	Aluminum	8.23	Co	Cobalt	0.0025	Eu	Europium	0.00012
Fe	Iron	5.63	Sc	Scandium	0.0022	Tb	Terbium	0.00011
Ca	Calcium	4.15	N	Nitrogen	0.0020	Lu	Lutetium	0.00008
Na	Sodium	2.36	Li	Lithium	0.0020	I	Iodine	0.00005
Mg	Magnesium	2.33	Nb	Niobium	0.0020	Tl	Thallium	0.000045
K	Potassium	2.09	Ga	Gallium	0.0015	Tm	Thulium	0.000025
Ti	Titanium	0.57	Pb	Lead	0.00125	Sb	Antimony	0.00002
H	Hydrogen	0.14	B	Boron	0.0010	Cd	Cadmium	0.00002
P	Phosphorus	0.105	Th	Thorium	0.00096	Bi	Bismuth	0.000017
Mn	Manganese	0.095	Sm	Samarium	0.00073	In	Indium	0.00001
F	Fluorine	0.0625	Gd	Gadolinium	0.00073	Hg	Mercury	0.000008
Ba	Barium	0.0425	Pr	Praseodymium	0.00065	Ag	Silver	0.000007
Sr	Strontium	0.0375	Dy	Dysprosium	0.00052	Se	Selenium	0.000005
S	Sulphur	0.026	Yb	Ytterbium	0.0003	A(r)	Argon	0.000004
C	Carbon	0.020	Hf	Hafnium	0.0003	Pd	Palladium	0.000001
Zr	Zirconium	0.0165	Cs	Cesium	0.0003	Pt	Platinum	0.000001
V	Vanadium	0.0135	Er	Erbium	0.00028	Te	Tellurium	0.000001
Cl	Chlorine	0.013	Be	Beryllium	0.00028	Ru	Ruthenium	0.000001
Cr	Chromium	0.010	U	Uranium	0.00027	Rh	Rhodium	0.0000005
Rb	Rubidium	0.009	Br	Bromine	0.00025	Os	Osmium	0.0000005
Ni	Nickel	0.0075	Ta	Tantalum	0.0002	Au	Gold	0.0000004
Zn	Zinc	0.0070	Sn	Tin	0.0002	He	Helium	0.0000003
Ce	Cerium	0.0067	As	Arsenic	0.00018	Re	Rhenium	0.0000001
Cu	Copper	0.0055	Ge	Germanium	0.00015	Ir	Iridium	0.0000001
Y	Yttrium	0.0033	W	Tungsten	0.00015			

SOURCE: After Brian Mason, 1966, and Konrad Krauskopf, 1968.

in spite of its heterogeneous aspect, the crust of the earth has a very simple composition. Notice that many of our common, cheap metals, such as chromium, nickel, zinc, copper, lead, and tin, are really rather rare and form only about 0.04 percent of the crust. How these elements are concentrated into mineable deposits will be discussed in later chapters.

Before discussing the important implications of these data, we should look at their origin. All abundance data are calculated from estimates of the amounts of the different rock types and their chemical analyses. The main problem is estimating the amounts of the rock types; however, geologic maps showing the distribution of rock types have been made for many parts of the earth's surface. In spite of this, such estimates are difficult, and different workers have made different estimates. The amounts shown in the tables are subject to some revision, but they show the general composition of the crust. Estimates of the composition of the whole earth or of the solar system are quite different.

DEFINITIONS

So far we have discussed *elements,* which *are materials that cannot be subdivided by ordinary*

chemical methods. The smallest unit of an element is the *atom.* The atom can be broken down into smaller particles such as electrons, neutrons, protons, etc., but except in the case of radioactive elements, this requires the large amounts of energy available in atom-smashing machines such as the cyclotron.

In nature, elements are combined to form *minerals, which can be defined as*

1. *naturally occurring, crystalline,*
2. *inorganic substances with*
3. *a definite small range in chemical composition and physical properties.*

Note that all three conditions must be met by a mineral.

Minerals are *crystalline* substances; that is, they *have an orderly internal structure (arrangement of atoms),* as opposed to such things as *glasses* which are *super-cooled liquids and have no internal order.* A *crystal is a solid form bounded by smooth planes which give an outward manifestation of the orderly internal structure.* (See Fig. 2–14.) Note the distinction between crystal and crystalline. Although all minerals are crystalline, they do not necessarily occur as geometric crystals.

The minerals group themselves naturally to form *rocks,* which can be defined as

1. *a natural aggregate of one or more minerals, or*
2. *any essential and appreciable part of the solid portion of the earth (or any other part of the solar system).*

Note that this definition does not require all rocks to be composed entirely of minerals. Some noncrystalline material is present in many quickly chilled volcanic rocks.

OXYGEN AND SILICON, BUILDING BLOCKS OF ROCK-FORMING MINERALS

Because the rocks of the crust are composed predominantly of eight elements, we would suspect at first glance that these eight elements, which must form most minerals, would occur in many different combinations. However, most rocks are composed of only a few combinations of elements—or minerals. The reason for this happy situation, which means that a beginner needs to learn only a few minerals to identify most rocks, can be understood from the data in Table 2–2. Rocks are composed of over 90 percent oxygen, by volume, because oxygen atoms are abundant and relatively large. Therefore, the number of possible combinations is controlled by the ways in which oxygen atoms can be arranged to build minerals. This is especially true because relatively few other atoms are readily available to form minerals. Therefore, although about 2000 minerals have been discovered, only 20 are common, and fewer than 10 form well over 90 percent of all rocks.

Thus the key to understanding the structure of minerals lies in how the atoms (more correctly called *ions,* because the atoms have gained or lost one or more electrons) in a mineral are arranged, i.e., how the other atoms

TABLE 2–2   Main elements in the continental crust.

Element	Weight (percent)	Atom (percent)	Volume (percent)
Oxygen	46.40	62.17	94.05
Silicon	28.15	21.51	.88
Aluminum	8.23	6.54	.48
Iron	5.63	2.16	.48
Calcium	4.15	2.22	1.19
Sodium	2.36	2.20	1.11
Magnesium	2.33	2.05	.32
Potassium	2.09	1.15	1.49
Totals	99.34	100.00	100.00

are fitted among the oxygen atoms. Clearly, the size of the atoms in question, compared to the size of the oxygen atom, will determine the structure of any mineral. This relationship is shown in two dimensions in Fig. 2–1. The number of oxygen ions that surround an element can be calculated from simple geometric relationships.

As can be predicted from the abundance data, most of the rock-forming minerals are composed largely of oxygen and silicon, generally with aluminum and at least one more of the "big eight." The sizes of oxygen and silicon are such that the small silicon ion fits inside four oxygen ions. This unit, which is the building block of the silicate minerals, is called the *SiO$_4$ tetrahedron*, because the *four oxygen ions can be considered to be situated on the four corners of a tetrahedron with the silicon ion at the center*. (See Fig. 2–2.)

The structures of the silicate minerals are determined by the way these SiO$_4$ tetrahedra are arranged. Two of these tetrahedra, if joined at a corner, can share a single oxygen ion. This forms a very strong bond. With this type of linkage, through sharing of oxygen, the structure of the silicate minerals is formed. The possible structures are shown in Fig. 2–3.

ATOMIC STRUCTURE

The sizes of ions determine the internal geometry of minerals and other crystalline substances; however, we must learn some features of atomic structure to understand how minerals are held together. This study will also provide an understanding of natural radioactivity, which is an important source of energy in the earth and will be discussed in later chapters.

Atoms are composed of smaller particles, and although physicists have discovered many such particles, only a few need to be discussed here. Most of the mass of

Small atom fits in space between 3 large atoms   Larger atom fits in space between 4 large atoms

FIGURE 2–1   The relative sizes of atoms determine how they can fit together. Shown here in two dimensions.

FIGURE 2–2   Three views of an SiO$_4$ tetrahedron. The large spheres are oxygen, and the small spheres are silicon.

	FORMULA OF SILICON-OXYGEN UNIT	NUMBER OF OXYGEN SHARED	EXAMPLE
1. Single tetrahedron	(SiO$_4$)	0	Olivine
2. Double tetrahedron	(Si$_2$O$_7$)	1	Epidote
3. Ring	(Si$_6$O$_{18}$)	2	Tourmaline
4. Chains	(SiO$_3$)	2	Pyroxene (augite)
5. Double chains	(Si$_4$O$_{11}$)	2 & 3	Amphibole (hornblende)
6. Sheets	(Si$_2$O$_5$)	3	Micas (muscovite, biotite)
7. Three-dimensional networks	(SiO$_2$)	4	Quartz, feldspars

FIGURE 2–3   Structures of silicate minerals.

an atom is in its nucleus. The *nucleus is composed of protons and neutrons* (except in the case of the hydrogen nucleus that is composed of only a single proton). *Protons are heavy and have a positive electrical charge.* The nucleus is surrounded by a cloud of the *much lighter, negatively charged electrons.* (See Fig. 2–4). An atom is mostly empty space because the diameter of the outer electron orbits, which is the size of the atom and averages nearly 0.00000002 centimeter, is between 20,000 and 200,000 times the diameter of the nucleus. The number of electrons equals the number of protons; hence, an atom is electrically neutral. The number of protons and electrons determines the chemical properties and so determines the identity of an element. *The number of protons (or electrons) is called the atomic number* of an element. In addition, an atom may have neutrons in its nucleus. *Neutrons have a mass slightly greater than that of a proton plus an electron, and are elec-*

trically neutral. *The number of neutrons in the atom of an element may vary, giving rise to isotopes.* Changing the number of neutrons in the atom of an element does not change the chemical properties, but it does change the mass of the atom. The *atomic mass number is the number of neutrons plus protons in the nucleus.* Most elements have naturally occurring isotopes, as shown in Table 2–3.

BONDING

The bonding of atoms to form minerals is done in several ways, most of which involve the electrons surrounding the nucleus. The electrons are not randomly distributed around the nucleus, but move in distinct orbits. Each orbit or shell contains a fixed number of electrons, as shown diagrammatically in two dimensions in Fig. 2–4. Most of the electron shells have subshells as indicated in Fig. 2–4 and Table 2–4. Up to seven shells are possible to

accommodate the number of electrons necessary to balance the number of protons in the nucleus. As the number of protons in the nucleus increases, the electron shells are filled. The first shell contains 2 electrons when filled, and the second, 8. The third shell can contain up to 18 electrons; the fourth, 32; the fifth, 32; the sixth, 9; and the seventh, 2. The electrons are distributed among the outer shells of any element in such a manner that the outer shell always contains eight or fewer electrons. Larger numbers can be accommodated only in interior shells.

The main principle involved in mineral bonding is to achieve a stable configuration with the outer shell filled. This can be accomplished in different ways. One possibility is for the atom to gain or lose electrons. An example of this is an atom with one electron in its outer shell losing this electron to another atom with seven electrons in its outer shell. *The resulting ions have opposite charges and so are mutually attracted.* (See Fig. 2–5.) *This is called ionic bonding,* and many minerals have ionic bonds.

Another way that the outer shell of an atom can be filled is by *two or more atoms sharing electrons.* In this way the outer shells of the atoms are filled even though the total number of electrons is not enough to satisfy both atoms individually. This is called *covalent bonding* and is shown in Fig. 2–6. Diamond is one of the few minerals with covalent bonding, and this accounts for the hardness of the diamond. Bonding of the silicate minerals is midway between ionic and covalent.

A third common type of bonding is the *metallic bond.* In a metal, *the atoms are closely packed so that the electrons are not owned by any particular nucleus.* This allows *the electrons to move rather freely through the metal* and so accounts for the good thermal and electrical conductivity of metals. In minerals,

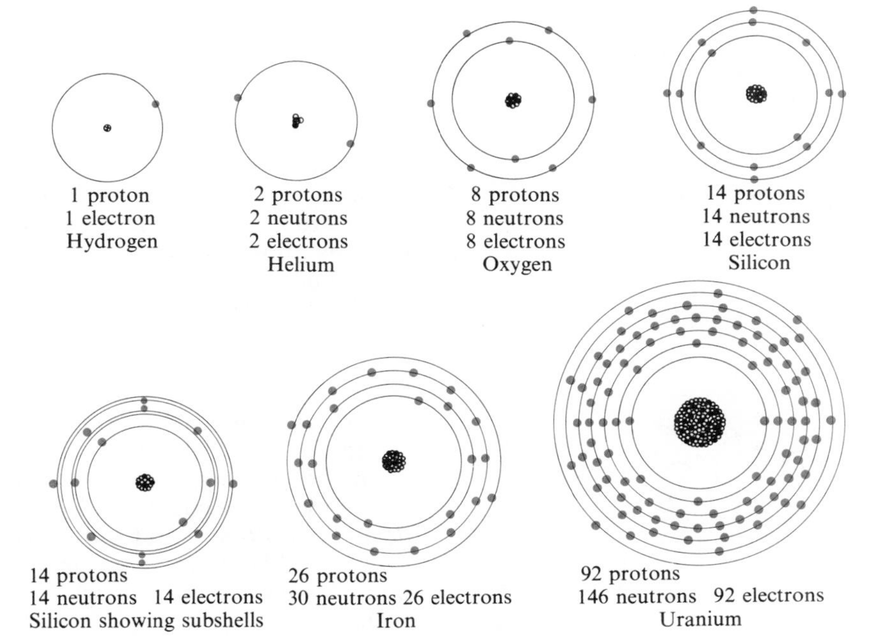

1 proton 1 electron Hydrogen	2 protons 2 neutrons 2 electrons Helium	8 protons 8 neutrons 8 electrons Oxygen	14 protons 14 neutrons 14 electrons Silicon

14 protons 14 neutrons 14 electrons Silicon showing subshells	26 protons 30 neutrons 26 electrons Iron	92 protons 146 neutrons 92 electrons Uranium

FIGURE 2–4 Diagrammatic representation of some common atoms. Although the electron shells are shown in only two dimensions here, the electrons actually move in spherical shells. The electron shells are complicated by subshells shown only for silicon. The nuclei contain almost all of the mass of the atoms, but the atomic radii are between 20,000 and 200,000 times the radii of the nuclei.

TABLE 2-3   Examples of naturally occurring isotopes.

Element	Symbol*	Atomic Number (= number of protons)	Nucleus		Atomic Mass Number (= number of protons + neutrons)	Electrons (= number of protons)	Remarks
			Protons	Neutrons			
Hydrogen	^1_1H	1	1	0	1	1	Common form
	^2_1H	1	1	1	2	1	Deuterium
Helium	^4_2He	2	2	2	4	2	Common form
	^3_2He	2	2	1	3	2	
Oxygen	$^{16}_8O$	8	8	8	16	8	Common form
	$^{18}_8O$	8	8	10	18	8	
Carbon	$^{12}_6C$	6	6	6	12	6	Common form
	$^{14}_6C$	6	6	8	14	6	
Potassium	$^{39}_{19}K$	19	19	20	39	19	Common form
	$^{40}_{19}K$	19	19	21	40	19	
Lead	$^{206}_{82}Pb$	82	82	124	206	82	
	$^{207}_{82}Pb$	82	82	125	207	82	
	$^{208}_{82}Pb$	82	82	126	208	82	
Uranium	$^{235}_{92}U$	92	92	143	235	92	
	$^{238}_{92}U$	92	92	146	238	92	

*The subscript is the atomic number and the superscript is the atomic mass number. Sometimes only the superscript is used because only the mass number can vary for a given element.

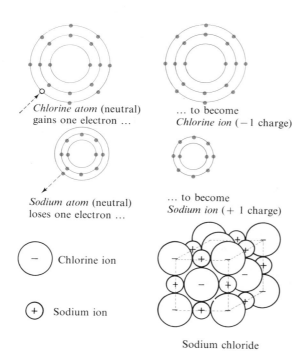

Chlorine atom (neutral) gains one electron ...

... to become Chlorine ion (−1 charge)

Sodium atom (neutral) loses one electron ...

... to become Sodium ion (+ 1 charge)

⊖ Chlorine ion

⊕ Sodium ion

Sodium chloride

FIGURE 2-5   Ionic bonding. Ions are formed by the completion of an electron ring by the gain or loss of electrons. The resulting ions are held together (bonded) by the attraction of unlike electric charges.

metallic bonds are found in some sulfides and in native elements, such as gold.

The size of atoms or ions is very important in determining the structure of minerals. Ions that are within about 10 percent of the same size may mutually substitute in many minerals. Some examples of ion pairs that can mutually substitute are silicon and aluminum, sodium and calcium, iron and magnesium, and gold and silver. (See Table 2–4 and Fig. 2–7.) The type of bond also affects possible substitution, and although potassium and gold are nearly the same size, they do not substitute for each other. Potassium forms ionic bonds, and gold forms covalent bonds with oxygen.

Our knowledge of the internal structure of minerals comes from X-ray studies. X rays have wavelengths in the same size range as atoms, and so are deflected by the

TABLE 2-4 Atomic and ionic radii of some elements. Note the general relationship between gain or loss of electrons and the difference between the atomic and ionic radii.

Element	Atomic Number	Number of Electrons per Shell and Subshell						Ion Formed	Number of Electrons Gained or Lost to Form Ion	Radii (Angstroms)	
		1st	2nd	3rd	4th	5th	6th			Ionic	Atomic
Oxygen	8	2	2·4					O^{-2}	2 gained	1.40	0.60
Sodium	11	2	2·6	1				Na^+	1 lost	0.97	1.86
Magnesium	12	2	2·6	2				Mg^{+2}	2 lost	0.66	1.60
Aluminum	13	2	2·6	2·1				Al^{+3}	3 lost	0.51	1.43
Silicon	14	2	2·6	2·2				Si^{+4}	4 lost	0.42	1.17
Sulfur	16	2	2·6	2·4				S^{-2}	2 gained	1.85	1.04
Chlorine	17	2	2·6	2·5				Cl^-	1 gained	1.81	1.07
Potassium	19	2	2·6	2·6	1			K^+	1 lost	1.33	2.31
Calcium	20	2	2·6	2·6	2			Ca^{+2}	2 lost	0.99	1.96
Titanium	22	2	2·6	2·6·2	2			Ti^{+3}	3 lost	0.76	1.46
								Ti^{+4}	4 lost	0.68	1.46
Iron	26	2	2·6	2·6·6	2			Fe^{+2}	2 lost	0.74	1.24
								Fe^{+3}	3 lost	0.64	1.24
Silver	47	2	2·6	2·6·10	2·6·10	1		Ag^+	1 lost	1.26	1.44
Gold	79	2	2·6	2·6·10	2·6·10	2·6·9	1	Au^+	1 lost	1.37	1.44

atoms in a crystal. This deflection occurs because the atoms in a crystal are packed in an orderly way. If the packing were not orderly, the X rays would be deflected in a random way. Instead, the regular pattern of the deflected X rays can be used to determine the internal structure of a mineral. (See Fig. 2–8.)

NATURAL RADIOACTIVITY

The material in this section is not directly related to mineralogy—the main topic of this chapter. The geological application of this section appears in later chapters. It is included here because it is directly related to atomic structure.

Not all nuclei are stable. *Radioactive nuclei disintegrate spontaneously, releasing energy* in the process. Understanding why and how this takes place was one of the great advances in physics during the first half of the twentieth century. In natural radioactivity, the unstable nucleus emits several types of high-energy particles and also releases energy in the form of electromagnetic waves similar to light energy.

The more important particles released in natural radioactivity are

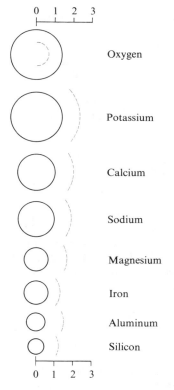

FIGURE 2–7 Comparison of ionic sizes of some common elements. Their atomic sizes are indicated by the dashed lines. The scale is in nanometers. (1 nanometer equals one-billionth of a meter (10^{-9} meter.)

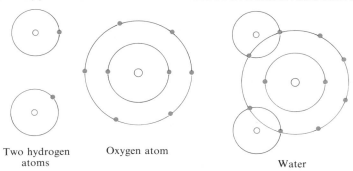

FIGURE 2–6 Covalent bonding, shown diagrammatically, results when electrons are shared by more than one atom.

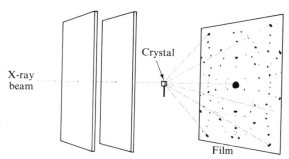

FIGURE 2–8 A beam of X rays is deflected by the atoms in a crystal into a regular pattern, thus revealing the internal structure of the crystal.

called *alpha* and *beta particles. Alpha particles are helium nuclei (ions) and so consist of two protons and two neutrons.* They are therefore positively charged. *Beta particles are electrons moving at velocities near the speed of light* and are negatively charged, as well as much smaller and less massive than alpha particles. The amount of energy released by emission of a particle depends on the mass and the velocity of the particle.

The other type of radiation in natural radioactivity is gamma radiation. *Gamma rays are energy in the form of electromagnetic waves.* Visible light, radio waves, and X rays are also energy in the form of electromagnetic waves and differ only in their frequency or wavelength. Gamma rays differ from X rays only in frequency; gamma rays have a somewhat higher frequency.

In radioactive decay, an atom is changed to an atom of another element. This means that the number of atoms of a radioactive element decreases with the passage of time. If the rate of disintegration is known, then by measuring the amounts of the parent and daughter elements, the age of crystallization of the mineral containing the parent element can be found. This is the principle of radioactive dating, and works only because the rate of disintegration is not affected by temperature, pressure, and so forth.

The rate of radioactive decay is expressed in terms of the half-life.

In one half-life, one-half of the radioactive atoms present will decay to the daughter element. In the next half-life, one-half of the remaining radioactive atoms will decay, and so on. (See Fig. 2–9.)

Only a few of the natural radioactive isotopes are of geologic importance. (See Appendix E for a list of the naturally occurring radioactive isotopes.) Some of these are useful in age determination, and others are sources of radioactive heating of the earth. Those that have proved most useful in dating are carbon 14, potassium 40, rubidium 87, uranium 235, uranium 238, and thorium 232. The requirements for dating are a reasonable rate of decay (half-life), the retention of daughter isotopes, and the existence of common minerals containing the parent

element. (Radioactive dating is covered in Chapter 16.) Those most important as a source of radioactive heating can be determined from consideration of abundance data and amount of energy released, both of which are shown in the tables. They are uranium 235, uranium 238, thorium 232, and potassium 40.

The data for the main heat-producing isotopes are shown in Table 2–5. Notice that because all of these isotopes have been decaying throughout geologic time, more of them were present in the early history of the earth and so produced more heat at that time. Note also that potassium 40 and uranium 235 both have much shorter half-lives than uranium 238 and thorium 232 and so are assumed to have been much more abundant at that time and thus produced much more heat during the early stages of the earth.

In the earth the energy released by radioactivity goes mainly into heat. Each kind of radiation behaves differently. The heavy, slow-moving alpha particles are stopped by a sheet of paper, but their impact is very destructive. They impart some of their kinetic energy to electrons and, being heavy, may disrupt whole atoms. Beta particles are much lighter

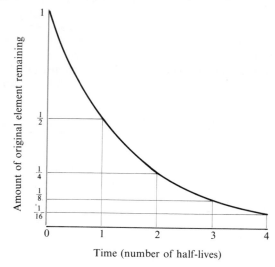

FIGURE 2–9 Rate of radioactive decay. During each half-life, one-half of the remaining amount of the radioactive element decays to its daughter element.

TABLE 2–5   Radioactive heat production in rocks.

Radioactive Isotope	Granitic Rocks		Basalt	
	Present	4500 Million Years Ago	Present	4500 Million Years Ago
Uranium 238	10.60	22.2	1.34	2.68
Uranium 235	0.46	37.5	0.06	4.74
Thorium 232	11.47	14.4	1.67	2.10
Potassium 40	3.30	40.1	0.73	8.89
Total	25.83	114.2	3.80	18.41

SOURCE: J. W. Winchester, "Terrestrial Heat Flow, Radioactivity, and the Chemical Composition of the Earth's Interior," *Journal of Geological Education* (December, 1966), Vol. 14, No. 5, pp. 200–204.
NOTE: At the time the earth formed 4500 million years ago, the shorter-lived radioactive elements were more abundant and so produced more heat than at present. The units used are 10^{-14} calorie per second per gram of rock.

and faster. They are about one hundred times more penetrating than alpha particles, but do about one hundred times less damage. Because they are too light to affect nuclei, they lose their kinetic energy by interaction with electrons. Gamma rays move with the velocity of light and have no mass. They are about one hundred times more penetrating than beta particles but do much less damage. Like beta rays, gamma rays only disrupt electrons. Heat is molecular motion, so the main effect of

radioactivity is to heat the surroundings by imparting motion to electrons and atoms. Other effects are damage to a crystal lattice and increase of potential energy by moving electrons from one shell to another. These less-important features are discussed below.

It should also be clear that only gamma rays have high energy and can penetrate very far, so most radiation detectors work mainly on gamma rays. Alpha rays are absorbed by about four inches of air, and beta rays by

about a yard of air, but gamma rays can travel over a half-mile in air. Less than an inch of rock will absorb alpha and beta radiation, but gamma rays can penetrate over 30 centimeters (1 foot) of rock. These points should be taken into account in prospecting for radioactive mineral deposits. Figure 2–10 shows a means of detecting radioactivity.

IDENTIFICATION OF MINERALS— PHYSICAL PROPERTIES

The definition of mineral suggests that minerals can be identified in several ways. The fact that minerals have reasonably fixed chemical compositions suggests that ordinary methods of chemical analysis can be used, and, historically, this approach was important in the development of both mineralogy and chemistry. The drawbacks to its current use are that a laboratory, as well as much skill and knowledge, is needed, and the relatively recent discovery that certain minerals having identical

A   B

FIGURE 2–10   The mineral uraninite. A. An ordinary photograph. B. An autoradiograph made by placing photographic film next to the mineral in a darkroom. The radiation from the uraninite exposes the film even though no visible light reaches the film. This can be done without a darkroom by wrapping the film in opaque paper and laying the flat mineral on the film. Exposure times are generally a few weeks. Photo from Ward's Natural Science Establishment, Inc., Rochester, N.Y.

chemical compositions have very different internal (crystal) structures. In this later case, the *polymorphs (many forms)* generally have very different characteristics, such as graphite and diamond, which are both composed of pure carbon. Proof that certain minerals were indeed polymorphs had to wait until the use of X rays to probe crystal structure was perfected. Another problem in the use of chemical analysis is that analysis of the all-important silicate minerals is an extremely difficult task, even in the best-equipped laboratories. Fortunately, this problem can be bypassed by using the easily determined physical properties to identify minerals, and this can be supplemented where necessary with spot chemical tests.

The *physical properties include such things as color, luster, hardness, weight (specific gravity), crystal form, fluorescence, taste, solubility, cleavage, magnetism, radioactivity, and many more.* The use of the more important physical properties follows.

HARDNESS

The *hardness* of a mineral *is its resistance to scratching.* As might be guessed, it is a difficult property to measure exactly because the amount of force, the shape of the scratches, and the relation of the surface tested to the crystal structure will all affect the measured hardness. In spite of this, the gross hardness is easily measured and is of great help in identification. Because surface alteration may change the apparent hardness, fresh surfaces only must be tested.

The hardness scale (Mohs) is based on ten minerals. The steps are not of equal value but are arbitrarily defined. The steps are reasonably equally spaced except that the step between corundum and diamond is very large.

1. Talc (Softest mineral)
2. Gypsum
3. Calcite
4. Fluorite
5. Apatite
6. Orthoclase
7. Quartz
8. Topaz
9. Corundum
10. Diamond (Hardest mineral)

A useful hardness scale is:

2.
—Fingernail
3.
—Copper cent (Use a bright shiny one, or you will only test the hardness of the tarnish.)
4.
5.
—Knife blade or glass plate
6. —File
7. —Quartz
8.

To determine the hardness of a mineral, you must find the softest mineral on the hardness scale that will scratch it. With a little practice, only a knife blade and your fingernail will be necessary to estimate the hardness. Note that two sub-

stances of the same hardness can scratch each other. A pitfall that can trap the neophyte is accepting any mark on the mineral as a scratch, because just as soft chalk leaves a mark on a hard blackboard, a soft mineral may leave a mark on a hard one. You can check this by trying to remove the mark with your moistened finger, by feeling the mark with your fingernail (even a tiny depression can be detected), and by looking very closely at the mark with a lens. The final test should always be to reverse the test and see whether the mineral can scratch the test mineral.

CLEAVAGE AND FRACTURE

Cleavage and fracture describe the way a mineral breaks. *Any irregular break is termed a fracture.* Many terms have been coined to describe fracture, but most, such as even, fibrous, splintery, or hackly, are self-explanatory. A common type of fracture, called *conchoidal, is the hollow, rounded type of break that occurs in glass.* (See Fig. 3–33)

A mineral has a *cleavage has a direction of weakness that, when broken, produces a smooth plane that reflects light.* (See Figs. 2–11 and 2–12.) The common minerals may have up to six such directions. Note that a cleavage is a direction; therefore, one cleavage direction will, in general, produce two cleavage surfaces. Cleavages can be recognized by their smoothness and by their tendency to form pairs or steps, as suggested in the sketches. A series of steps is more common in minerals with poor cleavage, that is, minerals in which the cleavage direction is only slightly weaker than any other direction in the mineral. Even in this case, each of the cleavage surfaces will be smooth and parallel and so will reflect light as a unit.

The angle between cleavage directions can be important in the

Two cleavages meeting at right angle

Two cleavages meeting at oblique angle

Perfect cleavage in one direction – Note two cleavage surfaces

Poor cleavage in one direction – Note step effect

FIGURE 2–11 Cleavage.

A

B

C

D

FIGURE 2–12 Minerals showing cleavage. A. Mica crystal with one cleavage. B. Galena, cubic cleavage. C. Calcite. Note the double refraction of the string behind the crystal. D. Serpentine asbestos with fibrous cleavage. Photos from Ward's Natural Science Establishment, Inc., Rochester, N.Y.

identification of some minerals as noted in Appendix A.

The planes of weakness that are the cleavage directions in a mineral are caused by the atomic structure of the mineral. Cleavages form along directions of weak bonding as shown in Fig. 2–13.

CRYSTAL FORM

All crystalline substances crystallize in one of six crystal systems. (See Figs. 2–14, 2–15, and 2–16.) If the mineral grows in unrestricted space, it develops the external shape of its crystal form; if it cannot grow its external shape, its crystalline nature can be determined only under the microscope or by X-ray analysis. Many external forms are possible in each of the systems; however, the system can be determined by the symmetry of the crystal. Crystallography is a fascinating subject, and more information can be found in the mineralogy references listed at the end of this chapter.

COLOR

Strictly speaking, color means the color of a fresh, unaltered surface, although in some cases the tarnished or weathered color may help in identification. For some minerals color is diagnostic, but many, such as quartz, may have almost any color, due to slight impurities.

STREAK

Streak is the color of the powdered mineral. To see the streak,

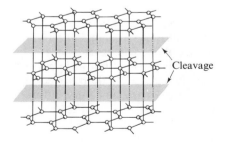

FIGURE 2–13 Structure of graphite. The bonds between the carbon sheets are weak, causing the cleavage in graphite. Graphite and diamond are both composed of carbon only; the differences between them are caused by the internal structure.

31

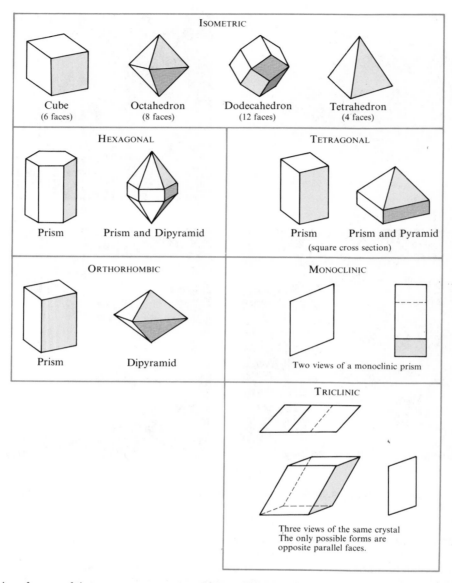

FIGURE 2–14   Examples of some of the more common crystal forms. They are shown and named for reference only.

the mineral is rubbed on a piece of unglazed porcelain called a streak plate. The color of the streak in many minerals is a more constant property than the color. The streak may be a very different color from the mineral color, and this, too, is of help in identification. Care should be taken, especially when working with small disseminated crystals, that only the mineral, and not the matrix, is rubbed on the streak plate. Many powdered minerals are white and are said to have no streak. Miner-

als with a hardness greater than the streak plate will, of course, have no streak.

LUSTER

Luster is the way a mineral reflects light. The two main types of luster are metallic and nonmetallic. The distinction is difficult to describe in words; either it looks like metal, or it does not. A few fall in between and can be called submetallic. Dozens of terms have been proposed to describe all of

the types of luster. A simplified outline of the most commonly used terms follows. Most terms are self-explanatory.

Metallic
   Bright
   Dull
(Submetallic) best to avoid if possible.
Nonmetallic
   Adamantine (brilliant luster, like a diamond)
   Vitreous (glassy)
   Greasy

FIGURE 2–15   Mineral specimens showing crystal forms. Parts A, Quartz, and B, Galena, from Ward's Natural Science Establishment, Inc., Rochester, N.Y.; parts C, Orthoclase (potassium feldspar), and D, Pyrite, from Smithsonian Institution.

Resinous
Waxy
Pearly
Silky
Dull or earthy

SPECIFIC GRAVITY

Specific gravity is a measure of the relative weight of a substance. It *is the ratio of the mass of a substance to the mass of an equal volume of water.* It is measured by weighing the substance in air and in water,

Specific gravity

$$= \frac{\text{weight in air}}{\text{weight in air} - \text{weight in water}}$$

This is a difficult measurement to make with most specimens. Esti-mations of specific gravity can be made fairly accurately by one who is experienced in handling miner-als. It is of help in identification, especially with the very heavy minerals.

The *density* of a substance is its mass per unit volume; when ex-pressed in grams per cubic centi-meter, density is numerically equal to specific gravity.

TASTE AND SOLUBILITY

Taste and solubility can be deter-mined by touching the specimen with the tongue. A soluble mineral will have the feel of a lump of sugar, and an insoluble mineral may feel like glass.

Reaction with dilute hydrochlo-ric acid is another solubility test.

Calcite will react, producing effer-vescence. Dolomite (see Tables A–1 and A–2 in Appendix A) must be powdered to increase the surface area before effervescence occurs. Dolomite will also react with hot dilute acid or with strong acid, both of which also increase the rate of chemical reaction.

Other physical properties may require special equipment to test. Examples are radioactivity (Geiger counter), magnetism (magnet), fluorescence (ultraviolet light).

ROCK-FORMING MINERALS

The main rock-forming minerals are *silicates,* minerals in which

33

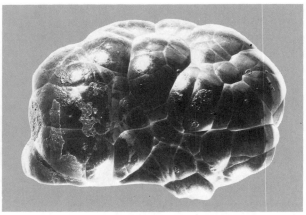

A

B

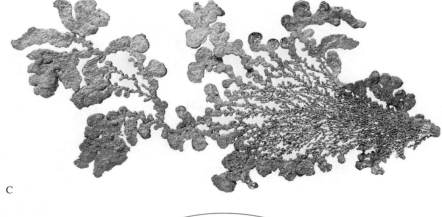

C

FIGURE 2–16 Minerals whose crystalline nature is not readily apparent. In most of the specimens shown, the individual crystals are too small to be seen with the naked eye. A. Agate. B. Hematite. C. Native copper. Part A is from Ward's Natural Science Establishment, Inc., Rochester, N.Y.; parts B and C are from the Smithsonian Institution.

elements such as iron, magnesium, sodium, calcium, and potassium are combined with aluminum, silicon, and oxygen. Their identification will be mentioned here and summarized in Appendix A. Their structures are shown in Fig. 2–3. Table 2–8 shows a common classification of minerals.

FELDSPARS

Feldspars are the most abundant minerals. (See Fig. 2–17.) They are composed of sodium, calcium, and potassium, combined with silicon, aluminum, and oxygen. There are two main types, and they are complicated by mixing.

Plagioclase feldspar is an example of a continuous mineral series. The high-temperature plagioclase is *calcic*, $CaAl_2Si_2O_8$; at low temperature *sodic plagioclase*, $NaAlSi_3O_8$, forms. A continuous series of intermediate plagioclase containing both sodium and calcium can form, depending on the temperature and composition of the melt. This is possible because sodium and calcium ions are about the same size. Aluminum

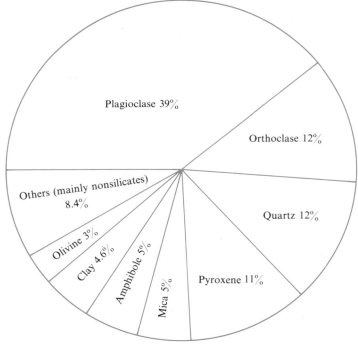

Plagioclase 39%

Orthoclase 12%

Others (mainly nonsilicates) 8.4%

Quartz 12%

Olivine 3%

Clay 4.6%

Amphibole 5%

Mica 5%

Pyroxene 11%

FIGURE 2–17 Relative abundances of minerals in the earth's crust. Data from Ronov and Yaroshevsky, 1969.

TABLE 2–8   Common minerals and their classification

Native elements	Gold, Au; graphite, C
Sulfides (and sulphosalts)	Pyrite, FeS_2; galena, PbS
Oxides and hydroxides	Hematite, Fe_2O_3; magnetite, Fe_3O_4; "limonite," $FeO(OH)·nH_2O$
Halides	Halite, NaCl; fluorite, CaF_2
Carbonates, nitrates, borates, and iodates	Calcite, $CaCO_3$; dolomite, $CaMg(CO_3)_2$
Sulfates, chromates, molybdates, tungstates	Gypsum, $CaSO_4·2H_2O$; anhydrite, $CaSO_4$
Phosphates, arsenates, and vanadates	Apatite, $Ca_5(F,Cl,OH)(PO_4)_3$
Silicates	Rock-forming minerals

and silicon, which are also nearly the same size, also mutually substitute. The microscope must be used to determine the relative amounts of sodium and calcium in plagioclase.

Potassium feldspar, $KAlSi_3O_8$, is more complicated. A number of varieties have been recognized, but their identification requires microscopic or X-ray techniques. The variety that generally forms in granitic rocks is commonly, although generally incorrectly, called *orthoclase;* and the variety in volcanic rocks, which crystallize at higher temperatures than granitic rocks, is called *sanidine.*

Limited amounts of sodium can be accommodated in potassium feldspar in spite of the differences in ionic sizes. This limited mixing is further complicated in that the temperature of formation further controls the amount of mixing. (See Fig. 2–18.)

At high temperatures, such as in volcanic rocks, all mixtures of sodium and potassium are possible and can be preserved by rapid chilling. Slow cooling will allow the potassium feldspar and the sodic plagioclase to unmix. In some rocks the unmixing is complete and two distinct crystals form; but in many, the two feldspar crystals are intergrown and form *perthite.*

Feldspars are identified by their light color, hardness (6), and two cleavages at nearly right angles. Striations on one cleavage direction will distinguish plagioclase (Fig. 2–19). Perthite will appear as mottled, irregular veining of different color (Fig. 2–20).

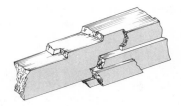

FIGURE 2–19   Plagioclase striations appear on only one of the two cleavage directions.

OLIVINE

Olivine, $(Mg,Fe)_2SiO_4$, is the simplest of the ferromagnesian or dark-colored minerals. The parentheses in the formula mean that magnesium and iron may be present in varying amounts because they are nearly the same size. Magnesium olivine forms at higher temperatures than the iron-rich types. Olivine generally forms small, rounded crystals that, because of alteration, may give deceptively low hardness tests. It can be recognized by its distinctive olive-green color.

PYROXENE

Pyroxene is the family name of a very large and complex group of minerals. Here we will consider only *augite,* $(Ca,Na)(Mg,Fe^{II}, Fe^{III},Al)(Si,Al)_2O_6$, the main rock-forming mineral of the family. The possible range in composition is suggested by its complex formula. It is generally recognized by its dark color, hardness (5–6), and two cleavages that meet at nearly right angles.

AMPHIBOLE

Amphibole is a family name of another very large and complex group of minerals. Here we will consider only *hornblende,* $Ca_2Na(Mg,Fe^{II})_4(Al,Fe^{III},Ti)(Al,Si)_8O_{22}(O,OH)_2$, the main rock-former in the group. Its composition is even more complex than that of augite. It is black, has a hardness of 5–6, and has two cleavages that meet at oblique angles, distinguishing it from augite.

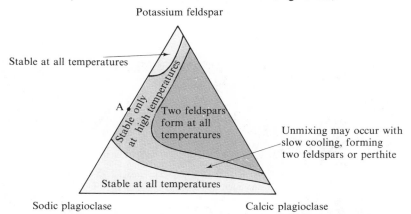

FIGURE 2–18   This triangular composition diagram shows the possible mixtures of the three feldspars and their temperature stability. This diagram is based on experimental data. Compositions on the left side of the diagram are common in the granitic rocks discussed in the next chapter. A feldspar composed of 50 percent sodic plagioclase and 50 percent potassium feldspar would fall on point A on the diagram.

GEMS

Gems are generally prized for their beauty, but must have other qualities as well. Most gems are minerals (pearls and amber are notable exceptions because they are organic), but many beautiful minerals are not gems. Rarity, and therefore expense, are certainly qualities of gems in the minds of many people.

Hardness is important to a gem. It must be hard in order to resist scratching, and it must also be tough enough to be durable and resist breakage. Many gems are cut, generally on cleavages, so that they have many faces or facets; others are shaped and polished (cabochons). Jade is a gem prized for its toughness; because of its toughness, it can be carved into beautiful objects. Diamonds are much harder than jade but not nearly so tough; diamonds will shatter if struck because they have good cleavage. Opal and turquoise are really too soft and too easily damaged to make good rings, as many owners will sadly testify.

Much of the beauty of gems results from their color, and especially how they affect light. The brilliance of many gems, such as diamond, comes from their high index of refraction, which is a measure of how much a light ray is bent on entering the crystal. Note that if the faces are cut at the correct angles, internal reflection occurs, giving the diamond its sparkle or brilliance. Commonly, too, such gems also have high dispersion so that the reflected light shows many colors. This occurs because each color or wavelength of light is bent a different amount in the gem; therefore, the entering white light is split into all the colors of the rainbow. Other gems, such as opal, also show a play of colors or iridescence, but the dispersion in them is due to minute fractures, or to intergrowths of tiny crystals.

CLASSICAL GEMS

Diamond. C. Hardness 10; the hardest mineral. Good cleavage. Very high index of refraction gives it fire and brilliance. Colorless, or pale shades of yellow, red, orange, green, blue, or brown. Most come from South America and Africa. They originate in disseminated magmatic deposits and are found in placer deposits (see Chapter 6). A number have been found in North America: in igneous rocks near Murfreesboro, Arkansas, and in surficial sediments and glacial deposits at many places.

Corundum. Al_2O_3. Hardness 9. No true cleavage but may be split easily in several directions. Corundum is a common mineral, found in metamorphic rocks and in placer deposits. Only clear

FIGURE 2–20 Cleavage surface of perthite, showing intergrowth of two types of feldspar. Photo from Smithsonian Institution.

corundum forms gems. Red varieties are *ruby,* and all other types are *sapphire.* Their colors come from traces of other metals. Star sapphire and ruby have minute inclusions arranged in certain places in the crystal lattice, giving the star effect.

Beryl. $Be_3Al_2Si_6O_{18}$. Hardness $7\frac{1}{2}$ to 8. Cleavage in one direction. Beryl is a common mineral found in many granites, especially pegmatites, and in felsites. Beryl is an ore of the rare element beryllium, and only clear beryl is a gem. Blue-green varieties are *aquamarine,* green varieties are *emerald,* and other names are used for other colors. The color of specimens from some localities can be changed by heating.

Topaz. $Al_2(SiO_4)(F,OH)_2$. Hardness 8. Cleavage in one direction. Yellow, pink, and blue-green. Occurrence is like that of beryl.

OTHER POPULAR GEMS

Jade. Two different minerals are called jade: *jadeite* $NaAl(Si_2O_6)$ and *nephrite,* an amphibole. Both have hardness about 6, no cleavage, and are green. Because they are very tough, they can be carved. Both are found in metamorphic rocks.

Turquoise. $CuAl_6(PO_4)_4(OH)_8 \cdot 2H_2O$. Hardness 6. No cleavage. Blue or green color. Found mainly as veins in altered volcanic rocks.

Garnet is a family of common, rock-forming minerals. Only a few are clear red gems.

Olivine is one of the common rock-forming minerals. *Peridot* is green, transparent olivine.

Feldspar is a family of common, rock-forming minerals. *Moonstone* is a milky-white, opalescent variety of orthoclase. *Amazonstone* is green microcline. Both are feldspars.

Quartz is a common, rock-forming mineral. The gem varieties are *rock crystal,* clear; *amethyst,* purple; *citrine,* yellow; *smoky quartz,* brown to black; and *rose quartz,* pink. Very finely crystalline types of quartz, such as *agate, chert,* and *chalcedony,* are used as semiprecious gems under an almost endless number of names. *Opal,* $SiO_2 \cdot nH_2O$, is also considered one of the quartz family, although opal is not crystalline and so is not really a mineral.

This list includes only the more popular gems, and to cover all gems and varieties would require an entire book.

MICA

The micas are easily recognized by their colors and perfect cleavage. The common micas and their formulas are

Biotite—black mica,
    $K(Mg,Fe)_3(AlSi_3O_{10})(OH)_2$
Muscovite—white mica,
    $KAl_2(AlSi_3O_{10})(OH)_2$

QUARTZ

Quartz, SiO_2, is generally clear or white but may have any color. It is recognized by its hardness (7) and lack of cleavage.

As in many other minerals, the color of quartz can be caused by very minor amounts of impurities. How this can occur is shown in the following diagrams of crystal defects (Fig. 2–21). In other cases color is caused by radiation as described earlier.

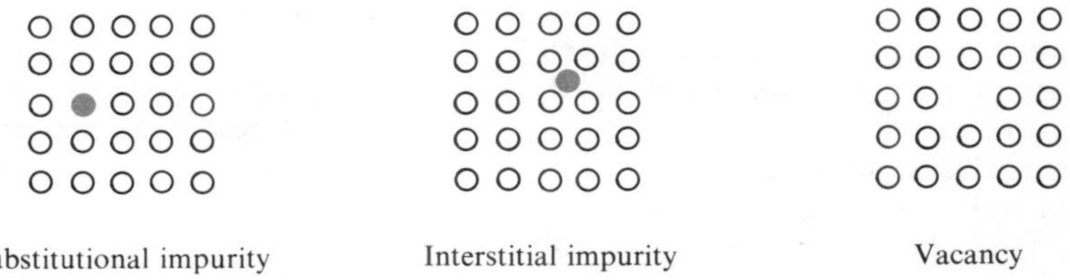

Substitutional impurity   Interstitial impurity   Vacancy

FIGURE 2–21   Some types of crystal defects shown diagrammatically.

SUMMARY

The crust of the earth is composed almost entirely of 8 elements: oxygen, silicon, aluminum, iron, calcium, sodium, magnesium, and potassium.

Elements are materials that cannot be subdivided by ordinary chemical methods. A little over 100 are known.

Atoms are the smallest units of an element. They are composed of electrons, neutrons, and protons.

Minerals are

1. naturally occurring, crystalline
2. inorganic substances with
3. a definite small range in chemical and physical properties.

Crystalline substances have orderly internal structures.

A *crystal* is a solid form bounded by smooth planes which give an outward manifestation of the orderly internal structure.

A *rock* is

1. a natural aggregate of one or more minerals *or*
2. any essential and appreciable part of the solid portion of the earth (or any other part of the solar system).

There are about 2000 minerals; only about 20 are common, and fewer than 10 form over 90 percent of all rocks.

Ions are atoms that have gained or lost electrons and so have an electrical charge.

Silicon and oxygen, the two most abundant elements, form the SiO_4 *tetrahedron* which is the basic building block of the rock-forming minerals.

The shape of the SiO_4 tetrahedron is determined by the relative sizes of oxygen and silicon ions, and consists of one small silicon ion surrounded by four large oxygen ions.

SiO_4 tetrahedra are linked by sharing oxygen ions, and the resulting arrangements determine the structures of silicate minerals.

The small number of rock-forming minerals is determined by the limited number of ways that SiO_4 tetrahedra can join together and how the other abundant ions can fit into these structures.

The *nucleus* of an atom contains most of the mass of an atom and is composed of protons and neutrons.

Protons are heavy and have a positive electrical charge.

Neutrons have slightly more mass than protons and are electrically neutral.

The nucleus is surrounded by a cloud of light, electrically negative *electrons.*

An atom is mostly empty space.

The number of electrons equals the number of protons, so an atom is electrically neutral.

The number of protons determines the identity of an element and is called the *atomic number* of an element.

The number of protons and neutrons is the *atomic mass number.*

The number of neutrons in the atom of an element may vary, giving rise to *isotopes.*

Ionic bonding is due to the attraction of unlike electrically charged ions.

Covalent bonding results when atoms complete their outer electron shells by sharing electrons.

In *metallic bonding,* the atoms are closely packed and the electrons are free to move throughout the metal.

Ions of similar size can substitute for each other in certain minerals.

In radioactive decay, an atom changes into an atom of another element. In this process, alpha, beta, and gamma radiation may be emitted.

Alpha particles are helium nuclei (ions) and so consist of two protons and two neutrons.

Beta particles are high-velocity electrons.

Gamma rays are electromagnetic waves somewhat higher in frequency than X rays.

Half-life is the time required for one half of the remaining radioactive atoms to decay. This rate is unaffected by temperature, pressure, etc., and so radioactivity can be used to determine age.

The main radioactive elements used in geologic dating are carbon 14, potassium 40, rubidium 87, uranium 235, uranium 238, and thorium 232.

The main radioactive sources of energy in the earth are uranium 235, uranium 238, thorium 232, and potassium 40.

Minerals are best identified by their physical properties, such as cleavage, fracture, crystal form, hardness, color, streak, luster, specific gravity, taste, solubility, magnetism, and fluorescence.

Cleavage is a smooth break caused by planes of weakness in a mineral.

Conchoidal fracture is the hollow, rounded type of break that occurs in glass.

Hardness is the resistance of a mineral to scratching.

Streak is the color of the powdered mineral.

Luster is the way a mineral reflects light. The main types are *metallic* and *nonmetallic.*

Specific gravity is the ratio of the mass of a substance to the mass of an equal volume of water.

The main rock-forming minerals are:

Feldspars
  plagioclase
    calcic plagioclase $CaAl_2Si_2O_8$
    sodic plagioclase $NaAlSi_3O_8$
  potassium feldspar (orthoclase, sanidine) $KAlSi_3O_8$.

All mixtures of calcic and sodic plagioclase are possible. Limited mixing of potassium feldspar with plagioclase is possible at high temperatures, but only a few of these mixtures are stable at ordinary temperatures. The unmixed feldspar produced by slow cooling of these mixes is called *perthite.*

Olivine $(Mg,Fe)_2SiO_4$
Pyroxene
   augite $(Ca,Na)(Mg, Fe^{II},Fe^{III},Al)$
   $(Si,Al)_2O_6$
Amphibole
   hornblende $Ca_2Na(Mg,Fe^{II})_4$
   $(Al,Fe^{III},Ti)(Al,Si)_8O_{22}(O,OH)_2$

Mica
   biotite (black mica) $K(Mg,Fe)_3$
   $(AlSi_3O_{10})(OH)_2$
   muscovite (white mica)
   $KAl_2(AlSi_3O_{10})(OH)_2$
Quartz, SiO_2

QUESTIONS

1 The four most abundant elements (weight percent) in order are ____ , ____ , ____ , and ____ .
2 Define mineral.
3 Distinguish between crystal and crystalline.
4 Why are there only eight important rock-forming minerals?
5 Can you relate the perfect cleavage in mica to its internal structure?
6 How are minerals distinguished from rocks?
7 Distinguish between silicon and silicate.
8 Why are the physical properties of graphite and diamond so different?
9 What factors determine whether one element can substitute for another in a mineral?
10 Why are silicon-oxygen tetrahedra the building blocks of most minerals?
11 Describe the units of which atoms are made.
12 What are ions, and what is ionic bonding?
13 What is an isotope?
14 Which are the geologically more important radioactive elements?
15 Define half-life. Why was radioactive heating more important in the early history of the earth than it is now?

SUPPLEMENTARY READINGS

Berry, L. G., and Brian Mason., *Mineralogy.* San Francisco: W. H. Freeman and Co., 1959, 612 pp.
Dietrich, R. V., and B. J. Skinner., *Rocks and Rock Minerals.* Somerset, N.J.: John Wiley & Sons, Inc., 1979, 319 pp.
Ernst, W. G., *Earth Materials.* Englewood Cliffs, N.J.: Prentice-Hall, Inc., 1969, 150 pp. (paperback).
Hurlbut, C. S., Jr. and C. Klein., *Dana's Manual of Mineralogy,* 19th ed. New York: John Wiley & Sons, Inc., 1977, 532 pp.
Loeffler, B. M., and R. G. Burns., "Shedding Light on the Color of Gems and Minerals," *American Scientist* (November-December 1976), Vol. 64. No. 6, pp. 636–647.
Rapp, George, Jr., *Color of Minerals.* ESCP Pamphlet Series, PS-6, Boston: Houghton Mifflin Co., 1971, 30 pp.

3 IGNEOUS ROCKS AND VOLCANOES

INTRODUCTION

The study of igneous rocks developed from the study of mineralogy. It began with a controversy started by Werner who, as noted earlier, contributed much to mineralogy. Werner was a proponent of Neptunism, the theory that rocks were precipitated in layers from a universal ocean. Thus he taught that all rocks are what would now be termed sedimentary rocks. He thought that granite, metamorphic rocks, and volcanic rocks were all precipitates from this universal ocean. The active volcanoes, such as those in Italy, were explained as caused by the burning of coal beds. The eventual downfall of Neptunism came from the proof that basalt, which we now know is the most common volcanic rock, is a volcanic rock and not a chemical precipitate. One reason basalt was thought to be crystallized from a solution was that it commonly forms six-sided prisms that resemble crystals such as those of quartz. The basalt prisms are the result of contraction during cooling (see Fig. 3–4).

Abraham Gottlob Werner (1749–1817) began teaching mineralogy at Freiberg in 1774. He was an extraordinary teacher, and, as a result, students came to him from all over Europe. It was his students who spread his theories far and wide because he published almost nothing. His aversion to writing was so great that in his later years he would not even write letters. His great success in mineralogy was because he was an organizer, and he organized what was known about minerals into a useful classification. His interest in organization was such that he was said by some to buy books so that he could arrange them in his library. Neptunism, in a sense, was an attempt to organize rocks into a simple, easily understood system. Probably this simplicity was the great appeal of Neptunism, for no further study was necessary. A rock's place in the onion-skin layers of the earth could be determined by merely identifying and naming it.

The Neptunists had a number of problems, such as what happened to the water in the universal sea. They had no answer to this question. Another problem was that enough was known about chemistry at the time to know that the volume of rocks assumed to have precipitated was much greater than could have been in solution in the sea. No one seems to have been troubled by this.

The main problem, however, was the origin of basalt. Unknown to the antagonists of the Neptunism controversy, the volcanic origin of basalt had been demonstrated in 1774, which was before Werner had taken up Neptunism. The story begins with Jean Étienne Guettard (1715–1786) who, after visiting Auvergne, an area in central France, recognized the ancient volcanoes there. (See Fig. 3–1.) He published this observation in 1752. In 1770, he published the idea that basalt is not lava; by this he meant that the rocks he had seen at Auvergne were different from the similar rocks with no remaining vestige of volcanoes that were called basalt. Ancient basalt is commonly found with sedimentary rocks, and erosion has removed the more obvious indications of its volcanic origin, such as volcanic cones. Thus, in a sense, Guettard was on both sides of the basalt and Neptunism questions. Nicholas Desmarest (1725–1815) was another Frenchman who also studied the Auvergne volcanoes. His studies began in 1763 and were published in 1774. He clearly showed that the lava of Auvergne is basalt. This should have stopped the controversy before it started, but these studies went unnoticed for years.

The opponents of the Neptunists were called Plutonists because they believed that igneous rocks originated deep in the earth, the home of Pluto, the god of the underworld. Plutonism was just one aspect of *uniformitarianism, the idea that present processes occur-*

ing on earth can be used to interpret ancient rocks. This idea will be expanded in later chapters. Uniformitarianism was the idea of James Hutton (1726–1797), a Scotsman. He first stated it in 1788 and then published a book *Theory of the Earth, with Proofs and Illustrations* in 1795. The book was written in a difficult style and so was not widely read. After Hutton's death in 1802, his friend Playfair wrote a readable book, *Illustrations of the Huttonian Theory.*

The defeat of Neptune did not occur because of the work of the Plutonists as much as it did from the studies of Neptunists. Two of Werner's former students visited Auvergne. The first was Leopold von Buch (1774–1853), a German, who was one of the most influential geologists of the time and certainly the most traveled. He visited central France in 1802 and recognized the volcanic nature of the basalt there, but he did not concede that Werner's theory was completely wrong. From that point on, he gradually abandoned more and more of the Neptunist ideas as his travels brought more evidence before him. The other visitor to Auvergne was a Frenchman, Jean Françoise D'Aubuisson de Voissins (1769–1819). He abandoned Neptunism in 1804 after

Asama-yama, one of Japan's most active volcanoes. Photo from U.S. Geological Survey.

A

B

FIGURE 3–1 The volcanoes of Auvergne, France. A. Chaine des Puys. B. Puy de Pariou in foreground and Puy de Dôme in Chaine des Puys. Photos courtesy Maurice Krafft, Équipe Vulcain.

this visit. The journey was suggested by the referees of a paper embracing Neptunism that he had published the year before. This was a dramatic change. Thus Neptunism was challenged while Werner was still alive, and soon after his death Plutonism was widely accepted.

One reason that this controversy went on so long was the limited means available to study rocks. The hand lens was the only tool for close examination. It worked reasonably well on rocks with large grains such as granite, but it could give little insight as to the nature of fine-grained rocks such as basalt. Chemistry was developing at this time so some analyses were available to aid the study of fine-grained rocks.

The microscope was available, but its adoption by geologists was slow. Microscopic study is now one of the most important ways that rocks are studied. The technique used is to slice chips of rocks to the thinness of paper so that they are nearly transparent, and to examine these thin sections, as they are called, by passing light through them. The first thin sections were made by William Nicol who also invented the Nicol prism, a device that polarizes light. These thin sections were of petrified wood and were described in 1831. No further use was made of thin sections for about 25 years until Henry Clifton Sorby came to the University of Edinburgh and saw Nicol's thin sections. He clearly saw their value in the microscopic study of rocks. Interestingly, his work was not quickly accepted because many geologists said in effect that mountains cannot be studied under the microscope. In 1862, Sorby chanced to meet the German Ferdinand Zirkel (1838–1912) while on an excursion in Germany. Zirkel's adoption of Sorby's microscopic study of rocks put the Germans in the forefront of this phase of geology and led to the general acceptance of microscopic study of rocks.

The only other common way that natural rocks are studied is by chemical analysis. This method is applied mainly to rocks too fine-grained for microscopic study. Chemical studies and classification of rocks probably reached a peak near the turn of the present century. They may again become important because modern methods of instrumental analysis may make chemical analysis of rocks as cheap and easy as microscopic study.

Experimental study of igneous rocks has contributed much to our understanding. Such studies began early in the eighteenth century but did not become important until much later. Fouqué and Michel-Lévy did significant work in France, starting in the 1880s. Perhaps the most important work was that of N. L. Bowen (1887–1956) who worked mainly at the Geophysical Laboratory of the Carnegie Institution in Washington, D.C. His work on basalt melts is described in this chapter. Much of this work was done at atmospheric pressure and so may not duplicate conditions within the earth where most rocks crystallize. After World War II, furnaces were developed that could melt rocks under pressure, so the later studies of Bowen and his coworkers came much closer to duplicating natural conditions.

ROCK CYCLE

The rocks of the crust are classified into three types according to their origin. Two of these types are formed by processes deep in the earth and, therefore, tell us something about conditions within the crust. They are

1. *Igneous rocks, which solidify from a melt or magma; and*
2. *Metamorphic rocks, which are rocks that have been changed*—generally by high temperature and pressure within the crust.

The third type, which records the conditions at the surface, is

3. *Sedimentary rocks, which are composed of the weathered fragments of older rocks that are deposited in layers near the earth's surface by water, wind, and ice.*

Much of geology is concerned with the interactions among the forces that produce these three rock types. The relationships are quite involved, but can be illustrated by the rock cycle (Fig. 3–2). The relative abundance of igneous rocks is shown in Fig. 3–3.

Igneous rocks are formed by the crystallization of magma. A *magma is a natural, hot melt composed of a mutual solution of rock-forming materials (mainly silicates) and some volatiles (mainly steam) that are held in solution by pressure.* Magmas may contain some solid material, but are mobile. Magma probably originates

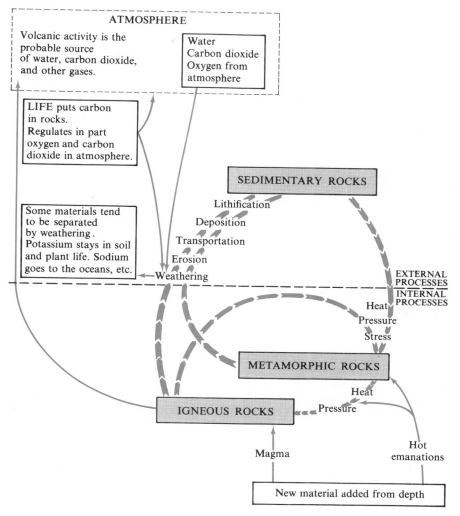

FIGURE 3–2 The middle of this diagram, showing the relationships among igneous, sedimentary, and metamorphic rocks, is what is generally considered the rock cycle. The upper and lower parts of the figure show how material is added to and subtracted from the rock cycle. The external processes of the upper half of the diagram are discussed in Chapters 7 through 12 of this book, and the internal processes in the lower half are discussed later.

near the bottom of the crust, but its origin is an important problem.

VOLCANOES

A great deal can be learned about igneous processes from the study of volcanoes. They are, after all, the only direct evidence we have for the existence of magma in the crust.

The active volcanoes today are of several different types, and the differences among them seem to depend on the composition of their magmas, more particularly on the behavior of the volatiles in the magmas. The most abundant volatile in magma is water, which escapes from volcanoes in the form of steam. Carbon dioxide is a common volcanic gas, but the sulfur gases (hydrogen sulfide and the oxides of sulfur), because of their strong odors, are the gases most easily noted near volcanoes. In addition, minor amounts of other gases, such as carbon monoxide, hydrochloric and hydro-

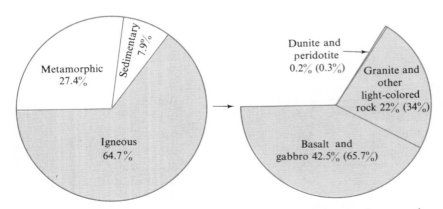

FIGURE 3-3 Relative abundance of igneous rocks in the earth's crust. For example, basalt and gabbro form 42.5 percent of all rocks and 65.7 percent of the igneous rocks. The igneous rock types are described later in this chapter. Data from Ronov and Yaroshevsky, 1969.

fluoric acid, ammonia, hydrogen, and hydrocarbons, are also released. Ordinary air also escapes from some volcanoes, especially those with porous rocks. Volcanic emanations are believed to have played a major role in the formation of the oceans and the atmosphere. (See Table 3-1.) Recent eruptions at Hawaii are estimated to release 1-2 percent gas during the early stages of eruption and about one-half percent during later stages. Laboratory studies suggest that, at depth, magmas may contain up to 5-8 percent dissolved water.

A simple analogy will illustrate the behavior of the water and other gases in volcanic magma.

TABLE 3-1 Composition of volcanic gas at Hawaii.

Gas	Volume (percent)
Water	70.75
Carbon dioxide	14.07
Hydrogen	0.33
Nitrogen	5.45
Argon	0.18
Sulfur dioxide	6.40
Sulfur trioxide	1.92
Sulfur	0.10
Chlorine	0.05
	99.65

SOURCE: After Shepherd, 1921 in G. A. Macdonald, "Physical Properties of Erupting Hawaiian Magmas," *Geological Society of America Bulletin* (August, 1963), Vol. 74, No. 8, pp. 1071-1078.

Most magmas are believed to contain at least a few percent water in solution. The water is dissolved in the magma just as carbon dioxide gas is dissolved in a bottle of soda pop. Most magma is able to dissolve this water because the magma is under considerable pressure due to the weight of the overlying rocks. The gas is dissolved in the soda pop as a result of the pressure inside the bottle. When the bottle is opened, the pressure is reduced and the gas begins to bubble out of the liquid. In the same way, when magma reaches the surface, the confining pressure is reduced and the gas, in this case mainly steam, is released.

If the magma has a low viscosity, the gas merely escapes or forms bubbles in the resulting rock. The tops of lava flows can commonly be distinguished from the bottoms by the presence of these bubbles at the top. (See Fig. 3-4.) In some cases, the bubble holes are filled by material deposited by the fluids, or are filled at a later time. The fillings are commonly silica materials such as opal or chalcedony, but many other minerals also occur here. Large filled holes of this kind are actively sought by rock collectors who cut and polish them.

If a fluid magma is erupted very suddenly, it may form a *froth of tiny bubbles* which quickly solidi-

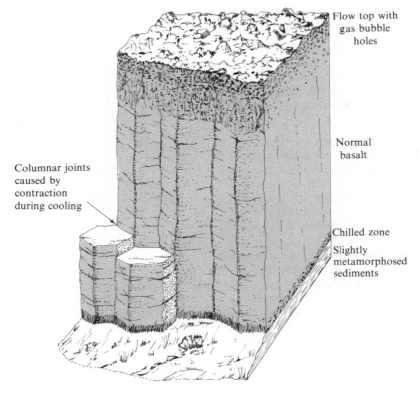

FIGURE 3-4 A section through a lava flow.

fies to *form the light, glassy rock pumice.* As a result of the many bubble holes, pumice is light enough to float. The formation of pumice can be likened to the opening of a bottle of warm soda pop or champagne, which results in the production of a froth. *A rock with coarser bubbles is called scoria.*

More violent events occur when more viscous magmas are erupted to the surface and the gas attempts to escape. In this case, the plastic rock is shattered by the pressure of the expanding gas. The gas may expand so rapidly that the rock explodes, blowing the fragments high in the air.

The fragments, or *pyroclasts,* produced in this way are named according to their size and shape (see Table 3–2). Pyroclasts smaller than 2 millimeters are called *ash grains,* those between 2 and 64 millimeters (2.5 inches) are termed *lapilli,* and those larger than 64 millimeters are called *blocks* if they are angular and *bombs* if they are rounded or show other evidence of having been melted or partly melted when they were ejected. The term *ash* implies burning, but, of course, these rocks do not result from burning. The term was first applied when volcanoes were thought to result from subterranean burning and so is an histori-

cal mistake. Unconsolidated deposits of these materials are called *tephra.* The rocks formed by this process are called *pyroclastic* ("fire-broken") *rocks.* Pyroclastic rocks are named according to the size of the fragments without reference to composition. *Consolidated ash is called tuff, consolidated lapilli is called lapilli-tuff, a pyroclastic rock composed of blocks is called pyroclastic breccia, and one composed of bombs is called agglomerate.* (See Fig 3–5.) The main problem with these rocks is distinguishing them from normal sedimentary

rocks formed by erosion of a volcanic area. The latter rocks are called *epiclastic tuff* or *epiclastic breccia,* or the like, depending on the size of the fragments. In general, field study is needed to distinguish between these two types of accumulation.

Volcanic bombs are formed by pyroclastic material that is still molten when thrown from the volcano. Again, the name is not really appropriate. Bombs are recognized by their shapes. A liquid assumes a streamlined shape while in flight. Most volcanic bombs are somewhat viscous when thrown

FIGURE 3–5   An outcrop of pyroclastic breccia in the Thomas Range, Utah. This rock is composed of coarse, angular fragments of volcanic rocks in a fine-grained matrix. The dark fragments are obsidian, and the light are pumice. Photo by M. H. Staatz, U.S. Geological Survey.

TABLE 3–2   Classification of pyroclastic material.

	Fragment	Unconsolidated Deposits	Rocks
General Terms	PYROCLAST—fragment ejected by volcanic action	TEPHRA	PYROCLASTIC ROCK
Pyroclasts >64 mm	BLOCK—angular, solid when ejected	BLOCK TEPHRA	PYROCLASTIC BRECCIA
	BOMB—rounded or other evidence suggests melted or partly melted when ejected	BOMB TEPHRA	AGGLOMERATE
Pyroclasts are between 2–64 mm	LAPILLUS	LAPILLI TEPHRA	LAPILLI TUFF
Pyroclasts <2 mm	ASH GRAIN	ASH	TUFF

NOTE: The term *epiclastic* is used if the rock or deposit is composed largely of material eroded from volcanic terranes, that is epiclastic tuff.

from a volcano and solidify during flight. Most spin during flight and so have an elongate shape. (See Fig. 3–6.) Some are still fluid when they hit the ground, and these flatten out.

Volcanic eruptions range from quiet to violent, as will be evident from the following descriptions. The examples used are currently active volcanoes and those that are recent enough that erosion has not greatly modified them. A volcano that has erupted in historical time is generally considered an active volcano; but only at one or two closely studied, well-instrumented volcanoes is it possible to predict eruptions, and even at these, warning comes just before eruption, and the place of eruption is known only generally. Thus, designating active volcanoes is difficult.

The rocks covering the surface of much of the western United States are volcanic, and many of these are millions of years old. This may seem old but is relatively young by geologic standards. Many of the original features of these rocks have been removed by erosion, but they are readily recognizable as volcanic. Less easily recognized are buried volcanic terranes. Such rocks are exposed by uplift and erosion in all parts of the earth, and the age of these rocks may be hundreds of millions

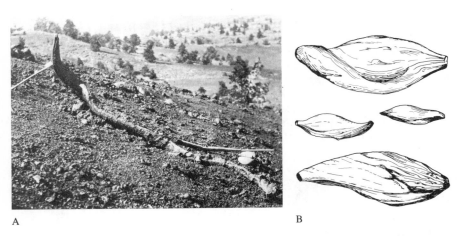

A   B

FIGURE 3–6 Volcanic bombs. A. An unusual volcanic bomb 4 meters (13 feet) long. Photo taken in 1902 in southern Idaho by I. C. Russell, U.S. Geological Survey. B. Typical volcanic bombs from Hawaii. Sizes range from under an inch to several feet. Many other shapes occur.

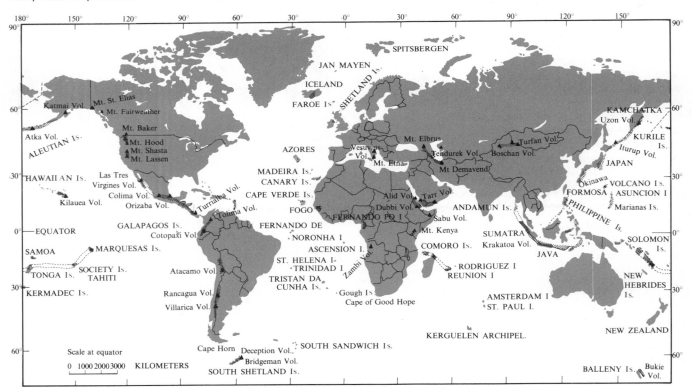

FIGURE 3–7 Volcanic areas of the world. Comparison with Fig. 14–35 will show the relationship with global structures. From G. A. Waring, U.S. Geological Survey Professional Paper 492, 1965.

of years. Thus study of present volcanoes will help in interpretation of ancient rocks—an important principle in geology.

Present-day volcanoes occur in relatively few areas, such as the rim of the Pacific Ocean, the Mediterranean Sea, and the Mid-Atlantic Ridge. Most of these areas are associated with sea-floor spreading. (See Fig. 3–7.)

It will be convenient to use rock names in the following paragraphs. Dark colored volcanic rock is called *basalt;* light colored, *felsite;* and intermediate between the two is *andesite.* The naming of

igneous rocks will be described later in the chapter.

The quietest eruptions are lava flows that come from dikes or dike swarms. The lava is basalt and apparently is very fluid. These quiet flows can produce great thicknesses and cover huge areas. An example is the Columbia River Basalt of southeast Washington and northeast Oregon that covers an area of over 77,750 square kilometers (30,000 square miles) and is over 1525 meters (5000 feet) thick at places. Thus this unit comprises at least 125,200 cubic kilometers (30,000 cubic miles) of ba-

salt. The source of this basalt field was four widely separated dike swarms. The very fluid basalt must have flowed like water because individual flows can be followed for over 160 kilometers (100 miles). (See Fig. 3–8.) Similar basalt fields are found at several places on the earth, such as the Deccan Plateau of India.

At other places similar basalt erupts from central vents, and large *shield volcanoes* form. They are called shield volcanoes because *the very fluid magma builds gentle slopes that resemble shields.* (See Fig. 3–9.) The Hawai-

FIGURE 3–8 Oblique aerial view looking east near the confluence of the Snake and Imnaha Rivers on the Oregon-Idaho border. The bedded rocks on the valley sides are basalt flows of the Columbia River Basalt. The lower parts of the valleys are cut in metamorphic rocks that underlie the basalt. Photo from U.S. Geological Survey.

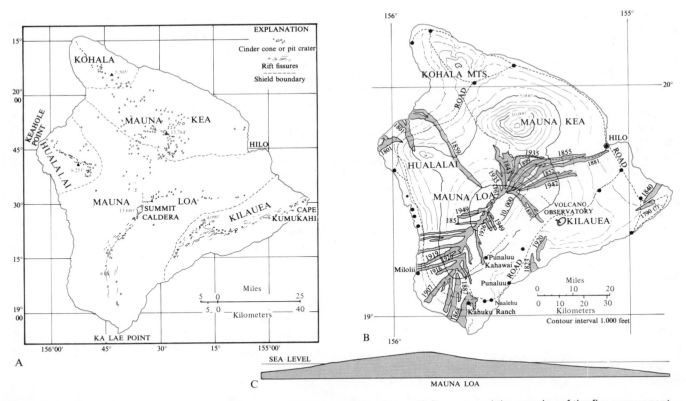

FIGURE 3–9   Maps of the island of Hawaii. A. Shows cinder cones, pit craters, rift fissures, and the margins of the five component volcanoes. B. Shows the historic lava flows through 1950 (color). C. Cross-section of Mauna Loa, a shield volcano, without vertical exaggeration. Parts A and B are from C. K. Wentworth and G. A. Macdonald, U.S. Geological Survey Bulletin 994, 1953, Part C is from K. O. Emery and others, U.S. Geological Survey Professional Paper 260A, 1954.

ian Islands are examples of this type of quiet eruption whose most violent displays are fountains of incandescent lava thrown up from lava pools in the craters. (See Figs. 3–10 and 3–11.) Despite the gentle slopes, the Hawaiian Islands rise a total of 8540 meters (28,000 feet) from the floor of the Pacific to 4204 meters (13,784 feet) above sea level.

The fluid basalts discussed so far are easily recognized by their ropy surfaces. (See Fig. 3–12.) However, even in Hawaii, not all flows are of this type. Some flows are more viscous; their surfaces harden during flowage and are broken into blocks by the traction of the fluid lava a few inches below the crust. These are the blocky lava flows. (See Fig. 3–13.) The terms *blocky* and *ropy,* although somewhat descriptive, are not too good, and many geologists use the Hawaiian words *aa* and *pahoe-*

hoe for blocky and ropy lava respectively. At some places the top of a flow may solidify, and the melted interior flows on, forming a lava tube or tunnel. Examples can be seen at many places, including the Modoc lavas in northeastern California and Craters of the Moon in Idaho.

Cinder cones probably result if the lava is more viscous than that which forms blocky flows. Pyroclastic material thrown out of the vent builds up a cone around the vent. A central depression around the vent remains at the summit. A *cinder cone is composed of tuff, breccia, ash, cinders, blocks, and*

FIGURE 3–10   Lava fountains 92 meters (300 feet) high on Kilauea, September 23, 1961. Note the trees burning around the active margin of the flow fed by the fountains. Photo by D. H. Richter, U.S. Geological Survey.

A

B

FIGURE 3-11   A. Oblique aerial view of the summit of Kilauea. The craters are believed to be collapse features. The inner crater, Halemaumau, has been the site of eruptions with lava fountaining and, at times, has been partially filled by lava forming a lava lake. This view is toward the north and shows the profile of 4204-meter (13,784-foot) Mauna Kea. Photo from U.S. Geological Survey. B. Looking southwest into Halemaumau at the collapse features formed during February 1960. An eruptive phase followed. The deep pit is about 230 meters (750 feet) below the rim. Photo by D. H. Richter, U.S. Geological Survey.

A

B

FIGURE 3-12   Ropy or pahoehoe lava. A. This flow near McCartys, New Mexico, moved toward the right. B. Cascade of a 1920 flow at Hawaii. Photos from U.S. Geological Survey: A by H. R. Cornwall, B by H. T. Stearns.

bombs. (See Fig. 3-14.) The slope is generally near 30 degrees, which is the steepest slope that can be made of loose material. Cinder cones are very common features and, because the loose material is easily eroded, are generally young features. They are rarely very high because these slopes are so unstable; such loose slopes are commonly tiring and frustrating to climb. Cinder cones are commonly made of basalt or andesite.

The steep-sided high volcanic mountains are generally *composite volcanoes,* so called because they *are made of a combination of pyroclastic material and lava.* Such volcanoes are generally made of andesite (intermediate composition) and combine explosive eruptions with outpouring of lava. Mt. Shasta and Mt. Rainier (Fig. 3-15) are examples. The lavas cover the pyroclastic material and form the skeleton that enables the mountain to grow large. The

lava is more viscous here, and the vents are blocked from time to time by congealing lava. These plugs cause the gas pressure to build up until the pressure exceeds the strength of the plug, and a greater or lesser explosion results. Under these conditions new vents may form on the sides of the volcano. Obviously, as the magma becomes more viscous, the explosive eruptions become more numerous and/or more destructive. Crater Lake, Oregon,

(Fig. 3–16) was formed by the explosive destruction of the top of a composite volcano similar to Mt. Rainier or Mt. Shasta. Interestingly, the volume of material blown out of Crater Lake is much smaller than the volume of the missing top of the mountain, suggesting that after explosive discharge of the magmatic gases the magma subsided. Most craters and calderas, as large craters are called, are formed by subsidence.

Krakatoa, between Java and Sumatra, has a somewhat similar history. It was a volcano 793 meters (2600 feet) above sea level that formed within a ring of islands marking the rim of an older volcano that lost its summit, probably in much the same way as Crater Lake formed. In 1883, the island of Krakatoa erupted explosively, destroying the island. The ash from this great explosion which rose high into the atmosphere was the cause of beautiful sunsets all over the world long after the eruption.

Lavas ranging in composition from andesite to felsite tend to be viscous, and they cause explosive eruptions. Plugs tend to block the vents, resulting in violent explosions of various kinds. (See Fig. 3–17.) One type has just been described. In 1902, Mt. Pelée on Martinique Island was in eruption, and a large spine formed in the vent, plugging it. This spine was pushed up about 458 meters (1500 feet), forming quite a spectacle. Finally, the pressure of the magmatic gases was enough to enable the gas-charged magma to break through. The resulting eruption was one of the most destructive ever witnessed, and led to the recognition of a very important type of pyroclastic rock. (Recall the earlier description of the formation of pumice.) At Mt. Pelée, the magmatic gas expanded so violently that instead of forming a frothy pumice, the expanding bubbles completely shattered the pumice, producing a mixture of hot gas and small glass fragments. Such a mixture is quite fluid because of the gas and can move very rapidly down even a gentle slope. When it comes to rest, *the hot glass fragments weld together, forming a welded tuff.* Eruptions of this type are called glowing clouds *(nuées*

A

B

FIGURE 3–13   Blocky or aa lava. A. Blocky lava flow and cinder cone. B. Steep front of a flow of viscous, blocky volcanic glass (obsidian). Photo from U.S. Forest Service.

FIGURE 3–14   Aerial view of Paricutin, a large cinder cone that began in a cornfield in 1943 and was active for about nine years. The lava flow spread from vents beneath the cone. Photo from Chicago Natural History Museum.

ardentes) or glowing avalanches. It was an eruption of this kind from Vesuvius that buried Pompeii in 79 A.D.

The 1902 eruption of Mt. Pelée was witnessed by a number of people. They reported that the hot cloud traveled at about 160 kilometers (100 miles) per hour. It killed all but about four of the perhaps 30,000 inhabitants of St. Pierre. The survivors were a prisoner in a dungeon, two people who were near the edge of the cloud, and one who was spared although all around him died. In the harbor all but two ships were sunk. Most of the witnesses were from those ships. They reported a sudden darkness and hot dust mixed with water which caused a short intense heat and a feeling of suffocation. Many of these people were badly burned.

Many welded tuffs closely resemble *obsidian* (volcanic glass) or felsite, and only recently has the abundance of welded tuffs been recognized. Their abundance on the surface of the continents may be as great as that of basalt. All welded tuffs are not associated with dome volcanoes. Much of Yellowstone National Park is covered by welded tuffs, and the geysers there probably get their heat from the cooling lava.

Underwater volcanic eruptions are common, and lavas sometimes flow into water bodies. This *causes rapid chilling of the lava, and causes the lava to subdivide into numerous interconnected lobes that look like a pile of pillows in cross-section.* The term *pillow lava* is applied to the resulting rocks. (See Fig. 3–18.) Such rocks are abundant on the ocean bottoms.

Erosion may remove the fragmental rocks from a volcano, leaving the lava that congealed in the central conduit as shown in Fig. 3–19.

Volcanic eruptions cannot yet be predicted; however, recent studies have detected the movement of magma at a few very

FIGURE 3–15 Mt. Rainier, a massive composite volcano in the Cascade Mountains, Washington. Photo from Washington State Department of Commerce and Economic Development.

FIGURE 3–16 Crater Lake, Oregon, fills a depression formed partly by explosive eruptions and mainly by subsidence of the magma. Wizard Island is a small volcanic cone with a crater at its summit. Photo from Oregon State Highway Department.

FIGURE 3–17 Novarupta Dome. Felsite dome in cinder cone at Katmai National Monument, Alaska. Welded tuffs are associated with this dome. Photo from U.S. Geological Survey.

FIGURE 3-18   Pillow lava at Depoe Bay, Oregon.

FIGURE 3-19   Shiprock, New Mexico, is probably part of an ancient volcano. Dikes radiating from it can be seen on the left skyline and the right foreground. Photo from New Mexico Department of Development.

closely studied, well-instrumented volcanoes. The movement is shown by many thousands of shallow, very weak earthquakes. The location of these earthquakes may ultimately reveal the magma plumbing system. It also appears that the magma moves upward to shallow magma chambers just before eruption. The latter hypothesis is suggested by the swelling of a volcano just before eruption. This is detected by very sensitive tilt meters. After an eruption the

magma apparently returns to the depths. This is indicated by the tilt meters and, at Hawaii, by the draining of the lava lakes that form in the craters there.

ENVIRONMENTAL HAZARDS OF VOLCANOES

The destruction caused by volcanoes is of two kinds—the direct destruction by material issuing from the volcano (see Figs. 3-20

and 3-21) and the indirect damage caused by mudflows and floods. The former has already been described in the discussion of explosive volcanoes, such as Mt. Pelée and Krakatoa. There is little defense against such rapid destruction except evacuation at the first indication of activity by the volcanoes. The first eruption may be of the glowing cloud (nuée ardente) type and may occur anywhere on the volcano. In such cases there may be no escape. In the quieter eruption of basaltic volcanoes, such as-those in the Hawaiian Islands, the fluid basalt moves at a few kilometers an hour and generally there is enough time to evacuate. This may seem of little value if your home and fields are inundated by either lava or air-borne ash, but at least you are alive and able to start over.

Indirect damage from volcanoes can also be swift and without warning. Many volcanoes are high mountains and even in the tropics are glacier covered. An eruption under a glacier can cause extensive melting that results in floods. Also many volcanoes, especially those that have been inactive for long periods, have large lakes in their depressed summit areas. Crater Lake in Oregon is an example. An eruption that breeches the lake, or an eruption into the lake, can cause the draining of the lake and so cause floods. The eruption of a submarine or shallow-island volcano such as Krakatoa can cause flooding by seismic sea waves.

Probably the largest areas inundated by volcanoes are those swept by mudflows. Although some mudflows are concurrent with the eruption, most occur at the next rainy season. Although the latter mudflows can be predicted, many lives are lost because many people refuse to leave their homes and fields. Mudflows are probably the most important way that pyroclastic volcanic rocks are distributed from volcanoes

and so are a normal geologic process. A pyroclastic eruption can deposit ash and coarser material on the slopes of the volcano and on other slopes downwind. This loose material is not stable and at the next rain becomes mud that can flow even on very gentle slopes. Rain commonly accompanies volcanic eruption, and, although we don't know why, it may be related to the large amounts of steam that are part of most eruptions. Thus mudflows can accompany eruptions. Pyroclastic eruptions commonly kill vegetation by covering low plants, and the falling debris strips the leaves from trees. Without vegetation, the runoff is increased, and the possibility of mudflows is greater.

Mount St. Helens in the Cascade Range in southern Washington erupted in 1980, causing wide-

FIGURE 3–20 In early 1973, an eruption began at Heimaey, an island that is part of Iceland. The cone in the left background is composed of pyroclastics which also partially cover the houses. Photo from Consulate General of Iceland.

A

B

FIGURE 3–21 Changes at Heimaey Island, Iceland, caused by volcanic eruption. A. 1960. B. 1973. Photo A, U.S. Air Force; B, Icelandic Geodetic Survey. Both courtesy of Thorvaldur Bragason.

spread damage. (See Fig. 3–22.) Its last previous eruption was in 1857, and at that time it had been active between about 1800 and 1857. Study of Mt. St. Helens and its deposits shows that it has erupted 15 times in the last 4500 years, and that a typical eruptive period lasts about 100 years, with 100–500 quiet years between eruptive intervals.

The eruption began with a number of shallow, mild earthquakes, and a week later, on March 27, steam was noted issuing from the summit. The earthquakes and eruptions continued, with rock and ice carried at times to heights of over 5 kilometers (3 miles). Fine rock material was carried many miles, and sulfur dioxide as well as steam issued from the crater.

During this time the crater enlarged, and an unusual type of earthquake activity called *harmonic tremors* revealed that magma was moving in the volcano. By late April, it was noted that the north side of the crater was rising and bulging northward. Apparently, a dome was being emplaced, and the uplift continued at over a meter per day.

A

B

C

D

FIGURE 3–22   1980 eruption of Mt. St. Helens. A. On April 10, before the eruption, the mountain was symmetrical and snow-covered. B. Eruption on May 18. C. Summit area after the eruption. D. Mudflow fills the Toutle River valley after the eruption. Photos from U.S. Geological Survey, A, B, and D by Austin Post.

The big eruption, which occurred on May 18, began with two closely spaced earthquakes that caused the bulge, now 125 meters (410 feet), to landslide. The landslide released the pressure on the dome or plug, and a blast of incandescent gas and ash resulted. The blast was estimated at 50 meters per second (160 feet per second), and it devastated an area of 500 square kilometers (190 square miles). It knocked down trees and seared them with 200° C (390° F) temperatures. Between two and three cubic kilometers (one-half and three-fourths cubic mile) of landslide and pyroclastic material moved up to 20 kilometers (12 miles) down stream valleys, filling some to depths of 60 meters (200 feet). After the initial blast, the ash and steam rose vertically to heights of as much as 19 kilometers (12 miles) and were carried hundreds of kilometers to the east. Fine ash several centimeters deep accumulated at many places downwind. A few hours after the blast, mudflows and debris flows fed by melting glaciers moved into river valleys, causing some flooding and warming of the river water. About 400 meters (1300 feet) was removed from the top of Mt. St. Helens by the blast, and the summit crater was enlarged to a diameter of about 2 kilometers (1.5 miles) and a depth of 1.5 kilometers (1 mile).

CRYSTALLIZATION OF IGNEOUS ROCKS

One of the most important concepts in igneous geology is the *reaction series* first described by *Bowen,* an American geochemist who studied silicate melts in the laboratory. *This series shows the sequence of minerals that form in the crystallization of a basaltic melt,* which is the most common type of volcanic rock. It also suggests how *the earliest-formed crystals might be separated from* *the melt or magma, forming rocks with a composition different from that of the original melt. Such a process is called* differentiation, and we will discuss its role when we consider the origin of magmas and of the various types of igneous rocks. Although Bowen originated the idea of a reaction series from studies of experimental melts, the general sequence of crystallization of most rocks was already known as a result of intensive microscopic studies. The reaction series is an oversimplification but is very useful. (See Fig. 3–23.)

The reaction series really consists of three series. The right-hand side is a *continuous series;* that is, *all compositions of plagioclase from entirely calcic to entirely sodic and all compositions in between occur.* The left-hand side is a *discontinuous series;* that is, *the changes from one mineral to the next occur in discrete steps.* Each of the minerals named in the left-hand series is actually a continuous series in composition of that mineral group. The lower series is only a sequence of crystallization. The arrangement shows the relative, but not exact, sequence of crystallization; calcic plagioclase and olivine tend to crystallize at the same time, and the lower series is last to crystallize. *It should be emphasized that Bowen's series holds for only* *some basaltic magmas,* but, as we shall see, basalt magma is very important in igneous geology.

The interpretation of the crystallization of a basaltic magma in terms of Bowen's series is as follows. In general, the first mineral to crystallize from the melt is olivine. In most cases, at the same time, or a little later, calcic plagioclase begins to crystallize. The discontinuous series continues with falling temperature; the olivine reacts with the melt to form pyroxene, which at the same time also starts to crystallize directly from the melt. Plagioclase continues to crystallize during falling temperature; the composition, however, continuously becomes more sodic. If cooling is slow enough, these two processes go on: the minerals on the left side react with the melt to form the next lower mineral, while that mineral is crystallizing from the melt, and the plagioclase continuously reacts with the melt to form more plagioclase that is more sodic in composition. This process ends when all the ferromagnesian minerals and plagioclase are formed. Then the potassium feldspar-muscovite-quartz series begins. This series may in part overlap the sodic part of the plagioclase series.

Another way to explain the discontinuous series is to consider what happens on heating one of

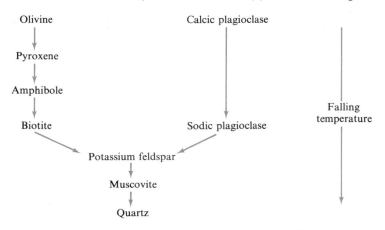

FIGURE 3–23 Bowen's reaction series. Shows the general sequence of mineral formation with falling temperature in a basaltic melt.

the minerals. When one of the discontinuous series minerals is heated, it does not melt at a single temperature the way ice or similar substances do. Instead, when an amphibole, for example, is heated, it begins to melt when a certain temperature, depending on its composition, is reached, and melting continues over a range of temperature. Examination of the remaining solid during this first stage of melting, by quick quenching to prevent reaction during slow cooling, would show that the amphibole has been transformed to a pyroxene and a liquid, just the reverse of what takes place during the slow cooling of a magma. On further heating, a temperature is reached at which further melting would occur. As before, melting would take place over a range in temperature, resulting in a melt and olivine. Finally, a temperature would be reached at which all the olivine would melt.

The evidence for the reaction series is abundant. Under the microscope, for instance, pyroxene crystals with jackets of amphibole are seen, recording a discontinuous reaction that was interrupted, probably by cooling too rapidly for the reaction to be completed. (See Fig. 3–24.) In differentiated rock bodies, slow cooling and low viscosity have combined to preserve some of the main points of the reaction principle. In such rock bodies, the early-crystallized olivine sank to the bottom and so was removed from contact with the main mass of the magma, allowing progressively lighter-colored rocks to form in the upper parts of the body. (See Fig. 3–25.)

Some geologists believe that such differentiation explains the great variety of igneous rocks, but this is probably an oversimplification. The reaction series applies to basaltic magmas, and the vast areas of basalt lava certainly indicate that basalt magmas are common in the earth's crust. But, as we will see later, rocks of granitic composition are also very abundant. The vast amount of granitic rocks and their volcanic equivalents suggests that these rocks also represent important magma types in the crust, because they are far too abundant to be derived from basalt by differentiation. Complete differentiation of basalt can produce, at the most, 15 percent granitic-composition rocks. This points out a major problem of geology— the origin of magmas. Understanding the crystallization of basalt gives some insight into the genesis of rock types as shown in the next section.

CLASSIFICATION

Igneous rocks are classified on the bases of composition and texture. The classification is an attempt to show the relationships among the various rock types; hence, it is an attempt at natural or genetic classification.

Composition, in the classification of igneous rocks, follows the reaction series. Most rock names came into use long before Bowen described the reaction series.

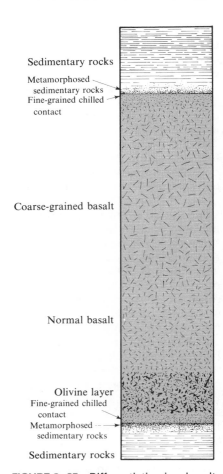

FIGURE 3–25 Differentiation in a basalt sill. A narrow column through a sill is shown here. The olivine layer develops by the sinking of the early-formed, heavy olivine crystals.

These names had been applied to the more common rocks. It is of more than passing interest that these common groups of rocks fall into the various steps in the differentiation of a basaltic magma according to Bowen's reaction series. This correlation suggests that the compositional part of igneous rock terminology may be genetic. This is shown in Fig. 3–26.

Because almost all coarse-grained rocks contain feldspar, the most common mineral family, the *type* of feldspar is the most important factor in the composition. Of secondary importance is the type of dark mineral which generally accompanies each type of feldspar. See classification diagram, Fig. 3–43.

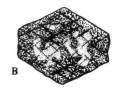

FIGURE 3–24 Evidence for the Bowen reaction series seen in microscopic study of igneous rocks. A. Augite crystal with remnants of olivine inside it. B. Hornblende crystal with augite remnants. In both cases the reaction was incomplete, thus preserving the remnants of the earlier mineral. Note the cleavage angles in augite and hornblende. The minerals shown here are typically about 3 millimeters (⅛ inch) in diameter, but may be much larger or smaller.

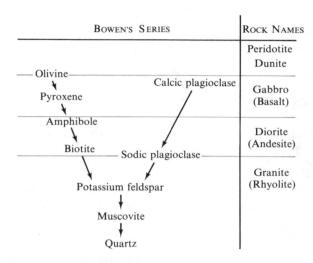

BOWEN'S SERIES	ROCK NAMES

Bowen's Series diagram showing:
Olivine → Pyroxene → Amphibole → Biotite, with Calcic plagioclase → Sodic plagioclase, leading to Potassium feldspar → Muscovite → Quartz

Rock Names: Peridotite / Dunite; Gabbro (Basalt); Diorite (Andesite); Granite (Rhyolite)

FIGURE 3–26 This diagram shows in a very general way how differentiation by separation of successively formed crystals could result in several rock types originating from a single basalt magma. (Volcanic rock types are in parentheses.) It also shows why minerals widely separated in the reaction series, such as quartz and olivine, rarely appear in the same rock.

In a similar way, the texture tells much about the cooling history of an igneous rock. *In igneous rocks the texture refers mainly to the grain size.* Rocks that cool slowly are able to grow large crystals; but quickly chilled rocks, such as volcanic rocks, are fine grained. To see the significance, we must consider the mode of occurrence of igneous rocks.

TEXTURE AND OCCURRENCE

Igneous rocks occur in two ways, either as *intrusive (below the surface)* bodies or as *extrusive (on the surface)* rocks. The ultimate source of igneous magma is probably deep in the crust or in the upper part of the mantle, and the terms *intrusive* and *extrusive* refer to the place where the rock solidified. Intrusive igneous rocks can be seen only where erosion has uncovered them. They are described as *concordant* if the contacts of the intrusive body are more or less parallel to the bedding of the intruded rocks, and *discordant* if the intrusive body cuts across the older rocks. (See Fig. 3–27.)

The largest discordant bodies are called batholiths. These are very large features, the size of mountain ranges such as the Sierra Nevada in California, which is nearly 644 kilometers (400

miles) long by about 80 kilometers (50 miles) wide. Many other ranges, both larger and smaller, also contain batholiths. (See Fig. 3–28.) Because batholiths are large and also because they probably were emplaced at least several thousand feet below the surface, they cooled very slowly. This slow cooling permitted large mineral grains to form; therefore, it is not suprising that batholiths are composed mainly of granitic rocks with crystals large enough to be easily seen. (See Fig. 3–29.) *Smaller bodies of coase-grained igneous rocks are termed* stocks. Coarse-grained igneous rocks are called *plutonic igneous rocks* because they were once believed to have crystallized at great depth.

Dikes are tabular, discordant intrusive bodies. They range in thickness from a few inches to several thousand feet, but generally are of the order of a few feet to tens of feet in thickness. (See Figs. 3–30 and 3–31.) They are

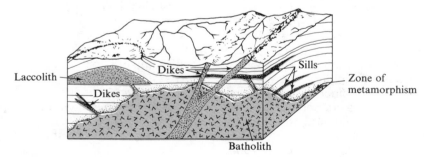

FIGURE 3–27 Intrusive igneous rock bodies. The laccolith and sills are concordant with the enclosing sedimentary beds, and the batholith and dikes are discordant. The heat from the batholith metamorphoses the surrounding rocks.

Laccolith • Dikes • Dikes • Batholith • Sills • Zone of metamorphism

FIGURE 3–28 Batholiths of western United States.

FIGURE 3–29   Granite has a coarse, interlocking texture formed by crystallization from a melt. Photo from Ward's Natural Science Establishment, Inc., Rochester, N.Y.

FIGURE 3–30   Dikes cutting through horizontal sedimentary rocks. The dikes are more resistant to erosion than are the sedimentary rocks. West Spanish Peak, Colorado, is in the background. Photo by G. W. Stose, U.S. Geological Survey.

grained, or glassy (see Fig. 3–33) if cooled so rapidly that no crystallization occurs.

Many igneous rocks are a mixture of coarse and fine crystals, and this texture is called porphyritic. (See Fig. 3–34.) Such a texture, in general, records a two-step history of the rock. The large crystals form as a result of slow cooling, perhaps in a deep magma chamber. Then the magma is moved higher in the crust to form a dike, sill, or flow. The remainder of the rock cools rapidly, resulting in a fine-grained or even glassy matrix surrounding the early-formed larger crystals. A rock with a coarse matrix and some much larger crystals may also be described as porphyritic. It is doubtful whether all such rocks have a two-step history as outlined here.

The temperature of basalt lava is in the range of 1000° C to 1200° C, felsite lava is about 900° C, and granitic rocks are believed to be emplaced at about 820° C. Advancing basalt flows at Hawaii have been measured at temperatures as low as 745° C. The cooling time of igneous rocks is not well known. Batholiths are believed to solidify in one to a few million years. Dikes, depending on size, probably solidify in several tens of thousands of years. Basalt lava is known to have hardened to a depth of over ten feet in two years. Rocks are poor conductors of heat; hence, intrusive rocks take a very long time to cool because the heat must be conducted away by the surrounding rocks. Lava flows lose their heat to the atmosphere more rapidly.

A summary of textural terms will be of help in learning to recognize igneous rock types:

Glassy—no crystals
Fine-grained—grains can be seen in sunlight with a good hand lens
Coarse-grained—crystals easily seen
Pegmatitic—crystals over one centimeter. (This texture is discussed later.)

generally much longer than they are wide, and many have been traced for miles. Most dikes occupy cracks and have straight, parallel walls. Because the intrusive rock is commonly resistant, dikes generally form ridges when exposed by erosion.

The concordant intrusive bodies are *sills* and *laccoliths*. They are very similar and are intruded between sedimentary beds. *Sills are thin and do not noticeably deform sedimentary beds.* (See Fig. 3–32.) *Laccoliths are thicker bodies and* uparch the overlying sediments, in some cases forming mountains.

Dikes and sills are small bodies compared to batholiths and have much more surface area for their volume; thus, these bodies cool much more rapidly and are commonly fine grained. Laccoliths are generally fine grained but, depending on their size, may be coarse grained.

The extrusive rocks result from lava flows and other types of volcanic activity. These rocks commonly cool rapidly and are fine

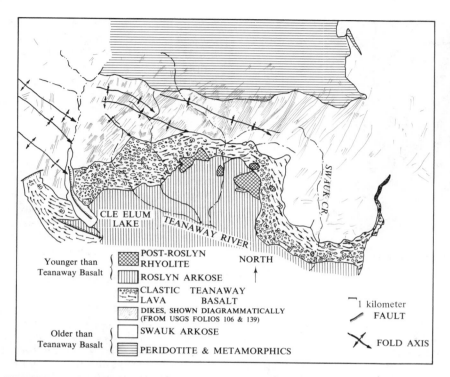

FIGURE 3–31 Teanaway dike swarm, Cascade Mountains, Washington. At places the dikes, shown diagrammatically here, comprise over 90 percent of the surface, recording much dilation. The dikes were the source of the Teanaway Basalt. From R. J. Foster, *American Journal of Science*, Vol. 256, 1958.

FIGURE 3–32 The black horizontal band is a sill. The inclined band is a crosscutting dike. The dark stripes in the sandstone are carbonaceous sediments. Taylor Glacier region, Victoria Land, Antarctica. Photo by W. B. Hamilton, U.S. Geological Survey.

Porphyritic—generally are named for the texture of the groundmass, giving such terms as *porphyritic basalt* for a dark, fine-grained rock with some larger crystals, and *porphyritic granite* for a granite with some much larger crystals.

by geologists working in small research submarines. At a few places the mid-ocean ridge rises above sea level, forming islands, such as Iceland, where basalt volcanism can be observed. The mid-ocean ridges are also areas of high heat flow. Heat flow measurements taken all over the earth reveal that at the mid-ocean ridges, the flow of heat energy from the interior is about twice the average of the rest of the ocean. This localized heat flow apparently is the cause of the mid-ocean ridges and their volcanic activity.

ORIGIN OF MAGMA AND PLATE TECTONICS

Igneous rocks must originate deep in the earth. To find out how and where magmas form, volcanic areas should be considered. The volcanic areas associated with plate tectonics are active geologically and are a good place to study.

At the mid-ocean ridges, new basaltic ocean floor is created. Submarine volcanism there has been observed and photographed

FIGURE 3–33 Obsidian or volcanic glass. The round, shelllike fractures are called conchoidal fractures. Photo from Ward's Natural Science Establishment, Inc., Rochester, N.Y.

FIGURE 3-34 Porphyritic texture. The light-colored feldspar crystals grew during slow cooling, and subsequent chilling probably produced the fine-grained dark matrix. Photo from Ward's Natural Science Establishment, Inc., Rochester, N.Y.

The ocean-floor basalt is about 4.8 kilometers (3 miles) thick, and this layer and the overlying sedimentary and volcanic rocks are called the *oceanic crust.* The layer below the crust is called the *mantle;* it is from this layer that the oceanic basalt must originate. The mantle is believed to be composed largely of the minerals olivine and pyroxene, which, depending on the amounts of each, form the rocks called *peridotite* or *dunite.* Seismic studies show that in parts of the upper mantle the rocks are soft because seismic waves are attenuated. The temperature and pressure at that level

are such that the mantle rocks are partially melted. At the places where the heat flow from the deep interior is high—the mid-ocean ridges—more of the rocks may be melted. Rocks expand when they are heated and when they are melted, and this expansion may be the reason for the mid-ocean ridges. The hot rocks take up more volume, so the upper mantle pushes the sea floor upward, forming the mid-ocean ridges. The mantle rocks below the ridges, although partially melted, are still quite rigid (Fig. 3–35).

The origin of the basalt magma that rises to the surface at the *mid-ocean ridges is probably the partially melted material from the upper mantle.* Rocks do not have a single melting point as simpler materials do. Ice or copper, for instance, at any given pressure, will each melt at a certain temperature. Many minerals and most rocks behave differently, and melting occurs over a wide temperature range. Partial melting of the mantle peridotite can form a magma of basaltic composition. A basalt magma formed in this way would be less dense than the rest of the mantle, and so would rise to cause submarine volcanism on the ocean floor.

The volcanic arc volcanoes erupt lava of andesite composition. This magma, too, may be caused by partial melting. At the volcanic island arc-trench areas, seismic studies show that the basaltic ocean crust and its overlying sediments are carried deep into the upper mantle (Fig. 3–36). In this case it is the basalt and the sediments that are partially melted at the higher temperature and pressure into which they are pushed in the upper mantle. The magma generated in this way may not always reach the surface and cause volcanic activity. These less fluid magmas may form intrusive bodies such as batholiths, as suggested in Fig. 3–36.

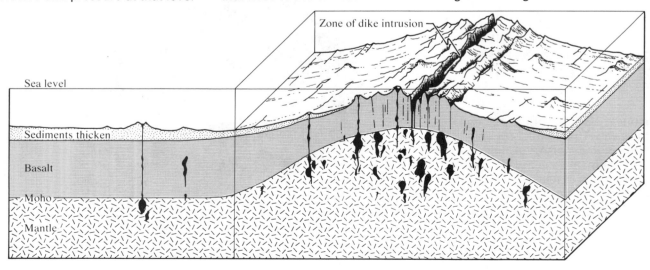

FIGURE 3–35 Ocean-floor basalt originates at mid-ocean ridge by partial melting of mantle rocks. Drawing by Robert Tope.

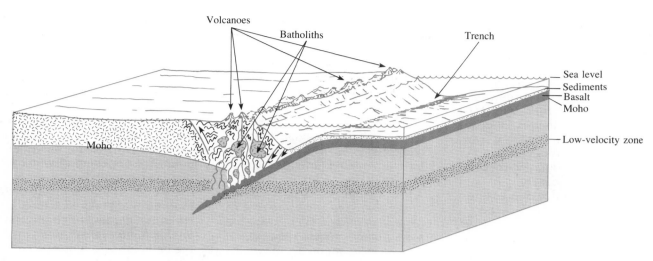

FIGURE 3–36 Batholiths and volcanoes form at a convergent plate boundary from partial melting of ocean-floor basalt and sedimentary rocks.

MOVEMENT OF MAGMA IN THE CRUST

Once the magma is generated, the next problem is: why does it rise in the crust? Several suggestions have been made. Rocks, like most substances, expand when they are melted. Thus, magma is less dense than the surrounding rocks and tends to rise, much as oil mixed with water rises and floats on the top.

Salt domes may illustrate intrusion caused by gravitational rise of lighter material. Salt domes occur in many parts of the world, and because some are associated with oil fields, they have been closely studied. Salt domes occur in sedimentary rocks; the salt originates from salt beds formed by the evaporation of sea water. Salt,

or halite as the mineral is called, has a specific gravity of 2.2. The specific gravity of sedimentary rocks increases with depth of burial because the weight of the overlying rocks compresses the deeper rocks. At the Gulf Coast of the United States the near-surface sedimentary rocks have a speciifc gravity of 1.9, at 610 meters (2000 feet) the specific gravity has increased to 2.2, and at 3050 meters (10,000 feet) it is 2.4. In this area the salt beds are very deep and thus are much lighter than the surrounding rocks. The salt of lesser density tends to rise and float to the top. This seems unlikely because we are used to considering rocks as strong, rigid materials. It can be easily shown, however, that rocks are not strong enough to support themselves. A moment's thought will show that rocks have finite strength and that the weight of overlying rocks will

exceed this strength at some depth. Below this depth rocks behave as viscous fluids. In the present case, salt behaves as a viscous fluid and the overlying sedimentary rocks are very weak. To start the formation of a salt dome, some initial irregularity is probably necessary, such as a high point on the salt bed, or a weakness or break in the overlying sediments. The rising salt may remain connected to the salt bed or, in some cases, become completely detached from the salt bed. The tops of many salt domes are mushroom-shaped. (See Fig. 3–37.) Some salt domes may have risen up to 13 kilometers (8 miles) in the Gulf Coast. The mechanics of salt dome formation are complex, and some downward movement of the surrounding sedimentary rocks may have occurred.

Magmas are less viscous than salt and, like salt, are less dense

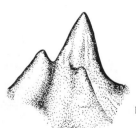

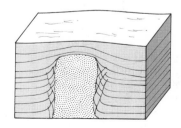

1525 m
1525 m

FIGURE 3–37 Salt domes showing some common shapes and the fractures caused in the surrounding rocks by their rise.

than the rocks at depth. Thus, the same mechanism, which may account for the rise of salt domes, may also account for the rise of some magmas. The internal structure of salt domes is somewhat similar to the internal structures of some intrusive igneous rocks, further suggesting a similarity. The heat of igneous rocks could weaken the surrounding rocks and thus assist the rise.

A possible source of pressure within the magma is the expansion of the volatiles. This was discussed earlier in connection with the surface manifestations of volcanism. In the present context, the volatiles in the newly formed magma would tend to expand as the magma rises because the confining pressure (weight of overlying rocks) is reduced. This might crack the enclosing rocks, allowing the magma to rise further.

Expanding gas bubbles might also be able to erode the enclosing rocks. This principle is used in a type of drilling and may be the mechanism by which some volcanic rocks reach the surface. Possible examples are the cylindrical, pyroclastic-filled conduits found in some volcanic areas.

Another possible way that magma can work its way through the crust is to react with the enclosing rocks. The magma could, by chemical reaction, dissolve or partially dissolve the rocks above the magma. Some melting or partial melting is also possible. For melting to occur, a similar amount of crystallization would have to occur elsewhere in the magma because magmas do not have enough excess heat energy to do much melting. Excess heat energy would have to be in the form of high temperature, but lava at the surface is generally very close to its melting point, indicating that, in general, magmas do not have excessively high temperatures. This process would tend to change the composition of the magma by assimilation of parts of the enclosing rocks. Differentiation

would probably also be going on in the magma.

EMPLACEMENT OF BATHOLITHS

The classic explanation of the emplacement of batholiths in the crust is *stoping.* The name comes from the mining term for the process of removing ore overhead. *The hot magma may cause blocks of the overlying enclosing rock to be broken off and fall through the melt.* Such a process requires a low-viscosity magma and would allow much contamination of the magma by reaction with the stoped blocks. Most batholiths contain many inclusions, probably of the enclosing rocks, giving some credence to this hypothesis. (See Figs. 3–38 and 3–39.)

Emplacement of batholiths is a major problem of geology. The great size of a batholith like the Sierra Nevada immediately raises the question; What happened to the large volume of rocks that the batholith displaced? Traditionally this question has been answered by two opposing points of view.

One school of thought holds that the batholith made room for itself by a combination of stoping and pushing aside the pre-existing rocks. The many included blocks, presumably of prebatholith rocks, found in some batholiths, and the zone of the deformed older rocks that surrounds most batholiths are the main evidence for this viewpoint. The opposing viewpoint is that the batholith was created more or less in place by the metamorphism of the pre-existing rocks that accompanied the mountain-building event during which the batholith was emplaced. This possibility is included in the rock cycle, Fig. 3–2. The evidence for this theory is the metamorphic rocks that surround most batholiths. In some cases these zones of high-grade metamorphic rocks are very wide, and in other cases the heat from the crystallizing magma is enough to account for the smaller halo of metamorphism.

It seems unreasonable to think that a huge batholith the size of the Sierra Nevada was all emplaced at the same time. Recent studies have shown that this batholith is composed of a number of

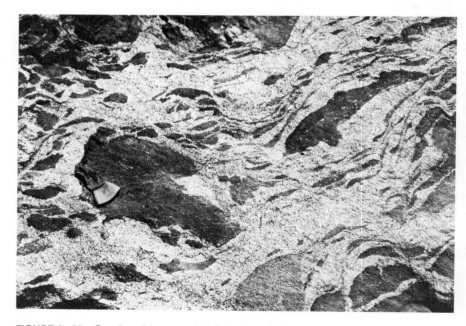

FIGURE 3–38 Granite with many dark inclusions in Yosemite National Park, California. The glove shows the scale. Photo by F. C. Calkins, U.S. Geological Survey.

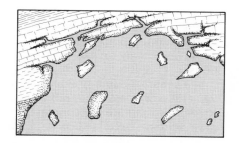

FIGURE 3–39 Stoping. Forceful injection of the magma breaks fragments from the roof that sink through the magma, reacting with it.

much smaller plutons. The finer-grained margins of these plutons that resulted from more rapid cooling indicate that at no time was the entire batholith molten. Radioactive dates indicate that the emplacement of the batholith took place over a few tens of millions of years. (See Fig. 3–40.)

Because of their large surface area, batholiths have traditionally been considered to be thick bodies extending deep into the crust. Some recent seismic and gravity studies suggest that batholiths may be relatively thin features. Figures 3–27 and 3–41 show these two interpretations of batholiths.

All of these data have been incorporated into one recent theory. This theory holds that the batholith is formed by the bubblelike rise of the detached pods of magma generated in the upper mantle or deep crust. These pods of magma form a thin batholith near the surface. At some places the magma reaches the surface and forms extensive volcanic rocks, and some batholiths intrude volcanic rocks that are similar in composition to the main batholith. These magma pods tend to spread out near the surface and so might account for the deformation that surrounds many batholiths. In passing through the crustal rocks, the magma would react with those rocks and could assimilate some of them by reaction and partial melting. In any case, the rocks through which the magma passes would be metamorphosed. Erosion of the thin layer of granitic rocks would expose an extensive metamorphic terrane similar to that found in many mountain ranges. Thus this theory would explain most of the data of batholiths and also explain the origin of extensive areas of metamorphic rocks. Like batholiths, these metamorphic terranes are difficult to explain because the known rise in temperature with depth is not enough to account for the amount of metamorphism.

PEGMATITE AND QUARTZ VEINS

The water and other volatiles (gases) in magma may also be involved in the formation of pegmatite and ore veins. When a deep intrusive body, such as a batholith, is emplaced, the rim is quickly chilled and solidifies first. This is commonly recorded by the fine-grained margins of an intrusive body. The volatiles are thus trapped inside the intrusive body; and as it solidifies, presumably from the margin inward, the volatiles are confined to a smaller and smaller volume in the center of the body. The residual fluid is rich in the elements that form the last minerals on the Bowen series, especially quartz, feldspar, and mica, as well as less common elements that do not fit into the rock-form-

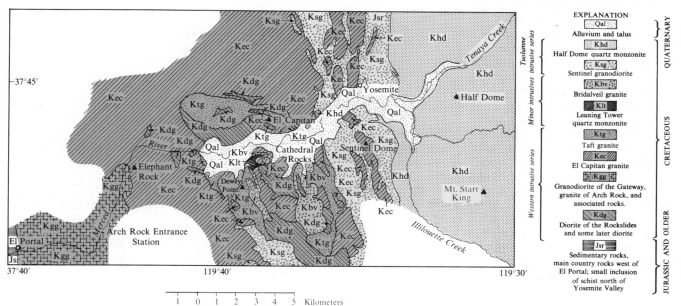

EXPLANATION

Qal
Alluvium and talus — QUATERNARY

Khd
Half Dome quartz monzonite

Ksg
Sentinel granodiorite — Minor intrusives intrusive series / Tuolumne

Kbv
Bridalveil granite

Klt
Leaning Tower quartz monzonite

Ktg
Taft granite — CRETACEOUS

Kec
El Capitan granite — Western intrusive series

Kgg
Granodiorite of the Gateway, granite of Arch Rock, and associated rocks.

Kdg
Diorite of the Rockslides and some later diorite

Jsr
Sedimentary rocks, main country rocks west of El Portal; small inclusion of schist north of Yosemite Valley — JURASSIC AND OLDER

FIGURE 3–40 Generalized geologic map of Yosemite Valley in the Sierra Nevada, California. The Sierra Nevada batholith is made of many smaller intrusive bodies as shown on this map. Granodiorite and quartz monzonite are types of granitic rocks included under the term *granite* in this text. From F. C. Calkins and D. L. Peck, California Division of Mines and Geology Bulletin 182, 1962.

63

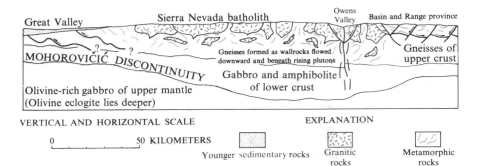

Gneisses formed as wallrocks flowed downward and beneath rising plutons

MOHOROVIČIĆ DISCONTINUITY

Gneisses of upper crust

Gabbro and amphibolite of lower crust

Olivine-rich gabbro of upper mantle (Olivine eclogite lies deeper)

VERTICAL AND HORIZONTAL SCALE

0   50 KILOMETERS

EXPLANATION

Younger sedimentary rocks   Granitic rocks   Metamorphic rocks

FIGURE 3–41 Hypothetical cross-section through the Sierra Nevada, based on the idea that the granitic rocks of the batholith moved through the metamorphic rocks to their present positions. This cross-section is based in part on seismic data. Compare with Fig. 3–27, which shows a thicker batholith. From Warren Hamilton and W. B. Myers, U.S. Geological Survey Professional Paper 554C, 1967.

ing minerals. This water-rich residual fluid eventually escapes, perhaps as a result of the increase in pressure that can result when the volatiles are concentrated in a small volume. The increase in pressure may cause cracks that are filled by this material (Fig. 3–42). Contraction due to cooling may also help to cause the cracks. The fluid has a low viscosity because of the volatiles, and this increases the mobility of the ions. As a result, very large crystals form. The resulting *very-coarse-grained rocks are called pegmatites,* and they may have the same composition as the enclosing magma or may contain unusual minerals. In the latter case the fluids are believed to concentrate the minor elements to form the unusual minerals. Pegmatites commonly occur as dikelike masses in or near intrusive bodies. They are actively sought because they contain large and often beautiful minerals, such as beryl, tourmaline, topaz, fluorite, all of the rock-forming minerals, and many more.

The residual fluids are probably also responsible for the formation of quartz-metal veins, but the process is not yet understood. The last fluid in a batholith would be rich in quartz and in the elements such as copper, lead, zinc, gold, silver, and sulfur, that do not fit into the crystal structures of the rock-forming minerals. A batholith may contain only a very tiny per-

centage of these elements, but the total amount of metal may be quite large. Quartz veins containing metals in the form of sulfide minerals are found around some intrusive bodies. (See Fig. 3–42.) These veins do not grade into pegmatites, so they must form by a somewhat different, although probably related, process.

In still other cases these fluids may cause widespread alteration of the intrusive body and/or the surrounding rocks. Much more study is needed to understand the actions of the magmatic volatiles that are believed to cause all of these diverse effects.

IDENTIFICATION AND INTERPRETATION

Identification of the coarse-grained igneous rocks will present no serious problems if one is able to identify the minerals and estimate their relative amounts. The first clue comes from the color of the rock which, following the charts (Figs. 3–43 and 3–44), will suggest which minerals should be looked for. In the use of the term *light color,* it should be understood that any shade of red or pink, no matter how dark, must be considered a light color. In the same way, any shade of green should be considered a dark color. Overall color should not be used

in place of good mineral determination, because exceptions are common enough to be troublesome (especially on examinations).

In a sense, the fine-grained, or volcanic, rocks are more difficult to identify than are the coarse-grained rocks because less data, in the form of identifiable minerals, are available; but because of this, the classification is less rigid. The procedure is similar in that color is used as the first step and, if no porphyritic crystals are present, the only step. If the porphyritic crystals can be identified, they are used in the same way as are the identifiable minerals in coarse-grained rocks; but because all potential minerals may not be present, color must be considered in deciding on a name. Care must be used not to mistake filled gas holes for porphyritic crystals. In many cases, it is necessary to have a chemical analysis in order to name a fine-grained rock accurately. Also, the volcanic rocks are not simply fine-grained equivalents of the coarse-grained rocks, as the chart might lead one to believe, but contain some small but distinct variations in composition, which chemical analyses will detect.

On the chart shown as Fig. 3–43, granite is used for most of the coarse-grained, quartz-bearing rocks. A number of other names are also in use for these rocks, depending on whether they contain one feldspar, two feldspars, or mixed feldspar (perthite). (See Fig.

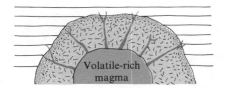

Volatile-rich magma

FIGURE 3–42 Cross-section of a crystallizing batholith showing origin of veins. Crystallization has proceeded inward from the fine-grained, quickly chilled margin. The volatiles are concentrated in the still molten center. The increased pressure causes cracks to form that are filled by the metal-bearing quartz-rich volatiles.

3–45.) Diorite generally has more dark minerals than does granite, and the feldspar is plagioclase. Quartz diorite is distinguished from granite in that it contains only plagioclase feldspar. The latter distinction, however, is not always easy to make.

Understanding the formation of a rock is much more important than simply naming it. It is necessary to discuss classification so that the meaning of rock names and the relationships among the various rock types are understood. Therefore, when studying a rock, try to interpret its origin, following mainly the discussion in the section on texture and mode of origin. Thus, the coarse texture of a granite implies that the specimen came from a large, slowly cooled, intrusive body that crystallized deep in the earth and has been exposed by uplift and deep erosion. A pumice sample must have come from a volcano, perhaps from an explosive eruption, and study of other nearby exposures may enable reconstruction of the volcano. When actual outcrops are studied, all of the information discussed here can be used in deciphering the origin and history of the rocks.

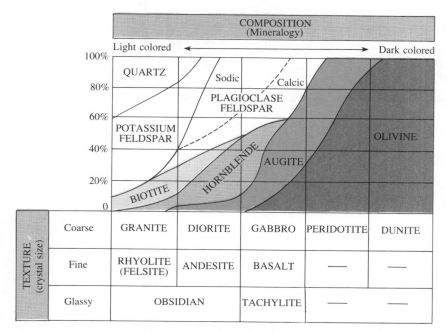

FIGURE 3–43   Chart showing the classification of the igneous rocks. Composition is indicated horizontally, and texture is indicated vertically. The upper part of the figure shows the range in mineral composition of each rock type.

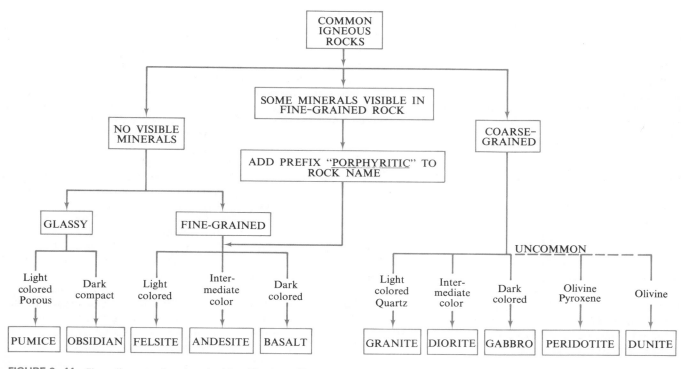

FIGURE 3–44   Flow diagram showing the identification of igneous rocks.

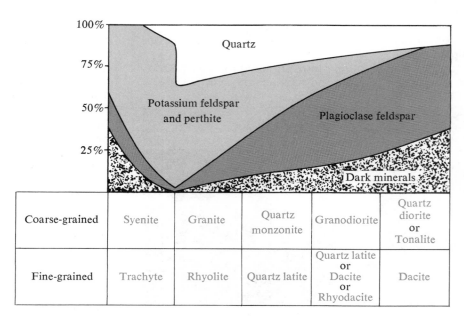

Coarse-grained	Syenite	Granite	Quartz monzonite	Granodiorite	Quartz diorite or Tonalite
Fine-grained	Trachyte	Rhyolite	Quartz latite	Quartz latite or Dacite or Rhyodacite	Dacite

FIGURE 3–45 Subdivision showing the names in use for the granitic rocks. The chart is an amplification of the left end of Fig. 3–43.

SUMMARY

The rocks of the crust are classified by origin into three types: *igneous,* solidified from magma; *metamorphic,* changed from another rock, generally by high temperature and pressure; and *sedimentary,* deposited in layers on the earth's surface by water, wind, and ice.

Magma is a natural, mobile, hot melt composed of rock-forming materials (mainly silicates) and volatiles (mainly steam) that are held in solution by pressure. It may or may not contain suspended solids.

In the sudden eruption of low viscosity magma, the expanding gases form a froth, resulting in the light, glassy rock *pumice.* Coarser bubble holes result in the rock *scoria.*

In eruption of more viscous magmas, the gas pressure shatters the rock into *pyroclastic* fragments: *volcanic ash* if sand-size or smaller, *lapilli* if larger but less than 64 millimeters (2.5 inches). Those larger than 64 millimeters (2.5 inches) are called *blocks* if angular and *bombs* if rounded. The *pyroclastic rocks* formed from these materials are *tuff*

if composed of ash, *lapilli tuff* if made of lapilli, *pyroclastic breccia* if composed of blocks, and *agglomerate* if made of bombs. *Tephra* is a general term for unconsolidated pyroclastic material.

Volcanic bombs are formed by pyroclastic material that is still molten when thrown out.

The quietest eruptions are very fluid basalt lava flows from dikes or dike swarms that can produce great thicknesses and cover huge areas.

A *shield volcano* is formed by eruption of very fluid basalt lava from a central vent.

Fluid basalt forms *ropy* or *pahoehoe* lava flows.

More viscous basalt lava breaks into blocks, forming *blocky* or *aa* lava flows.

Cinder cones are formed by pyroclastic material thrown out of a vent by the eruption of still more viscous magma.

Composite volcanoes are the very high volcanic mountains made of a combination of pyroclastic material and lava. If the lava is so viscous as

to plug the vent, the pressure of the volatiles in the magma may build up until an explosion results. The more viscous the magma, the more numerous or destructive the explosions.

Very explosive eruptions of hot gas and shattered glass fragments are called *nuées ardentes (glowing clouds, glowing avalanches);* this fluid pyroclastic material forms the rock *welded tuff* when the hot glass fragments come to rest and weld together. Welded tuffs are relatively abundant and may or may not be associated with dome volcanoes.

Pillow lava forms when lava flowing into water bodies is chilled relatively rapidly.

Eruptions cannot yet be predicted, but magma movements are detected by many shallow, weak earthquakes, by tilt-meter measurements, and so forth.

Volcanic destruction is of two kinds: direct damage by the material erupted; and indirect damage by mudflows of pyroclastic material, and by floods from melted glaciers and breached lakes.

Bowen's reaction series shows the sequence in the crystallization of a basalt magma.

Separation of early-formed crystals may form rocks with a composition different from the original melt; this process is called *differentiation.*

In a *continuous mineral series,* the full range of compositions of that mineral group occurs in sequence from one end member to another.

In a *discontinuous mineral series,* the change is from one mineral group to another in separate, discrete steps.

With falling temperature, Bowen's reaction series has the following sequence of crystallization: a discontinuous series of olivine to pyroxene to amphibole to biotite, concurrently with a continuous series of calcic to sodic plagioclase, followed by the sequence potassium feldspar, muscovite, and, lastly, quartz.

Igneous rocks are classified by composition and texture. The composition classification follows the reaction series. Texture refers to the grain size and indicates the mode of occurrence and cooling history of the rock.

Intrusive igneous rocks solidify below the surface of the earth; *extrusive* igneous rocks solidify on the surface.

Plutonic igneous rocks are coarse-grained rocks that crystallized deep in the earth.

Intrusive igneous rocks are *discordant* if they cut across the bedding of the rocks they intrude, and *concordant* if they are more or less parallel to the bedding.

Discordant intrusives are *batholiths,* very large, coarse-grained bodies; *stocks,* smaller, coarse-grained bodies; and *dikes,* tabular, usually parallel-walled, linear, fine-grained bodies.

Concordant intrusives are *sills,* thin, usually fine-grained bodies intruded between sedimentary layers without noticeable deformation; and *laccoliths,* thicker, fine- to coarse-grained bodies which up-arch the sedimentary layers above them.

Extrusive rocks result from lava flows or other volcanic activity. They are fine-grained or, with very rapid cooling, glassy.

The textural terms are:

Glassy—no crystals

Fine-grained—grains can be seen in sunlight with a good hand lens

Coarse-grained—crystals easily seen

Pegmatitic—crystals over one centimeter

Porphyritic—coarse crystals in a fine-grained or glassy matrix, or some very coarse crystals in a coarse-grained rock. Porphyritic texture may record a change in conditions during the cooling history of the rock.

The mid-ocean ridges are areas of basalt volcanism. The basalt apparently originates from partial melting of upper mantle rocks.

The andesite magma of volcanic island arcs probably originates from partial melting of the descending slab of ocean-floor basalt and its overlying sediments.

Magma probably rises because the melted rock expands and is less dense than the surrounding rocks.

Salt domes from buried salt beds are also less dense than the surrounding rocks, and may illustrate another case of gravitational rise of lighter material.

Expansion of the volatiles in magma or erosion by expanding gas bubbles may help the magma to rise.

Most large batholiths are composed of a number of smaller plutons.

Traditionally batholiths have been thought to extend deep into the crust and to have been emplaced by *stoping* and pushing aside the pre-existing rocks.

Stoping is a process by which blocks of pre-existing rocks above the magma fall through the melted magma, allowing the magma to rise.

Another theory is that batholiths may be relatively thin features. They may be emplaced by the bubblelike rise of detached pods of magma, which metamorphose the rocks through which they pass.

Quartz veins and very coarse-grained *pegmatite* masses may result from the residual fluid in magma, which is composed of trapped volatiles, elements to form the last minerals of Bowen's series, and less common elements which do not fit into the common rock-forming minerals and therefore may form unusual minerals.

In general, coarse-grained igneous rocks are identified by identifying minerals and estimating their relative amounts. The type of feldspar is important.

Fine-grained igneous rocks are identified primarily by color or chemical analysis.

Study of the texture and composition of the rock enables interpretation of its formation.

QUESTIONS

1 Define igneous rock.
2 Describe the formation of pumice.
3 Describe the types of volcanoes. Include how they are formed and which is steepest and why.
4 What hazards do volcanoes pose?
5 Describe the crystallization of a basaltic melt.
6 Describe the crystallization of plagioclase in a magma.
7 What happens when a crystal of calcic plagioclase is melted?

8   Is the classification of most igenous rocks genetic? Discuss.
9   How can a sill be distinguished from a flow?
10 What cooling history does a porphyritic texture imply?
11 What is a plutonic igneous rock?
12 What is a batholith; what is the usual composition and what problems do batholiths present?
13 Name a concordant igenous body.
14 In what type of occurrence (sill, flow, batholith, etc.) would you expect to find glassy, fine-grained, porphyritic, and coarse-grained igneous rocks?
15 Describe the formation of ore veins.
16 Igneous rocks are classified on the bases of _____ and _____.
17 What minerals are most likely to be visible in a felsite porphyry?
18 How does pegmatite form?
19 What the main volcanic gases?
20 What is the probable origin of the mid-ocean ridge basalt?
21 What is the probable origin of andesite at volcanic island arcs?

SUPPLEMENTARY READINGS

General

Dietrich, R. V., and B. J. Skinner, *Rocks and Rock Minerals.* Somerset, N.J.: John Wiley & Sons, 1979, 319 pp.

Ernst, W. G., *Earth Materials.* Englewood Cliffs, N.J.: Prentice-Hall, Inc., 1969, 150 pp. (paperback).

Volcanoes

Aspinall, W. P., H. Sigurdsson, and J. B. Shepherd, "Eruption of Soufrière Volcano on St. Vincent Island, 1971–1972," *Science* (July 13, 1973), Vol. 181, No. 4095, pp. 117–124.

Bullard, F. M., *Volcanoes of the Earth.* Austin, Tex.: University of Texas Press, 1976, 579 pp.

Crandell, D. R., and D. R. Mullineaux, *Volcanic Hazards at Mount Rainier, Washington.* U.S. Geological Survey Bulletin 1238, 1967, 26 pp.

Crandell, D. R., and D. R. Mullineaux, *Potential Hazards from Future Eruptions of Mount St. Helens Volcano, Washington.* U.S. Geographical Survey Bulletin 1382-C, Washington, D.C.: Superintendent of Documents, 1978, pp. C1-C26 (reprint).

Decker, Robert, and Barbara Decker, "The Eruptions of Mount St. Helens," *Scientific American* (March 1981), Vol. 244, No. 3, pp. 68–80.

Decker, Robert, and Barbara Decker, *Volcanoes.* San Francisco: W. H. Freeman and Co., 1981, 244 pp.

Eaton, G. P., and others, "Magma beneath Yellowstone National Park," *Science* (May 23, 1975), Vol. 188, No. 4190, pp. 787–796.

Heiken, Grant, "Pyroclastic Flow Deposits," *American Scientist* (September–October 1979), Vol. 67, No. 5, pp. 564–571.

Kittleman, L. R., "Tephra," *Scientific American* (December 1979), Vol. 241, No. 6, pp. 160–177.

Macdonald, G. A., *Volcanoes.* Englewood Cliffs, N.J.: Prentice-Hall, Inc., 1972, 510 pp.

Miller, C. D., *Potential Hazards from Future Eruptions in the Vicinity of Mount Shasta Volcano, Northern California.* U.S. Geological Survey Bulletin 1503, Washington, D.C.: Superintendent of Documents, 1980, 43 pp.

Moore, J. G., "Mechanism of Formation of Pillow Lava," *American Scientist* (May–June 1975), Vol. 63, No. 3, pp. 269–277.

Peck, D. L., T. L. Wright, and R. W. Decker, "The Lava Lakes of Kilauea," *Scientific American* (October 1979), Vol. 241, No. 4, pp. 114–128.

Sheets, P. D., and D. K. Grayson, eds., *Volcanic Activity and Human Ecology.* New York: Academic Press, 1979, 644 pp.

Williams, R. S., Jr., and J. G. Moore, "Iceland Chills a Lava Flow," *Geotimes* (August 1973), Vol. 18, No. 8, pp. 14–17. Describes the use of seawater in an attempt to save a fishing port from the hazards of a lava flow.

4 WEATHERING AND SEDIMENTARY ROCKS

The study of sedimentary rocks began very early. Da Vinci around 1500 and Steno in 1668 noted that sedimentary rocks had originally been deposited in horizontal layers, that the fossils that they contain were once living animals, and that seas had once covered much of what is now land. These were major observations, and had they been generally recognized, geology would have developed much earlier than it did. During the eighteenth century a number of attempts were made to classify and organize the sedimentary layers. Study of the sedimentary layers and their contained fossils ultimately led, in the early nineteenth century, to a geologic time scale and to deciphering the geologic history of the earth.

The study of sedimentary rocks themselves, their composition and formation, came much later. Study of sedimentary rocks may have lagged behind study of igneous rocks because they seemed to present fewer problems than controversial rocks such as basalt. Study of how sedimentary rocks form probably started in earnest when uniformitarianism was generally accepted. Uniformitarianism, as was described in the last chapter, in this case means that to understand ancient rocks, we must study how similar rocks are forming today. Uniformitarianism is central to all geology, and so it is a theme to which we will return in later chapters. It began with James Hutton, late in the eighteenth century, but it did not become widely accepted until it was popularized by Charles Lyell (1797–1875). Lyell wrote a textbook that was first published in 1830 and went on through many editions. He extended and illustrated uniformitarianism and was responsible for its acceptance.

Even though the scene was set for study of the sedimentary rocks with the acceptance of uniformitarianism and the development of microscope techniques, study of sedimentary rocks developed slowly. The microscope was used first on igneous and metamorphic

rocks as described in Chapter 3. Study of these rocks gave insight into internal processes of the earth, and so was an important research area. The practical geologists of the nineteenth century were engaged in mining, and so they, too, were mainly interested in igneous and metamorphic rocks. A few geologists did do important work on sedimentary rocks, but it took a new economic incentive to give the study of sedimentary rocks its rightful place in geology. Petroleum is found in sedimentary rocks, and its importance in the twentieth century encouraged detailed study of all aspects of sedimentary rocks.

Sedimentary rocks are formed at or near the surface of the earth. They are composed of rock fragments, weathering products, organic material, or precipitates. Most are deposited in beds or layers. They comprise a very small volume of the earth, only about 8 percent of the crust. (See Fig. 4–1.) In spite of their small volume, however, they cover about 66 percent of the surface. They are formed in a number of different ways, but because the raw materials come from weathering, we will begin there.

WEATHERING

PROCESSES

Mechanical weathering breaks up a pre-existing rock into smaller fragments, and *chemical weathering,* acting on these small fragments, *rearranges the elements into new minerals.*

Mechanical weathering is done mainly by the forces produced by the expansion of water due to freezing, and to a lesser extent by such things as forces of growing roots (see Fig 4–2), worms and other burrowing animals, lightning, expansion and contraction caused by heating and cooling, and human activity. The most important of these methods is the re-

sult of the 9 percent expansion that water undergoes when it freezes. Tremendous forces are produced in this way, and their efficacy is shown by the thick layers of broken rock that form in areas that go through a daily freeze-thaw cycle many times a year. (See Fig. 4–3.)

Another type of mechanical weathering is caused by the expansion of rock when erosion removes the weight of the overlying rocks. The outermost layer of the rock may expand enough so that it breaks from the main mass, forming cracks or joints parallel to the surface. (See Fig. 4–4). This process is called *exfoliation* and is best displayed in massive rocks such as granite. Granite bodies that have few cracks or joints may be formed into domes in this way (Fig. 4–5). This may also occur in boulders released by frost action. In this case layers like onion skin may form. (See Fig. 4–6.) However, most weathering of boulders in this way is caused by chemical weathering.

The products of mechanical weathering include everything from the huge boulders found beneath cliffs to particles the size of silt. The erosion and transportation of these fragments will be discussed in Chapters 7 through 11. These fragments can form rocks such as sandstones, and the shape of the fragments in such a rock can reveal some of the history of the rock. Thus if the fragments are sharp and angular, they probably

Sandstone and shale in New Mexico. Photo by E. F. Patterson, U.S. Geological Survey.

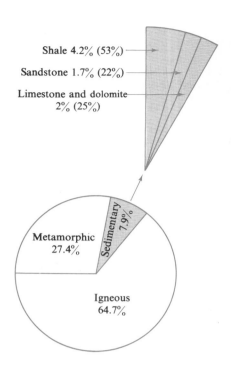

Shale 4.2% (53%)

Sandstone 1.7% (22%)

Limestone and dolomite
2% (25%)

Metamorphic
27.4%

Sedimentary
7.9%

Igneous
64.7%

FIGURE 4–1 Relative abundance of sedimentary rocks in the earth's crust. For example, shale forms 4.2 percent of all rocks, and 53 percent of the sedimentary rocks. Data from Ronov and Yaroshevsky, 1969.

FIGURE 4–2 Sidewalk and curb have been moved by tree roots.

FIGURE 4–4 Cracks or joints parallel to the surface are believed to be caused by expansion. The expansion occurs as erosion removes the overlying rocks. The granitic rocks in this view crystallized at least a mile beneath the surface. Jointing of this type on a larger scale is conspicuous in many areas of granitic rocks. Photo by G. K. Gilbert, U.S. Geological Survey.

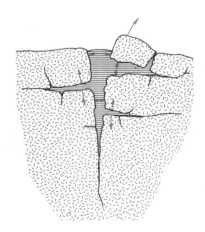

FIGURE 4–3 Frost wedging. Water in crack freezes from top down because ice is less dense than water (floats) and because the water in a crack is cooled most at the surface. This restricts the water in the crack. Because ice occupies more volume than water, further freezing creates forces due to this expansion. These forces cause further cracking of the rock.

were buried quickly; but if the fragments are rounded, they were abraded during long transportation. (See Fig. 4–7.)

Water is important in most types of weathering. The role of water is largely determined by its structure. The water molecule is composed of an oxygen atom and two smaller hydrogen atoms that are attached to the oxygen about 105 degrees apart, as shown in Fig. 2–6. The hydrogen has a slight residual positive charge,

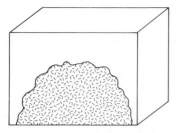

A. Batholith formed at a depth of several kilometers

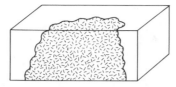

B. Uplift and erosion expose the batholith

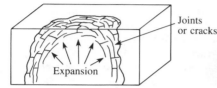

Joints or cracks

Expansion

C. The release of the pressure due to the weight of the overlying rocks allows expansion that causes joints or cracks to develop

D. Half dome in Yosemite National Park

FIGURE 4–5   Exfoliation domes. The photo by F. C. Calkins, U.S. Geological Survey.

and the oxygen a similar weak negative charge. Because of these charges the molecules tend to form in groups of two to eight. More groups are formed near the freezing point than in liquid water, and because the groups take up more space, water expands on freezing. These residual charges also make water an excellent solvent. *Removal of material in solution by ground water is called leaching.*

The rate of chemical weathering depends in general on the temperature, the surface area, and the amount of water available. Except in cold or very dry climates, it generally keeps up with mechanical weathering, and the two can be separated only in concept. Mechanical weathering provides the large surface area necessary for chemical activity to take place at the surface of the earth. (See Fig. 4–8.)

In following chemical weathering, we will see what happens to the elements in the main rock-forming minerals. The main reactions involved are *oxidation, hydrolysis,* and *carbonation. Oxida-*

tion is reaction with oxygen in the air to form an oxide. Hydrolysis is reaction with water (*hydration* is the addition of water molecules to a mineral), and *carbonation is a reaction with carbon dioxide in the*

air to form a carbonate. The carbonation reaction begins with the uniting of carbon dioxide (CO_2) and water (H_2O) to form carbonic acid (H_2CO_3), or

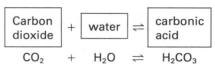

| Carbon dioxide | + | water | ⇌ | carbonic acid |

$$CO_2 + H_2O \rightleftharpoons H_2CO_3$$

This acid plays an important role in many weathering reactions. The air in soil, because of biologic processes, contains up to 10 percent carbon dioxide, much more than the small amount in the atmosphere. In all of these reactions, substances are added so that the total volume is increased. The details of how these reactions take place are not well understood and are an important area of current research in soil science, a relatively new branch of science. The reactions take place slowly and are probably influenced by the composition of the water films that surround the particles and by the organic materials present (particularly the acids). The importance of organic materials in

FIGURE 4–6   Spheroidal weathering. Expansion of surface layers can produce spherical masses with onion-skin layers. Such weathering probably begins on joint cracks; and although splitting-off of the layers is mechanical weathering, the expansion of the layers is probably caused largely by chemical weathering. This photo of basalt in Puerto Rico by C. A. Kaye, U.S. Geological Survey.

73

WATER AND MECHANICAL WEATHERING

Water behaves in some unusual ways, and its properties determine the climate of our earth. Here we will consider only one aspect of water, the underlying reason for its expansion on freezing and the reason for its other unusual properties. Most materials expand when heated and contract when cooled. Water does too, but with a very important exception. Water contracts upon cooling but reaches its maximum density at 4°C. Further cooling causes water to expand. This very unusual behavior is generally accepted as commonplace by most of us because we have come to expect ice cubes to float in a drink. The expansion is caused by what are called *hydrogen bonds.* Hydrogen bonds also account for water's very high capacity to absorb heat, its high boiling and freezing temperatures, the very large amounts of heat needed to cause a phase change (as from water to ice), and its ability to dissolve ionic substances.

A water molecule is composed of one oxygen and two hydrogen atoms, with the two hydrogen atoms 105 degrees apart and imbedded in the oxygen. (See Fig. 1.) This is a polar molecule because the two hydrogen atoms are positively charged, and negative charges exist on the opposite side of the oxygen atom. These

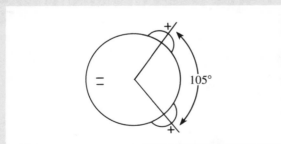

FIGURE 1

separated charges give polar compounds their properties. In water, the negative charges are at two places on the oxygen atom. The bond between those negative places and the positive

hydrogen atoms of other water molecules (the hydrogen bond) is stronger than similar bonds in other polar substances.

Water molecules tend to link together on hydrogen bonds. The motion of the molecules, which is proportional to their temperature, determines the number of bonds. As water cools, more hydrogen bonds form because there is less molecular motion. As molecular motion decreases, the molecules come closer together, so the water becomes denser as the volume decreases. This continues until the water is cooled to 4°C.

At 4°C, the water begins to take on the ice structure. Hydrogen bonds among water molecules begin to form the six-sided (hexagonal) shape of ice crystals. Note in Fig. 2 that the ice

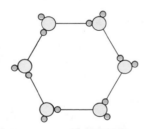

FIGURE 2

structure is quite open, and so the molecules in this structure occupy more space than they did as warmer water. Thus, as the open ice structure begins to form at 4°C, the density of water decreases, and the water expands. The expansion of water as it turns into ice is the main cause of most mechanical weathering. As ice cools further, it behaves normally and contracts.

At 4°C, the density of water is 1.0000 g/cm^3, and its specific volume is 1.0000 cm^3/g. At 0° C, the density of ice is 0.9170 g/cm^3, and its specific volume is 1.0905 cm^3/g. Thus, between 4°C and 0°C, the expansion is about 9 percent. Most of this expansion occurs at 0°C, where freezing occurs.

FIGURE 4-7 Sand grains showing different degrees of roundness. The rounded sand grains were probably shaped by abrasion during longer transportation than the angular grains. If the fragments are of different materials, the softer will be rounded first.

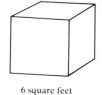

6 square feet
864 square inches
(6 x 12 x 12)
(5,574 square centimeters)

1,728 in.^{2}
(11,148 cm^2)

3,456 in.^{2}
(22,297 cm^2)

6,912 in.^{2}
(44,593 cm^2)

FIGURE 4-8 Surface area versus particle size. Shows how mechanical weathering increases the surface area and thus helps to increase chemical weathering. A cubic foot is shown in each case.

weathering and soil formation is covered in the next section.

Air pollutants such as carbon dioxide and sulfates, released by burning fossil fuels and industrial processes, cause accelerated weathering of building stone and other materials in cities. The ultimate fate of many pollutants is not known, but it appears that some are absorbed and decomposed by soil.

The rainwater involved in most weathering is not pure water. In addition to carbon dioxide, it contains dissolved salts. Atmospheric water vapor, in general, will not condense to form clouds or rain without a nucleus around which to condense. Dust particles provide some of these nuclei, but most are tiny crystals of salts from the ocean. Spray from ocean waves evaporates, and the tiny crystals of salts formed in this way are carried away by the wind. The salts formed when seawater is evaporated are described in a later section. Near the ocean, rainwater has the same chemical composition as seawater, but is very much

diluted. Farther inland, the composition is variable because the winds transport the various salts from the ocean differently. The chemical composition of rainwater does have an effect on the clays produced by weathering, but little more can be said on the complex subject of clay mineralogy here. However, the salts in rainwater accumulate in lakes that have no outlet to the sea, such as Great Salt Lake.

Chemical weathering is concentrated at edges and corners of rocks and so tends to make originally angular rocks more spheroidal (Fig. 4-9). *Spheroidal weathering* may continue to remove layers as in Fig. 4-6). The cause of

such weathering is the expansion that occurs during chemical weathering.

Chemical weathering is most intense on grain surfaces, and because of the increase in volume during chemical weathering, this trait can cause disaggregation of a rock. This process may account for fragmentation of rocks in areas where mechanical weathering by frost action does not occur. Laboratory experiments show that expansion and contraction from daily heating and cooling are not enough to cause disintegration, even on a desert.

Climate affects the topography produced by weathering. Dry areas where mechanical weathering predominates generally have bold, angular cliffs (see Fig. 4-10), and humid areas have rounded topography. Limestone, which is made of the soluble mineral calcite, is readily dissolved by carbonic acid. Thus it is rapidly weathered and tends to form valleys in humid regions, but it is a resistant rock and forms bold outcrops in dry areas. In warm, moist climates, weathering may extend to depths of over 122 meters (400 feet). Maximum weathering probably takes place just below the ground surface—the depth at which fence posts tend to break.

Chemical weathering can be defined also as changes taking place near the surface that tend to restore the minerals to equilibrium with their surroundings. The primary minerals form, in most cases, at elevated temperature and pressure and are in equilibrium under those conditions.

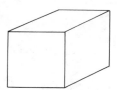

FIGURE 4-9 Spheroidal weathering occurs because weathering is more intense on edges and corners.

When these minerals are exposed at the surface, they are in conditions far from those under which they formed; therefore, they react to form new minerals which will be in equilibrium. One can then predict that the highest-temperature minerals will probably be affected first in chemical weathering, but of course many other factors such as the bonding within the mineral will influence its weathering. It is fortunate, then, that Bowen's reaction series (Fig. 3–23), which shows the sequence of mineral formation with falling temperature, also shows the relative stability to weathering. Olivine is the first to be affected and quartz the least affected, although biotite is easily weathered. Organic soil acids can, in some cases, upset these relationships.

The weathering of potassium feldspar will illustrate chemical weathering. An overall reaction can be written as

FIGURE 4–10 Angular topography in an arid area. Weathering has accentuated minor differences among the sedimentary layers. Bryce Canyon National Park, Utah. Photo from Union Pacific Railroad.

$$\underset{\text{2 potassium feldspar}}{\boxed{\text{2 potassium feldspar}}} + \underset{\text{carbonic acid}}{\boxed{\text{carbonic acid}}} + \underset{\text{water}}{\boxed{\text{water}}} \rightarrow \underset{\text{clay}}{\boxed{\text{clay}}} + \underset{\text{potassium carbonate}}{\boxed{\text{potassium carbonate}}} + \underset{\text{4 silica}}{\boxed{\text{4 silica}}}$$

$$2\ KAlSi_3O_8 + H_2CO_3 + H_2O \rightarrow Al_2Si_2O_5(OH)_4 + K_2CO_3 + 4\ SiO_2$$

The feldspar is broken down by a process in which hydrogen replaces potassium. The hydrogen can come from water, from carbonic acid, or from the hydrogen ions that surround plant roots. The first two sources are inorganic, and the last is an example of biological weathering. The potassium released in the latter way is taken up by the plant. The clay mineral shown in the illustrative reaction is not the only clay mineral that can form. Potassium and magnesium clays are also common and form in some weathering environments. Weathering reactions of other feldspars are similar, differing only in that sodium and calcium are involved in place of potassium.

The weathering of the ferromagnesian minerals is also similar, and magnesium-bearing clays may form in this way. The ferromagnesian minerals are more easily weathered because the iron is easily oxidized, breaking down the mineral.

The products of chemical weathering of the common minerals can be summarized as follows:

Quartz—little chemical weathering, sand grains produced.
Feldspars—
   Potassium feldspar ($KAlSi_3O_8$).
      K by carbonation produces K_2CO_3.
      Al,Si,O by hydration produce clay ($Al_2Si_2O_5(OH)_4$).
      SiO_2 is released as soluble silica and colloidal silica.
   Plagioclase feldspars—produce the same end products except that $CaCO_3$ and Na_2CO_3 are produced instead of K_2CO_3.
Muscovite—produces the same products as potassium feldspar.
Ferromagnesian minerals—depending on composition, produce the same products plus iron oxides — hematite Fe_2O_3 (red iron oxide) or limonite

FeO(OH) (yellow iron oxide). Magnesium forms soluble magnesium carbonate, and some is used in producing clay minerals.

The products of weathering and their ultimate disposition are summarized in Table 4-1. (See also Fig. 4-11.)

From this brief outline of chemical weathering it is possible to make some interpretation of sedimentary rocks. As might be expected, most sandstones are composed largely of quartz because it is the mineral most resistant to weathering. Some sandstones contain much feldspar, and such rocks formed under conditions

that did not permit chemical weathering to complete its job. This situation could have developed several ways. Accumulation might have been too rapid to permit complete weathering, or the climate may have been too dry or too cold. Certainly there are other possibilities, but this serves as an example of how sedimentary rocks can be interpreted.

SOILS

Soils are of great economic importance. They can be said to be the bridge between terrestrial life and the inanimate world. *Soil, as the*

TABLE 4–1 Weathering products and their disposition.

Weathering Product	Disposition
Clay	Forms shale.
Silica	Forms chert and similar rocks.
Potassium carbonate (K_2CO_3)	Some is transported to ocean, some is used by plant life, and some is adsorbed on or taken into certain clays.
Calcium carbonate ($CaCO_3$)	Forms limestone.
Sodium carbonate (Na_2CO_3)	Dissolves in ocean.
Iron oxides	Form sedimentary deposits of iron oxide.
Magnesium	Some replaces calcium in limestone to form dolostone, and some is transported to ocean.

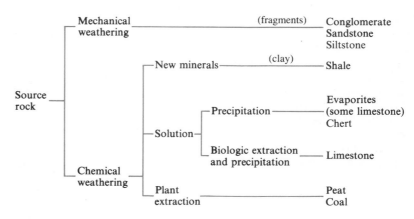

FIGURE 4–11 Simplified chart showing origins of sedimentary rocks.

term is used here, is the material that supports plant life, and so supports all terrestrial life. Only lichen and primitive plants, such as silver sword in Hawaii, can get nutrition directly from rock without soil. The development of soil is a complex interplay of weathering and biologic processes.

The clay minerals produced by chemical weathering are very important in soil. Clay minerals are tiny plates, or sheets; these sheets have a small negative charge on their surfaces. Water molecules also have their charges separated. The positive sides of water molecules are attracted to and adhere to the clay minerals, so water is held in the soil where plants can use it. Other positively charged ions such as potassium and calcium also adhere to clay particles. Plant roots are surrounded by positively charged hydrogen ions, and these ions are exchanged for the water and other ions on the clay minerals. This is an important process in the growth of plants (Fig. 4–12).

The importance of soil to life is shown by the composition of some foods. Steak contains about 12.5 percent phosphorus, iron, and calcium by weight. Thus each pound of steak contains about two ounces of rock material obtained from soil by plants. A much larger amount of rock had to be weathered to release these two ounces. Plants use much weathered rock material. For example, an acre of alfalfa yields about four tons of alfalfa per year. To produce this alfalfa, over two tons of rock must be weathered to obtain the phosphorus, calcium, magnesium, potassium, and other elements in the alfalfa.[1] This corresponds to about 0.178 millimeter (0.007 inch) of rock weathered. The average rate of erosion is estimated to be about 1 inch per 9000 years, so this is about 5.5 times the annual erosion rate. This "uses" soil at a very

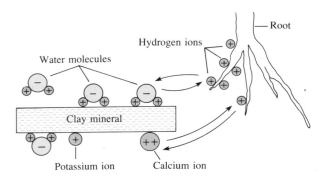

FIGURE 4–12 Clay minerals attract water ions and other ions that are used by plants.

high rate, so fertilizers must be added. Thus soils can be used faster than they form, a fact not always taken into account in agricultural economics.

There are many different types of soils. At first it was thought that the parent rock determined the soil type. We now know that climate is more important than parent rock; but the situation is complex, and other factors, especially the type of vegetation, are involved. Soils can be described as mature and immature. On hillslopes, erosion may prevent soils from reaching maturity and the transported soil may accumulate in valleys. (See Fig. 4–13B.) Many soils form on surface materials that have been transported. It apparently takes at least a few hundred years for a soil to develop. For this reason soil conservation is very important. The range in rate of soil formation is great. In 45 years, 35.6 centimeters (14 inches) of soil formed on

pumice from the eruption of the volcano Krakatoa. At the other extreme, in many places no soil has formed on glacially polished surfaces in about 10,000 years. (See Fig. 10–9.)

In most soils three layers, or *horizons,* can be recognized. They are the topsoil, the subsoil, and the partly decomposed bedrock. In addition, in some areas, especially forests, a layer of organic material may form at the surface. (See Figs. 4–13 and 4–14.) The material that gives the upper parts of many soils their dark color is *humus.* Humus is formed by the action of bacteria and molds on the plant material in the soil. Chemical weathering releases some of the materials used by plants, and humus also provides food for plants.

Soils form as a result of the weathering processes just discussed and the effects of the organic material present, mainly humus. One of the most important processes in the formation of soils

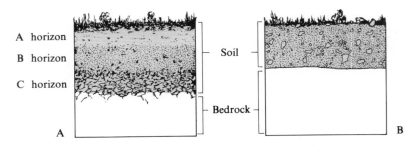

FIGURE 4–13 Soil profiles. A. Residual soil with well-developed horizons or layers. B. Transported soil without layering.

---

[1]W. D. Keller, "Geochemical Weathering of Rocks: Source of Raw Materials for Good Living," *Journal of Geological Education* (February, 1966), Vol. 14, No. 1, pp. 17-22.

is the leaching of material from the topsoil and the deposition of this material in the subsoil. Water reacts with humus to form acid that is very effective both in leaching and in causing chemical weathering. Earlier, the similar role of carbonic acid, formed by carbon dioxide and water, was discussed. Carbon dioxide is also formed in soil from the decay of organic material.

We can now consider a few of the more important types of soils. In the United States the soil of the humid east is different from that of the drier west. The dividing line is approximately at the 64-centimeter (25-inch) annual rainfall line and runs almost north-south near 97° longitude. (See Fig. 4–15.) The characteristics of the soils about to be discussed are summarized in Table 4–2.

The soil that predominates in eastern United States is called *pedalfer* (pedon, *Greek for soil, and* Al, Fe, *the symbols for aluminum and iron*). Humus is well developed in the temperate, moist climate. The abundant rainwater becomes strongly acid by reaction with the humus and leaches the topsoil. Aluminum and iron leached from the topsoil are deposited in the subsoil, giving it a brown (limonite) color. Soluble materials such as calcite or dolomite are leached from this type of soil and are generally removed by the ground water. These soils are quite acid, and clays are well developed. Ground water carries the clay to the subsoil so that topsoil may be somewhat sandy, due to resistant quartz fragments. Pedal-

FIGURE 4–14 Soil profile revealed in a pit near Ladson, South Carolina. Photo by H. E. Malde, U.S. Geological Survey.

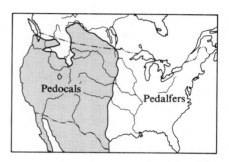

FIGURE 4–15 Map of major soil types in the United States. The dividing line is approximately at the 64-centimeter (25-inch) annual rainfall line except in the northern Rocky Mountains where low temperature affects the soil type.

TABLE 4–2 Generalized summary of soil types. See text for details.

	Temperate Humid [>64-cm (25-in.) Rainfall]	Temperate Dry [<64-cm (25-in.) Rainfall]	Tropical Savanna (Heavy Seasonal Rainfall)	Extreme Arctic, Desert (Hot or Cold)
Climate				
Vegetation	Forest	Grass-brush	Grass and tree	Almost none, so no humus develops
Typical Area	Eastern U.S.	Western U.S.	—	—
Soil Type	Pedalfer	Pedocal	Laterite	
Topsoil	Sandy; light colored; acid.	Commonly enriched in calcite; whitish color.	*(Zones not developed)* Enriched in iron (and aluminum); brick red color.	No real soil forms because there is no organic material. Chemical weathering is very slow.
Subsoil	Enriched in aluminum, iron, clay; brown color (limonite).	Enriched in calcite; whitish color.	All other elements removed by leaching.	
Remarks	Extreme development in conifer forests because abundant humus makes ground water very acid. Produces a light gray soil because of removal of iron.	*Caliche* is the name applied to the calcite-rich soils.	Apparently bacteria destroy the humus so no acid is available to remove iron. Tropical rain forests develop soils similar to pedalfers.	

fers generally are best developed under forests. Conifer forests produce a thick layer of litter on the forest floor and so develop more humus. Therefore, the soil is very acid and its development is accentuated; the resulting soil is light gray in color. It is in the pedalfers that calcite, generally called agricultural lime or limestone, is spread on the soil to combat its acidity.

In the drier west, less humus develops so the limited amounts of ground water are much less active chemically. Chemical weathering is also much slower in this climate so there is less clay than in pedalfers. Under these conditions leaching is not complete, and the soluble materials, especially calcite, are deposited in the subsoil because the ground water evaporates before it can remove them. Such soils are called *pedocals* (pedon *for soil, and* cal *for calcite*). *The calcite forms as whitish material in the soil, called* **caliche** (Spanish for *lime*). Caliche may accumulate in the topsoil as well as in the subsoil. This can occur when the ground water moves upward through capillary openings because of the evaporation at the surface. Pedocals tend to develop under grass- or brush-covered areas.

Soil formation in tropical climates is not fully understood, in part because few areas have been well studied. The soil in rainforests is somewhat similar to the pedalfers. The soil of the grass- and tree-covered savannas, however, is much different. Apparently because of the seasonal heavy rainfall, leaching is the dominant process. Everything, including silica and clays, is leached from the soil and carried away by ground water except iron and aluminum, which form hydrated oxides. Apparently, at the high temperatures, bacteria destroy the humus so that the ground water is not acid and cannot remove iron and aluminum. The soils produced in this way are called *laterites*. (See Fig.

4–16.) Laterite is from the Latin word for brick, and its reddish-brown color is similar to that of a brick. Laterites do not form distinct layers as other soils do. As might be expected, they are poor soils on which to grow crops, as shown by the many unsuccessful attempts to develop new farmlands in emerging tropical countries. Most laterites are iron-rich; however, in some areas they are mainly aluminum. Such a *mixture of hydrated aluminum oxides is called* **bauxite** and is a valuable ore. Bauxite probably forms only from the weathering of aluminum-rich rocks that have little iron, such as some feldspar-rich igneous rocks. At some places laterites are mined for iron ore and, at a few places, for their nickel or manganese; here the parent rock probably was rich in these elements.

In the extreme climates, such as arctic and desert, very little chemical weathering occurs, and there is almost no organic material; therefore, no real soil can develop.

TRACE ELEMENTS IN THE SOIL AND HEALTH

Some elements present in the environment in only trace amounts have a large influence on health. A few obvious examples of good and bad effects of trace elements are widely known. The relationship between dental caries and fluorine in the water supply, and selenium poisoning of cattle by "loco weed" *(Astragalus racemosus)* are known to almost everyone.

Although some trace elements, such as iodine, apparently move through the atmosphere, most come from the soil and enter the life cycle either through water supply or plants. Although climate is more important in soil formation than the type of bedrock, the bedrock does influence the trace elements in the soil. Because most of these elements are metals, mining areas are commonly enriched in one or more of them. Many other areas are also either enriched or deficient in some or many of the trace elements. The amount of chemical weathering seems important, and both arid and tropical areas are generally deficient. Not all trace elements are beneficial, as will be shown. Insecticides, sprays, and fertilizers also add materials to plants and soils.

Deciding which elements should be considered trace elements is not a simple decision. Most living

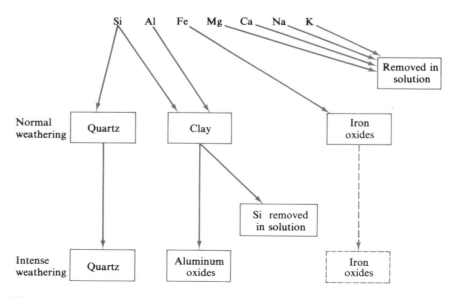

FIGURE 4–16   Simplified summary of weathering showing how laterites form.

tissue is composed of hydrogen, carbon, nitrogen, oxygen, sodium, magnesium, phosphorus, sulfur, chlorine, potassium, calcium, and in those animals with hemoglobin, iron. In general the trace elements regulate enzyme actions. The elements that do this are calcium, magnesium, iron, manganese, cobalt, copper, zinc, and molybdenum. Zinc seems to be necessary for growth and to reach maturity. Obviously there is overlap between the two lists. In addition, there is another group of elements called *age elements* because they accumulate during life. This group includes beryllium, aluminum, silicon, titanium, vanadium, chromium, nickel, arsenic, selenium, tin, lead, silver, cadmium, barium, gold, and mercury. The role of this last list is not clearly understood, and future studies may show that some of them belong in the other lists. Excess or deficiency of some of these can cause great changes or even death. Some can cause tumors and so may cause some cancers. The role of metals in both the cause and treatment of cancer is an area of current research. Some of the evidence showing the effect of trace elements on health is comparison of maps showing trace elements in soils with maps showing either mortality or morbidity of disease. Such evidence is subject to various interpretations and so is somewhat controversial. In the following paragraphs, a number of examples of the effects of trace elements will be described.

Fluorine aids in reducing dental caries, and many municipalities add it to drinking water for this reason. Geographic areas where teeth are good also differ from areas where teeth are poor in having higher calcium, molybdenum, aluminum, titanium, and calcium-magnesium ratio, and lower copper, manganese, barium, and strontium. The evidence that fluorine reduces caries is excellent, but in spite of this, many people object to adding it to their water.

Low-fluorine areas have higher rates of both osteoporosis and aortic calcification, so the benefits of fluorine may go far beyond sound teeth. In addition, other studies suggest that vanadium, boron, strontium, barium, lithium, and titanium may prevent caries, and cadmium, lead, copper, zinc, and chromium may increase them.

In Georgia, highland areas with soils containing many trace elements have lower cardiovascular death rates for middle-aged white males than coastal-plain areas where soils are deficient in trace elements. (See Fig. 4–17.) In general, areas of hard water have lower rates of cardiovascular disease.

Goiter is well known to be caused by a deficiency in iodine. We avoid the enlarged neck of goiter by adding iodine to salt in places where iodine is low. Iodine is in seawater, so those living near the sea or eating much seafood generally get enough. Mountain areas and recently glaciated areas tend to be low in iodine. The distribution of iodine is erratic, so many exceptions can be found.

Cancer may also be related to trace elements. Low-iodine areas have a suspicious relationship to breast cancer. Cancer is less common near the Gulf of Mexico than in the northern states, but this might be related more to climate than to trace elements. Diabetes is less common in the tropics, and insulin doses are smaller there; again this may be climatic.

Cattle fed forage low in selenium have a very high incidence of muscular dystrophy, which can be cured by selenium injection. The soils of part of the northern great

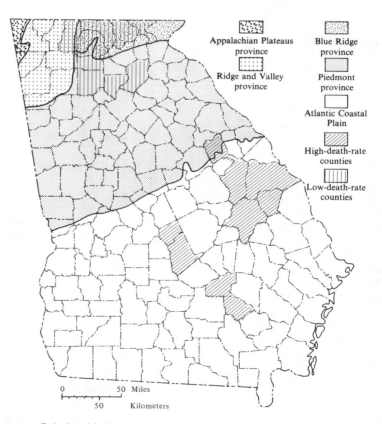

FIGURE 4–17 Relationship between cardiovascular disease and geologic provinces in Georgia. The areas with both high and low death rates for cardiovascular disease among white males of ages 35–74 are shown. From U.S. Geological Survey Professional Paper 574C, 1970.

plains are rich in selenium; some plants, such as loco weed, concentrate the selenium, and cattle eating it are poisoned. Apparently the selenium enrichment of this plant is seasonal or otherwise variable, because ranchers report conflicting observations. Thus, both too little and too much selenium are detrimental to animal life.

In a Florida study, salt-sick disease in cattle and anemia in children were related to deficiencies in iron and copper in the soil. Addition of these two elements to diet promptly caused recovery. In addition, cobalt aided in copper retention, and molybdenum accentuated copper deficiency and also interfered with phosphorus metabolism.

Anthrax, another cattle disease, is controlled by soil type. Apparently the spores live in organic-rich calcium soils, and the disease reaches cattle after such infected soils dry out.

Other studies suggest that some soils in western United States contain toxic amounts of vanadium, molybdenum, and selenium. Some of the peat used in gardens contains high amounts of uranium, and some natural fertilizers add cadmium and zinc to soils. Mercury is high in many soils. All of these points have a bearing on claims of the value of foods raised on organic versus chemical fertilizer, and apparently every case must be considered individually. Thus some food fads may unknowingly be based on trace elements. Medicinal plants in many cases also depend on trace elements for their value.

Lithium is sometimes used in the treatment of certain types of mental illness. Cities with trace amounts of lithium in their water supply have fewer admissions to mental hospitals than those that do not, according to one preliminary study.

SEDIMENTARY ROCKS

LITHIFICATION

The products of weathering generally are eroded, transported, and deposited before they are transformed into sedimentary rocks. These processes are described in Chapters 7 through 11.

The transformation of a sediment into a rock is called lithification. Several processes are involved. *Cementation is the deposition by ground water of soluble material between the grains* and, as might be expected, is most effective in coarse-grained, well-sorted, permeable rocks. The main cementing agents are

Calcite—recognized by acid test. (Do not mistake the effervescence of cement for the overall effervescence of limestone.)
Silica—generally produces the toughest rocks.
Iron oxides—color the rock red or yellow.

Other cements, such as dolomite, are possible. In certain poorly sorted rocks, clay, which generally colors the rock gray or green-gray, may be thought of as forming the cement.

Compaction is an effective lithifier of fine-grained rocks such as shale and siltstone. Compaction *generally comes about from the weight of overburden during burial,* and the reduction in volume due to compaction and squeezing out of water may amount to over 50 percent in some shales.

Recrystallization probably is important in producing chert and in transforming calcite-bearing mud into limestone. Recrystallization *produces an interlocking texture.*

CLASSIFICATION AND FORMATION

As with the igneous rocks, it is more important to interpret the formation of these rocks than merely to name them. Knowing the origin of the materials that compose a sedimentary rock and understanding the origin of its sedimentary features will permit such interpretation.

According to the way they were formed, sedimentary rocks can be classified as marine, lacustrine (lake deposited), glacial, eolian (wind deposited), fluvial (river deposited), and so forth. They can be classified also by type as limestone, chert, quartzose sandstone, and so forth, or by their mode of origin as clastic, chemical precipitate, or organic. In practice, these are blended to produce a practical classification as follows.

Clastic Sedimentary Rocks *are composed of rock fragments or mineral grains broken from any type of pre-existing rock.* (See Fig. 4–18.) They are subdivided according to fragment size. Commonly, sizes are mixed, requiring intermediate names such as sandy siltstone. They are recognized by their clastic texture. The fragments originate from mechanical weathering.

Composition is not used in the general classification of clastic sedimentary rocks because the composition may be affected by several factors. The examples of this are almost endless, but a few simple ones will illustrate. A quartz sandstone may be formed

Clastic texture
Note open spaces.

Interlocking texture. Developed in chemical precipitates such as some limestones and in many igneous and metamorphic rocks.

FIGURE 4–18 Recognition of textures is very important in the identification and interpretation of rocks.

if the source area is mainly quartz (perhaps an older quartz sandstone), or a quartz sandstone may be the result of prolonged weathering, leaving only quartz fragments. Large amounts of feldspar in a sandstone may imply rapid deposition and burial before chemical weathering could decompose the feldspar, or it might imply a cold climate in which

chemical weathering is very slow. In many studies composition is used to subdivide sandstones as indicated in Table 4–3.

The actual formation of shale is somewhat more complex than indicated in Table 4–3. Much of the clay that gives a shale its fissility apparently develops or is mechanically reoriented after deposition. This is suggested by the lack of

geologically young shale. Young fine-grained clastic rocks are mainly mudstones.

Nonclastic Sedimentary Rocks are formed by chemical precipitation, by biologic precipitation, and by accumulation of organic material. These processes extract specific materials from their surroundings, generally seawater, and precipitate these substances, forming rocks. The rocks are classified mainly by composition. (See Table 4–4.) As with the clastic rocks, these rocks are commonly mixed, both among themselves and with the clastic rocks.

Limestone is composed of calcite. It is recognized by effervescence with dilute hydrochloric acid. It is generally of biologic origin and may contain fossils. (See Fig. 4–22A.) A rock composed mainly of fossils or fossil fragments is called *coquina*. (See Fig. 4–22B)

Oolites are tiny spherical grains formed by layers of calcium carbonate deposited around a small sand grain or shell fragment. Rocks composed of these grains are called *oolitic limestones.*

Limestone can form in many ways as shown in Table 4–5. Most limestone probably originates from organisms that remove calcium carbonate from seawater. The remains of these animals may accumulate to form the limestone directly, or they may be broken and redeposited. Calcite can also be precipitated chemically. The

TABLE 4–3   Clastic sedimentary rocks.

Size	Sediment	Rock
> 2 mm	Gravel	*Conglomerate*—generally has a sandy matrix. May be subdivided into *roundstone* and *sharpstone conglomerate* (*breccia* is a synonym) depending on the fragment shape. (See Fig. 4–19.)
$\frac{1}{16}$–2 mm	Sand	*Sandstone*—recognized by gritty feel. (See Figs. 4–20 and 4–21.) Generally designated as coarse, medium, or fine if well sorted. They have also been subdivided on the basis of composition. The more important types are *Quartzose sandstone* (mainly quartz) *Arkosic sandstone (arkose)*—over 20 percent feldspar. *Graywacke*—poorly sorted, with clay or chloritic matrix.
$\frac{1}{256}$ – $\frac{1}{16}$ mm	Silt	*Siltstone (mudstone)*—may be necessary to rub on teeth to detect grittiness, thus distinguishing it from shale.
< $\frac{1}{256}$ mm	Clay	*Shale*—distinguished from siltstone by its lack of grittiness and its *fissility* (ability to split very easily on bedding planes). Rocks composed of clay that lack fissility are called *claystone*.

A   B

FIGURE 4–19   Conglomerate. A. Rounded pebbles in a sandstone matrix. B. Sharpstone conglomerate (breccia). The angular pebbles in this rock were probably not transported far compared with the pebbles in A. Photos from Ward's Natural Science Establishment, Inc., Rochester, N.Y.

FIGURE 4–20   Sandstone. The bedding is too thick to show in this specimen. Photo from Ward's Natural Science Establishment, Inc., Rochester, N.Y.

83

A   B

FIGURE 4-21   A. Sandstone outcrop in New Mexico. Note geologist at base of cliff for scale. B. Closeup of a specimen of the same sandstone. Photos by J. R. Stacy, U.S. Geological Survey.

TABLE 4-4   Nonclastic sedimentary rocks.

Rock	Principal Texture	Principal Composition	How Recognized
Limestone	Fine	Calcite $CaCO_3$	Effervesces with dilute hydrochloric acid. Generally of biologic origin and may contain fossils. *Coquina* is limestone composed mainly of fossils or fossil fragments.
Chalk	Fine	Calcite $CaCO_3$	Soft white limestone formed by the accumulation of microscopic shells. Effervesces with dilute hydrochloric acid. Compare with diatomite.
Dolomite	Fine	Dolomite $CaMg(CO_3)_2$	Effervesces with dilute hydrochloric acid after scratching (to produce powder). May have irregular holes.
	Fine	Silica SiO_2	Hardness 6 or 7. Similar to chalcedony. *Flint* is black or dark gray.
Gypsum	Fine to coarse	Gypsum $CaSO_4 \cdot 2H_2O$	Hardness 2. One good cleavage and two poorer cleavages.
Rock salt	Fine to coarse	Halite $NaCl$	Cubic crystals and cleavage may be visible. Salty taste.
Diatomite	Fine	Silica SiO_2	Soft white rock formed by the accumulation of microscopic shells composed of silica. Distinguished from chalk by lack of effervescence.
Coal	Fine to coarse	Carbon C	Soft, black, burns.

amount of carbon dioxide in seawater controls the amount of calcite that can remain in solution. If the amount of carbon dioxide is reduced by warming the water, as would occur in shallow tropical water, calcite may be precipitated. Marine plants also remove carbon dioxide from seawater and aid in calcite precipitation. Bacteria can also aid chemical precipitation of calcite by making the water more alkaline. Decaying organic material can increase the amount of carbon dioxide and so retard precipitation of calcite. Limestone is presently being formed by chemical precipitation on the shallow Bahama Banks where the factors discussed are favorable.

Chalk is soft, white limestone formed by the accumulation of the shells of microscopic animals. The shells are composed of calcium

A   B

FIGURE 4-22 Limestone. A. Fossiliferous limestone. A fine-grained white limestone with abundant fossil fragments. B. Coquina, a variety of limestone composed almost entirely of fossil fragments. Photos from Ward's Natural Science Establishment, Inc., Rochester, N.Y.

TABLE 4-5 Origins of the more important nonclastic sedimentary rocks.

| Rock | Biologic | Chemical | | Remarks |
		Evaporite	Other	
Limestone	x	x Rare	x	
Dolomite		x Rare		Forms mainly by replacement
Chert	x		x Precipitation from volcanic silica	Most biologic chert recrystallizes from opal
Gypsum		x		
Rock salt		x		
Chalk	x			
Diatomite	x			
Coal	x			

carbonate, and chalk is recognized by effervescence with acid.

Dolomite is composed mainly of the mineral dolomite. It is recognized by effervescence with dilute hydrochloric acid after scratching (to produce powder); it will also react (without scratching) with concentrated or with warm dilute hydrochloric acid. Dolomite is generally formed by replacement of calcite, presumably soon after burial. Some of the calcium atoms in calcite are replaced by magnesium atoms. The reduction in volume in this replacement may produce irregular voids and generally obliterates fossils.

Chert is microscopically fine-grained silica (SiO_2). It is equivalent to chalcedony (see Mineral Table, Appendix A). It is unfortunate that many names have been applied to the fine-grained varieties of silica. These names were based on minor differences, such as color and luster, that cannot be substantiated by modern methods, and there is little agreement among mineralogists on their naming. Both chert and chalcedony may contain opal and fine-grained quartz. Dozens of names have been applied to chalcedonic silica; of these, a few may be useful, e.g., agate for banded types

and flint for dark gray or black chert.

Chert originates in several ways. Some may precipitate directly from seawater in areas where volcanism releases abundant silica. Most comes from the accumulation of silica shells of organisms. These organisms remove dissolved silica from seawater to form their shells or skeletons. These silica remains come from diatoms, radiolaria, and sponge spicules, and are composed of opal. Opal is easily recrystallized to form chert. Thus much chert is recrystallized, making the origin difficult to discern. In recrystallization the silica may replace other materials, and fossils replaced by silica are common. If this occurs in limestone, beautifully preserved fossils with delicate features intact can be recovered by dissolving the limestone with acid. Chert may form either beds or nodules.

Gypsum forms from the evaporation of seawater. Rocks formed in this way are termed *evaporites;* they are discussed below. The identification of gypsum is described in Appendix A.

Rock salt is composed of halite. It is recognized by its cubic cleavage and its taste. (See Appendix A.) It is deposited when restricted parts of the sea are evaporated.

Diatomite is a soft, white rock composed of the remains of microscopic plants. Because the remains are composed of silica, it is distinguished from chalk by lack of effervescence.

Coal is formed by the accumulation of plant material.

Evaporite deposits are formed by the evaporation of seawater. (See Table 4-6.) Gypsum and rock salt are the main rocks formed in this way. When seawater is evaporated at surface temperatures, such as in a restricted basin, the first mineral precipitated is calcite. Dolomite is the next mineral precipitated, but only very small amounts of limestone and dolomite can be formed in this way. Evaporation of a kilometer col-

TABLE 4–6 Composition of seawater.

Element	Percent	Origin	Rock Formed
Chlorine	0.019	Volcanoes	Rock salt (halite)
Sodium	0.011	Chemical weathering	Rock salt (halite)
Magnesium	0.001	Chemical weathering	Dolomite
Sulfur	0.0009	Volcanoes	Gypsum
Calcium	0.0004	Chemical weathering	Limestone (calcite)
Potassium	0.0004	Chemical weathering	Potassium salts

umn of seawater would only produce a few centimeters of limestone and dolomite. After about two-thirds of the water is evaporated, gypsum is precipitated; and when nine-tenths of the water is removed, halite forms. During the last stages of precipitation, potassium and magnesium salts form. Thick beds of rock salt imply evaporation of large amounts of seawater.

The rocks just described are the most important, but by no means the only, sedimentary rocks. Any mixture of types is possible and, because any weathering product may form a sedimentary rock, endless variety is possible. Some less abundant types include economically important deposits, such as iron oxides, phosphorous rocks, bauxite (aluminum ore), and potash.

Shale, limestone, and sandstone make up over 99 percent of all sedimentary rocks. Because feldspars are the most abundant min-erals, their main weathering product, shale, is the most abundant sedimentary rock. Limestone is next in abundance, and sandstone is last. (See Fig. 4–1.) The abundance of shale is somewhat less than is predicted from the abundance of clay-forming silicate minerals, suggesting that some clay is deposited in the deep sea basins. Clay particles are very small and sink slowly; they can be carried thousands of miles by gentle currents. Table 4–7 summarizes the more common sedimentary rock types.

FEATURES OF SEDIMENTARY ROCKS

The most noticeable feature of an outcrop of sedimentary rocks is the *bedding,* which *records the layers in the order of deposition with the oldest at the bottom.* In some instances the beds are too thick to show in a small outcrop, and, for the same reason, many hand specimens do not show obvious bedding.

Study of the types of fossils and of sedimentary features can tell much about the environment of deposition and, in some cases, the postdepositional history. *Mud cracks* on bedding planes record periodic drying and suggest shallow water and, perhaps, seasonal drying (Fig. 4–23). *Ripple-marks,* too, suggest shallow water with some current action, but are also known to form in deep water. Detailed study of ripple-marks can show type and direction of current. (See Figs. 4–24 and 4–25.) Current action can also cause localized scouring or cutting to produce *cut-and-fill features.* (See Fig. 4–26.) Another type of current feature is *cross-bedding.* (See Figs. 4–27 and 4–28.) Sand dunes likewise exhibit cross-bedding. These features can also be used to recognize beds that have been overturned by folding or faulting. (See Fig. 4–29.) Currents can also align mineral grains and fossils and so impart linear internal structures to rocks that can be used to determine the current direction.

If a mixture of sediments is suddenly deposited into a sedimentary basin, then the large fragments sink faster than the small ones. This does not violate Galileo's famous experiment, because here we are concerned with falling

TABLE 4–7 Sedimentary rocks. This simplified chart shows some of the relationships among common sedimentary rocks.

Sediment (size)	Composition					
	Quartz	Feldspar	Rock Fragments	Volcanic* Rock Fragments	Calcite	Clay
	Quartzose+	Arkosic+	Graywacke+	Pyroclastic	LIMESTONE	
Gravel (> 2 mm)	CONGLOMERATE BRECCIA			BRECCIA AGGLOMERATE	COQUINA	
Sand (1/16 – 2mm)	SANDSTONE			TUFF	CALCARENITE	
Silt (< 1/256 mm)	SILTSTONE (MUDSTONE)				MICRITE CALCILUTITE CHALK	SHALE

*See Table 3–2.
+Use these terms as adjectives: i.e., arkosic sandstone.

A

B

FIGURE 4–23   A. Mud cracks at Death Valley, California. B. Ancient mud cracks preserved in a sandstone in New Mexico. Photos from U.S. Geological Survey; A by J. R. Stacy, B by E. F. Patterson.

Current direction

Current ripple mark – does not show top and bottom of beds

Wave ripple mark – indicates top and bottom of beds. Formed by back-and-forth slosh

FIGURE 4–24   Ripple-marks form where currents can act on sediments. Ripple-marks form at right angles to the current direction, and asymmetrical ripple-marks indicate which way the current flowed. To see why only wave ripple-marks can be used to determine top and bottom of beds, turn the page upside-down.

bodies in a viscous medium in which velocity of descent is controlled mainly by the size. The type of bedding produced by this kind of sedimentation is called *graded bedding* (Fig. 4–30); we can demonstrate this experimentally by putting sediments of various sizes in a jar of water, shaking, and then allowing the contents to settle. Graded beds can form in many ways, such as the sudden influx of sediment by a storm or seasonal flow of rivers. Their formation at the base of the continental slope is described in the next section.

The *sorting* of a sediment *is a measure of the range in size of its fragments.* If the fragments that compose a sediment are all similar in size the sediment is well sorted. The sorting of a sediment determines other textural features. The *porosity* of a sedimentary rock *is a measure of its empty, or void, space,* and its *permeability is a measure of the interconnections of the pore spaces.* These properties become quite important when one is concerned with recovering oil or ground water from an underground reservoir. The amount of pore space in sedimentary rocks is quite varied but surprisingly large. If we consider packing of uniform spheres (and most sand grains are rounded), we find

that the closest possible packing contains 26 percent void space, and the loosest, 48 percent. (See Fig. 4–31.) The permeability will vary depending on the amount of pore space, the sorting, the cementation, and the size of the particles. A shale may have over 30 percent porosity, but the surface tension of water will prevent flowage through the tiny openings.

Slump features as shown in Fig. 4–32 form if sediments are deposited on a slope or if they are tilted while still soft. Another type of deformation of soft rock is the vertical movement of sediment shown in Fig. 4–33.

DEPOSITION

Most sedimentary rocks are deposited in oceans, and these marine rocks, easily dated by their contained fossils, record much of the earth's history. In the present section a few of the more important marine environments will be considered.

Clastic Rocks   One of the interesting observations of marine sedimentary rocks is that at many places individual sedimentary units, such as sandstones, can be

FIGURE 4–25   Oscillation ripple-marks, Erie County, Ohio. Note the presence of ripple-marks in successive layers. Photo by D. F. Demarest, U.S. Geological Survey.

87

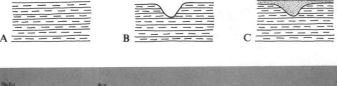

D

FIGURE 4–26   Scouring produced by localized currents. A. Initial deposition. B. Channel cut by current. C. Deposition is resumed, filling the channel. Such a feature records an interruption in deposition and can be used to distinguish top and bottom. D. Channel filled with conglomerate. Note that the conglomerate is more resistant to weathering than the sandstone. Channel is about ten feet deep. Daggett County, Utah. Photo by W. R. Hansen, U.S. Geological Survey.

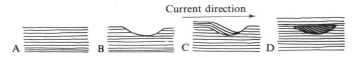

Current direction

FIGURE 4–27   Development of cross-beds. A. Initial deposition. B. Channel cut by current. C. Channel filled by deposition from one side. D. Normal deposition after channel is filled. Cross-beds can also be used to distinguish top and bottom.

traced for hundreds of miles. At other places the beds are irregular and have limited horizontal extent. To see how this comes about we will consider how marine sedimentary rocks are formed. We will begin with the clastic rocks (sandstones and shales). As will be seen, the origin of the fragments that compose these rocks is weathering and erosion. These fragments are delivered to the ocean by rivers.

Because most marine sediments consist of material delivered by rivers, most sedimentary rocks form near the continents. Sampling and drilling in the oceans show this to be true. If we wish to interpet the thick accumulations of old sedimentary rocks exposed in today's mountain ranges, we should study the continental shelves and the continental slopes. (See Fig. 4–34.)

The *continental shelf* is part of the continent, and its extent is determined by seismic studies of the crustal structure. However, the 100-fathom, or 183-meter (600-foot), depth line is generally close to the edge of the continental shelf. The continental shelf is almost flat; it has an average slope of about 1.9 meters per kilometer (10 feet per mile). The break in slope between the shelf and the continental slope generally begins

A

B

FIGURE 4–28   Cross-bedding. A. In river-deposited coarse sandstone. Photo by H. E. Malde, U.S. Geological Survey. B. In wind-deposited sandstone in Zion National Park, Utah. Photo by H. E. Gregory, U.S. Geological Survey.

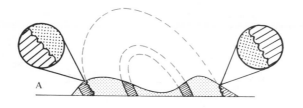
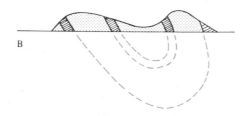

FIGURE 4–29   Alternate interpretations of the structure of outcrops of folded sedimentary rocks. The ripple-marks in A show that it is the correct interpretation. In the absence of ripple-marks or other sedimentary features that indicate top and bottom, it would not be possible to choose between the two interpretations.

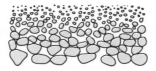

FIGURE 4–30   Graded bedding. Produced by the more rapid settling of coarse material than fine material.

Close packing   Open packing

FIGURE 4–31   The packing controls the porosity, as illustrated here in the idealized case of spheres.

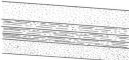
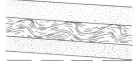

FIGURE 4–32   Soft sediments can slide down even very gentle slopes, developing slump features.

FIGURE 4–33   A. Photo of deformed bedding. B. Diagram showing how deformation of soft rock forms the structure shown in photo.

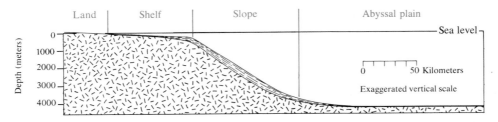

FIGURE 4–34 Continental margin showing typical continental shelf, continental slope, and abyssal plain. The vertical scale is exaggerated.

at a depth of about 122–152 meters (400–500 feet), and the range is from near sea level to 458 meters (1500 feet). The average depth of the shelves is near 61 meters (200 feet). The continental slope extends on the average to a depth of 3.2 kilometers (2 miles) or more, with a slope of two to six degrees. This is a very steep slope compared to the mountain ranges of the earth.

The continental shelves are covered with clastic sediments for the most part. One would expect that, in general, the coarsest material should be found near shore or near river mouths and the finer silt and clay should be farther from the source. (See Fig. 4–35.) This is probably true in a very general way, but the many departures suggest that the situation is not simple. (See Fig. 4–36.) One very obvious reason is that during the recent glacial time the oceans were lower because much water was held in the glacial ice. Apparently, too, tidal and other currents are fairly effective in moving and sorting clastic material on the relatively shallow shelves. This type of reworking of the sediments may possibly account for the uniform beds extending over large areas that are found on the present continents. In any event, pebbles and conglomerate are not uncommon far out on the shelf.

The present coastal plains on the Atlantic and Gulf coasts are underlain by what are apparently continental shelf deposits. Uplift has exposed the landward parts of these sedimentary accumulations, and seismic studies and deep drilling indicate that they extend onto

and thicken on the present continental shelf. On the Gulf Coast these deposits are several miles thick, and their features and fossils indicate shallow-water deposition. The latter suggests slow subsidence.

Much less is known about the structure and formation of the *continental slopes.* Limited sampling shows the surface, at least, to be covered mainly with fine sediments, presumably swept off the shelves. A more important

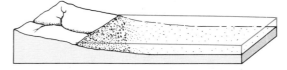

FIGURE 4–35 Idealized diagram showing coarse material being deposited near the shore and finer material further out.

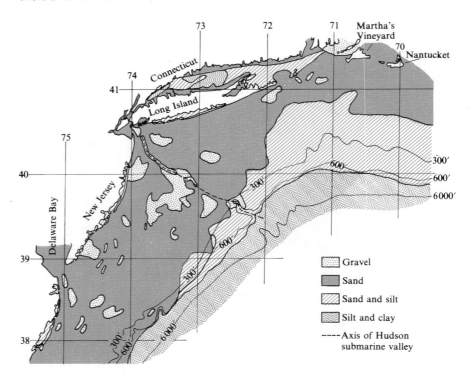

Gravel
Sand
Sand and silt
Silt and clay
----Axis of Hudson submarine valley

FIGURE 4–36 Bottom sediments on part of the continental shelf in northeastern United States. The many departures from the idealized case shown in Fig. 4-35 are probably caused by bottom currents and by deposition when sea level was lowered during glacial times. Adapted from F. P. Shepard and G. V. Cohee, *Bulletin,* Geological Society of America, Vol. 47, 1936.

process in the building of the slopes may be submarine landslides or mudslides. Following earthquakes, submarine telephone cables may be broken. The cables do not all break at the time of the earthquake but at later times, farther down the slope. In some cases many cables have been broken and the time of breaking is accurately known. This, plus the burial of some of the cables reported by the repair ships, strongly suggests that submarine landslides move down the continental slopes.

Another feature of continental shelves and slopes is the *submarine canyons.* These submarine canyons have many of the features of river canyons, and some are larger than such river canyons as the Grand Canyon. The origin of such features has been the subject of lively speculation, and not everyone is yet satisfied with the explanations offered. At first they were thought to have formed by ordinary river erosion at a time when sea level was lower than it is at present. Many of the submarine canyons have heads near the mouths of large rivers, making this a reasonable hypothesis. However, many of them extend to depths of over 3.2 kilometers (2 miles), and sea level changes of this magnitude in the recent past seem impossible. The submarine landslides mentioned above offer the most likely explanation. Such landslides consist of sediment-laden water. Because they are heavier than the surrounding water, these landslides or mudflows move along the bottom and are capable of eroding the soft bottom material. *These muddy waters are called turbidity currents.* They can move down very gentle slopes and can transport fairly large pebbles. Landsliding of unconsolidated material on the continental slopes can account for the canyons there; and turbidity currents initiated by stream discharge, storms, or other currents can account for the submarine canyons

on the continental shelves. As well as explaining the canyons, these processes can also account for the building of the continental slopes, for seismic studies show the slopes to be composed of sedimentary rocks.

Turbidity currents and submarine landslides can also account for the smooth ocean bottoms at the foot of the continental slope at many places. (See Fig. 4–37.) These smooth surfaces are apparently similar to the fans that develop where rivers enter broad valleys. These submarine fans contain pebbles and shallow-water shells at some places. The fans do not occur everywhere at the foot of the continental slope.

To further evaluate the possible existence and importance of turbidity currents in the geologic past, we will look at their depositional features. Turbidity currents, because of their density and speed, can transport large fragments. These larger fragments and the main mass of fine material are deposited fairly rapidly. The larger, heavier fragments settle first, forming graded beds. The eroding ability of turbidity currents suggests that they may erode or otherwise deform the top of the underlying bed. This is shown in Fig. 4–38.

Carbonate Rocks The carbonate rocks are limestone and dolomite. Limestones are being deposited at present in warm shallow seas, and most of the features of ancient limestones suggest that they, too, were formed in such an environment. Calcite, the mineral of which limestone is composed, is soluble in water containing dissolved carbon dioxide. To precipitate calcite, the amount of dissolved carbon dioxide must be reduced because calcite is only slightly soluble in pure water. Warm water can dissolve much less of any gas, including carbon dioxide; this is one reason why limestones form in warm water. Marine plants are also common in warm water, and

they extract carbon dioxide from the water as part of their life process. This, too, aids in the precipitation of calcite. Many lime-secreting organisms live in this environment, and their shells and other structures also accumulate on the sea floor. Fossils of organisms that lived in warm shallow waters are present in most limestones, showing that this is the environment in which most limestones formed. A very few limestones form by chemical precipitation in restricted parts of the oceans and in lakes.

Dolomite is not, so far as we know, depositing anywhere except at rare restricted parts of the ocean where evaporation occurs. Yet there are many thick and very extensive dolomite beds that formed in the geologic past. Study of these ancient dolomite beds reveals that they are limestones that have been replaced by dolomite. Apparently, either soon after formation or during burial, if limestone is in contact with water containing magnesium, the magnesium reacts with calcite, $CaCO_3$, and forms dolomite, $CaMg(CO_3)_2$.

Geosynclines and Plate Tectonics It was noted long ago by geologists that many mountain ranges are formed of bent and broken layers of sedimentary rocks. These sedimentary rocks had been deposited as flat layers in shallow seas. We know that the seas were shallow because the rocks have features that commonly form in shallow water, such as mud cracks, and the fossils that they contain are organisms that live in shallow waters. The mountain ranges began, then, as elongate basins of deposition, and these areas of sedimentation were named *geosynclines (geoclines).* The sedimentary rocks in geosynclines are much thicker than nearby rocks of similar age. Geosynclinal mountain ranges are commonly near the margins of continents, and so some of them may have formed in the continental shelf-

FIGURE 4–37 Turbidity current in Lake Ontario at mouth of Niagara River. The turbidity current is caused by the sediment-bearing river water. Note that part of the river water is carried by long-shore currents.

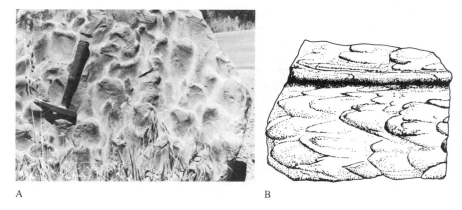

A   B

FIGURE 4–38 Depositional features of turbidity currents. A. Photo. The bottom of a bed is shown. B. Diagram. The lobe-like casts point in the opposite direction to current movement. The feature that crosses the specimen is the cast of a groove, probably caused by a boulder carried by the turbidity current. The specimen shown is about two feet in diameter.

continental slope area just described.

The sediments of continental shelves tend to be sandstones near the coast, with shale or limestone accumulating farther from shore, although many exceptions are found (Fig. 4–39). The rocks that underlie the continental slope are generally clastics, shale and sandstone, commonly intermixed.

Few or no volcanic rocks are found in the continental shelf-slope area.

In some areas of most geosynclinal mountain ranges, volcanic rocks are found interbedded with sedimentary rocks. This suggests that in these cases convergent plate boundaries may be involved. Volcanic island arcs form where oceanic plates collide with conti-

nental plates and so are also at the margins of continents (Fig. 4–40). The volcanic borderlands can be the source of both volcanic rocks and clastic sediments. Geosynclinal clastic sediments containing volcanic rocks generally are very poorly sorted and so have a wide variety of sizes and types of fragments. Such rocks contrast with the sediments on

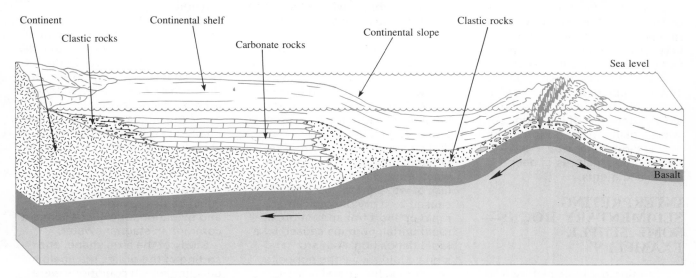

FIGURE 4–39 Development of thick sections of sedimentary rocks at a continental shelf and slope, and much thinner layers of sediments and volcanic rocks on the ocean floors and at mid-ocean ridges.

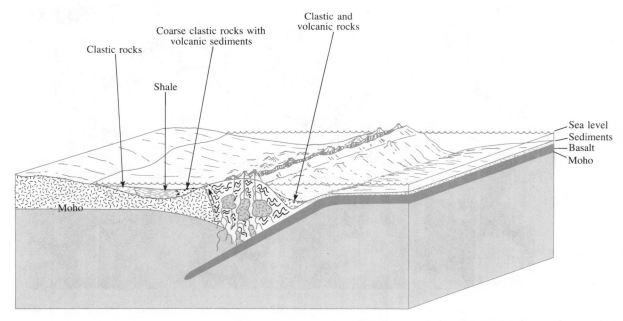

FIGURE 4–40 Development of thick sections of sedimentary and volcanic sedimentary rocks at a convergent plate boundary.

93

continental shelves that tend to be well sorted. It is later plate movements that are believed to crumple the geosynclinal sediments.

In the deep ocean, far from land, the sediments consist mainly of wind-carried dust, largely of volcanic origin, and the fallen shells of tiny animals that live near the surface. Sediments of this type that are composed of more than 30 percent biologic material are called oozes. (See Fig. 4–41.) The rate of accumulation of oozes is very slow compared with other sedimentary rocks. It is estimated that it requires between 1000 and 10,000 years to deposit one millimeter. The volcanic dust settles so slowly that most of it is oxidized while sinking to the bottom, and so it is termed *red clay*. Fig. 4–42 shows some features of deep ocean sediments.

INTERPRETING SEDIMENTARY ROCKS— SOME SIMPLE EXAMPLES

Interpreting geologic history from the study of sedimentary rocks is an important activity carried on by many geologists. These studies are conducted at many levels, both in the field and in the laboratory. The few examples given here may help you to interpret the rocks in your area. These examples are simple, but a little imagination will reveal the many other possibilities and the many complexities that are possible. The following examples will be limited to the clastic sedimentary rocks, especially sandstones.

In studying a hand specimen of sandstone, the composition of the grains may reveal the rock type that was eroded to form the sandstone. The amount of weathering influences the mineralogy of the grains so that the grains of rock that could reveal the source are, in general, only present if very little weathering has occurred. Such a lack of weathering could be caused by a dry or cold climate, rapid or short transportation, or rapid burial, perhaps caused by a rapid deposition. A quartz sandstone could reveal the opposite conditions. In general, a quartz sandstone implies complete weathering, and a feldspathic sandstone (arkose) implies less weathering and/or rapid burial. The silt in shale may also reveal the degree of weathering or reworking. The shale associated with quartz sandstone generally has mainly quartz in its silt fraction. Shales that are rapidly deposited with little weathering contain much feldspar in their silt fraction.

The color of a sedimentary rock may also reveal information about its formation. Red color, red beds, reveals oxidation because the red color comes mainly from hematite (red iron oxide). Seasonal drying is generally necessary as in an arid climate, the general interpretation given to red beds. Black color in sedimentary rocks is generally caused by carbon of organic origin. This implies that organisms were deposited faster than they could be destroyed by bacteria and oxidation. Such situations are common in stagnant water.

Study of the size, shape, and sorting of the grains may help in deciding which conditions were most important in the formation of

FIGURE 4–41 Radiolarian ooze from a depth of 5205 meters (17,066 feet), about 970 kilometers (600 miles) south of Hawaii. Photo from D. R. Horn, Lamont-Doherty Geological Observatory.

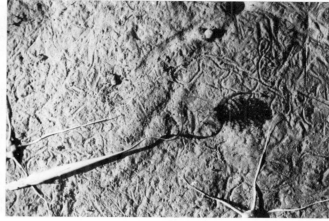

A   B

C

FIGURE 4–42   Ocean bottom photographs. A. Hydrographer Canyon off the east coast of the United States. Soft coral and other bottom organisms at a depth of 723 meters (2400 feet). B. Hydrographer Canyon showing brittle stars and other organisms. C. Deep sea bottom in Romanche Trench. Photos courtesy of R. M. Pratt.

the sandstone. Large grains suggest nearness to the source or short distance of transportation. Angular fragments, unless very resistant to abrasion, have not traveled far. The sorting can reveal, among other things, the mode of transportation. Wind-transported sediments are very well sorted, and glacially deposited rocks have an extreme size range. All of these features can be seen with a hand lens, but the complete study of a hand specimen in the laboratory requires the use of the microscope and other instruments.

Much more can be learned at the outcrop of the rock layer. Here the bedding can be seen. The

thickness and uniformity of the beds can reveal much about their origin. The irregular bedding of river deposits contrasts with the regularity of many shallow-water marine sedimentary rocks. The many features of sedimentary rocks, such as ripple-marks, cross-beds, and graded bedding, can also be studied at the outcrop. These features and their interpretation were described earlier. Fossils can also be collected at the outcrop. The environment of deposition can, in many cases, be determined from study of fossils, because the types of animals that live near shore are different from those in deep water; they also dif-

fer between warm and cold water. Perhaps the most important use made of fossils is to determine the age of the sedimentary layers. The deformation, if any, of the beds can also be seen in many outcrops.

The next level of study of a sedimentary layer is to trace its areal extent by mapping its outcrops. By combining the interpretations made at each outcrop it is possible to reconstruct the basin in which the beds were deposited. An example of such an interpretation is shown in Fig. 4–43. In this example, the source area for the Morrison Formation is revealed from the current directions shown by

95

the cross-bedding, the increase in thickness near the source, and the coarseness of the grains near the source. Note also that the Morrison Formation is exposed in only a small part of the basin, and everywhere else it has either been removed by erosion or is covered by younger rocks. Most geologic interpretations must be made on the basis of such fragmentary evidence.

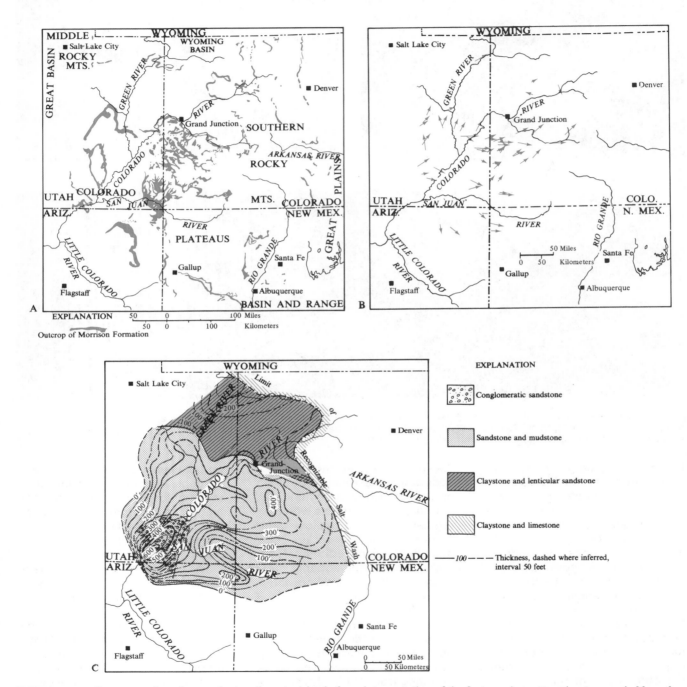

FIGURE 4–43 Reconstruction of an ancient sedimentary basin from interpretation of the features in scattered outcrops. A. Map of outcrops of the Morrison Formation. B. Current direction inferred from cross-bedding in the Salt Wash Member of the Morrison Formation. The arrows show the down-slope direction of cross-beds. Length of the arrows is proportional to the consistency. C. Thickness and sediment type in the Salt Wash Member. Note that the coarser sediments are closer to the source area inferred from both thickness and cross-bedding. From L. C. Craig and others, U.S. Geological Survey Bulletin 1009E, 1955.

SUMMARY

Sedimentary rocks are composed of rock fragments, weathering products, organic material, or precipitates; and most are deposited in *beds* or layers. They compose only about 8 percent of the earth's crust, although they cover about 75 percent of the surface.

Mechanical weathering breaks up a rock into smaller fragments; *chemical weathering* rearranges the elements into new minerals.

Mechanical weathering is caused mostly by the forces produced by the expansion of freezing water, and to a lesser extent by growing roots, burrowing animals, lightning, and people.

Leaching is the removal of material in solution by ground water.

Chemical weathering is controlled by temperature, surface area, and amount of water available. Mechanical weathering provides the large surface area needed by chemical weathering.

Chemical weathering increases volume by adding substances. *Oxidation* is reaction with oxygen in air to form an oxide; *hydrolysis* is reaction with water; and *carbonation* is reaction with carbon dioxide in air to form a carbonate. Carbonation involves uniting carbon dioxide with water to form carbonic acid.

Dry climates tend to have bold cliffs because of the predominance of mechanical weathering. Humid areas have rounded topography. Maximum weathering probably occurs just below ground surface.

Bowen's reaction series, in general, shows relative stability to chemical weathering; olivine is first to be affected and quartz least affected, although biotite is easily weathered.

Quartz—very little weathering, sand grains produced.

Feldspars and muscovite—potassium, calcium, and sodium form carbonates by carbonation. Some potassium carbonate goes to ocean, some is used by plants, some goes to clays. Calcium carbonate forms limestone. Sodium carbonate dissolves in ocean. Aluminum, silicon, and oxygen produce clay which forms shale. Silicon and oxygen produce silica which forms chert and similar rocks.

Ferromagnesian minerals produce same products plus iron into *hematite* (red iron oxide) or *limonite* (yellow iron oxide), and magnesium into some clay minerals or magnesium carbonate which replaces calcium in limestone to form dolomite.

Most sandstones are composed largely of resistant quartz fragments; sandstones with much feldspar formed under conditions that did not permit complete chemical weathering.

Soil is the material that supports plant life. Climate is more important than parent rock type in determining soil type.

Most soils have three layers or *horizons: topsoil, subsoil,* and partly decomposed bedrock. *Humus,* dark material in the upper part of soil, is formed by action of bacteria and molds on plant material. Soils form by weathering processes, leaching, and carbonation, and effects of an acid formed by water and humus.

Soils in humid areas are called *pedalfers* (*pedon,* Greek for soil, and *Al, Fe* for aluminum and iron). Pedalfers are quite acid because of abundant rainfall and humus. Aluminum and iron are leached from topsoil and deposited in subsoil.

Soils in drier areas [less than approximately 64 centimeters (25 inches) of precipitation annually] are *pedocals* (*pedon* for soil and *cal* for calcite). Leaching is not complete, and calcite is deposited rather than removed by solution in ground water. Such calcite deposits are called *caliche.*

Laterites are formed in tropical savannas; everything is leached away except iron and aluminum which form reddish brown hydrated oxide deposits. *Bauxite* (aluminum ore) is such a deposit, weathered from an aluminum-rich parent rock.

Trace elements in soils, which enter the life cycle through water supplies or plants, have a large influence on health.

Lithification is the transformation of sediment into a rock. *Cementation* is deposition of soluble material, such as calcite, silica, and iron oxides, between the grains by ground water. *Compaction* lithifies fine sediments by weight of overburden during burial. *Recrystallization* produces the interlocking texture of chert and limestone.

Clastic rocks are composed of fragments or minerals broken from any type of pre-existing rock; clastic rocks are generally subdivided by fragment size, not by composition, except for some sandstone types.

Conglomerate—fragments over 2 mm; may be subdivided into *roundstone* or *sharpstone conglomerates (breccia)* if fragments rounded or angular respectively.

Sandstone—fragments between $\frac{1}{16}$ and 2 mm. Some compositional types are *quartzose sandstone* (mainly quartz), *arkosic sandstone* or *arkose* (over 20 percent feldspar), and *graywacke* (poorly sorted, clay or chloritic matrix).

Siltstone (mudstone)—fragments $\frac{1}{256}$ to $\frac{1}{16}$ mm, gritty.

Shale—fragments less than $\frac{1}{256}$ mm (clay), not gritty, has *fissility* (ability to split on bedding planes). Nonfissile clay rocks are *claystone.*

Nonclastic rocks are formed by chemical or biologic precipitation, and by accumulation of organic material.

Limestone is composed of calcite, in some cases from chemical precipitation but usually of biologic origin. If composed mainly of fossils it is called *coquina.*

Chalk is soft, white limestone formed by accumulation of the calcium carbonate shells of microscopic animals.

Dolomite is formed by replacement of calcite with dolomite.

Chert is fine-grained silica; some precipitated directly from seawater, most from accumulation of silica shells of organisms.

An *evaporite* is a deposit formed by evaporation of seawater. The precipitation sequence is calcite, dolomite, gypsum, rock salt, and lastly potassium and magnesium salts. *Gypsum* and *rock salt (halite)* are the main rocks formed this way.

Diatomite is soft, white rock composed of silica remains of microscopic plants.

Coal is formed by accumulation of plant material.

Shale, limestone, and sandstone make up over 99 percent of all sedimentary rocks. Shale is most abundant because of the abundance of feldspars.

Bedding of sedimentary rocks records the layers in order of deposition with the oldest at the bottom.

Mud cracks record periodic drying and can show top and bottom of deformed beds.

Ripple-marks can show type and direction of currents, and wave ripple-marks can show top and bottom.

Cut-and-fill features are caused by localized scouring, and can show top and bottom.

Cross-bedding, another current feature, also shows top and bottom.

Graded bedding is formed by larger fragments settling faster than smaller fragments; bottom is indicated by the larger fragments.

Porosity is a measure of the amount of void space in a sediment; *permeability* is a measure of the interconnections of the void spaces.

Continental shelves, the almost flat portions of the continent below sea level, are generally covered with clastic sediments.

Continental slopes, the steep slopes between the continental shelves and the deep ocean platforms, are mainly covered with fine sediments.

Slopes and shelves have some large *submarine canyons.* These may be caused by *turbidity currents,* underwater landslides consisting of heavy, sediment-laden water which can erode the soft bottom material.

Many mountain ranges are formed of bent and broken layers of sedimentary rocks. These sedimentary rocks are much thicker than nearby rocks of the same age. *Geocline* or *geosyncline* is the name given to the elongate basin in which the rocks accumulated.

Geosynclines (geoclines) generally form at continent-ocean borders and may form at continental shelf-slopes or at convergent plate boundaries.

In the deep ocean, the sediments are composed of slowly settled, wind-carried dust, largely of volcanic origin, and fallen shells of tiny animals. Sediments that have more than 30 percent biologic material are called *oozes.*

Sedimentary rocks are important in the interpretation of geologic history through study of grain size and shape, depositional features, size and shape of depositional area, fossil study including dating, and deformation of the area.

QUESTIONS

1 Describe two methods of mechanical weathering.
2 Describe two methods of chemical weathering.
3 List the common minerals in the order of their stability to chemical weathering.
4 In what climate is chemical weathering most effective and why?
5 In what climate is mechanical weathering most effective and why?
6 Why is quartz a common mineral in sandstones?
7 What effects would you expect the increased carbon dioxide in industrial areas to have on stone buildings?
8 Why is mature soil zoned? Describe the zones.
9 Discuss the importance of climate and rock type on the soil produced.
10 List the trace elements that are harmful and those that are helpful to humans.
11 What is the origin of chert?
12 What are the common cements in sandstones?
13 Clastic sedimentary rocks are classified on the basis of _____.
14 What rocks would form from the weathering of a granite in a cold climate? In a warm, moist climate?
15 How can top and bottom of the beds be determined in a group of vertical sedimentary beds?
16 How might sediments deposited in shallow water differ from those deposited in deep water?
17 What features could be used to distinguish turbidity-current-deposited sediments?

18 What geologic history might be inferred from each of the following rocks?
   a. A feldspar-rich sandstone.
   b. A well-rounded conglomerate.
   c. A sandstone composed almost entirely of quartz.
   d. A limestone with abundant fossils.

19 What kinds of fossils would you expect to find in rocks deposited in the following environments?
   a. Tropical shallow water.
   b. Deep water.
   c. Temperate near shore.
   d. Marsh.

SUPPLEMENTARY READINGS

General

Dietrich, R. V., and B. J. Skinner, *Rocks and Rock Minerals.* Somerset, N.J.: John Wiley & Sons, 1979, 319 pp.

Ernst, W. G., *Earth Materials.* Englewood Cliffs, N.J.: Prentice-Hall, Inc., 1969, 150 pp. (paperback).

Sedimentary Rocks

Freeman, Tom, *Field Guide to Layered Rocks.* ESCP Pamphlet Series, PS-3. Boston: Houghton Mifflin Co., 1971, 43 pp.

Heezen, B. C., "The Origin of Submarine Canyons," *Scientific American* (August 1956), Vol. 195, No. 2, pp. 36–41. Reprint 807, W. H. Freeman and Co., San Francisco.

Hsü, K. J., "When the Mediterranean Dried Up," *Scientific American* (December 1972), Vol. 227, No. 6, pp. 26–36. An excellent article on the interpretation of climate and environment from sedimentary rocks and other geologic features.

Laporte, L. F., *Ancient Environments.* Englewood Cliffs, N.J.: Prentice-Hall, Inc., 1968, 116 pp. (paperback).

Millot, Georges, "Clay," *Scientific American* (April 1979), Vol. 240, No. 4, pp. 108–118.

Pettijohn, F. J., and P. E. Potter, *Atlas and Glossary of Primary Sedimentary Structures.* New York: Springer-Verlag, 1964, 370 pp.

Potter, P. E., and F. J. Pettijohn, *Paleocurrents and Basin Analysis.* Berlin: Springer-Verlag, published in United States by Academic Press, Inc., New York, 1963, 296 pp.

Reading, H. G., ed., *Sedimentary Environments and Facies.* New York: Elsevier North Holland, Inc., 1978, 557 pp.

Zenger, D. H., "Dolomitization and Uniformitarianism," *Journal of Geological Education* (May 1972), Vol. 20, No. 3, pp. 107–124.

Weathering

Boyer, R. E., *Field Guide to Rock Weathering.* ESCP Pamphlet Series, PS-1. Boston: Houghton Mifflin Co., 1971, 37 pp.

Foth, Henry, and H. S. Jacobs, *Field Guide to Soils.* ESCP Pamphlet Series, PS-2. Boston: Houghton Mifflin Co., 1971, 38 pp.

Gauri, K. L., "The Preservation of Stone," *Scientific American* (June 1978), Vol. 238, No. 6, pp. 126–136.

Keller, W. D., "Geochemical Weathering of Rocks: Source of Raw Materials for Good Living," *Journal of Geological Education* (February 1966), Vol. 14, No. 1, pp. 17–22.

Trace Elements and Health

Cannon, H. L., and D. F. Davidson, *Relation of Geology and Trace Elements to Nutrition.* Geological Society of America, Special Paper No. 90, 1967, 68 pp.

Gillie, R. B., "Endemic Goiter," *Scientific American* (June 1971), Vol. 224, No. 6, pp. 92–101.

Mertz, Walter, "The Essential Trace Elements," *Science* (September 18, 1981), Vol. 213, No. 4514, pp. 1332–1338.

Scherp, H. W., "Dental Caries: Prospects for Prevention," *Science* (September 24, 1971), Vol. 173, No. 4003, pp. 1199–1205.

Shacklette, H. T., H. I. Sauer, and A. T. Miesch, *Geochemical Environments and Cardiovascular Mortality Rates in Georgia.* U.S. Geological Survey Professional Paper 574C, Washington, D.C.: U.S. Government Printing Office, 1970, 39 pp.

Spencer, J. M., "Geology, Health, and Phosphorus," *Journal of Geological Education* (May 1974), Vol. 22, No. 3, pp. 93–96.

Van Ness, G. B., "Ecology of Anthrax," *Science* (June 25, 1971), Vol. 172, No. 3990, pp. 1303–1307

5 METAMORPHIC ROCKS

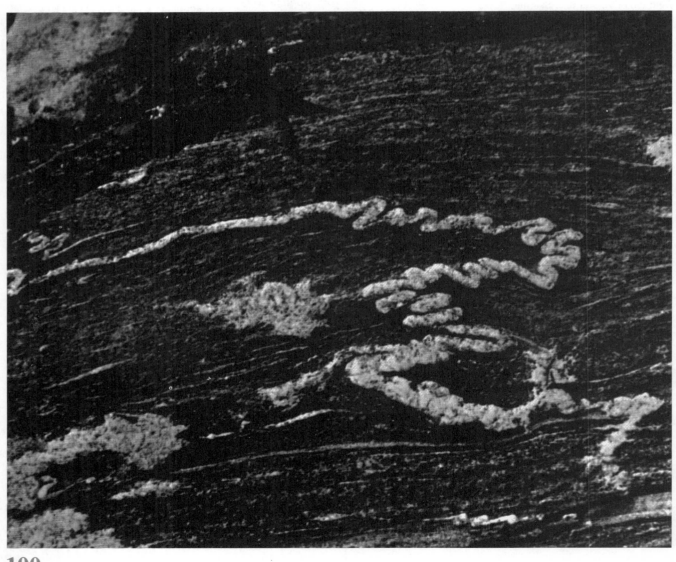

INTRODUCTION

Metamorphic rocks are rocks of any type or composition that have been changed in either mineral composition or texture.

The study of metamorphic rocks developed later than the study of igneous rocks, although the same methods of study are used for both rock types. Metamorphic rocks are the more difficult of the two to study because of their complex structures and their variations in composition and mineralogy. Hutton was the first to recognize clearly that rocks could undergo changes, although his concept of metamorphism was crude in the light of later knowledge. By the middle of the nineteenth century, French geologists had established the two main types of metamorphism, contact metamorphism and regional metamorphism (to be described).

Even before Werner, the idea developed that metamorphic rocks are the oldest rocks. Their complex structures and their locations in the cores of many mountain ranges and under the oldest sedimentary rocks suggest their antiquity. We now know that although the oldest rocks are metamorphic rocks, not all metamorphic rocks are old. The discovery of deformed fossils in metamorphic rocks showed this as early as 1814 when De Charpentier found fossils in mica schist in the Alps. This discovery helped to convince some Neptunists that Werner's age assignment was wrong. In 1881 Reusch found fossils in metamorphic rocks in Norway, and this discovery accelerated the study and understanding of metamorphic rocks. At the present time, the mapping of metamorphic areas is aided by the use of radioactive dating. Mapping complex structures without some means of recognizing the ages of the rocks is a very difficult undertaking.

Around the turn of the century, studies of the intensity or grade of metamorphism were made from field mapping and microscopic study. Studies were made by Barrow in Scotland in 1893, Becke and Grubenmann in the Alps in 1904, Goldschmidt at Oslo in 1911, and Escola in Finland in 1915. These men established the principles involved in the mineralogical changes that occur in metamorphism. Refinements of this work, especially using the results of more recent laboratory studies, is an ongoing type of research.

AGENTS OF METAMORPHISM

Metamorphic rocks are rocks that have been changed while in the solid state, either in texture or in mineral composition, by any of the following: heat, pressure, directed pressure (stress), shear, or chemically active solutions. Because any rock of any type or composition can be subjected to any or all of the above agents, the variety of metamorphic rocks is endless. Note that the changes that take place in the lithification of a sedimentary rock are, according to the definition, metamorphic. Such changes, however, are not considered metamorphic, but are arbitrarily excluded.

The changes that occur during metamorphism are the result of an attempt to reestablish equilibrium with the new conditions to which the rock is now subjected. Again note the similarity to weathering. Weathering occurs when rocks that formed deep in the earth are subjected to surface conditions. However, when surface-formed rocks, such as shales, are metamorphosed, the changes that occur proceed in the opposite direction and higher-temperature minerals are formed.

Many factors promote or retard metamorphic reactions. Large surface area promotes chemical reaction so fine-grained rocks react faster than coarse-grained rocks. Glass is less stable than crystalline material and so also reacts faster.

Heat is an important agent in causing metamorphism. It is well known that as one goes deeper into the earth in mines or drill holes, the temperature increases. Thus, deep burial alone can cause the temperature to rise, but most metamorphic rocks probably form at places where the temperature rise with depth is greater than normal.

Pressure also is the result of burial. The pressure is due to the weight of the overlying rocks. At the increased temperature and pressure deep in the earth, rocks lose much of their strength and so behave plastically. This causes the pressure due to the weight of overlying rocks to act uniformly in all directions, just as the pressure due to the weight of the water in a swimming pool acts in all directions.

Directed pressure (stress) is also commonly present during metamorphism. The effect of this is to fold the more or less plastic rocks. Stress may deform rocks, and flattened or stretched pebbles or fossils record some of this deformation.

Shear results when the rocks are broken and moved by directed pressure, as in faults. In metamorphic rocks the movement is commonly distributed on many closely spaced shear surfaces. (See Fig. 5–1.)

Chemically active solutions are very important in metamorphic reactions. Even if temperature and

Complexly folded quartz vein in mica schist. Potomac River gorge near Washington, D.C. Photo by J. C. Reed, Jr., U.S. Geological Survey.

pressure are high enough to cause a metamorphic reaction to occur, some water is generally necessary to allow the reaction to take place. Without water to act as a catalyst, the reaction may actually be too slow to occur. Water is generally present in most rocks, and only a thin film on the grain boundaries is necessary. The water may contain other materials that cause or promote the reactions. It can also bring in or remove elements from the rock. Small amounts of water may also greatly reduce the strength of rocks undergoing metamorphism.

The hot waters that cause low-grade metamorphism have apparently been found in some areas in the quest for geothermal power. Hot springs are relatively common and are spectacular at places such as Yellowstone National Park. Recent study of the geothermal area near the Salton Sea in southern California shows that the hot waters there have caused metamorphism in the surrounding sediments.

TYPES OF METAMORPHISM

Metamorphic rocks can be subdivided into three genetic groups, although there is gradation among the groups.

1. *Thermal* or *contact metamorphic rocks.* These rocks generally are found at the margins of intrusive igneous bodies such as batholiths. The rocks are termed *hornfels.*
2. *Regional metamorphic rocks* are so-called because they generally occupy large areas. These rocks form deep in the crust, and their presence at the surface reveals much uplift and erosion. *The rocks generally have recrystallized under stress so that the new minerals grow in preferred orientation. These rocks are said to be foliated,* and the most common are *gneiss* (pronounced "nice") and *schist.*

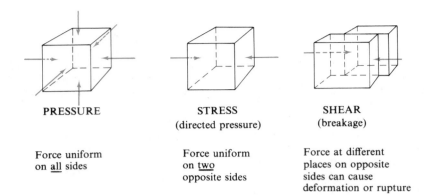

FIGURE 5–1 Pressure, stress, and shear.

3. *Dynamic metamorphic rocks* result from breaking and grinding without much recrystallization.

FORMATION AND IDENTIFICATION OF METAMORPHIC ROCKS

For classification and identification, metamorphic rocks can be subdivided into two textural groups:

1. Foliated—having a directional or layered aspect.
2. Nonfoliated—homogeneous or massive rocks.

NONFOLIATED METAMORPHIC ROCKS

Because the nonfoliated rocks are the simpler, we will consider them first. There are two types of nonfoliated metamorphic rocks. The first type consists of *thermal or contact metamorphic rocks called hornfels* (singular). They generally occur in narrow belts around intrusive bodies and may originate from any type of parent rock. Sometimes these are called *baked rocks;* the name is appropriate, because their formation is quite similar to the baking of clay pottery in a kiln. They are generally fine-grained, tough rocks that are difficult to identify without microscopic study unless the field rela-

tions are clear. They range from being completely recrystallized with none of their original features preserved to being slightly modified rocks with most of the original features preserved. (See Fig. 5–2.) In some of these rocks new minerals grow. At the early stages of growth the irregular growths produce what are called *spotted rocks.* Calcite-bearing contact rocks are generally quite spectacular, containing large crystals of garnet and other minerals.

The second type of nonfoliated metamorphic rock develops if the newly formed metamorphic minerals are equidimensional and so do not grow in any preferred orientation. The best examples of this are the rocks that result when monomineralic rocks such as limestone, quartz sandstone, and dunite are metamorphosed. In the case of limestone, which is composed of calcite, no new mineral can form; thus, the calcite crystals,

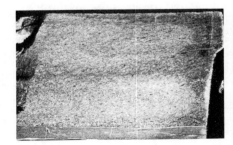

FIGURE 5–2 Hornfels, showing bedding and graded bedding. This polished specimen is just over two inches in diameter. Photo by F. C. Armstrong, U.S. Geological Survey.

which are small in limestone, grow bigger, develop an interlocking texture, and become *marble*. (See Fig. 5–3.) Marble can be distinguished from limestone only by its larger crystals and lack of fossils; being composed of calcite, both effervesce in acid. Marble may have any color; pure white calcite rocks are more apt to be marble than limestone. Impure limestones, as just noted, develop new minerals, particularly garnet. Dolomite marble also occurs.

Because quartz sandstone similarly can form no new minerals, the quartz crystals enlarge and intergrow to form *quartzite*. Quartzite is distinguished from chalcedony, chert, and opal by its peculiar sugary luster. The term quartzite is best reserved for the metamorphic rock, although quartz sandstone is called quartzite by some and the metamorphic rock is called metaquartzite.

Both marble and quartzite may form under stress, and so the crystals may be aligned. However, because the crystals are equidimensional, this preferred orientation can be detected only by microscopic examination. Thus, although we call these rocks nonfoliated, some may have a hidden foliation.

So far we have been concerned with *progressive* metamorphism, which is metamorphism caused by increased temperature, pressure, and other agents of metamorphism. The next example is one of *retrogressive* metamorphism, which is metamorphism at a temperature lower than that of the rock's original formation. Dunite is composed mainly of olivine, which is the highest-temperature mineral in Bowen's series. When it is subjected to the same metamorphic conditions that produce the rocks discussed earlier, it too is metamorphosed, but at a temperature lower than that of its formation. The olivine of dunite is changed to the mineral *serpentine,* and the metamorphic rock produced is called *serpentinite*. In other instances, instead of serpen-

FIGURE 5–3   Marble showing coarse interlocking texture. Photo from Ward's Natural Science Establishment, Inc., Rochester, N.Y.

tine, *talc* is the metamorphic mineral that forms. If nonfoliated, the talc rock is called *soapstone;* if foliated, it is called *talc schist.* Slight differences in the composition of the parent rock or of the water solutions that cause this type of metamorphism probably determine which mineral is formed.

FOLIATED METAMORPHIC ROCKS

Foliated rocks are the more common of the metamorphic rocks (see Fig. 5–4) and are the product of regional metamorphism. Regional metamorphism generally takes place in areas undergoing

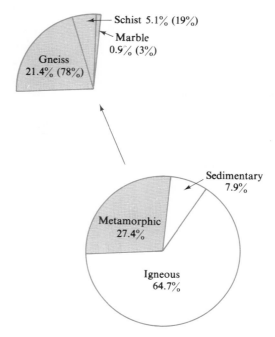

FIGURE 5–4   Relative abundance of metamorphic rocks in the earth's crust. For example, gneiss forms 21.4 percent of all rocks and 78 percent of the metamorphic rocks. Data from Ronov and Yaroshevsky, 1969.

103

deformation, and these very active areas are generally exposed in the cores of mountain ranges. Regional metamorphic rocks are produced by the same stress that makes mountain structures. The forces that fold and fault the shallow rocks exposed on the flanks of mountain ranges can provide the stress fields under which the deep-seated metamorphic rocks, now exposed in the uplifted cores of mountain ranges, were recrystallized.

One example of the rocks produced by increasing metamorphic grade will show most of the foliated metamorphic rock types. These same rock types will form when many other parent rocks are metamorphosed; the differences in original chemical composition will change the relative amounts of the metamorphic minerals only somewhat.

Shale is composed mainly of clay. During lower-grade metamorphism, the clay is transformed to mica, and at higher grades of metamorphism, to feldspar. Which micas and which feldspars form depends on the bulk chemical composition. Compare this with chemical weathering and note that weathering is, in a sense, retrogressive metamorphism. Because mica is a flat, platy mineral, it tends to grow with its leaves perpendicular to the maximum stress, forming a preferred orientation. (See Figs. 5–5, 5–6, and 5–7.) Microscopic examination of some foliated rocks shows that the bedding planes are slightly displaced in the plane of the foliation. This suggests that foliation, at least at some places, forms in shear directions. The shear direction is in general at about 45 degrees to the maximum stress in uniform materials. Thus foliation may form perpendicular to the direction of maximum stress, or at about 45 degrees to that direction.

The sequence in the metamorphism of a shale is given in Table 5–1. The rock names are applied for the overall texture of the rock and not strictly for the min-

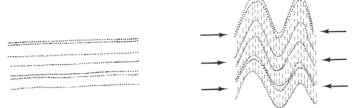

Original flat-lying sedimentary rocks.

Mica, shown diagrammatically by short dashes, grows perpendicular to the forces that may fold the rocks during metamorphism. Note that the foliation may form at an angle to the bedding.

FIGURE 5–5 Development of foliation.

FIGURE 5–6 Foliation and bedding in folded rocks near Walland, Tennessee. The rock type is slate. Compare with Fig. 5–5. Photo by A. Keith, U.S. Geological Survey.

eral transformations which occur over a range in temperature and pressure. (See Figs. 5–8, 5–9, and 5–10.) Most *schists* contain feldspar, but the name *gneiss* is reserved for rocks with much more feldspar than mica. Other rocks that have the same general bulk chemical composition, thus producing the same metamorphic rocks, are certain pyroclastic and volcanic sedimentary rocks, many sandstones, arkose, granite, and rhyolite. The coarse-grained rocks

in this list are rarely the parents of slate or fine-grained schist but, in general, remain more or less unaffected until medium-grade metamorphism is reached. Occurrence of slate in areas where the non-shale rocks are not metamorphosed suggests that slate may be formed by a mechanical reorientation of clay particles during folding, together with limited recrystallization.

Only one other group of regional metamorphic rocks must be

TABLE 5-1 Metamorphism of shale.

Sedimentary Rock	Low-grade Metamorphism	Medium-grade Metamorphism	High-grade Metamorphism
Shale ⟶ Clay	⟶ **Slate** ⟶ Clay begins to be transformed into mica. Mica crystals are too small to see, but impart a foliation to the rock. May also form by mechanical rearrangement of clay alone.	⟶ **Schist** ⟶ Mica grains are larger so that the rock has a conspicuous foliation.	⟶ **Gneiss** Mica has transformed largely to feldspar, giving the rock a banded or layered aspect.

FIGURE 5-7 Typical exposure of slate on Prudence Island, Rhode Island. Bedding is nearly vertical and is shown by color differences. Foliation is perpendicular to the bedding. Photo by J. B. Woodworth, U.S. Geological Survey.

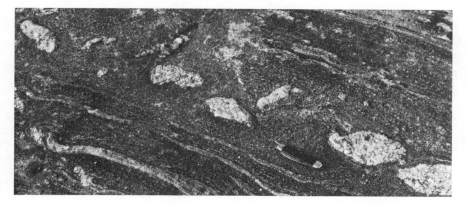

FIGURE 5-8 Outcrop of foliated rock. The rock is an impure marble, and the foliation is parallel to the bedding. The light-colored pods above the knife are quartzite, and the pods formed when a sandstone bed broke brittlely when the rocks were folded. Photo by J. A. Donaldson, Geological Survey of Canada, Ottawa.

considered among the most common metamorphic rocks. These rocks are formed by the metamorphism of basalt, certain pyroclastic and volcanic sedimentary rocks, gabbro, graywacke, and some calcite-bearing or dolomitic sedimentary rocks. The sequence is given in Table 5-2.

Dynamic metamorphism results when rocks are broken, sheared, and ground near the surface where the temperature and pressure are too low to cause any significant recrystallization. Thus, these rocks are commonly associated with fault zones, but there are all gradations between these rocks and ordinary schist, especially lower-grade schist formed from coarse-grained rocks. The fine-grained, ground rock in a fault zone is called *fault gouge* or *fault breccia* if it has little cohesion, and similar cohesive rocks are generally called *mylonite*. As the amount of recrystallization of such rocks increases, they become gradational with schist and gneiss. The origin of these rocks is clear when they are encountered in the field, but they are very difficult to identify in hand specimen.

An unusual type of dynamic metamorphism can occur at impact craters caused by meteorites.

The origin, texture, and mineralogy of the common metamorphic rocks are summarized in Figs. 5-11, 5-12, and 5-13.

INTERPRETATION OF METAMORPHIC ROCKS

Study of a metamorphic rock should reveal what the parent or original rock was, and under what conditions or in what environment it was metamorphosed. At some places, a layer of metamorphic rock can be followed or mapped to areas where it is not metamorphosed. In this way, the parent rock can be determined, as well as the changes that take place as the

FIGURE 5–9 Garnet-mica schist. The large garnet crystals grew during metamorphism. Foliation does not show clearly in the orientation pictured. Photo from Smithsonian Institution.

TABLE 5–2 Metamorphism of basalt.

Parent Rock	Low-grade Metamorphism	Medium- and High-grade Metamorphism
Basalt ⟶	Greenschist ⟶	Amphibolite
		(Amphibole schist)
	Fine-grained, foliated	Coarse-grained, foliated
	Chlorite (green "mica")	Plagioclase
	Green amphibole	Dark amphibole
	Quartz	

degree of metamorphism changes. Such studies, coupled with laboratory experiments, can yield much information about conditions and processes deep in the crust. That is, if we know the conditions of formation of a mineral or, better, an assemblage of minerals, when we find rocks composed of such minerals, we know the conditions under which the rocks formed. The problem is complicated by the diversity of natural rocks in that they contain many more elements, which are the variables in experiments, than can be easily studied in experiments. It is also not possible to reproduce the effects of geologic time in the laboratory, nor it is generally possible to detect the effects of solutions or volatiles that may have catalyzed the reactions in natural rocks. Thus, although we are very rapidly gaining an understanding of metamorphic processes, many problems remain.

To interpret metamorphic rocks fully, one must know whether they contain equilibrium mineral assemblages or whether conditions changed before the metamorphic reactions were complete; whether material was added or subtracted during metamorphism; and whether the rocks were subjected to one or more periods of metamorphism. One can think of many more possibilities that must be considered: for example, if the reactions take place during deep burial, will the rocks revert to something like their original state during slow uplift? No doubt some such effect may operate in some instances, but most of the chemical reactions involved go very slowly in the reverse direction. If this were not so, nothing but weathering products would be visible at the surface.

METAMORPHIC GRADE

The concept of *metamorphic grade* was just introduced in discussing the formation of meta-

A

B

FIGURE 5–10 Gneiss. A. The foliation is shown by discontinuous layers of dark- and light-colored minerals. B. The layers in this gneiss are more irregular. Photos from Ward's Natural Science Establishment, Inc., Rochester, N.Y.

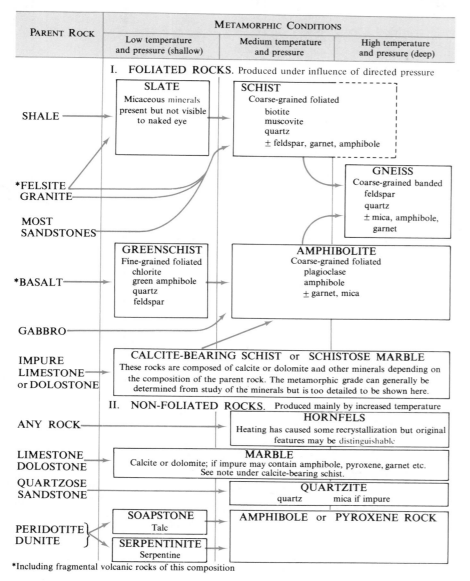

PARENT ROCK	METAMORPHIC CONDITIONS		
	Low temperature and pressure (shallow)	Medium temperature and pressure	High temperature and pressure (deep)

I. FOLIATED ROCKS. Produced under influence of directed pressure

SHALE

SLATE
Micaceous minerals present but not visible to naked eye

SCHIST
Coarse-grained foliated
biotite
muscovite
quartz
± feldspar, garnet, amphibole

*FELSITE
GRANITE

MOST SANDSTONES

GNEISS
Coarse-grained banded
feldspar
quartz
± mica, amphibole, garnet

*BASALT

GREENSCHIST
Fine-grained foliated
chlorite
green amphibole
quartz
feldspar

AMPHIBOLITE
Coarse-grained foliated
plagioclase
amphibole
± garnet, mica

GABBRO

IMPURE LIMESTONE or DOLOSTONE

CALCITE-BEARING SCHIST or SCHISTOSE MARBLE
These rocks are composed of calcite or dolomite and other minerals depending on the composition of the parent rock. The metamorphic grade can generally be determined from study of the minerals but is too detailed to be shown here.

II. NON-FOLIATED ROCKS. Produced mainly by increased temperature

ANY ROCK

HORNFELS
Heating has caused some recrystallization but original features may be distinguishable

LIMESTONE DOLOSTONE

MARBLE
Calcite or dolomite; if impure may contain amphibole, pyroxene, garnet etc.
See note under calcite-bearing schist.

QUARTZOSE SANDSTONE

QUARTZITE
quartz   mica if impure

PERIDOTITE DUNITE

SOAPSTONE
Talc

AMPHIBOLE or PYROXENE ROCK

SERPENTINITE
Serpentine

*Including fragmental volcanic rocks of this composition

FIGURE 5–11 A generalized chart showing the origin of common metamorphic rocks.

TABLE 5–3 Index minerals and grade of regional metamorphism developed in shale.

Chlorite	Low Grade
Biotite	↑
Almandine garnet	
Staurolite	
Kyanite	
Sillimanite	High Grade

A line on a map showing the first occurence of each of the index minerals is an *isograd* and is the boundary between mineral zones. That is, the first occurrence of biotite is the boundary between the chlorite zone and the biotite zone.

morphic rocks. Metamorphic grade is an interpretation based on field and laboratory evidence and should be further examined. The first attempts to map and study *grade of metamorphic rocks* consisted of mapping mineral zones. This was done by restricting the study to a single parent rock type, such as shale, and recording the first occurrence of certain *index minerals* as the metamorphic grade increased. (See Table 5–3 and Fig. 5–14.) This simple method has been im-

proved by considering assemblages of several minerals and by comparing such equilibrium assemblages in rocks of many compositions.

METAMORPHIC FACIES

A *metamorphic facies* can be defined as a group or assemblage of minerals that formed in response to an environment. Using mineral assemblages is an obvious improvement over index minerals in determining that the rock reached

equilibrium. Equilibrium mineral assemblages are determined from field studies. At each place studied, rocks of a given composition are traced from areas of little or no metamorphism to areas of intense metamorphism. The minerals in the rock change as the metamorphic conditions or environment changes. Mineral assemblages found at many places, especially if in the same sequence, are assumed to be equilibrium assemblages.

Once equilibrium assemblages or metamorphic facies are recognized, the next step is to learn under what conditions or environments such an assemblage forms. Laboratory experiments are performed so that the environment in which the facies forms can be discovered. In this way, the conditions deep in the earth's crust where metamorphism occurs are learned. The result of many of these laboratory studies is that the amount of water vapor in the rocks greatly influences which mineral assemblage forms.

The relationships among the common metamorphic facies are shown in Fig. 5–15, and the facies and the minerals in each facies are shown in Table 5–4. Note that the facies are generally named for minerals, but that rocks of different compositions, even though metamorphosed under the condi-

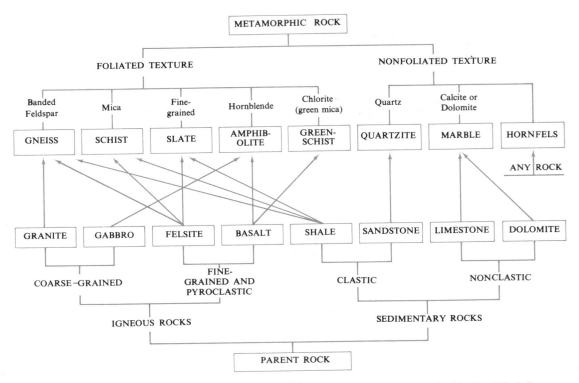

FIGURE 5–12 Origin of common metamorphic rocks. Not all possible parent rocks are shown in this simplified diagram.

tions or environment, may not contain that mineral. The facies was named from studies of rocks of some composition, shale, for example, and later work determined what minerals formed in basaltic rocks when subjected to the same environment. Thus, sanidinite facies rocks need not contain sanidine.

ORIGIN OF GRANITE

As shown in Fig. 5–16, at the highest grades of metamorphism some melting may occur, and so metamorphic processes may merge with igneous activity. Many of the small lenses of granitic rocks found in high-grade gneiss probably originate from such partial melting. (See Fig. 5–17.) The origin of larger bodies of granite, such as batholiths, in high-grade metamorphic terranes is a problem. Some show chilled, fine-grained margins and other features that are clearly igneous. Others have completely gradational contacts and could be metamorphic rocks or could have intruded the gneiss

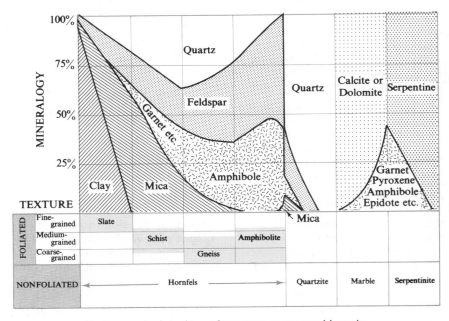

FIGURE 5–13 Texture and mineralogy of common metamorphic rocks.

just after the gneiss formed and was still hot. Some of the latter granites are probably the result of metamorphism, but the isotope composition of most batholiths indicates an origin in the lower crust or upper mantle.

METAMORPHIC ROCKS AND PLATE TECTONICS

The active areas of plate tectonics—mid-ocean ridges and conver-

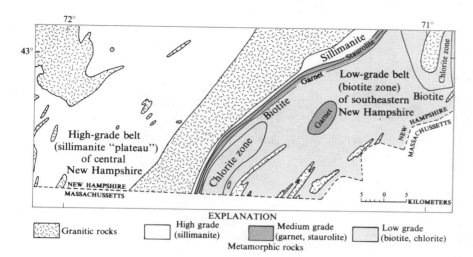

FIGURE 5–14   Metamorphic grade as interpreted from metamorphic minerals in shaly rocks in southeastern New Hampshire. From Warren Hamilton and W. B. Myers, U.S. Geological Survey Professional Paper 554C, 1967.

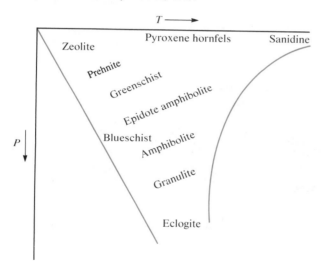

FIGURE 5–15   General relationships among metamorphic facies.

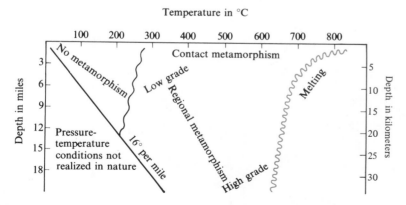

FIGURE 5–16   Diagram showing the conditions under which metamorphism occurs. Pressure is proportional to depth. Biotite is believed to crystallize between 200°C and 300°C and sillimanite between 600°C and 700°C, or very near the melting point of granite.

gent plate boundaries—are also places where metamorphism occurs. At mid-ocean ridges the rocks near the crest are metamorphosed, as shown in Fig. 5–18. The crest is an area of active basalt volcanism, so it is not surprising that the nearby rocks are subjected to metamorphic conditions. Volatiles, especially water, from the volcanic activity and the heat that generates the magma are probably important causes of this metamorphism.

Where an oceanic plate collides and moves downward under a continental plate, the resulting igneous and metamorphic rocks are easier to study because they are above sea level. The most common metamorphic rocks are the regional metamorphic rocks, and they are formed at such places. The heat and volatiles that cause the metamorphism are apparently associated with the rising magma that forms the volcanoes and the batholiths, as shown in Fig. 5–19. Near the slab of descending ocean floor is a zone of high pressure–low temperature metamorphism, shown in Fig. 5–19. This limited region is where the glaucophane schist facies rocks form. These rocks commonly contain blue-colored minerals, such as glaucophane, and are called *blueschist* by some geologists.

109

FIGURE 5-17 The small, irregular occurrence of massive granitic rock that cuts across the foliation may have formed by partial melting. The granitic rock is localized in a minor fault zone where the foliation is disturbed. Dannemora Mountain, New York. Photo by A. W. Postel, U.S. Geological Survey.

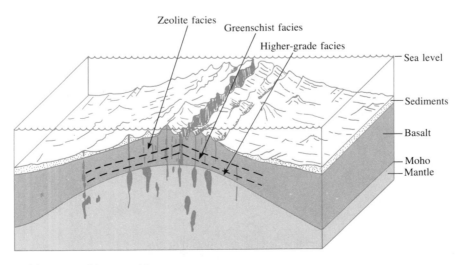

FIGURE 5-18 Metamorphism at a mid-ocean ridge.

TABLE 5-4 Typical mineral assemblages formed in metamorphic facies for rocks of shale and basalt composition. This table shows how elements are rearranged into different minerals in different metamorphic environments.

Metamorphic Facies	Rocks of Shale Composition												Rocks of Basaltic Composition													
	Clay	Chlorite	Stilpnomelane	Muscovite	Biotite	Garnet	Sillimanite	Cordierite	Andalusite	Sanidine	Orthopyroxene	Quartz	Zeolites	Smectite	Prehnite	Pumpellyite	Chlorite	Epidote	Lawsonite	Amphibole	Garnet	Orthopyroxene	Clinopyroxene	Omphacite	Quartz	Plagioclase
Zeolite	x	x										x	x	x												
Prehnite-pumpellyite	x	x	x									x			x	x										
Greenschist		x		x								x					x	x		(Actinolite) x						(Albite) x
Blueschist (glaucophane schist)		x		x		(Spessartine) x						x						x	x	(Glaucophane) x				x		
Epidote amphibolite		x		x	x	(Almandine) x						x						x		(Hornblende) x	(Almandine) x				x	(Albite) x
Amphibolite				x	x	x	x					x								(Hornblende) x	x				x	(Andesine) x
Pyroxene hornfels					x			x	x			x										x	x		x	(Labradorite) x
Granulite					x	x	x					x									x	x	x		x	(Labradorite) x
Sanidinite						x	x			x	x	x										x	x		x	(Labradorite) x
Eclogite	None known																				(Pyrope) x			x		

111

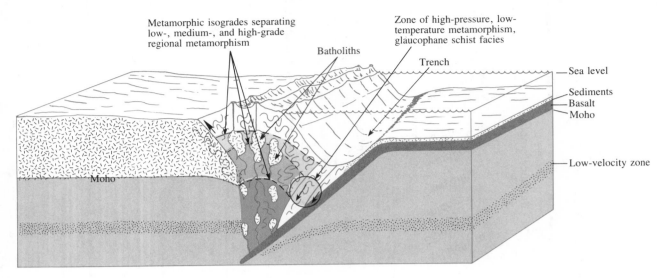

FIGURE 5-19 Metamorphism at a convergent plate boundary.

SUMMARY

Metamorphic rocks are rocks which have been changed, either in texture or mineral composition by heat and pressure, usually as a result of burial, directed pressure (stress), shear, or chemically active solutions, especially water.

There are three genetic groups with gradation among them:

1. *Thermal* or *contact metamorphic rocks,* called *hornfels,* generally found at margins of intrusive igneous bodies.
2. *Regional metamorphic rocks* occupy large areas, and reveal much uplift and erosion after their formation. They are usually *foliated* (the new minerals have preferred orientation because of recrystallization under stress). The most common are *gneiss* and *schist.*
3. *Dynamic metamorphic rocks* result from breaking and grinding without much recrystallization.

Textural groups are (1) *foliated*—directional or layered aspect and (2) *nonfoliated*—homogeneous or massive.

There are two types of *nonfoliated* rocks:

1. Thermal or contact metamorphic rocks, called *hornfels,* fine-grained, tough *baked rocks* in narrow belts around intrusive bodies.
2. The development of equidimensional new minerals, usually from monomineralic parent rocks. Limestone into *marble;* quartz sandstone into *quartzite;* dunite into *serpentinite* if serpentine is the new mineral, *soapstone* if talc is the new mineral (foliated talc rock is *talc schist*).

Progressive metamorphism is caused by increased pressure, temperature, and so forth, as in cases of marble and quartzite. *Retrogressive* metamorphism is at lower temperature and so forth than parent rock's formation; the olivine in dunite is the highest temperature mineral in Bowen's reaction series, and it is metamorphosed at a temperature lower than that of its formation.

Foliated metamorphic rocks are caused by the stresses that fold

mountains in regions undergoing deformation. Rocks are named largely for texture and not strictly for minerals. Examples are given in Tables 5-1 and 5-2.

Dynamic metamorphism occurs in fault zones or, rarely, at meteorite impact craters. Noncohesive ground rock in fault zone is *fault gouge* or *fault breccia.* Similar cohesive rocks are *mylonite.* These grade into schist and gneiss.

Interpretation of metamorphic rocks gives valuable insight into earth processes and history.

Metamorphic grade is determined by mapping the first occurrence of *index minerals* in metamorphic rocks of a given composition.

A *metamorphic facies* is a group or assemblage of minerals that formed in response to an environment.

Metamorphic processes may merge with igneous as in the origin of some granite bodies.

Much metamorphism is associated with the igneous processes that occur at mid-ocean ridges and convergent plate boundaries.

QUESTIONS

1 Define metamorphic rock.
2 What are the types of metamorphism, and which agents are most important in each?
3 What is foliation, and how does it form?
4 What is contact metamorphism, and what types of rocks are formed?
5 How does a shale differ from a slate?
6 What is the low-grade metamorphic equivalent of a basalt?
7 Serpentinite forms from metamorphism of _____.
8 Soapstone forms from metamorphism of _____.
9 How is the metamorphic grade established?
10 How might the field relationships differ between an igneous and a metamorphic granite?

SUPPLEMENTARY READINGS

General

Dietrich, R. V., and B. J. Skinner, *Rocks and Rock Minerals.* Somerset, N.J.: John Wiley & Sons, 1979, 319 pp.

Ernst, W. G., *Earth Materials.* Englewood Cliffs, N.J.: Prentice-Hall, Inc., 1969, 150 pp. (paperback).

Metamorphic Rocks

Dietrich, R. V., "Migmatites—A Résumé," *Journal of Geological Education* (September 1974), Vol. 22, No. 4, pp. 144–156.

Fyfe, W. S., F. J. Turner, and John Verhoogen, *Metamorphic Reactions and Metamorphic Facies,* Chapter 1, "Concept of Metamorphic Facies," pp. 3–20; Chapter 7, "Mineral Assemblages of Individual Metamorphic Facies," pp. 199–239. Geological Society of America, Memoir 73, 1958.

Higgins, M. W., *Cataclastic Rocks.* U.S. Geological Survey, Professional Paper 687, 1971, 97 pp.

Mason, Brian, *Principles of Geochemistry,* 3rd ed., Chapter 10, "Metamorphism and Metamorphic Rocks, pp. 248–282. New York: John Wiley & Sons, Inc., 1966.

6 MINERAL RESOURCES

Many of the materials that form the crust of the earth are useful to us. The term *mineral resources is used to mean natural resources that come from the earth.* Note that the term *mineral* used in this sense does not follow the definition of mineral used earlier because many of the materials discussed in this chapter, strictly speaking, are not minerals. The most important mineral resources are metals and energy-producing materials, and this chapter will be limited mainly to these. Many other mineral products are mined, however, such as sand, gravel, building stone, clay, diatomite, and limestone.

Scientific understanding of how economic mineral deposits form developed very slowly. In 1556, Agricola's book *De Re Metallica* was published, and, as noted earlier, it was an important, useful book that helped the miners of that time. Agricola recognized that the veins that contain the ore minerals were deposited by solutions in channels that formed later than the surrounding rock. In these observations he was far ahead of his time; however, no further progress was made for almost 200 years. In that interval a number of fanciful ideas developed, such as veins are the twigs of a golden tree whose roots are at the center of the earth, and veins develop as a result of celestial influences, especially the sun's rays.

By the middle of the eighteenth century, more scientific study was progressing, particularly in Germany at the Mining Academy at Freiberg where Werner was active. In the nineteenth century, two important theories of vein formation were put forward. The first is that veins are deposited by hot water solutions, and the second is that solutions moving through rocks leach the metals that are then deposited in veins. Apparently both processes cause ore deposits. The late nineteenth century was a time of great interest in mining, and many geologists contributed to our understanding of mineral deposits. In the early

twentieth century, much of this work was used by Waldemar Lindgren to formulate a classification of mineral deposits. The classification used in this chapter is adapted from Lindgren. Like most geologic features, mineral deposits are related to plate tectonics, and this realization has aided in both exploration for and understanding of mineral resources.

This chapter is concerned with the origin of mineral resources, and few new principles are involved. Before discussing formation of mineral deposits, we will consider some of their unique features and their importance.

DISTRIBUTION AND IMPORTANCE

Mineral resources are not spread evenly around the continents. This creates the have and the have-not nations. A mine or other mineral deposit is a rare feature, geologically speaking.

The importance of mineral deposits to man cannot be overemphasized. The wealth of nations is dependent on their natural resources. Although not always emphasized, mineral deposits have played a very important role in history. In ancient times, wars were fought and civilizations prospered or failed because of gold and silver deposits. In recent times, metals and energy-producing materials, such as iron, coal, and petroleum, have determined the course of history.

EXHAUSTIBILITY

Every mineral deposit has finite dimensions. When the ore is mined out, the deposit has been used. Thus all mineral deposits are exhaustible, and there is no way to replace them. This is not news to anyone who has traveled through the West and seen the many ghost towns. The importance of the exhaustibility of mineral deposits is seen when rates of mining are considered. In the last fifty years more metal has been

mined than in all preceding time. Thus, we are using our natural resources at an unprecedented rate. Unless new sources are found, we face the prospect of nations decaying like the ghost towns of the West. This drastic possibility will be averted by increased exploration (prospecting), using new, sensitive chemical, physical, and biologic methods; by finding substitutes, such as plastics; and by developing new refining methods that are effective with very low-grade ores. In any case, it seems quite evident that we will have to pay more for our metals in the future.

An *ore is anything that can be mined at a profit.* Therefore, what is ore in the United States may not be ore in the Canadian Arctic or Central Australia because of the cost of mining and the distance to market. It also follows that what is worthless today may be ore tomorrow because of changing economic conditions or technological changes.

Water, both surface and underground, and soil are the most im-

Strip mining coal in Ohio. Photo from U.S. Bureau of Mines and Marion Power Shovel Co.

115

portant mineral resources. Life would not be possible without both. These two basic resources are different from all other mineral resources in that they are replenishable, although the time required to replenish them may be long by human standards.

RESOURCES AND POPULATION

The mineral resources of the earth are finite, and once mined, no more new metal can be had. This is one aspect of the spaceship earth analogy. Some metals, such as aluminum, will probably never run out, but some others, especially silver and mercury, and perhaps lead, zinc, gold, and tin may very well run out.

Because metal mines are not evenly distributed, no industrial country is self-sufficient in all mineral resources. Obviously, the industrial countries are using mineral resources in much greater quantities than the underdeveloped countries. The undeveloped or nonindustrial countries are striving to become industrial countries and raise their standards of living. Thus the rate of use of minerals is bound to increase, and the increase will be rapid because most of the earth's population lives in the undeveloped countries.

Some examples of metal usage will show the magnitude of the problem. The per capita use of iron in the United States is about one ton, and for the total world population the figure is 0.17 ton. The United States' usage is about six times that of the world average, or, said another way, iron production would have to be increased six times if the whole earth is to have the same standard of living as the United States. It is estimated that the population of the earth will double by the year 2000. To keep the same world per capita use of iron, production will have to be doubled; or if everyone is to have the United States' stan-

dard of living, production must be increased 12 times. Of course, not only iron but also all other mineral resources will be needed in similar ratios to achieve these goals. These projections are probably too high for a number of reasons, such as export of manufactured goods by the industrial countries. On the other hand, per capita consumption in the United States and elsewhere has been rapidly accelerating, suggesting that the projections may be too low. It seems clear that the potential demand for metals in the year 2000 will be at least several times that of today. The question is, Can this potential demand be met?

The United States, like most industrial countries, is not self-sufficient in many mineral resources; it must import large quantities. A question of great importance is, What is a fair price to pay an undeveloped country for its unreplenishable mineral resources? All people look forward to the day when they will have a standard of living like ours. This will be difficult to achieve if their resources are gone. On the other hand, they cannot develop or even use their resources without capital, and their only ready way to get capital is to sell their resources. It takes many millions of dollars to find a resource, and many more to develop it, and, of course, much technical know-how. The historical background to the present world situation shows how undeveloped countries have been exploited in the past. The colonial system developed with the industrial revolution. The mother country was generally an industrial country, and the colonies supplied the mother country with raw materials. The political breakdown of this system began with the American Revolution and was pretty well completed after World War II with the formation of many new independent countries in Africa and Asia. Most of the colonialism of today is economic colonialism, and revolves around the

question of the price paid for mineral resources. Prices may have to increase, and this may change the standards of living of the whole world. The constant upward revision of the royalties paid for oil in the Near East is an example of the kinds of change that seem likely to occur. Anyone who does not believe the importance of mineral resources in world politics needs only to study the history of Europe for the last 200 years or so. The territorial settlements after each war involved mainly the important mineral resource areas, such as the iron of Alsace-Lorraine, the coal of the Saar, and the lead-zinc of Poland.

Waste Disposal—''Garbage is a resource for which we have not yet found a use.'' The amount of material discarded is alarming. In the United States it is about 5.5 pounds (2.5 kilograms) per person per day—almost twice what it was in 1920. Much of the increase is due to modern packaging and to disposable items. Another aspect of the spaceship earth idea is that we can never really throw anything away; we can only store it in a different way. Most cities either bury or burn their garbage—processes that have been in use since before recorded history. Some cities have tried to make fertilizer out of garbage but have been unable to sell it. Several cities are compacting trash and either shipping it away or using it locally for land fill. Experiments are underway using trash as fuel to run electric generators and converting garbage to petroleum.

One way to eliminate part of the disposal problem is to recycle as much trash as possible. Bottles and aluminum cans are currently being recycled by consumers. Commercially reclaimed metals include about 40 percent of our lead and 25 percent of our copper. About three million tons of iron are discarded each year in the United States. Recycling trash can greatly reduce the need to find and mine metal ores, and so lessen one problem discussed earlier. In this sense, one can look at dumps and automobile wrecking yards as national resources although they still may be odorous eyesores.

It is estimated that the United States each day produces enough waste to cover 400 acres to a depth of 10 feet. The cost to government for collection and disposal is over $3000 million per year, a sum only exceeded by funds for schools and roads.

PLATE TECTONICS AND METAL DEPOSITS

Plate tectonics created the rocks of the earth's crust and in many cases caused their later deformation, so it is not surprising that metal deposits are also related to plate tectonics. At mid-ocean ridges new oceanic crust is created from material coming up from the earth's mantle. At several places on mid-ocean ridges, oceanographers have observed the formation of metallic sulfide ores. The sulfide minerals apparently originate with the basalt magma and are deposited by hot water. The hot water is seawater that circulates through the basalt and overlying sediments. The circulating water is heated by the basaltic magma, and the hot water rises to the ocean floor, carrying the sulfides with it. Hot brines [56°C (133°F)], containing large amounts of dissolved metals, have been found in the Red Sea above the mid-ocean ridge. At several places on the East Pacific Rise off the coast of Mexico, geologists and oceanographers in small research submarines have seen these hot springs emitting very hot water [380°C (716°F)] that is black with sulfides. Rapid cooling causes the sulfides and other material to deposit in cones 1.5 to 6 meters (5 to 30 feet) high that resemble small volcanic cones. One observer remarked that they

"looked like a row of factory chimneys belching black smoke." Some of the springs are cooler and emit clear water, and a deepwater fauna unlike any other yet discovered was found around these springs. The animals are giant tube worms, clams, and crabs. The copper sulfide deposits on the island of Cyprus are associated with basaltic rocks like those on ocean bottoms, and the deposits are believed to have formed at a mid-ocean ridge.

At convergent plate boundaries where the ocean crust dives deep into the mantle, magmas are generated by the resulting melting. Metal deposits are associated with these magmas, and the metals occur in rough sequence with depth of origin or distance from the volcanoes (Fig. 6–1). The metals may originate in the ocean crust that is melted or may come from the mantle rocks through which the magma moves. Many of the mineral deposits of western North and South America are believed to have formed in this way. Although the magmas formed at convergent plate boundaries may be enriched, they do not contain enough metal to be ore, so other enrichment processes are necessary to form metal deposits.

FORMATION OF MINERAL DEPOSITS

Mineral deposits occur at places where unusual conditions caused the concentration of some elements far in excess of their normal abundances. Table 6–1 shows the average amounts of some elements in the earth's crust. Copper, for instance, has an average abundance of 0.0055 percent, but copper ore in the United States must average about 0.5 percent copper. Thus, copper must be concentrated by at least 90 times its aver-

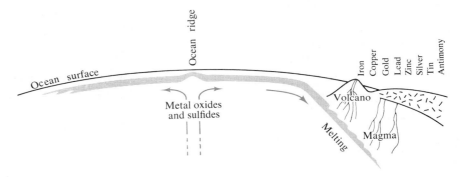

FIGURE 6–1 In plate tectonics, material rises from the earth's interior at ocean ridges, moves outward, and then downward. The rising material contains metals that may form deposits on sea floors and may be concentrated by magmatic processes where the descending sea floor is melted.

TABLE 6–1 Concentration factors for some common metals. These concentration factors are approximate because the amount of metal in ore may change with location and economic conditions.

Metal	Percent In Earth's Crust	Percent In Ore	Concentration Factor
Mercury	0.000008	0.2	25,000
Lead	0.00125	5.0	4,000
Nickel	0.0075	1.0	133
Copper	0.0055	0.5	90
Iron	5.63	30.0	6
Aluminum	8.23	38.0	5

age amount to form copper ore. Many mines produce more than one metal, thus reducing the percentage of any one element that must be present to make the deposit economic. In any case, to form a metal mine, an element must be concentrated at least several to a few thousand times its average abundance. We will now see how this can be done. (See Tables 6–2 and 6–3.)

DEPOSITS ASSOCIATED WITH IGNEOUS ROCKS

Many mines are associated with igneous rocks, and we will consider them first. The different magma types, such as granite and gabbro, each have typical metals that are associated with them.

Disseminated Magmatic Deposits are the simplest of the magmatic deposits. *The valuable mineral is disseminated or scattered throughout the igneous body.* In the diamond deposits of South Africa, for example, the diamonds are disseminated in an unusual rock, somewhat similar to peridotite. (See Fig. 6–2.) In this example, the valuable mineral is unique

to this rock type and, because of its great value, can be mined. In the other types of deposits, the ore mineral or minerals are concentrated rather than disseminated within the body.

Crystal Settling was already described in the discussion of differentiation of magma. (See Figs. 3–25 and 6–3.) Heavy, early-crystallized minerals sink to the bottom of the igneous body. Many chromite and/or magnetite deposits form in this way, generally in gabbro bodies. (Magnetite crystallizes early under some conditions and late under others.) In some cases, the segregation of the early-formed ore minerals is effected by squeezing the liquid magma away, leaving the early-formed crystals.

Late Magmatic Deposits *form when the ore-forming constituents of the magma are among the last to crystallize.* The metalliferous material is heavy; it tends to settle to the bottom of the igneous body and to crystallize there and in the interstices between the earlier-formed minerals. At the late stage when these deposits form, as pointed out earlier, the magma may be water-rich. In other cases late deformation may force the metalliferous fluid into cracks (see Fig. 6–4).

PEGMATITE

Pegmatite bodies also form late in the crystallization of a magma. These extremely coarse-grained rocks may form as dikes or irregular pods. (See Figs. 6–5 and 3–

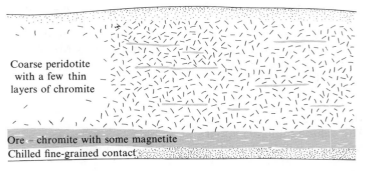

Coarse peridotite with a few thin layers of chromite

Ore – chromite with some magnetite
Chilled fine-grained contact

FIGURE 6–3   Chromite layers form by settling of early-formed chromite crystals. The parent igneous rock is peridotite, dunite, or gabbro. Commonly the igneous body has a more irregular shape than shown.

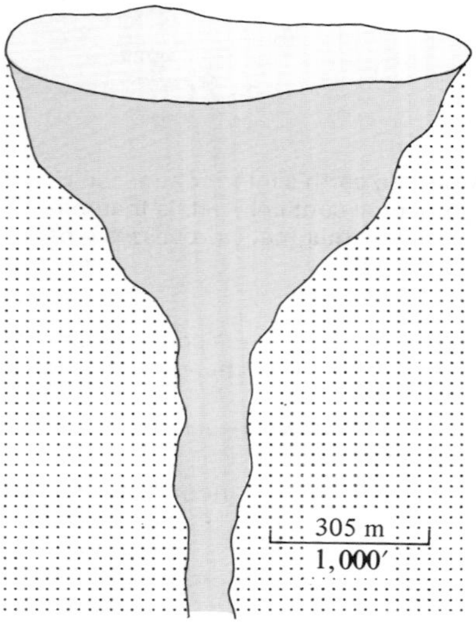

TABLE 6–2   Genetic classification of mineral deposits.

Type	Example
Magmatic	
Disseminated	Diamonds—South Africa (Fig. 6–2)
Crystal settling	Chromite—Stillwater, Montana (Fig. 6–3)
Late magmatic	Magnetite—Adirondack Mountains, New York (Fig. 6–4)
Pegmatite	Beryllium and lithium—Black Hills, South Dakota (Fig. 6–5)
Hydrothermal	Copper—Butte, Montana (Fig. 6–6)
Contact-metamorphic	Lead and silver—Leadville, Colorado (Fig. 6–8)
Sedimentary	
Evaporite	Potassium—Carlsbad, New Mexico
Placer	Gold—Sierra Nevada foothills, California (Fig. 6–10)
Weathering	Bauxite—Arkansas (Fig. 6–12)

FIGURE 6–2   South African diamond pipe. Diamonds are scattered through more or less vertical, cylindrical bodies of an unusual igneous rock somewhat similar to peridotite.

305 m
1,000′

42.) Just how the large crystals form is not known, but the water- and volatile-rich solutions must permit the growth of large crystáls. Not all pegmatites are igneous, for many are found in metamorphic terranes far from any igneous body. They are sought for their large crystals of mica, which are used in industry, and for feldspar, quartz crystals, and gems. They are also the main

TABLE 6-3   Geologic occurrence of some metals.

Metal	Main Ore Minerals	Principal Type Occurrence	Example
Aluminum	Bauxite $Al_2O_3 \cdot nH_2O$	Weathering	Arkansas
Antimony	Stibnite Sb_2S_3	Hydrothermal	Yellow Pine, Idaho
Beryllium	Beryl $Be_3Al_2Si_6O_{18}$	Pegmatite	Black Hills. S.D.
Cadmium	By-product of zinc ores	Hydrothermal (in sphalerite)	Missouri (Tri-State Region)
Chromium	Chromite $Fe(Cr,Fe)_2O_4$	Magmatic	Stillwater, Montana
Cobalt	Cobaltite $CoAsS$ Smaltite $CoAs_3$	Hydrothermal	Zaire
Copper	Bornite Cu_5FeS_4 Enargite Cu_3AsS_4 Chalcocite Cu_2S	Hydrothermal	Bingham, Utah Butte, Montana
Gold	Tellurides $AuTe_2$ Native gold Au	Hydrothermal Placer	Homestake, S.D. Mother Lode. Cal.
Iron	Hematite Fe_2O_3 Magnetite Fe_3O_4	Sedimentary and Weathering	Lake Superior Region
Lead	Galena PbS	Hydrothermal	Missouri (Tri-State Region) Coeur d'Alene, Id.
Manganese	Oxides MnO_2	Weathering	Chamberlain, S.D.
Mercury	Cinnabar HgS	Hydrothermal	New Almaden, Cal.
Molybdenum	Molybdenite MoS_2	Hydrothermal	Climax, Colo.
Nickel	Sulfides NiS	Magmatic	Sudbury, Ontario
Platinum	Native platinum Pt	Magmatic Placer	Ural Mountains, Russia
Silver	Sulfides Ag_2S By-product of zinc-lead mine	Hydrothermal	Comstock, Nev. Coeur d'Alene. Id.
Tin	Cassiterite SnO_2	Hydrothermal Placer	Malaya
Titanium	Ilmenite $FeTiO_3$ Rutile TiO_2	Magmatic	Adirondack Mtns., New York
Tungsten	Scheelite $CaWO_4$ Wolframite $(Fe, Mn)WO_4$	Contact-metamorphic	Small deposits in Nev. and Cal.
Uranium	Pitchblende U_3O_8 Uraninite UO_2 Carnotite $K_2(UO_2)_2(VO_4)_2 \cdot 3H_2O$	Sedimentary (Carnotite)	Colorado Plateau
Vanadium	Carnotite $K_2(UO_2)_2(VO_4)_2 \cdot 3H_2O$ Roscoelite $K(V,Al)_2(OH)_2AlSi_3O_{10}$	Sedimentary (Carnotite)	Colorado Plateau
Zinc	Sphalerite ZnS	Hydrothermal	Missouri (Tri-State Region)

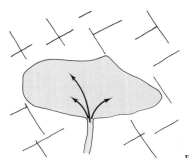

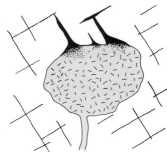

A. Injection of magma

B. After partial crystallization, the magma body is deformed and the late fluid is squeezed into newly-formed cracks.

FIGURE 6-4 An example of segregation of late magmatic fluid. In this case, the late fluid is separated from the rest of the magma; in other cases, the late fluid may sink and fill the spaces between the early crystals near the bottom of the igneous body.

FIGURE 6-5 Pegmatite in the Black Hills of South Dakota. The large, light-colored crystals, such as the one the hammer is on, are spodumene, a lithium ore. Photo by J. J. Norton, U.S. Geological Survey.

source of columbium, tantalum, the rare earths, beryl (the main beryllium mineral), and lithium. Many other minerals occur in pegmatites, including some high-temperature sulfides.

HYDROTHERMAL DEPOSITS

Hydrothermal, or *hot water, deposits* are the most common type of metal deposit. Their possible origin from the late magmatic fluids was discussed in Chapter 3. They grade from high-temperature to low-temperature deposits. Typical hydrothermal deposits contain sulfide minerals, such as pyrite, chalcopyrite, and galena, in quartz veins. (See Figs. 6-6 and 6-7.) Generally they are associated with batholiths, and the veins are both in the batholith and the surrounding rocks. (See Fig. 3-42.) However, many hydrothermal deposits have no nearby outcropping igneous rocks and the origin of the depositing solutions is not known. In such cases the ores may have been deposited by circulating ground water, as mentioned later in connection with geothermal power. In these instances, the metals might come from deep magmas or might be leached from other rocks by the circulating ground water. Examples of hydrothermal deposits include copper at

Butte, silver at the Comstock Lode, lead-zinc at Missouri, and gold at the Mother Lode in California.

CONTACT METAMORPHIC DEPOSITS

Contact Metamorphic Deposits form in the zone of metamorphism that surrounds an intrusive

FIGURE 6-6 Underground mining, Kelley Mine, Butte, Montana. Some of the veins in the granitic rock being drilled can be seen. Photo from Anaconda Co.

body. The magma and its fluids react with the wall rock and the ground water and replace the wall rock with ore minerals. These deposits are commonly found in limestone near intrusive bodies because the limestone is especially reactive with the fluids. (See Fig. 6–8.) Many deposits formed in this way contain high-temperature sulfide minerals similar to those in high-temperature hydrothermal deposits. Some iron deposits, a few fairly large, formed in this way: Iron Springs, Utah, is an example.

SEDIMENTARY DEPOSITS

Sedimentary processes can also concentrate valuable elements.

Evaporites, or chemical precipitates, such as salt beds, especially those containing potassium salts, are sought actively. Such beds come from the evaporation of sea-

FIGURE 6–7 Aerial view of open pit copper mine at Bingham Canyon, Utah, about 32 kilometers (20 miles) southwest of Salt Lake City. This is the largest copper mine in the United States. Photo by Don Green, courtesy of Kennecott Copper Corp.

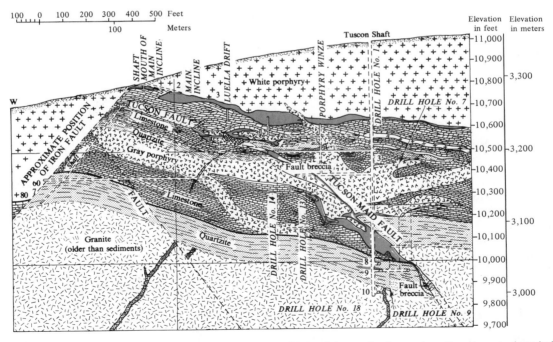

FIGURE 6–8 Cross-section of part of the Leadville district, Colorado. Most of the ore bodies replace limestone and are believed to have originated from the porphyry. Some veins are present, and the ore bodies along the fault show that it acted as a passage for some of the ore fluids. Replacement ore bodies generally have complex shapes. See Fig. 6–9. From F. A. Aicher, U.S. Geological Survey Professional Paper 148, 1927.

121

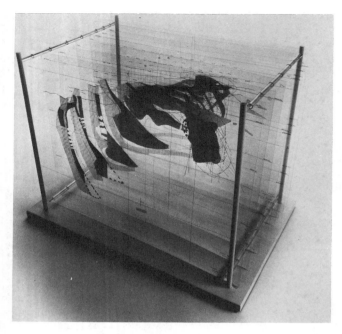

FIGURE 6–9 A model of a mine made by drafting successive cross-sections on transparent material. The complex shapes of ore bodies can be studied by this technique. Photo from Anaconda Co.

water or of a salt lake. Because seawater contains only 3.5 percent dissolved material, mostly common salt, several thousand feet of seawater must evaporate to produce 30 meters (100 feet) of salt. Many salt deposits, however, are much thicker than this. The evaporation is thought to occur in shallow, slowly sinking basins, partially dammed by sand bars, with fresh brine from the ocean flowing over the sand bars, perhaps during storms. Recently it has been suggested that salts could accumulate in deep basins with high evaporation and stagnant bottoms. The chemistry of these brines is complex, but the general sequence of precipitation is calcium and magnesium carbonate, gypsum, salt (halite), and finally, potassium and bromine salts. Salt beds are mined in the area of an ancient basin in Michigan, Ohio, and New York. Potassium salts are mined near Carlsbad, New Mexico.

At other times and places in the geologic past, generally in restricted basins, deposits of phos-

phate and iron-rich rocks have occurred. Rather specialized conditions are probably necessary to form such rocks. The phosphate deposits of Idaho, Wyoming, and Montana formed in such a restricted sea.

Placer Deposits *form by the concentration of heavy mineral fragments by rivers.* A heavy, or more properly, a dense mineral fragment is moved less easily than a lighter fragment of the same size. For this reason a river tends to separate, or sort, its transported load. The denser fragments are the first to be deposited and tend to be concentrated on the insides of curves, at places where the gradient is less, and in similar places.

Minerals concentrated in this way must be dense and resistant to weathering. Examples include gold, diamond, and platinum. Gold placers were very important in the opening of the west, especially in California in 1849. (See Fig. 6–10.) Placer deposits are "poor man's deposits" because very little equipment is necessary

to prospect for them or to exploit them. This easily-obtained wealth started the westward movement. Ocean waves can also produce placers, and beach sands containing valuable heavy minerals are mined at some places.

Deep Ocean Deposits have recently attracted much attention. Bottom photographs and dredge samples have revealed that parts of the floor of the Pacific Ocean are covered by manganese nodules. (See Fig. 6–11.) These interesting rocks have formed by the precipitation of manganese and iron dissolved in seawater onto sand grains and pebbles on the sea bottom. This precipitation is so slow that 1000 to 100,000 years are required to deposit a layer one millimeter thick. The deposition occurs in the form of oxides and hydroxides. These nodules average about 24 percent manganese and 14 percent iron, with small amounts of nickel, cobalt, copper, and other elements. Some contain 2 percent or more copper, nickel, or cobalt. The elements come from chemical weathering of the continents, by leaching of submarine basalt, and from waters associated with volcanic areas such as mid-ocean ridges.

Because of the large areas involved, these nodules constitute a very valuable resource, and techniques for their mining are being developed. However, the nodules also contain large amounts of silica (SiO_2), and current metallurgical techniques do not permit the separation of manganese from silica profitably.

DEPOSITS CONCENTRATED BY WEATHERING

Weathering processes also contribute to mineral wealth. Weathering and the development of soil are described in Chapter 4. Soils are surely our most valuable mineral commodity. In addition, some tropical soils (laterites) are mined for their aluminum (bauxite), iron, or other elements. (See Fig. 6–12.)

Weathering processes are also active on outcropping mineral deposits. Their effect on many deposits is great and, unless understood, can be very misleading. If an ore or its weathering products are insoluble, the ore will be enriched by weathering unless the weathered material is removed by erosion. This is especially true in mineral deposits in soluble rocks such as limestone. (See Fig. 6–13.) This kind of enrichment has fooled many prospectors and investors who thought that they had a good mine on the basis of near-surface assays, but found the ore too low-grade to be economical when a shaft was sunk. Gold, lead, copper, silver, and mercury enrichments of this type are very common.

At sulfide deposits, the weathering processes can become very complex. The sulfide minerals are oxidized to become soluble sulfates. Pyrite, which is present in almost every case, weathers to sulfuric acid and soluble iron sulfates, all of which aid in the solution of other sulfide minerals. Iron, silver, and copper sulfides are dissolved in this way. The soluble sulfates trickle slowly downward, enriching the lower part of the deposit. The main zone of enrichment occurs at the water table. Here the chemical environment differs, mainly in its lack of oxygen, and precipitation as sulfides occurs. Again, the shrewd prospector should not be fooled by the near-surface shows unless they, by themselves, are rich and large enough to mine.

Weathering of mineral deposits does help the prospector. The oxidation of the ever-present pyrite leaves a surface residue of iron oxides, especially hematite and limonite. These red-brown caps clearly mark the outcrops of sulfide veins, and so they are the prime target of the prospector. (See Fig. 6–14.)

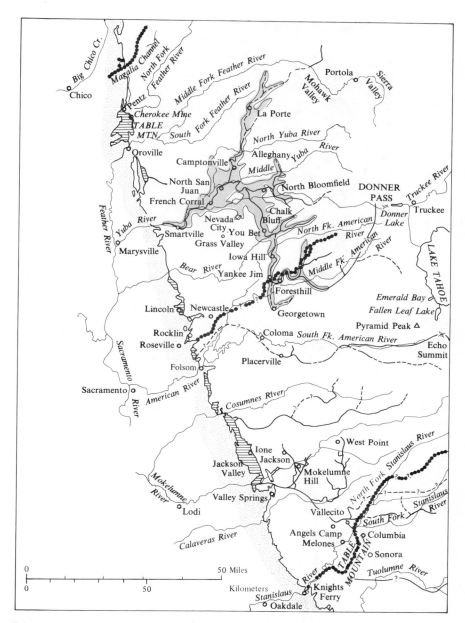

FIGURE 6–10 Map showing the location of placer gold deposits in the Sierra foothills in California. The ancient rivers along which these placers accumulated were in different locations than the present rivers. The older of the rivers are shown by heavy lines and the younger by dotted lines. These rivers flowed into a sea indicated by the stippled area. From P. C. Bateman and Clyde Wahrhaftig, California Division of Mines and Geology Bulletin 190, 1966.

ENERGY RESOURCES

Modern civilization depends on energy. At first our only use of energy was in the form of food. Later we used wood fires, and the industrial revolution required the use of coal. The fossil fuels, coal and petroleum, are the main energy sources at present, with nuclear power developing very rapidly.

The energy sources of the earth are:
Solar radiation
Food
Fossil fuels
Water power

123

FIGURE 6–11 Manganese nodules on the floor of the Atlantic Ocean. Photo courtesy of R. M. Pratt.

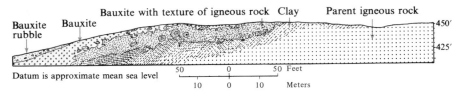

FIGURE 6–12 Cross-section of the Pruden bauxite mine, Arkansas. The bauxite is the weathering product of the aluminum-rich igneous rock. From Mackenzie Gordon, Jr., and others, U.S. Geological Survey Professional Paper 299, 1958.

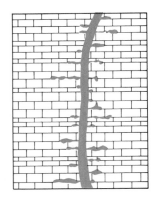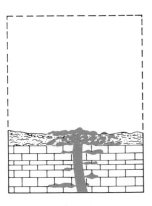

FIGURE 6–13 Residual concentration by solution of the limestone that encloses the deposit. The ore minerals must be insoluble.

Direct solar energy
Tidal energy
Geothermal energy
Nuclear energy

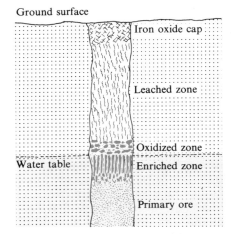

FIGURE 6–14 Weathering of a sulfide vein deposit. Rainwater percolating to the water table leaches metals from the vein and redeposits them near the water table, creating an enriched zone. The reddish iron oxide cap helps to attract the attention of prospectors.

The amount of energy that the earth receives from the sun is many times more than the amount used by humans. If a practical way to use it could be devised, no other source would be needed. It should be remembered that the earth re-radiates the energy received from the sun. If it did not, the temperature of the earth would increase, and geologic history shows that the average temperature has remained about the same for at least 500 million years. Solar energy is our most important energy source. It is used by plants, providing the food that supports all of the life on the earth. In the past, plants in the form of wood provided an important source of energy. A very small amount of solar energy is preserved in the form of fossil fuels such as coal and petroleum.

FOSSIL FUELS

Coal and *petroleum* are called *fossil fuels because they originate from once-living material.*

Coal comes from plant material. To form coal, plant material must accumulate. However, under most conditions, even with abundant growth of vegetation as in a tropical rainforest, plant material does not accumulate; instead, it decays and is attacked by bacteria as

124

soon as it falls. Most coal beds seem to have developed in swamps, where, as soon as a fallen tree becomes waterlogged, it sinks to the bottom and ceases, or nearly ceases, to rot, especially if the water is stagnant as it is in many swamps. The plant material at the bottom of such a swamp becomes *peat.* The peat ultimately becomes coal, largely due to compaction from the weight of overlying sediments. The first stage in this process produces *lignite,* and further pressure produces *bituminous coal* and finally, in rare cases, *anthracite.* (See Fig. 6–15.) In this process the original peat may be compressed to less than one twenty-fifth of its original thickness. To produce a reason-

ably pure coal, little other sediment can be accumulating in the swamp at the same time.

Petroleum also comes from organic material, but the process is not so clear as it is with coal. Oil and gas come from marine sediments. Dark organic-rich shale is the probable source. Such sediments are accumulating today in the Black Sea where the bottom muds contain up to 35 percent organic material compared with about 2.5 percent in ordinary marine sediments. In normal marine sediments the organic material is oxidized, but in stagnant bottom water it can accumulate. The situation is very similar to the formation of coal. As with coal, appar-

ently, deep burial, some heat, and much time are necessary to produce gas and oil. Once formed, the petroleum must migrate to a permeable rock from which it can be extracted.

Petroleum and natural gas move very slowly in the shale source rocks; they are not economically recoverable by drilling until they have migrated to more permeable reservoir rocks. Thus, to have an oil field, it is necessary to have *source rocks, reservoir rocks, and a trap,* generally an impermeable cover. Because petroleum is lighter than water, it rises to the top of a reservoir. Several types of traps are shown in Fig. 6–16. Not all oil forms in marine rocks. The oil shales of the Rocky

Peat bog

Peat

Lignite

Bituminous—bright coal

Coal areas of the United States

Bituminous—cannel coal

Anthracite

Bituminous—splint coal

Coke

FIGURE 6–15 Types of coal. Courtesy U.S. Bureau of Mines.

125

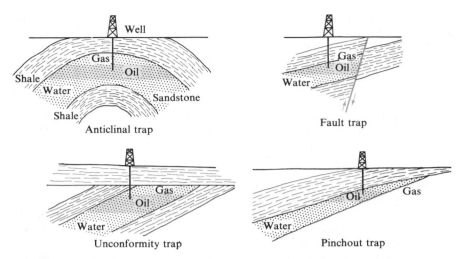

FIGURE 6-16 Several types of oil and gas traps. The terms *anticline*, *unconformity*, and *fault* are discussed in later chapters.

Mountain area, which contain a large percentage of our petroleum reserves, formed in fresh-water lakes, probably by the same process. In oil shale, the petroleum material is disseminated through the shale and has not migrated to a reservoir rock from which it can be easily recovered. The petroleum is recovered by heating the rock, which is a more expensive process than production from an oil well. Oil shale of marine origin occurs in the area from Michigan and western Pennsylvania to Oklahoma, as well as in Alaska and elsewhere.

Although the fossil fuels are our present main source of energy, their future is not bright. The earth's coal supplies are expected to last for two or three centuries, but petroleum will run out in 70 to 80 years. At the present time, about 60 percent of the world's and 67 percent of the United States' industrial energy comes from oil and gas. These figures suggest that many changes are due in the relatively near future.

Natural gas is the cleanest of fossil fuels, and it will probably be the first to run out. Coal, which is the most widely used, releases sulfur and mercury. Sulfur is a major problem in some areas where it causes atmospheric pollution. Use of low-sulfur coal or removal of sulfur from the smoke is required in some cities. Low-sulfur coal is also needed in the iron and steel industries. It is estimated that 2700 kilograms (3000 tons) of mercury are also released from burning coal each year. This is about equal to the amount lost in industrial processes, and over ten times the amount released by natural weathering.

WATER POWER

Water power is an attractive power source because of the lack of atmospheric pollution. Damming rivers, however, spoils some of their aesthetic value, and the lakes formed by the dams will ultimately fill with silt. Transporting the power from a hydroelectric plant requires transmission lines. In spite of these problems, water power will probably be important in Africa and South America because these continents lack large supplies of coal but have large unused amounts of water power. Water power could supply the earth with energy at the present rate of usage, but because of the silting of the lakes, it could only do this for one or two centuries.

SOLAR RADIATION

Solar radiation as a source of energy is attractive because no fuel is used. The disadvantages are that large areas are required and that only certain areas such as deserts receive enough sunlight to make such systems feasible. These disadvantages are offset somewhat because desert areas are largely unused for any other purpose, although many ecologists would disagree because they feel that deserts, like mountain, forest, and beach areas, should be preserved for recreation, ecologic, and aesthetic reasons.

The relatively cloudless deserts of southwestern United States receive about 0.8 kilowatt per square meter for the middle six to eight hours of the day most of the year. A black surface can absorb most of this energy, but a black body is also an excellent radiator. Recently, surface coatings have been developed that allow the sun's radiation to pass through and be absorbed, but these surface films are reasonably opaque to the lower wavelength radiation from the collector. The collector would heat molten salts that would maintain a fairly constant temperature during overnight operation. A large-scale experiment will be necessary to prove the practicality of such a system.

TIDAL ENERGY

Tidal energy can only supply a small amount of our energy needs, and no really practical method has been devised to utilize it. The advantages are no fuel, no waste, no pollution, and minimum ecologic and scenic damage.

GEOTHERMAL POWER

Geothermal power is relatively new in this country but has been used in Italy since 1904. Basically, it is simply the use of steam escaping from the earth to generate power. Hot springs and geysers are fairly common, but at only a few places does steam escape from the earth continuously. Yellowstone National Park is a thermal area where super-heated water or steam could be obtained by drilling. Electric power is being generated in this way at The Geysers in north central California. (See Fig. 6–17.) The water escaping from these thermal areas is mostly rainwater that has been heated, probably by contact with a cooling magma. Isotope studies show that very little of the water is primary water from a magma. In most areas, these thermal waters contain much dissolved material, a fact that is both an asset and a major problem. After being used to generate power, the waters can be evaporated by solar heat to harvest some of the salts. This is a clear gain. The problems are the disposal of the raw or partially harvested waters, and the corrosion and deposition that these waters cause in pipes. It is interesting to note that at some places these waters contain appreciable amounts of metals such as lead, copper, and iron. Thus these hot waters may closely resemble the hydrothermal solutions that form mineral veins. As we learn more about such hot waters and how to harness them, they may become an important power source. We know that the earth's interior is hot, and perhaps some day we can use deep holes to obtain heat from which substantial power can be generated.

The number of volcanic areas where geothermal steam is available is small. The life of a steam well is also limited. The best current estimates suggest that about one-fifth of our present energy needs could be met for about fifty years by geothermal power.

At a few places geothermal power can be obtained from hot, dry rocks. Water pumped into such places is heated to steam, which can be used to generate power. This steam contains much less dissolved material and so

FIGURE 6–17 Geothermal steam escaping from wells at The Geysers, California. Photo by Pacific Gas and Electric Co.

greatly reduces one of the problems of geothermal power.

NUCLEAR ENERGY

Nuclear energy may be the energy source of the future. At present, its use is growing. However, several problems must be overcome to insure its use. The most important are disposal of radioactive wastes, safety, and development of reactors that use the more plentiful ores. Other serious obstacles are public acceptance of nearby nuclear reactors, disposal of unused heat, and assurance that fissionable products of breeder reactors will be used for peaceful purposes.

The elements useful for producing atomic power are uranium and thorium. Thorium occurs mainly in hydrothermal veins. Uranium occurs in hydrothermal veins, pegmatites, and in some sedimentary rocks.

Nuclear energy can be released by fusion and fission. Fusion is the process used in the hydrogen bomb, and so far, no method has been found to release this energy slowly. It remains to be seen if this will ever become a practical power source.

Fission reactors are in current use. Some of these reactors are burners in the sense that they consume their fuel; others are converters or breeders. The difference is of supreme importance for the future. Uranium is the main fuel used in reactors. Natural uranium is a rare element and occurs in three isotopes. Uranium 238 is 99.283 percent, uranium 235 is 0.711 percent, and uranium 234 is 0.006 percent of natural uranium. Of these, uranium 234 is so minor it can be neglected. Uranium 235 is the only naturally fissionable isotope and is used in burner reactors. Almost all of the present reactors are of this type, and they burn uranium 235. There is only enough uranium 235 to last about a century at the present rate of development. It is possible to build breeder reactors that will convert the abundant uranium 238 and thorium 232 into fissionable isotopes, but some uranium 235 is needed for these breeder reactors. Such breeders must be developed soon, or the uranium 235 will be used up and nuclear energy will not be available—and there seems to be no other energy source for the distant future. Breeder reactors can produce the materials used in nuclear weapons so control of their by-products is a serious problem.

Radioactive waste products and other very dangerous wastes present special problems. Some have been dumped at sea in special concrete and/or steel containers, but their potential danger to the whole ocean makes this practice not acceptable to many. The waste from an atomic reactor is about the same amount as the total fuel. It must be held for about 600 years before it is safe. Salt may be very good material in which to bury atomic wastes. It is stable in earthquakes, has a high melting point, and is a good nuclear shield. For these reasons salt domes are currently being investigated as burial sites for radioactive waste.

THE WEEKEND PROSPECTOR

If you want to try prospecting, do it as recreation and not as a way to become rich and famous. Prospecting is fun and may get you into areas that are both beautiful and remote. If you decide to look for gold, remember that almost everyplace has been prospected and mined at least three times: at the time of the initial gold rush; during the period immediately after the rush, when followers worked the low-grade deposits; and later, during the 1930s, when many jobless people fled to the gold fields.

With the recent rise in the price of gold, both amateur and professional gold-seekers have been at work. But gold is difficult to find, and new major discoveries are very unlikely to be made; thus the weekend gold prospector should search near or at those places where gold has been found in the past.

A weekend prospector is not likely to find gold veins in rocks but can recover placer gold. Any river that passes through the old gold camps is likely to have some placer gold. At many places, dredges have worked the rivers, but small placer deposits may still be found. Any place that the river slows is where heavy materials such as gold are likely to be dropped by the river; the outsides of curves and behind obstructions such as boulders are simple examples.

To test such places, the gold pan is an indispensable tool. A shovelful of gravel, preferably from the deepest level of the gravel because the gold is among the first minerals deposited, is panned. The gold pan is swirled around so that the heaviest grains fall to the bottom and the lighter ones are allowed to escape at the sides. This process is continued until only the densest minerals remain in the pan. These minerals are called "black sands" because most, but by no means all, of the dense minerals are black. If any gold is present, it should be obvious in the black sand.

Diving is becoming a common means of recovering gold. It is not a new method of prospecting because it was used by the "forty-niners" in California. Wet suits, scuba gear, and other equipment have become common and have helped to popularize diving for gold. A diver can, using suction equipment, recover the gravel and, hopefully, the gold that is lodged in cracks, joints, and other places. Such cracks and other irregularities on the stream bed,

especially if they are across the flow, are natural traps for gold and other heavy minerals.

Placers are found not only on present rivers but also on ancient river channels. These and other gold deposits are known in some dry areas and are not currently economical because of the lack of water.

Gold, of course, is not the only objective of prospectors. In searching for any commodity, one must be able to identify its minerals and the common rocks with which it is associated. A common approach to prospecting is to use some simple test for the presence of a particular mineral, and, once it is found, trace it uphill to its source. Panning is a classical test; some others are listed later. If the source is a vein deposit that has been exposed to weathering, it may be quite obvious when it is approached. Weathered veins are generally marked by the red-yellow and brown oxides and hydrous oxides of iron, and in dry regions at some places by brilliantly colored minerals.

The gold pan can be used to find deposits of other minerals such as *monazite*, which, because it is dense, is concentrated with the black sand. Monazite is a rare-earth mineral and is radioactive because it contains some thorium. A radiation counter, such as a Geiger counter, will detect black sands containing monazite; and so, in some cases, one can prospect for gold with a Geiger counter.

Modern prospectors use many diverse methods, including geochemical, geophysical, and geobotanical studies. Some of the methods are quite sophisticated, but many are simple. Geochemical tests include wet chemical tests that can be performed in the field, as well as instrument methods, such as spectrograph, X-ray spectroscopy, and radio activation. The idea is that if there is a mineral deposit concealed beneath the soil, its presence can be detected by traces of metals that have "leaked" from the deposit.

Some geophysical methods are also available to the amateur prospector. The use of "black-light" to detect *scheelite*, an ore of tungsten, is an example. Magnetic minerals can be found using the *dip needle*, a sort of compass in which the needle is free to move in the vertical plane. Radiation counters, such as Geiger and scintillation counters, are also in common use, but one must realize that most radiation cannot pass through more than a few inches to a few feet of rock. More specialized geophysical methods are seismic, gravity, and electrical studies.

Plants can also be used in prospecting. Some plants systematically gather certain elements and concentrate them in some part, such as the leaves. In this case, chemical tests of the leaves would reveal the prospect. Some plants have extensive root systems, so they may sample a large volume of soil. Some plants grow best where certain elements are in the soil, and so their presence reveals the element.

The references that follow give more detail on the methods described here. State publications and publications of the U.S. Geological Survey will provide information on mining districts. Many state surveys give various types of help to prospectors. Do not trespass; be sure you know which areas, generally federal land, are open to prospecting. A knowledge of mining law and how to stake a claim can also be obtained from the library.

REFERENCES

H. L. Cannon. "Botanical Prospecting for Ore Deposits," *Science* (1960), Vol. 132, No. 3427, pp. 591–598.

W. B. Clark. "Diving for Gold in California," *California Geology* (November, 1980), Vol. 33, No. 11, pp. 243–249. A publication of California Division of Mines and Geology, Sacramento.

H. E. Hawkes. *Principles of Geochemical Prospecting.* U.S. Geological Survey Bulletin 1000-F. U.S. Government Printing Office, Washington, D.C., 1957, pp. 225–355.

A. H. Lang. *Prospecting in Canada: 3rd ed.* Canada Geol. Survey Economic Geology Series No. 7, 1956, 401 pp.

H. B. Mertie, Jr. "The Gold Pan; a Neglected Geological Tool," *Economic Geology* (1954), Vol. 49, No. 6, pp. 639–651.

H. M. Nakagawa and H. W. Lakin. *A Field Method for the Determination of Silver in Soils and Rocks.* U.S. Geological Survey Professional Paper 525-C. U.S. Government Printing Office, Washington, D.C., 1965, pp. C172–C175.

P. K. Theobald, Jr. *The Goldpan as a Quantitative Geologic Tool.* U.S. Geological Survey Bulletin 1071-A. U.S. Government Printing Office, Washington, D.C., 1957, 54 pp.

S. V. Victorov, Y. A. Vostokova, and D. C. Vyshivkin. *Short Guide to Geo-Botanical Surveying.* New York: Macmillan Company, 1964, 158 pp.

J. F. Walker. *Elementary Geology Applied to Prospecting,* 5th ed. British Columbia Department of Mines, 1955, 196 pp.

F. N. Ward, H. W. Lakin, F. C. Canney, and others. *Analytical Methods Used in Geochemical Exploration by the U.S. Geological Survey.* U.S. Geological Survey Bulletin 1152. U.S. Government Printing Office, Washington, D.C., 1963, 100 pp.

U.S. Geological Survey. *Suggestions for Prospecting,* 1967 0-240-815. U.S. Government Printing Office, Washington, D.C., 24 pp.

SUMMARY

Mineral resources are natural resources that come from the earth, such as metals, energy-producing materials, sand and gravel, building stone, limestone, and so forth.

A mineral deposit is a geologically rare feature; once used it cannot be replaced except for ground water and soil, which require a very long replacement time. Special conditions concentrate the valuable material.

Ore is anything that can be mined at a profit. Economic conditions or technological changes affect which deposit is considered ore. Mineral resources greatly affect the course of history.

Many mineral deposits are associated with plate tectonics. Sulfide minerals of many metals are deposited at some mid-ocean ridges. The igneous rocks associated with convergent plate boundaries commonly contain metal ores.

Disseminated magmatic deposits—the valuable mineral is scattered throughout an igneous body.

Crystal settling—heavy, early-crystallized minerals sink to the bottom of magma in a crystallizing igneous body.

Late magmatic deposits—ore-forming minerals are concentrated by being among the last minerals to crystallize in a magma.

Pegmatite deposits—bodies of late-forming, very large crystals. (Some pegmatite bodies are metamorphic.)

Hydrothermal deposits—these deposits may be from late magmatic fluids, and commonly are metal sulfide minerals in quartz veins. The origin of some hydrothermal-depositing solutions may be circulating ground water and leached metals.

Contact-metamorphic deposits—deposits around intrusive bodies in the zone of metamorphism formed when magma reacts with the surrounding wall rocks (in many cases limestone) and ground water.

Evaporites—chemical precipitates

formed by the sedimentary processes in the evaporation of seawater.

Placer deposits—a sedimentary process in which heavy mineral fragments are sorted and concentrated by rivers and ocean waves.

Deep-ocean deposits—manganese and iron oxides and hydroxides are precipitated from seawater into nodules. It is not yet economic to separate the manganese from the accompanying silica.

Weathering processes provide our most valuable resource, soil. Also, tropical soils (*laterites*) are mined for *bauxite* (aluminum ore), iron, and so forth. Weathering processes also cause local zones of enrichment on other types of mineral deposits, some of which may be deceiving concerning the deposit's value.

The energy sources of the earth are:

Solar energy—sufficient if practical way to use it directly can be devised.

Food—solar energy through plants.

Fossil fuels—small amount of solar energy is stored in coal and petroleum. Supplies greatly limited.

The main *energy resources* are the fossil fuels which originate from once-living material.

Coal forms from accumulation of plant material under conditions which prevent its rot, probably in stagnant swamps. The sequence is *peat,* to *lignite,* to *bituminous coal,* to, in rare cases, *anthracite.*

Petroleum comes from organic-rich marine shales. To be easily recoverable, the petroleum has to migrate to a reservoir rock.

Water power—eventually limited by silting behind dams.

Solar radiation—advantage is no fuel used. Disadvantage is large areas are required; practicality uncertain.

Tidal energy—no practical method yet.

Geothermal power is the use of steam from hot spring and geyser areas to generate power. It is limited to a small number of geothermal steam areas, and steam well's life is limited.

Nuclear energy is produced from thorium (mainly from hydrothermal veins) and uranium (from hydrothermal veins, pegmatites, and in some sedimentary rocks).

Nuclear power is the likely energy source of future. Breeder fission reactors will need to be developed to use the more abundant uranium 238 before the necessary, less abundant uranium 235 is used up in the current burner reactors. Disposal of radioactive wastes is a very difficult problem.

QUESTIONS

1 What are the similarities and differences between mineral resources and other types of natural resources?

2 What types of mineral deposits form at mid-ocean ridges?

3 What types of mineral deposits are directly associated with igneous rocks? Might some of these be found in sedimentary or metamophic rocks?

4 Is the gold pan still a useful tool for the modern prospector?

5 How does weathering form mineral deposits?

6 What types of mineral deposits are associated directly with sedimentary rocks?

7 What parts of the earth are relatively unexplored for mineral resources?

8 Name an ore of lead _____.

zinc _____.

mercury _____.

iron _____.

copper _____.

9 Discuss whether the earth's mineral resources are sufficient for the next hundred years.

10 List the earth's energy sources, and discuss the future of each.

11 What are the problems facing us in disposing of our waste material?

SUPPLEMENTARY READINGS

Axtmann, R. C., "Environmental Impact of a Geothermal Power Plant," *Science* (March 7, 1975), Vol. 187, No. 4179, pp. 795–803.

Barnea, Joseph, "Geothermal Power," *Scientific American* (January 1972), Vol. 226, No. 1, pp. 70–77.

Bonatti, Enrico, "The Origin of Metal Deposits in the Oceanic Lithosphere," *Scientific American* (February 1978), Vol. 238, No. 2, pp. 54–61.

Brobst, D. A., and W. P. Pratt, eds.,

United States Mineral Resources. U.S. Geological Survey Professional Paper 820, Washington, D.C.: U.S. Government Printing Office, 1981, 722 pp.

Cameron, E. N., ed., *The Mineral Position of the United States 1975–2000.* Madison, Wis.: The University of Wisconsin Press, 1973, 100 pp. (paperback).

Cheney, E. S., "The Hunt for Giant Uranium Deposits," *American Scientist* (January–February 1981), Vol. 69, No. 1, pp. 37–48.

Committee on Geological Sciences, National Research Council, National Academy of Sciences, *The Earth and Human Affairs.* San Francisco: Canfield Press, 1972, 138 pp. (paperback).

Dick, R. A., and S. P. Wimpfen, "Oil Mining," *Scientific American* (October 1980), Vol. 243, No. 4, pp. 182–188.

Down, C. G., and J. Stocks, *Environmental Impact of Mining.* New York: Halsted Press, 1977, 371 pp.

Flowers, A. R., "World Oil Production," *Scientific American* (March 1978), Vol. 238, No. 3, pp 42–49.

Harte, John, and Mohamed El-Gasseir, "Energy and Water," *Science* (February 10, 1978), Vol. 199, No. 4329, pp. 623–634.

Hatheway, A. W., and C. R. McClure, eds., *Geology in the Siting of Nuclear Power Plants.* Reviews in Engineering Geology, Vol. IV, Boulder, Colo.: Geological Society of America, 1979, 256 pp.

Heiken, Grant, et al., "Hot Dry Rock Geothermal Energy," *American Scientist* (July–August 1981), Vol. 69, No. 4, pp. 400–407.

Hekinian, R., et al., "Sulfide Deposits from the East Pacific Rise Near 21°N," *Science* (March 28, 1980), Vol. 207, No. 4438, pp. 1433–1444.

Hubbert, M. K., "The Energy Resources of the Earth," *Scientific American* (September 1971), Vol. 224, No. 3, pp. 60–70.

Keyfitz, Nathan, "World Resources and the Middle Class," *Scientific American* (July 1976), Vol. 235, No. 1, pp. 28–35.

Lang, A. H., *Prospecting in Canada,* 4th ed. Canada Dept. Mines and Technology Survey, Geological Survey of Canada, Economic Geology Series No. 7, 1969, 308 pp.

Menard, H. W., "Time, Chance, and the Origin of Manganese Nodules", *American Scientist* (September-October 1976), Vol. 64, No. 5, pp. 519–529.

Menard, H. W., "Toward a Rational Strategy for Oil Exploration," *Scientific American* (January 1981), Vol. 244, No. 1, pp. 55–65.

Park, C. F., Jr., *Earthbound: Minerals, Energy, and Man's Future.* San Francisco: Freeman, Cooper and Co., 1975, 279 pp.

Skinner, B. J., *Earth Resources,* 2nd ed. Englewood Cliffs, N. J.: Prentice-Hall, Inc., 1976, 152 pp. (paperback).

Teller, Edward, *Energy from Heaven and Earth.* San Francisco: W. H. Freeman and Co., 1979, 287 pp.

Theobald, P. K., S. P. Schweinfurth, and D. C. Duncan, *Energy Resources of the United States.* U.S. Geological Survey Circular 650, Washington, D. C.: U.S. Government Printing Office, 1972, 27 pp.

Titley, S. R., "Porphyry Copper," *American Scientist* (November–December 1981), Vol. 69, No. 6, pp. 632–638.

7 DOWNSLOPE MOVEMENT OF SURFACE MATERIAL

A landscape, to James Hutton, was not a given thing, shaped once and forgotten, but rather a page from a continuing biography of the planet.

Loren Eiseley
Darwin's Century
Chapter IV

The outer few miles of the earth are called the crust. We will define the term later. Although the crust is the only part of the earth accessible to direct observation, studying the earth by observations in the crust is like trying to read a book by its covers. Actually, the problem is even greater than this because 71 percent of the surface is covered by oceans.

Classical geology developed from observations made on the continents. One of the most fundamental ideas or laws to have been developed in *uniformitarianism,* which says that *the present is the key to the past.* In other words, any structure in old rocks must have been formed by processes now going on upon the earth. Although this simple statement of uniformitarianism has limitations, it was very important to the development of geology. Uniformitarianism was not always accepted. We see erosion going on around us, note the material carried by muddy rivers, and conclude that rivers eroded the valleys in which they flow. Not too many years ago some geologists believed that valleys, especially deep valleys such as Yosemite, were formed by great earthquakes which split open the earth and that the rivers then flowed into them simply because they were the lower areas. This theory is called *catastrophism.*

Uniformitarianism implies slow processes going on over great lengths of time. Rapid processes, such as earthquakes, landslides, and volcanic eruptions, do occur, but, in general, do so at widely separated times so that their average rate is slow.

EROSION—THE CONSTANT SURFACE PROCESS

The main process going on today on the earth's surface is clearly erosion and is shown by gullies on hillsides, landslides, and similar examples. The agents of erosion, of which rivers are far and away the most important, are constantly wearing down the continents; nevertheless, we still have mountains. The present rate of erosion is such that, theoretically, all topography above sea level will be removed in about 12 million years. However, we know that erosion and deposition have been going on for a few thousand million years. Therefore, something must interrupt the work of the rivers and uplift the continents. Thus, much of geology is concerned with the struggle between the external forces of erosion and the internal forces which cause uplift.

Some familiar demonstrations of the internal forces which cause uplift are earthquakes, volcanoes, and folded and tilted sedimentary rocks that are now high on mountains but were originally formed as flat beds at or below sea level. These internal processes will be discussed later. Physical or dynamic geology is the study of this battle between the internal force of uplift and the external force of erosion. We will consider these two forces separately to eliminate confusion, but remember that they go hand in hand.

The underlying cause of erosion is downslope movements under the influence of gravity; the agents that cause such movements are water (rivers, and so forth), ice (glaciers), and wind. The material that is moved by these agents comes from the weathering processes discussed earlier.

The sun's energy is the ultimate source of the energy used in erosion, as shown in Fig. 7-1. This diagram also shows that erosion is a necessary byproduct of the distribution of the sun's energy around the earth. Without this distribution of energy by the oceans and the atmosphere, the equatorial regions would become hotter and the poles colder because most of the sun's energy falls near the equator.

IMPORTANCE OF DOWNSLOPE MOVEMENT

All of the surface material of the earth tends to move downslope under the influence of gravity. *The surface material is more or less weathered and is generally termed overburden to differentiate it from bedrock.* These downslope movements shape the earth's surface and so must be understood to interpret landscapes. The downslope movement generally ends at a river or stream that carries the material away. Thus, rivers and downslope movements work together to shape the landscape. A small example of this, discussed in the following chapter on rivers, shows that although a river downcuts to form a valley, the shape of the valley sides is determined by

Urban landslide in Oakland, California. The slide developed after heavy rains. Only a small part of the slide is shown.

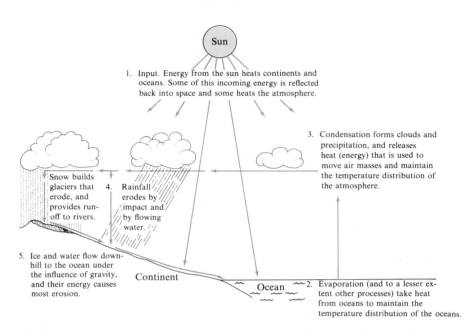

FIGURE 7-1 The water cycle and the earth's energy balance. Incoming energy from the sun is distributed around the earth by the circulation of the oceans and the atmosphere. The water cycle is part of the energy exchange between the oceans and the atmosphere, and it is the immediate source of the energy used in most erosion.

downslope movements, which widen the narrow cut made by the river. (See Fig. 8–11.) The efficiency of downslope movements is easily shown by analysis of slopes. Such analysis shows that most of the earth's surface slopes less than 5 degrees and that very little of the earth's surface slopes more than 45 degrees. Many forms of downslope movement are slow processes. However, when slopes are disturbed by buildings, roads, or irrigation, more rapid movements may be triggered, causing much damage.

FIGURE 7-2 Urban landslide in Oakland, California. This slide developed after heavy winter rains. Construction on the top of the hill may have altered the infiltration of rainwater. Only a small part of the slide is shown in the photo.

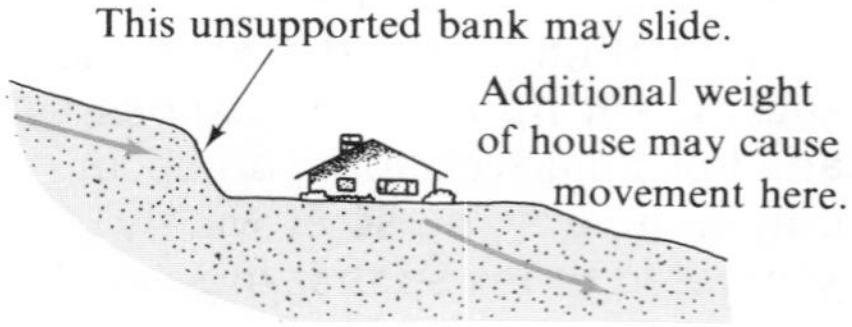

(See Fig. 7–2.) Thus, knowledge of these processes is important when any type of engineering project is considered on even moderately sloping ground. (See Fig. 7–3.)

TYPES OF DOWNSLOPE MOVEMENT

Downslope movements may be rapid or very slow; and they may involve only the very topmost surface material, or they may be massive movements that involve the total thickness of overburden or even bedrock. The processes at work on any slope depend on many factors. On steep slopes no overburden generally can develop; therefore, the dominant process on such slopes is *rockfall*, and the slope that develops at the foot is called a *talus*. (See Fig. 7–4.) On less steep slopes—how much less depends on the climate which controls the weathering processes—overburden may develop, and this material may move downslope in a number of ways which will be discussed. Thus weathering, which is controlled by the climate, determines the type of downslope movement, thereby causing the different aspects of the topography in a humid area and a dry area. (See Figs. 7–5 and 7–6.) The shapes of slopes and the theories of how these shapes are developed will be discussed in Chapter 8. The structure and rock types on a hill may also control the erosion processes. Differences in rate or kind of weathering can determine the amount of material available for downslope move-

This unsupported bank may slide.

Additional weight of house may cause movement here.

FIGURE 7-3 Hillside construction can cause earth movements if not properly planned.

GRADUALISM AND RARE, RAPID EVENTS

All geology textbooks describe uniformitarianism as implying slow processes that act over long periods of time. On the other hand are the rare, rapid, and large-scale geologic events that capture our attention in newspapers and television reports. Both are clearly normal geologic processes; the link between them is the immense length of geologic time.

Some examples of rare, rapid, and, at places, large-scale geologic events are earthquakes, volcanic events, landslides, and floods; these are rapid and catastrophic. In the early days of geology, one of the big arguments was between the uniformitarianists (who were gradualists) and the catastrophists. The uniformitarianists believed, for example, that river valleys were carved by the slow erosion by the river. The catastrophists believed that river valleys were formed by rending of the earth by huge earthquakes or the like. They used valleys such as the Grand Canyon and Yosemite (described in Chapter 10) as examples.

The problem of gradual versus rapid geologic events disappears if the results are averaged over a long period of time. Gretener uses the example of the Flims slide in the Swiss Alps. This is the largest known landslide, and it occurred in prehistoric time. About 12 billion (12,000 million) cubic meters of material moved in that slide. The area is drained by the Rhine River, and the Rhine deposits about a million cubic meters of sediment in Lake Constance each year. Thus, if all of the slide reached the Rhine, in a few to about ten thousand years the whole slide would be removed. So if, in this part of the Alps, such slides occur every few thousand years, which is very likely, they could account for much of the erosion of the area. Such an event would be considered catastrophic and rare by human life spans, but could be normal in terms of geologic time. Even though the slides are huge, if they occur only at long intervals, the erosion rate is, on the average, slow. Gradualism can be either a very slow process acting over a long time, or big, rapid events that occur at long intervals.

Other, less spectacular examples of rapid geological events are floods and the retreat of cliffs, especially coastal cliffs. In most cases, these events occur at long intervals when judged by human life spans. That floods are normal events on rivers is shown in Chapter 8. In Chapter 9, the point will be made that flash floods in deserts are rare, but they are the main cause of erosion there. In Chapter 13, earthquakes are described; they can be rapid, catastrophic events. The overall movement on the San Andreas fault in California is estimated at several hundred miles. The biggest well-documented movement on that fault is 20 feet that occurred in 1906 in the San Francisco earthquake. Clearly the total displacement is caused by many earthquakes, each with small movement.

REFERENCES

P. E. Gretener, "Significance of the Rare Event in Geology," *Bulletin,* American Association of Petroleum Geologists, 1967, Vol. 51, pp. 2197–2206.

Donald H. Zenger, "The Role of Rapid Events in Earth History," *Journal of Geological Education,* 1970, Vol. 18, pp. 42–43.

ment, and so the underlying bedrock structure may be etched in relief.

EROSION BY RAIN

Falling rain erodes surface material in two ways: by impact and by runoff. The impact of raindrops loosens material and splashes it into the air (see Fig. 7–7), and on hillsides, some of this material falls back at a point lower down the hill. Raindrops hit the ground with a velocity of about 9 meters (30 feet) per second. The resistance of the air controls this velocity. Two and five-tenths millimeters (0.1 inch) of rain amounts to about 11 tons of water on each acre, and the energy of this water is about 75,000 joules (55,000 foot-pounds) per second. Ninety percent of this great amount of energy is dissipated in the impact. Ninety percent of the impact splashes are to a height of 0.3 meter (1 foot) or less, and the lateral splash movement is about four times the height. Because impact erosion is most effective in regions that have little or no grass or forest cover, it is most effective in desert areas (or newly plowed fields) that are subjected to sudden downpours such as thunderstorms. Splash erosion accounts for the otherwise puzzling removal of soil from hilltops where there is little runoff. It can also ruin soil by splashing up the light clay particles that are then carried away by the runoff, leaving the silt behind.

The rainwater that does not infiltrate into the ground runs down the hillside and erodes in the same ways that river water erodes. This water may gather in lower places and cut gullies. The amount of erosion caused in these ways depends on the steepness of the slope and on the type and amount of vegetative cover. The effects of such erosion are best seen in areas where the vegetation has been removed by fires or humans.

MASSIVE DOWNSLOPE MOVEMENTS

The massive types of downslope movement that involve appreciable thicknesses of material are creep, earthflow, mudflow, and landslide.

A number of names are applied to these movements as described in the next few pages. They are termed *flows if they move as viscous fluids*, *slides if they move over the surface*, and *falls if they travel in the air in free fall*.

Creep is the slow downslope movement of overburden and, in some cases, bedrock. It is recognized by tilted poles and fence posts if the movement is great enough. Commonly, in such cases, trees will be unable to root themselves, and only shrubs and grass will grow on the slope. In other cases, where the creep is slower, the trunks of trees will be bent as a result of the slow movement. (See Figs. 7–8 and 7–9.)

If the amount of water in the overburden increases, an earth-

FIGURE 7–4 A talus develops at the foot of a steep slope as a result of mechanical weathering as shown in this view of Bryce Canyon National Park, Utah. Photo from Union Pacific Railroad.

HUMID   ARID

Shale
Sandstone
Shale
Limestone
Shale
Sandstone
Shale

FIGURE 7–5 Idealized erosional forms in humid and arid areas.

136

FIGURE 7–6 The Appalachian Mountains in North Carolina. Compare the rounded hillsides shown here with the angular topography of arid areas shown in Fig. 7–4. Photo by A. Keith, U.S. Geological Survey.

flow results. In this case a tongue of overburden breaks away and flows a short distance. An earthflow differs from creep in that a distinct, curved scarp is formed at the breakaway point. (See Fig. 7–10.)

With increasing water, an earthflow may grade into a *mudflow. The behavior of mudflows is simi-*

FIGURE 7–7 Erosion by raindrop. From U.S. Department of Agriculture, 1955 Yearbook, *Water.*

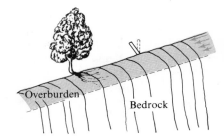

FIGURE 7–8 The effect of creep on fence posts and trees. This is caused by the slow downslope movement of the overburden.

lar to that of fluids. Rain falling on the loose pyroclastic material on the sides of some types of volcanoes produces mudflows, and this is a very important mode of transportation of volcanic material. Another type of mudflow is common in arid areas. In this case, a heavy thunderstorm produces large amounts of runoff in the drainage area of a stream. The runoff takes the form of rapidly moving sheets of water that pick up much of the loose surface material. Because

these sheet floods flow into the main stream, all of this muddy material is concentrated in the main stream course. As a result, a dry stream bed is transformed very rapidly into a flood. This flood of muddy material is a mudflow, and it moves very swiftly, in some cases with a steep, wall-like front. The thick mud can carry large boulders, and such a mudflow can cause much damage as it flows out of the mountains. Eventually, loss of water (generally by percolation into the ground) thickens the mudflow to the point that it can no longer flow. (See Fig. 7–11.)

Another type of mudflow, called *solifluction, occurs especially when frozen ground melts from the top down, as during warm spring days in temperate regions or during the summer in areas of permafrost.* (See Fig. 7–12.) *In this case the surface mud flows downslope.* Solifluction causes many problems in construction, especially in the far north areas of permafrost. *Permafrost areas, as the name implies, are areas where the deeply frozen ground does not completely thaw during the summer.*

FIGURE 7–9 Creep in slate at Resurrection Bay, Alaska. Photo by F. H. Moffit, U.S. Geological Survey.

FIGURE 7–10 Earthflows on a farm in eastern Washington. Note the scarps at the breakaway points. The left flow should probably be classified as a mudflow. Photo from U.S. Department of Agriculture, Soil Conservation Service.

FIGURE 7–11 Mudflow in the San Juan Mountains, Colorado. Its source can be seen on the skyline, and it ends in a lake. The trees growing on the mudflow show its age. Photo by W. Cross, U.S. Geological Survey.

Another type of movement of soil material, called *frost heaving,* is associated with freezing and thawing. In this case *the expansion of water as it freezes pushes pebbles upward through the soil or overburden.* This phenomenon is well known to farmers in regions such as New England where each spring fields are covered with pebbles that were not visible in the fall. In the Middle Ages peasants thought that the stones grew in the soil just as crops do. Frost heaving is a complex process, not yet fully understood. The small expansion of water during freezing is not enough to account for the several inches of uplift of pebbles or fence posts common in areas of frost heaving. Most of the soil water is in the pore spaces of the fine material. When this water freezes, it expands and so pushes nearby pebbles upward. During the process of freezing, soil moisture migrates to these pockets of fine material and forms lenses of ice; it is this additional volume of ice that causes the several inches of uplift. When melting occurs, the surface tension of the water in the fine sediments pulls the fine material together, leaving the larger pebbles at nearly the height to which the expansion pushed them.

In very cold areas, frost heaving causes *stone rings* or *polygons.* (See Fig. 7–13.) In such areas, the extreme cold causes contraction cracks to form at the surface. In summer, meltwater fills the cracks, and frost heaving proceeds as outlined above. Pebbles rise to the surface, forming the polygons.

Landslides are rapid slides of bedrock and/or overburden. Two types involve bedrock: the *landslide* or *rockslide,* and the *slump.* *Rockslides develop when a mass of bedrock breaks loose and slides down the slope* (Fig. 7–14). In many cases the bedrock is broken into many fragments during the fall, and this material behaves as a fluid, spreading out in the valley. It may even flow some distance uphill on the opposite side of the valley if the valley into which it falls is narrow enough. Such landslides are sometimes called *avalanches,* but this term is best reserved for snowslides. Landslides are generally large and destructive, involving millions of tons of rock. As suggested in the diagrams, landslides are apt to develop if planes of weakness, such as bedding or jointing, are parallel to a slope, especially if the slope has been undercut by a river, a glacier, or people. Thus, landslide danger can be evaluated by geologic study. *Slides involving overburden are called debris slides.*

Slumps tend to develop in cases where a strong, resistant rock overlies weak rocks. Note in Fig. 7–15 the curved plane of slippage,

138

the reverse tilt of the resistant unit that may provide a basin for a pond to develop, and the rise of the toe of the slump, which is most pronounced where the underlying rock is very weak and can flow plastically. Unlike rockfalls, slumps develop new cliffs nearly as high as those previous to the slump, setting the stage for a new slump. Thus slumping is a continual process, and generally, in areas of slumping, many previous generations can be seen far in front of the present cliffs. (See Fig. 7–16.)

Liquefaction Slides associated with the 1964 Alaskan earthquake show two other factors. (See Figs. 7–17, 7–18, and 7–19.) Slides may occur under water and create destructive waves, commonly called *tidal waves,* but better called *seismic sea waves* because they have nothing to do with tides. (See Chapter 13.) The Alaskan slides also illustrate sliding by a peculiar type of clay that behaves as a fluid, or liquefies, when disturbed. This type of clay is termed *quick clay;* it is generally formed from the very fine products of glacial erosion and so is found mainly at northern latitudes. These clays contain much water, commonly 50 percent or more by weight, and the tiny clay platelets are randomly oriented. Because most quick clays were deposited in the ocean, the interstitial water is saline. The salt ions in this water help to keep the clay particles stuck together, but if fresh ground water moves through the clay, flushing out the salt, the clay becomes unstable. Vibrations may then squeeze the water out, bringing the clay platelets into contact and making the mixture a fluid. (See Fig. 7–20.)

Another example will show how liquefaction occurs. Sediments that are not compacted, that is, have open packing (see Fig. 7–21A), may liquefy if they are nearly saturated with water and the water cannot drain away. If we

A

B

FIGURE 7–12   Effects of permafrost. A. Solifluction probably caused the smooth slopes in the middle ground of this view of the Sierra Nevada in California. B. Melting of permafrost caused the distortion of this building in Alaska. Photo A from U.S. Forest Service, B by T. L. Péwé, U.S. Geological Survey.

assume that a sediment is composed of uniform spheres, when it is in open packing, it has almost 48 percent open space, and this space is filled with water. Vibration, as in an earthquake, can cause the spherical grains to move into close packing (see Fig. 7–21B). In close packing there is only 26 percent open space, so almost half of the water (22 percent by volume of the original sediment) is ejected from the sediment. If this water cannot drain away, the sediment liquefies or becomes a

mud with no strength, so it flows like a fluid. Of course, few sediments are composed of uniform spheres, but many sand deposits can liquefy if they meet the other requirements—open packing and saturation. Many stream-deposited sediments meet these requirements.

TRIGGERING

Massive downslope movements may be triggered in many ways.

139

A

B

FIGURE 7-13   Stone polygons formed by frost action. A. Close-up view in Alaska. Photo by H. B. Allen, U.S. Geological Survey. B.Oblique aerial view near Churchill, Manitoba. Photo from Department of Energy, Mines and Resources, Ottawa. Canadian Government Copyright (A 1426–66).

A

FIGURE 7-14   Landslide near Hebgen Lake, Montana. A. The slide was triggered by an earthquake and came from the left, leaving a large scar on the mountain side. The debris traveled uphill on the right side of the valley. The slide dammed the Madison River, creating the lake in the foreground. The light-colored area on the slide is a cut made to drain the lake partially. Photo by J. R. Stacy, U.S. Geological Survey. B. This cross-section of the slide was drawn looking in the opposite direction from A. From J. B. Hadley, U.S. Geological Survey Professional Paper 435, 1964.

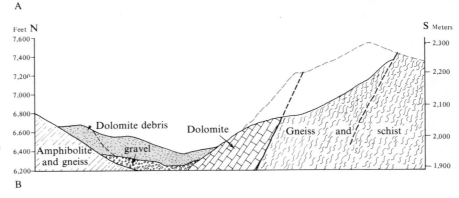

B

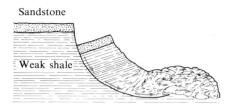

Sandstone

Weak shale

FIGURE 7–15 Slump block showing curved plane of slippage and reverse tilt of the slump block.

FIGURE 7–16 Slump formed in a weak shale in South Dakota. The slump is about 140 meters (450 feet) wide. Photo by D. R. Crandell, U.S. Geological Survey.

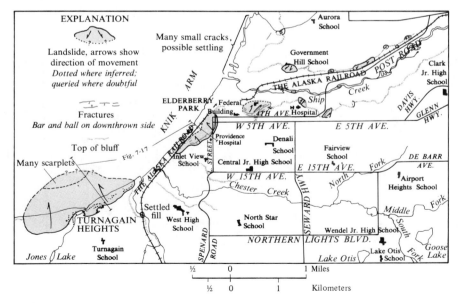

FIGURE 7–17 Map of Anchorage, Alaska, showing the location of slides caused by the Good Friday Earthquake of 1964. Cross-sections of the L Street and Turnagain Heights slides are shown in Fig. 7–18. From Arthur Grantz and others, U.S. Geological Survey Circular 491, 1964.

141

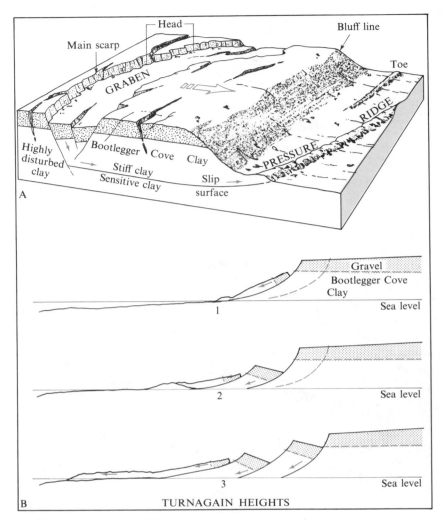

FIGURE 7–18 Diagrams showing A., L Street slide, and B., probable development of Turnagain Heights landslide. See Fig. 7–17 for locations. A graben is a down-dropped block. A from W. R. Hansen, U.S. Geological Survey Professional Paper 542–A, 1965; B after Arthur Grantz and others, U.S. Geological Survey Circular 491, 1964.

A   B

FIGURE 7–19 Landslides at Anchorage, Alaska, caused by the 1964 earthquake. See Fig. 7–17 for locations. A. Looking west at Government Hill School. Note the slumping. B. Aerial view looking south at the hospital on Fourth Avenue. Photos by W. R. Hansen, U.S. Geological Survey Professional Paper 542–A, 1965.

142

FIGURE 7–20 A dramatic example of quick clay. On the left an undisturbed sample of quick clay supports 11 kilograms (24 pounds). On the right another sample of the same quick clay pours like a liquid after being stirred. Photo from C. B. Crawford, National Research Council of Canada.

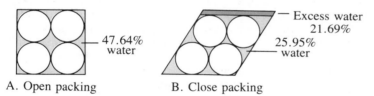

A. Open packing   B. Close packing

FIGURE 7–21 Liquefaction can occur if saturated open-packed sediments are deformed, causing close packing.

FIGURE 7–22 Rockslide caused by undercutting during construction near Quicksand, Kentucky. Photo by J. W. Huddle, U.S. Geological Survey.

Conical pile of dry sand
Angle of repose is about 31°.

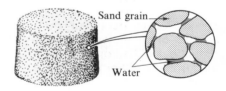

Moist sand can form nearly vertical walls.

Expanded view showing how surface tension of the pore water holds the sand together.

FIGURE 7–23 The effect of water on the cohesion of sand.

The most common are (1) undercutting of the slope (see Fig. 7–22), (2) overloading of the slope so that it cannot support its new weight and, hence, must flow or slide, (3) vibrations from earthquakes or explosions that break the bond holding the slope in place (see Figs. 7–14 and 7–18), and (4) additional water. The addition of water is generally seasonal, and this is why many newly made cuts stand until the following spring. The effect of the water is twofold: it adds to the weight of the slope, and it lessens the internal cohesion of the overburden. Although its effect as a lubricant is commonly considered to be its main role, this effect actually is very slight. The main effect is lessening the cohesion of the material by filling the spaces between the grains with water. A simple experiment with sand will illustrate this point. Dry sand can be piled only in cones with slopes slightly over 30 degrees. Damp sand, however, can stand nearly vertically because the small amounts of moisture between the grains tend to hold the grains together by surface tension. (See Fig. 7–23.) Additional water completely fills the intergrain voids; thus, there is no surface tension to hold the grains together, and a mud forms that flows outward as a fluid.

Certain clay minerals that occur in some soils increase in volume when water is added to them. Adding water to slopes composed of such swelling clays can trigger landslides.

Recognition of any of the types of mass wasting can be of great economic importance when contemplating any construction. The history of many housing developments suggests that even house or lot buyers could benefit by the ability to recognize them. Generally, mass movements are best detected by observation from a distance followed by detailed close study. Generally, a slope in motion will have few trees and may have a bumpy, hummocky appearance. Because some types of glacial deposits also have this appearance, one should know how to recognize glacial topography, which will be discussed in Chapter 10.

STABILIZING SLOPES

Landslides can be stabilized in a number of ways. The most important point to make is that all of these methods are done much more economically during the initial construction rather than as a first-aid measure later on. In many cases it is possible, with very minor redesign of any engineering project, to avoid potential landslide areas. One very common misconception is that landslides can be stabilized with planting. The author has personally met a number of homeowners in landslide areas in California who felt that growing either grass or ice plant would stabilize a landslide area. This is a complete misconception, and there are numerous examples of very large trees, even forests, moving as parts of landslide blocks. Many landslides move at depths greater than the normal root depth of most trees.

The main methods of stabilizing landslides are excavation, drainage, construction of retaining walls or similar structures, the use of rock bolts, and the injection of grout (cement) into the slide. Highways avoid landslides by bridging over the unstable areas. Most common of all these methods is to drain the landslide. Generally this is done by drilling holes into the slide and providing perforated pipes through which the wa-

ter can escape. In some cases this is coupled with an impervious layer over part of the landslide to prevent the entrance of surface water into the slide. In a few cases the slides are actually triggered by blasting to stabilize the area. The quick clays (see Fig. 7–20) of Anchorage were subjected to both blasting and vibration in an effort to stabilize them after the 1964 earthquake. Some clay soils can be stabilized by adding chemicals.

LANDSLIDES AND THE URBAN ENVIRONMENT

Downslope movements range from rapid, dramatic events that kill hundreds and cause millions of dollars damage all the way to slow, almost imperceptible creep. Downslope movements cost hundreds of millions of dollars each year in direct damage, and by delays because of blocked roads and railroads; but, perhaps, more important to the average person, they commonly cause personal financial disaster if a new home is destroyed.

It is difficult to protect oneself from financial loss due to landslides. Landslides are typically excluded from homeowners' insurance, as is earthquake damage. Many policies also exclude damage due to subsidence. These coverages are available, but generally at high cost, and should be investigated by the homeowner.

In many cases the sale of a home is voided if knowledge of hazardous conditions is concealed, but, in general, real estate sales are *caveat emptor*—let the buyer beware. Getting the opinion of a consulting geological engineer is an approach in some areas. The problem with this approach used to be finding a competent person or firm; but licensing laws in several states have made this easier, although many licensed engineers who can legally give such advice know very

little about geology. The real problem is that consultants are vulnerable to law suits if their advice is in error; and with the recent rise in malpractice suits in all professional fields, consultants find it almost impossible to get insurance against such suits. Another problem the consultant faces is that he might also be sued by a landowner whose property loses value because of an inaccurate report that it has hazardous geologic conditions. It is difficult to give a geologic opinion on the stability of a single lot because much larger areas have to be studied to determine the geologic conditions. This means that consultants have to charge almost as much to examine a single lot as they do for a whole tract. Drilling and laboratory tests may be needed as well. Thus, the individual cannot do much to protect himself, but must look to government agencies for protection.

A few cities have laws designed to protect the home buyer. The Los Angeles building codes are commonly cited as examples for the rest of the nation. Los Angeles is endowed with many natural attractions but with many problems as well. Legislation dealing with earthquake hazards and air pollution also began in Los Angeles and has served as a model for the rest of the country. The landslide problem became acute during the very heavy rains of 1952, in part because population pressures and the desire for property with a view led to building homes in the foothills surrounding the city. The January 1952 rains caused about $7.5 million damage in Los Angeles. The building codes enacted before the following winter rains virtually eliminated high, steep cuts and fills, required permits for all hillside grading, and required inspection and geological reports. In addition, drainage has to be conducted to streets and away from cut and fill slopes, and erosion must be controlled by plantings or other devices.

Proper *compaction* is necessary to prevent fills from settling. Figure 7–24 shows subsidence of an area in San Francisco that was built on filled ground over an old swamp. The city has raised the grade of the streets so that the streets are halfway up what was the first floor of the houses.

FIGURE 7–24 Subsidence. The street was regraded and raised after the area subsided so that the garage of this house is unusable.

SUMMARY

Erosion is the main process going on on the earth's surface. Rivers are the most important agents. Internal forces uplift the surface so that erosion does not completely level the continents. The underlying cause of erosion is downslope movements under the influence of gravity.

Overburden is the more or less weathered surface material, as opposed to *bedrock.*

On steep slopes no overburden generally develops. *Rockfall* occurs, producing a slope called a *talus* at the bottom.

Weathering, controlled by climate, determines the amount of overburden that develops and the type of downslope movement, thereby causing the different aspects of topography in humid and dry areas. Structure and rock types also control erosion processes.

Falling rain erodes by *impact,* moving material by splashing, and by *runoff* of the rainwater that does not infiltrate the ground.

Types of downslope movement are *flows* if they move as viscous fluids, *slides* if they move over the ground, and *falls* if they travel in the air in free fall.

Creep is the slow downslope movement of overburden and, in some cases, bedrock.

With more water in the overburden, an *earthflow* results, forming a distinct curved scarp at the breakaway point.

With increasing water, a *mudflow* results. Rain on loose pyroclastic material on a volcano, or a heavy thunderstorm in an arid area can produce mudflows.

Solifluction is a type of mudflow that occurs when frozen ground melts from the top down so that the surface mud flows downslope.

Frost heaving results from the expansion of freezing water which

pushes pebbles upward through the soil. In cold regions, frost heaving forms stone rings or polygons.

Landslides are rapid slides of bedrock and/or overburden. *Landslides* or *rockslides* involve bedrock; a mass of bedrock breaks loose and slides down the slope. The material is often fragmented and behaves like a fluid in the valley, and the slide is often large and destructive. Although landslides are sometimes called *avalanches,* that term is best reserved for snowslides. Slides involving overburden are called *debris slides.*

Slumps are also rapid slides of bedrock and/or overburden. They tend to develop where a strong, resistant rock overlies weak rocks, and slip down a curved plane, tilting the top unit back, and creating a new cliff that will itself slump.

The most common triggers to massive downslope movements are

145

(1) undercutting of the slope by rivers or humans, (2) overloading the slope so that it cannot support its new weight, (3) vibrations from earthquakes or explosions, and (4) additional water which adds to the weight of the slope and lessens the internal cohesion of the material by filling the pore spaces.

Quick clay is a type of clay that behaves as a fluid when disturbed. A saturated, open-packed sandstone may liquefy if disturbed.

Landslides can cause great damage. Landslide potential should be evaluated before engineering projects are done. The main methods of stabilization are draining, excavating, construction of retaining walls or similar structures, rock bolts, injection of grout into slide, or controlled triggering by blasting.

QUESTIONS

1 Erosion tends to level the land surfaces. Why aren't the continents worn flat?
2 Describe the slow means of downslope movement.
3 Describe the rapid means of downslope movement.
4 Of what importance is raindrop impact in erosion?
5 How might rapid mass movements be triggered? What is the role of water in triggering?
6 In buying a hillside home, how can you protect yourself against possible landsliding?
7 Describe two ways that liquefaction can occur.

SUPPLEMENTARY READINGS

Ellison, W. D., "Erosion by Raindrop," *Scientific American* (November 1948), Vol. 179, No. 5, pp. 40–45. Reprint 817, W. H. Freeman and Co., San Francisco.

Kerr, P. F., "Quick Clay," *Scientific American* (November 1963), Vol. 209, No. 5, pp. 132–142.

Sharp, R. P., and L. H. Nobles, "Mudflow of 1941 at Wrightwood, Southern California," *Bulletin,* Geological Society of America (May 1953), Vol. 64, No. 5, pp. 547–560.

Sharpe, C. F. S., *Landslides and Related Phenomena.* New York: Columbia University Press, 1938, 137 pp.

Voight, Barry, ed., *Rockslides and Avalanches.* Part 1, "Natural Phenomena." New York: Elsevier, 1978, 834 pp.

8 RIVERS AND DEVELOPMENT OF LANDSCAPE

In the space of one hundred and seventy-six years the lower Mississippi has shortened itself two hundred and forty-two miles. This is an average of a trifle over one mile and a third per year. Therefore, any calm person, who is not blind or idiotic, can see that in the Old Oölitic Silurian Period, just a million years ago next November, the Lower Mississippi River was upwards of one million three hundred thousand miles long, and stuck out over the Gulf of Mexico like a fishing rod. And by the same token any person can see that seven hundred and forty-two years from now the Lower Mississippi will be only a mile and three quarters long, and Cairo and New Orleans will have joined their streets together, and be plodding comfortably along under a single mayor and a mutual board of aldermen. There is something fascinating about science. One gets such wholesale returns of conjecture out of such a trifling investment of fact.

Mark Twain
Life on the Mississippi
Chapter 17

Rivers are agents of erosion, transportation, and deposition. That is, they carve their own valleys and carry the eroded material downstream where it is either deposited by the river or delivered into a lake or the ocean. Rivers are the most important agents in transporting the products of erosion; they drain vastly larger areas than do glaciers, and, even in their infrequent times of flowage on deserts, they probably are able to carry more material than the wind can. Thus, it is the rivers that receive and transport the products of weathering that are fed into them by runoff from precipitation as well as by downslope movements. Weathering, downslope movement, and rivers work together to shape the landscape. Although the geologic role of rivers has been known for a long time, rivers are complex systems with many variables, and much remains to be learned about them. Even more factors are involved in the development of landscape,

and this is one of the active fields of research in geology.

Before Hutton's time very little thought was given to erosion and landforms. A few writers did note the erosion and deposition done by rivers, repeating the observation of the Nile made by the ancient Greek Herodotus about 500 B.C. One reason for the lack of interest was that most people believed in a strict interpretation of the Bible. In 1654 Bishop Ussher, from a study of Genesis, said that the earth was created on the night before Sunday, October 23, 4004 B.C., and that the Flood had begun December 7, 2349 B.C. A few thousand years is not enough time for a slow process like erosion to modify the earth's surface to any great extent. As a result, most people thought that the earth was either in the perfect condition in which it had been formed by the Creator, or that it had been modified by the Flood or by earthquakes. In the eighteenth century the viewpoint changed somewhat,

and near the end of that century Hutton argued that the slow process of erosion was accompanied by deposition, thus perpetuating the land. Change instead of destruction was Hutton's concept; a benign Deity, instead of a wrathful God, was a current theme. This was the intellectual climate in which modern geology developed.

In the nineteenth century, little more was done in studies of erosion until the last half of the century. The intervening years were active ones for geology, and this was when the geologic time scale was developed, a topic covered in Chapter 16. The scene then shifted to the United States. After the Civil War the West was opening up, and part of the westward interest was geologic exploration. Note that this was the same time that Africa was being explored as described in Chapter 1. On May 24, 1869, at Green River Station, Wyoming, John Wesley Powell, a one-armed Civil War veteran, and nine companions started on an epic

Erosion of plowed field by runoff of rainwater. Photo from U.S. Department of Agriculture, Soil Conservation Service.

boat trip that was to take them more than one thousand miles through the Grand Canyon to Virgin River, Arizona. Powell became a national hero and returned several times to study the Colorado Plateau. From these geological studies, Powell and Grove Karl Gilbert described the erosion and the work of the rivers. The Colorado Plateau, with its flat-lying sedimentary beds, open vistas, and deep canyons, is an excellent place to appreciate the effectiveness of erosion. The work of these men was organized by William Morris Davis in the years between 1880 and 1934 into what he called the cycle of erosion. This concept and the criticism of it will be discussed in this chapter.

FLOW OF RIVERS

Rainfall and other forms of precipitation are the source of river water. Only a small fraction of the total precipitation forms the rivers, but the amount of water is great. The fate of most of the rest of the precipitation is described in Chapter 12. The energy of rivers comes from the downward pull of gravity acting on the river water. In physics this is called *potential energy* and is equal to the mass times the acceleration of gravity times the height (mgh). Table 8–1 shows the tremendous amount of energy theoretically available from rivers. Only a small fraction of this energy is actually used to do geologic work. If this were not so, the landscapes of the earth would be far different. To see where the potential energy of rivers goes, and to gain some insight into river processes, the velocity of river flow must be considered.

The average velocity of river flow is less than 8 kilometers (5 miles) per hour, and most rivers are in the range from a fraction of a kilometer per hour to about 32 kilometers (20 miles) per hour. [Twenty-four kilometers (15 miles) per hour is the fastest ever mea-

TABLE 8–1 Calculation of approximate amount of energy available from rivers and the theoretically possible average velocity of rivers.

Potential energy = mass × acceleration of gravity × height

$$P.E. = mgh$$
$$mg = weight$$
$$P.E. = weight \times height$$

Average elevation of continents = 2755 feet (See Chapter 15)
Average runoff of continental U.S. = 1,200,000 million gallons per day (See Chapter 12)

1 gallon of water weighs 8.35 pounds
$$P.E. = 1,200,000,000,000 \times 8.35 \times 2755$$
$$= 2.76 \times 10^{16} \text{ foot-pounds per day}$$

(The total energy used in the U.S. in 1960 was about 3.1×10^{19} foot-pounds, or about 1000 times more.) If all of the river's energy were used to heat the water, it would only rise about 3.5°F.

The velocity can also be calculated:
Potential energy = kinetic energy
$$mgh = \tfrac{1}{2} mv^2$$
$$v^2 = 2gh$$
$$= 2 \times 32 \times 2755$$
$$= 176,000$$
$$v = 418 \text{ feet per second}$$
$$= 286 \text{ mph}$$

sured.] These velocities are far less than the 460 kilometers (286 miles) per hour average velocity (Table 8–1) predicted on the basis of the potential energy of rivers. Apparently the energy of rivers goes largely into heating the water by friction along the bed and banks and by turbulence. It is this friction and turbulence, however, that enable a river to erode and transport, as will be described in the next section. (See Fig. 8–1.)

The flow of a river is determined by many factors. The *discharge, or flow, can be defined as the*

amount of water passing a given cross-section of the river in a time unit (usually given in cubic feet per second, cfs). The discharge, then, is equal to the area of the cross-section times the velocity. The cross-section is determined by the shape of the river's channel, and the velocity is determined by a number of factors, the chief of which are the roughness, shape, and curving of the channel and the slope of the river. The amount of water available for run-off is also a factor, and this is determined by such climatic factors as precipitation, evaporation, vegetation, and permeability of soil and bedrock. On any given river, all of the factors that control the flow vary from place to place and with season.

The channel is probably the most important feature of a river. However, it is difficult to study because it is generally obscured by the water. In recent years, many soundings have been taken, and much has been learned about channel shape and, as will be seen in the next section, the changes in shape that occur with changes in flow. The shape of a channel greatly influences drag or friction. In general, the smaller the wetted perimeter, the less the drag. This suggests that if this were the only factor, rivers in eroding their channels might tend to form semicircular cross-sections because this shape has the least perimeter for its area. However, many other fac-

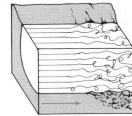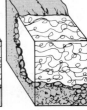

Smooth channel   Rough channel

FIGURE 8–1 Idealized diagram showing flow of water. The lines represent paths of water particles. Where the channel is smooth, the water flows in straight paths; but where the channel is rough, the flow is turbulent. Turbulence increases with increasing velocity.

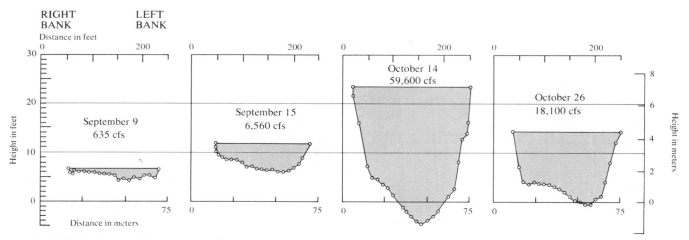

FIGURE 8-2 Changes in channel of San Juan River near Bluff, Utah, during flood in 1941. Note the changes in elevation of both the water surface and the channel depth. After L. B. Leopold and Thomas Maddock, Jr., U.S. Geological Survey Professional Paper 252, 1953.

tors influence channel shape and actual channels are quite irregular. (See Fig. 8-2.) At flood stage, rivers are higher than their channels and spill over into their valleys. This greatly increases the wetted perimeter, and the resulting drag slows the flow in the flooded valley, reducing the potential damage to the inundated area. The size and amount of the material transported by the river and the erodibility of the banks have a great influence on channel shape. (These factors will be described in the next section.) Figure 8-3 shows the distribution of velocity and turbulence in an idealized river channel both on a straight reach and on curves. A river changes in a systematic way from its headwaters to its mouth. Downstream the discharge increases because a larger area is drained; and to accommodate this increased flow, the width and depth both increase, as does the velocity. (See Fig. 8-4.)

The slope of a river also changes systematically from headwaters to mouth as shown in Fig. 8-5. The vertical scale in this figure is greatly exaggerated. A typical river has a slope of several hundred feet per mile near its headwaters and just a few feet per mile near its mouth. The lower 966 kilometers (600 miles) of the Mississippi River has a slope of less

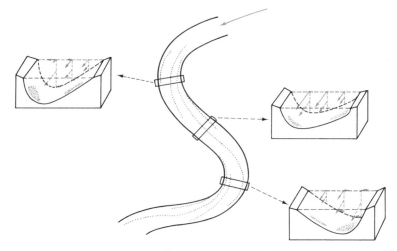

FIGURE 8-3 Distribution of velocity and turbulence in a river channel. In the cross-sections, the arrows show the direction and magnitude of the velocity. The surface velocity is indicated by two components, and the path of the surface water is shown by the dotted lines in the map view. The areas of most turbulence are indicated by stippling in the cross-sections.

than 9.3 centimeters per kilometer (one-half foot per mile). A steep, rapidly flowing river generally has a slope of 5.68 to 11.37 meters per kilometer (30 to 60 feet per mile). A slope of one degree is 17.5 meters per kilometer (92.4 feet per mile). Notice that the profile (Fig. 8-5) is not smooth but has irregularities, especially where tributary streams join. These irregularities are caused by abrupt changes in discharge and load. This suggests that the overall concave upward shape of the long profile is also the result of a balance between

discharge, load, and slope, as well as other factors. A brief study of river processes in the next section will give insight into this balance.

EROSION AND TRANSPORTATION

The main geologic work of rivers is transportation. The load is delivered to the river by downslope movement of material on the valley sides, by erosion by the river itself, and by tributary streams.

The load of a river can be considered in three parts: *dissolved load, suspended load, and bed load.* The total amount of material delivered by the rivers of the world to the oceans is estimated to be about 40,000 million tons per year. Human activities, especially farming and construction, contribute much of this load.

The dissolved load comes from chemical weathering and from solution by the river itself. Unlike the rest of the stream's load, it has no effect on the flow of the stream. The dissolved load of the world's rivers is estimated at 3900 million tons per year, just under 10 percent of the total load. Table 8-2 shows the average dissolved load. About half of the dissolved load is bicarbonate ion (HCO_3^-) that results from the solution of carbon dioxide in water, as described in weathering in Chapter 4.

The suspended load consists of the fine material that, once in the water, settles so slowly that it is carried long distances. As the flow of a river increases, some of the finer bed load becomes suspended load so that the distinction can be made only at a given moment. Suspended material is the cause of muddy river water. Because clay particles settle very slowly, they are carried relatively long distances. Silt settles much more rapidly; it is carried as suspended load only by turbulent water which has enough upward currents to keep the silt from settling. Thus clay is carried farther than silt, resulting in sorting of sediment by rivers. The suspended load is estimated to be almost two-thirds of the total load carried by rivers and amounts to over 25,000 million tons per year for all of the world's rivers.

The bed load is the material that is moved by rolling and sliding

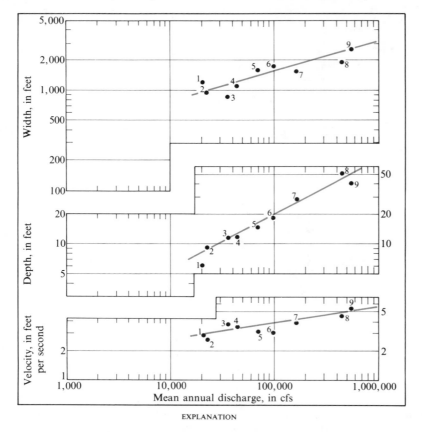

EXPLANATION

1. Missouri River at **Bismarck**, N. Dak. 4. Missouri River at Kansas City, Mo. 7. Mississippi River at St. Louis, Mo.
2. Missouri River at Pierre, S. Dak. 5. Missouri River at Hermann, Mo. 8. Mississippi River at Memphis, Tenn.
3. Missouri River at St. Joseph, Mo. 6. Mississippi River at Alton, Ill. 9. Mississippi River near Vicksburg, Miss.

FIGURE 8-4 Graphs showing the increase in width, depth, velocity, and discharge downstream in the Missouri River and lower Mississippi River. From L. B. Leopold and Thomas Maddock, Jr., U.S. Geological Survey Professional Paper 252, 1953.

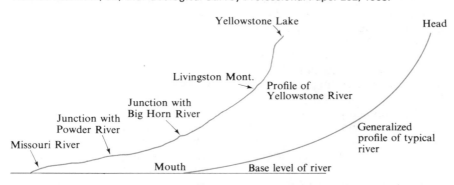

FIGURE 8-5 Longitudinal river profiles. Vertical scale is exaggerated. Yellowstone River profile from U.S. Geological Survey Water Supply Paper 41, 1901.

TABLE 8-2 Average composition of dissolved material in river waters of the world.

	Parts Per Million
Bicarbonate (HCO_3)	58.4
Calcium	15.0
Silica (SiO_2)	13.1
Sulfate (SO_4)	11.2
Chlorine	7.8
Sodium	6.3
Magnesium	4.1
Potassium	2.3
Nitrate	1.0
Iron	0.7
	119.9

SOURCE: From *Data of Geochemistry,* sixth edition, Chapter G, "Chemical Composition of Rivers and Lakes," D. A. Livingstone, U.S. Geological Survey Professional Paper 440-G, 1963.

along the bottom. It, too, is somewhat sorted by size; also the particles are worn by abrasion, making the pebbles and boulders in a typical stream bed subspherical. In addition to abrasion and solution, the bed-load pebbles are at times cracked and broken by impacts during movement. Almost all of the movement of most bed loads occurs during the generally very limited times of high flow. Thus the floods that occur seasonally, or even less commonly, mark times that bed loads are moved. In some rivers the ratio of maximum to minimum flow is a few hundred to one. In the summer or fall, it is difficult to believe that the large boulders in the stream bed, particularly a small stream, are part of the bed load and are being moved by the stream. (See Fig. 8–6.) One must remember that the largest boulders may move only during periods of very great flow that may occur a few times each century. However, if one visits such a stream during the spring when it is filled with melt water, the audible clicks when a large boulder rolls or slides into its neighbor can be heard. The bed load of most streams is about a quarter or less of the total load. The maximum size particle that can be moved by a river is proportional to the square of the velocity. Thus doubling the velocity increases by four times the size of particle that can be moved. In general, doubling the velocity increases the amount of material in transport between four and eight times. The movement of bed load during times of high flow is shown by the changes in size and shape of the channel. (See Fig. 8–2.)

Rivers erode by several processes which were necessarily discussed in describing transportation. If parts of the channel are soluble, erosion by solution will occur. *Abrasion occurs as a result of the transported material rubbing against the sides and bottom of the channel.* A common type of abrasion results in cylindrical holes, called *potholes,* worn in

bedrock, apparently by stones that are spun by turbulent eddy currents. (See Fig. 8–7.) In plunge pools below waterfalls, much of the erosion is caused by the moving water itself. The impact and

drag of the water erode and transport material. As noted earlier, the size of particle that can be moved in this way is proportional to the square of the velocity; however, velocities great enough to move

FIGURE 8–6　A stream at low water level, showing part of the bed load that is moved during times of high discharge. Photo from California Division of Beaches and Parks.

FIGURE 8–7　Potholes in a river bed cut in granitic rock. The potholes are formed during times of much greater flow. Photo by Mary Hill, courtesy California Division of Mines and Geology.

sand, and even coarser material, are necessary to cause erosion of the smooth surface of consolidated clay. The rate of erosion is slow on many rivers and may be noticeable only after major floods. (See Fig. 8–8.)

The rate of erosion of the continents is determined by measuring the amount of material carried by rivers. The average rate of erosion in the United States, determined in this way, is about 6 centimeters (2.4 inches) per thousand years. Of the major river basins the Columbia River basin is eroding at the rate of 3.8 centimeters (1.5 inches) per thousand years, the lowest rate; the Colorado River at 16.5 centimeters (6.5 inches) per thousand years; and the Eel River in northern California at between

1 and 2 meters (40 and 80 inches) per thousand years, the largest rate. The rate of erosion depends on a complex relationship of many factors, such as rainfall, evaporation, and vegetation, but note that the maximum rate of erosion occurs in areas with about ten inches of rainfall per year. (See Fig. 8–9.)

Erosion tends to deepen, lengthen, and widen a river valley. The ways that a river can deepen its channel were just discussed. *At the head of a stream any downcutting lengthens the valley. This process is called* headward erosion *and is the way that streams eat into the land masses.* (See Fig. 8–10.) A stream valley is widened by two methods. The valley-side processes, such as creep and landsliding, widen the valley. (See Fig.

8–11.) Lateral cutting by the stream itself is the other process.

Lateral cutting is most pronounced on the outsides of curves where the valley sides may be undercut. (See Fig. 8–12.) Thus a curving stream widens its valley, and straight stretches of streams longer than ten times the width are not common. Clearly, any irregularity on a stream bottom or bank can deflect the flow and cause a curve to develop. Once formed, a curve tends to migrate, thus widening the valley. Many rivers have *distinctive symmetrical curves called* meanders. *(See Fig. 8–13.)*

The migration of meanders is shown in Fig. 8–14, which also shows how the meanders may intersect, causing *oxbow lakes.* (See

A

B

C

D

FIGURE 8–8 These two pairs of photographs were taken nearly one hundred years apart and show that little erosion has occurred along the Colorado River in that interval. The early photographs were taken on John Wesley Powell's expeditions. A and B. Sockdologer Rapid, Grand Canyon National Park, Arizona. C and D. Schoolboy Rapids, Split Mountain Canyon, Dinosaur National Park, Utah. The changes in the foreground may have occurred in a single flood. Photo A, 1872, and C, 1871, by J. K. Hillers; B and D, 1968, by H. G. Stephens, all from U.S. Geological Survey.

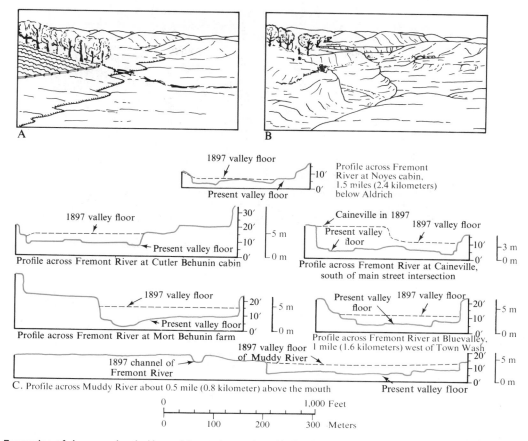

1897 valley floor

Present valley floor

Profile across Fremont River at Noyes cabin, 1.5 miles (2.4 kilometers) below Aldrich

1897 valley floor

Present valley floor

Profile across Fremont River at Cutler Behunin cabin

Caineville in 1897

1897 valley floor

Present valley floor

Profile across Fremont River at Caineville, south of main street intersection

1897 valley floor

Present valley floor

Profile across Fremont River at Mort Behunin farm

Present valley floor   1897 valley floor

Profile across Fremont River at Bluevalley, 1 mile (1.6 kilometers) west of Town Wash

1897 channel of Fremont River

1897 valley floor of Muddy River

Present valley floor

C. Profile across Muddy River about 0.5 mile (0.8 kilometer) above the mouth

0   1,000 Feet

0   100   200   300 Meters

FIGURE 8–9 Examples of river erosion in Henry Mountains region, Utah, since 1897. Climatic change is the probable reason for this erosion. Parts A and B show Pleasant Creek at Notom. The creek in part B (present day) is 6 meters (20 feet) deep. Part C shows a number of cross-sections of the Fremont River. From C. B. Hunt, U.S. Geological Survey Professional Paper 228, 1953.

FIGURE 8–10 The two streams in the foreground of this oblique aerial view are lengthening by headward erosion. The bigger one may ultimately capture the stream in the middleground. San Bernardino Mountains, California. Photo by J. R. Balsley, U.S. Geological Survey.

FIGURE 8–11 Widening of a stream valley by a small slide. In this case a forest fire has reduced the vegetation and thus caused the slide. Photo from U.S. Forest Service.

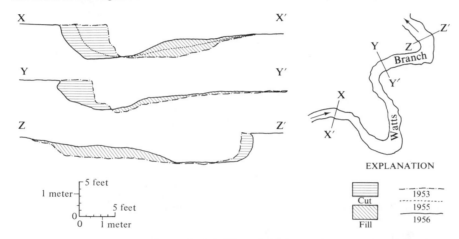

FIGURE 8–12 Erosion and deposition on Watts Branch near Rockville, Maryland, from 1953 to 1956. From M. G. Wolman and L. B. Leopold, U.S. Geological Survey Professional Paper 282C, 1957.

FIGURE 8–13 Meanders on the Green River in Utah. Cut-off meanders and old meander scars show the movement of the river on the floodplain. Looking north (upstream) toward the Book Cliffs. Photo from U.S. Geological Survey.

Fig. 8–15.) Meanders are not formed in the same way as are the irregular curves on a river. The curves that form meanders are not common curves, such as half circles. The meander-forming curve is the one that distributes the river's loss of energy most uniformly. Any other curve would tend to concentrate erosion and deposition at some point on the curve. (The development of river valleys by all of these methods of erosion will be described later.)

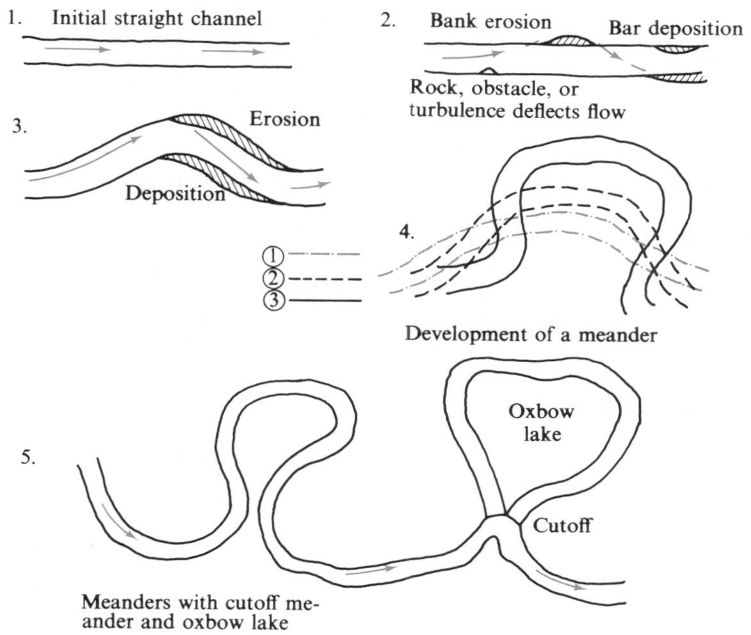

THE RIVER AS A SYSTEM

Rivers do not occur singly in nature. Each has tributaries and each tributary has smaller tributaries, and so on, down to the smallest rill. All of these parts function together to form a *river system,* or *drainage basin.* A change in any part of the system generally affects all of the other parts. One obvious illustration of this balance is that all the tributary streams meet the main stream at the level of the main stream. This accordance of junctions is too universal to be due to chance. (This accordance is not found in glaciated areas where the stream levels have been changed by glacial erosion.)

The main work of a river is to transport debris, and river systems regulate themselves so that they are able to transport the debris delivered to them. This is generally accomplished by changing the slope, by either erosion or deposition, so that at any given point for the discharge and channel characteristics the velocity is such that the debris is transported. Thus a change in any part of the system will affect the balance of the rest of the system. *A stream that has just the right gradient to be able to transport the debris delivered to it by its own erosion, its tributaries, downslope movement of material, and rain wash is called a graded stream.* This ability is a long-term average so that a given stretch of river at one season may be eroding its bed, and at another time it may be building up its bed. These changes in regimen vary from year to year with such climatic influences as amount and intensity of precipitation.

The long profile of a river (see Fig. 8–5) was mentioned earlier, and the reason for its shape can now be appreciated. The slope near the head must be steep for

FIGURE 8–14 Development of meanders.

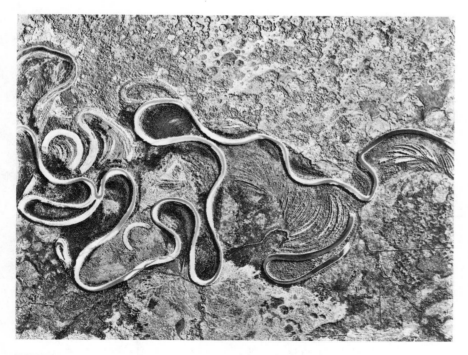

FIGURE 8–15 Meanders and oxbow lakes formed by cut-off meanders, Hay River, northwestern Alberta, Canada. Photo from Geological Survey of Canada, Ottawa.

the river to be able to do its work of eroding and transporting with the small discharge it has available. Farther downstream the velocity and discharge both increase and particle size decreases as does the number of banks (see Fig. 8–16), all of which permit the river to do its work with a lesser slope, even though the total load is increased. Thus the concave upward slope of a river enables the most uniform rate of work per unit of length and at the same time minimizes the work done in the whole system.

In most river systems the profile is more complicated because the ability of a river to erode and transport depends largely on the velocity, the amount of flow, the gradient, and the shape and roughness of the channel. At the point where a tributary joins the main stream, the regimen of the stream is abruptly changed by the additional flow and the increased debris from the tributary. The main stream reacts to this change by altering its gradient so that it can handle the new material. These changes are superimposed on the simple profile.

The limit to which the river can erode is, of course, the elevation of its mouth. This limit is called the base level of the river. (See Fig. 8–5.) For most big rivers the base level is sea level, but for some that flow into lakes, it is the elevation of the lake. For tributary streams, the base level is the elevation at which they join the main stream.

DEPOSITION

River-deposited sedimentary rocks (fluvial rocks) are relatively rare because rivers deliver most of their loads to lakes or to the ocean. River deposits in the form of deltas develop where rivers enter standing water; even in this case, however, much of the river's load may be reworked by waves and offshore currents, thus be-

FIGURE 8–16   The number of banks decreases downstream, reducing the friction.

coming marine or lacustrine (lake-deposited) rocks. Deposition occurs because the velocity of river water is slowed so abruptly on entering standing water that the bed load is dropped. The dissolved and suspended loads may be carried farther out. (See Figs. 8–17 and 8–18.) A *delta* can build outward into the ocean only if the waves and currents are not strong enough to move the deposited material.

Alluvial fans are similar to deltas and are formed where streams flow out of mountains into broad, relatively flat valleys. In this case,

the flatness of the valley slows the stream, causing deposition. In addition, especially in arid areas, the mountain stream may seep into the ground in the broad valley. Because the stream constantly shifts its position on the fan, the fan develops a cone shape (Fig. 8–19).

When the flow of a river fills its bed up to the top of the banks, any increase in discharge will cause flooding. The floodwater spills over the river banks, filling part or all of the valley. A river is doing its maximum erosion and transportation at the time of flood, and as the floodwater spreads out, its velocity is quickly reduced, causing *deposition near the banks of the river. Such deposits are termed levees* (or natural levees, to distinguish them from levees constructed for flood protection). (See Fig. 8–20.) Although the levees help to keep the river within its banks, the area between the levee and the side of the river valley may be lower than the river at

FIGURE 8–17   The Nile delta photographed from *Gemini IV.* The dark area is cultivated land. The Mediterranean Sea is to the left; the Sinai Peninsula is in the middleground. The Nile has a number of distributaries, and this accounts for the large cultivated area. Photo from NASA.

flood stage; this area is subject to inundation when the flooding river crests over the levee top. Because the river valleys are prime agricultural areas and because the mechanism of levee building provides a continuing supply of new soil, many people accept periodic flooding and live in the valleys of the large rivers. These people should expect spring flooding, for it is the normal, not the abnormal, behavior of the river. To some extent they can be protected by upstream flood-control dams, which regulate somewhat the flow during flood stage, but this system is expensive and not always effective as discussed later.

Rivers may also build up their beds by deposition. In some cases, a river may deposit material during a long period of time so that its meandering causes deposition covering an entire valley. Such a process produces a smooth surface. Later, the river may, because of some change in its regimen, cut into this surface and so form one or more lower *terraces* (Fig. 8–21). Terraces can also be formed by lateral cutting of bedrock by the river. One common cause of terrace building is associated with the recent glaciation. In areas where glaciers used to be active in the mountains, much debris of glacial erosion was rather suddenly released to the rivers when the glaciers melted. Because the rivers could not transport the material as rapidly as it was delivered, they deposited it in their valleys. As conditions returned to normal, they downcut into these deposits and formed terraces, most of which, in time, will be removed by erosion. As is shown in a later section, the study of terraces can tell much about the history of a river.

Braided streams are an unusual type of stream that forms under some conditions, especially if the load is large and coarse for the slope and discharge, and the banks are easily eroded. In a *braided stream* the channel divides and rejoins numerous times. (See Fig. 8–22.) Apparently the stream deposits the coarser part of its abundant load in order to attain a steep enough slope to transport the remaining load. This deposition forces the stream to broaden and so erode its banks, adding to the abundant load, per-

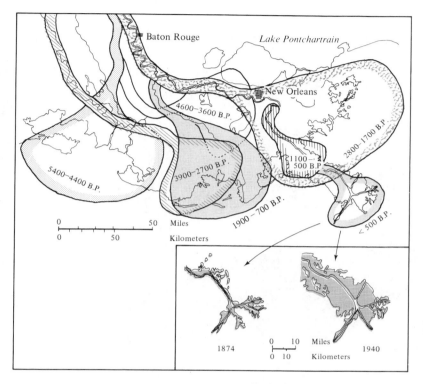

FIGURE 8–18 Development of the Mississippi River delta. The upper drawing shows the ages of subdeltas in years before present (B.P.). Upper drawing after Kolb and van Lopik, 1966, and lower drawings after Scruton, 1960.

FIGURE 8–19 Alluvial fans at canyon mouths, Mojave Desert, California. Photo by J. R. Balsley, U.S. Geological Survey.

159

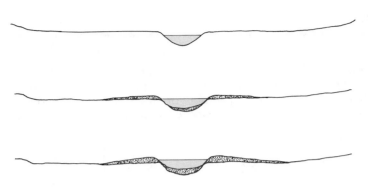

FIGURE 8-20 The development of levees. Levees form when silt is deposited beside the river during floods. If, as is generally the case, deposition also occurs on the river bed, the seriousness of later floods increases. The vertical scale is greatly exaggerated.

FIGURE 8-21 Terraces cut in river-deposited material by the Snake River. Teton Mountains on the skyline. Photo from Union Pacific Railroad.

FIGURE 8-22 Braided river. Vertical aerial photograph of North Platte River, Nebraska. Photo from U.S. Geological Survey.

haps choking the channel. The places where two channels join have increased flow and reduced drag. At such places some of the bed load is moved, but this reduces the slope and deposition begins again; however, the load is moved in the process, albeit only a short distance.

RIVER VALLEYS

River valleys can be described as narrow or broad, although all intermediate steps between the two exist. Narrow valleys are not much wider than the river channel, and broad valleys are much wider than the river. Another way to describe river valleys is youthful, mature, or old age (Fig. 8–23.).

Youthful, narrow valleys, which have V-shaped cross-sections, are generally found where rivers are actively downcutting. On many rivers they occur near the headwaters, but narrow valleys can form anywhere. Many, but far from all, are downcutting, nongraded rivers. Some narrow valleys occur in resistant rocks that slow the lateral cutting of a river; they commonly have rapids and waterfalls where resistant rocks slow the downcutting. Lakes may also form in narrow valleys, but, like falls and rapids, they are only temporary features. (See Figs. 8–24, 8–25, 8–26, 8–27, and 8–28.)

Valley widening begins, in general, after a stream has reached grade and is no longer rapidly downcutting. This occurs first, on most rivers, near the mouth where the increased flow allows the river to erode more efficiently. Wide floodplains characterize the lower parts of many rivers. (See Fig. 8–13.) In areas of weak rock, widening is also common. Wide valleys are generally fairly flat in cross-section. Meanders are common features of wide valleys, especially in areas with uniform banks composed of fine sediment. Meanders may alternate with areas of irregular curves. In a mature valley the meander belt occupies most of the width of the valley. An old-age valley is much wider than its meander belt. (See Fig. 8–23.) Another feature, familiar to fishermen, is the alternation of riffles and pools on most rivers. The riffles form on alternate sides of the river and are composed of the coarser part of the bed load. (See Fig. 8–29.)

River valleys can be used in the interpretation of geologic history. Evidence of uplift can be found in

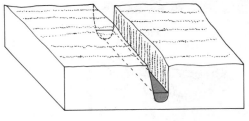

(Extreme youth) River has cut down rapidly so that downslope movements on the valley sides cannot keep up. Characterized by canyons, rapids, waterfalls, and, in some cases, lakes.

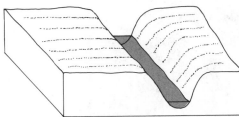

(Youth) Characterized by canyons, V-shaped valley, falls, rapids, and, in some cases, lakes.

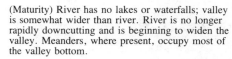

(Maturity) River has no lakes or waterfalls; valley is somewhat wider than river. River is no longer rapidly downcutting and is beginning to widen the valley. Meanders, where present, occupy most of the valley bottom.

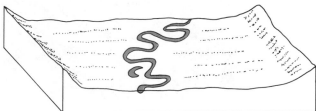

(Old age) River is in a broad valley. Valley is much wider than meander belt.

FIGURE 8–23 The development of river valleys. Valleys may be described as youthful, mature, or in old age.

161

FIGURE 8–24 An extremely narrow canyon with nearly vertical walls cut in sandstone. Downcutting was apparently faster than downslope movements could shape the valley sides. Photo from California Division of Beaches and Parks.

FIGURE 8–25 The Yellowstone River in Yellowstone National Park, with falls and rapids. Photo from National Park Service.

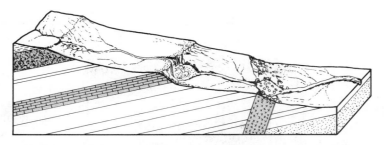

FIGURE 8-26   Block diagram showing how resistant rocks cause falls and rapids.

FIGURE 8-27   The Snake River Valley on the Oregon-Idaho border is a narrow, V-shaped valley. Oblique aerial photograph from U.S. Geological Survey.

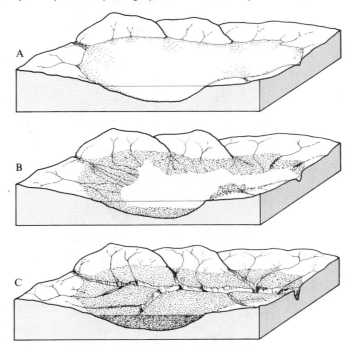

FIGURE 8-28   Three stages in the life of a lake. A. Initial lake. B. Lake has been partially filled by stream deposition. C. Lake is completely filled, and stream has cut a channel across the fill.

163

FIGURE 8-29   A narrow, V-shaped valley in which the river has alternating pools and riffles. Photo from U.S. Forest Service.

individual river valleys because river valleys widen as erosion progresses. Thus, if a river is cutting a V-shaped, youthful valley within a much wider valley, a change has

occurred causing the river to downcut actively. (See Fig. 8–30.) This change could be due to an uplift, a climatic change, or an increase in the flow of the river from

a stream capture (discussed below) or other cause. If no further changes occur, the new valley will widen as erosion progresses, removing the evidence. If a new uplift occurs, a new V-shaped valley will be formed. This is one way river terraces are produced (Fig. 8–31). If depression rather than uplift occurs, a river valley will be the site of deposition instead of erosion.

At some places uplift apparently is shown by rivers that have narrow, V-shaped valleys and a meandering map pattern. This may happen if uplift occurs slowly enough that the river that was meandering in a broad valley maintains its meandering course, although downcutting. (See Fig. 8–32.)

DRAINAGE PATTERNS

The development of the drainage pattern is an important part of the erosional history of an area. A few of the ways that streams develop can be observed in nature, but because geologic processes are slow, most of the development must be imagined. Initially, the runoff flows down the irregularities in the slope in many small rills. The flowing water erodes these rills; the ones with greater flow are eroded most. These larger rills become deeper gullies,

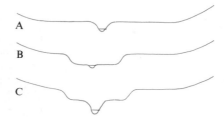

FIGURE 8–31   Successive uplifts have formed a terraced valley. A. An uplift caused the river initially in a broad valley to downcut, forming a V-shaped valley. B. Further erosion widened the new valley. C. A new uplift caused renewed downcutting. Ultimately erosion will destroy the older terraces, and it will not be possible to decipher this history.

FIGURE 8–30   A new valley is being cut into an already existing broad valley in this oblique aerial view of the Snake River on the Oregon-Idaho boundary. Photo from U.S. Geological Survey.

FIGURE 8–32 Entrenched meanders on the Grande Ronde River in Washington. Broad valley features like meanders in narrow, V-shaped valleys as shown here suggest a change in the regimen of the river, such as a slow uplift. Photo from Washington State Department of Commerce and Economic Development.

These conditions, however, are probably not commonly found in nature. One can also envision an uplift slow enough that the initial streams are merely entrenched. Thus uplift may be slow or rapid, or may accelerate or slow down, and each case will affect the erosional development of an area in a different way. The landscapes developed by these differing conditions will be described in the next section.

The effect of bedrock structure on stream pattern can be seen in many places. Areas of uniform bedrock commonly have branching streams in a pattern similar to the veins in a leaf. (See Fig. 8–37.) In other areas, the joints, or cracks in the rocks, may influence the stream pattern, and rectangular patterns commonly develop in this way. At other places, alternating hard and soft rocks may produce a trellis pattern. (See Fig. 8–38.) The initial slope of an area also affects the stream pattern, as shown in Fig. 8–39.

At other places there seems to be no relationship between river courses and geologic structures. At such places the rivers may cut across the upturned edges of resistant rock units that form high ridges. One way that such situations can develop is that the river is *superposed,* as shown in Fig. 8–40. The drainage pattern formed on flat-lying, uniform rocks; when the region was uplifted, the streams downcut through the overlying uniform rocks into the deformed rocks below. Eventually all, or almost all, of the overlying rocks in which the stream pattern formed are removed by erosion. The result is a stream pattern that cuts across the rock structures.

DEVELOPMENT OF LANDSCAPE

Development of landscapes is even more complex than the development of river valleys and

and ultimately one of them will become the master stream. The process of gully development can be seen in many excavations (see Fig. 8–33) and in fields where plowing has removed the natural vegetation that prevented erosion. (See Fig. 8–34.) The rills develop by downcutting and by headward erosion. The rill with the most water will downcut fastest, and this will accelerate headward erosion by this rill and all of its tributary rills. The *lengthening by headward erosion enables this rill to capture the flow of the rills it intersects. This increases the area drained by the main rill and accelerates its*

erosion. The process of stream capture, called **stream piracy,** continues, and the rill develops first into a gully and finally into a master stream. (See Figs. 8–35 and 8–36.)

Stream patterns develop in many ways. Most stream patterns are probably in part inherited from an earlier development and in part dependent on the underlying rock structure and changes in elevation. Other factors that influence drainage development include climate, runoff, and vegetation. The preceding description tacitly assumes an initially fairly smooth surface that was rapidly uplifted.

FIGURE 8–33   Rills developed in an excavation in glacial deposits in Iowa. Photo by W. C. Alden, U.S. Geological Survey.

FIGURE 8–34   Gullies cut in a sandy slope in a two-year period in Wisconsin. Photo by W. C. Alden, U.S. Geological Survey.

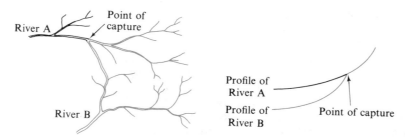

FIGURE 8–35   A common type of stream piracy. Rapidly downcutting minor tributary of River B captures the headwaters of River A. The gradients of the two rivers near the point of capture are shown in the right diagram.

FIGURE 8–36 An example of stream capture is shown in this photo of the southern part of the Arabian Peninsula taken from *Gemini IV.* The Gulf of Aden is in the background. Wadi Hadramaut in the foreground clearly flows toward the right because it is wider there. Many of the tributary streams flow toward the left, and their flow is abruptly reversed where they join Hadramaut. Such reversals are common where rivers have been captured. Photo from NASA.

FIGURE 8–37 Branching stream pattern (dendritic pattern) developed in massive rocks. From U.S. Geological Survey, Mount Mitchell quadrangle, North Carolina-Tennessee.

drainage patterns, and all of the factors discussed above are involved. Thus it is no wonder that much controversy exists. We will begin with the simplest case and use it to illustrate other possibilities, which will show why controversy exists.

The simplest viewpoint on the development of landscape was an American idea, and it had a pronounced influence on the history of geology. The cornerstone of modern geology is the doctrine of *uniformitarianism, which, in the present context, states that the* *present processes at work on the earth, given enough time, caused the present landforms.* This principle, although recognized by a few earlier men, was widely adopted near the end of the eighteenth century. Geology, however, developed mainly in Western Europe during the nineteenth century; and most geologists, especially the island-dwelling English, thought that ocean waves rather than rivers were responsible for most erosion. The opening of the American West following the Civil War provided a vast new area for geologic study. The great rivers and canyons, such as the Colorado, and the many mountain ranges at various stages of dissection, together with the dry climate that bared great vistas, made recognition of huge amounts of erosion simple. (See Fig. 8–41.)

Late in the nineteenth century, William Morris Davis of Harvard codified the earlier studies into what he termed the *cycle of erosion.* Davis visualized every *landscape going through a series of changes from initial uplift to complete leveling by weathering and erosion.* His emphasis was on the forms produced, and he paid little attention to the processes involved. He clearly recognized that an area rarely goes through the complete cycle, but, rather, the cycle generally is restarted by new uplifts. This accounts for the fact that there are very few examples of areas that have gone through the complete erosion cycle. In the *youthful stage, a newly uplifted area is relatively flat with deeply*

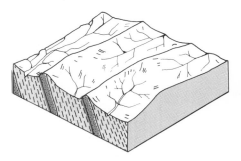

FIGURE 8–38 Alternating hard and soft rocks produce a trellis drainage pattern.

entrenched streams (Figs. 8–42 and 8–43). As erosion proceeds, the interstream areas are reduced in size as a result of downcutting by the rivers and widening of the valleys by downslope movement of material. *When no flat interstream areas exist, the stage has progressed to* maturity. (See Figs. 8–42 and 8–44.) This is the stage of maximum relief. *As the interstream areas are rounded and lowered, the area passes into* old age (Fig. 8–42). In the ultimate stage, *when even these gentle interstream areas are worn almost flat, the area is called a* peneplain ("almost a plain," sometimes spelled *peneplane*). Flat eroded areas such as peneplains are rare features on the earth's surface, and this fact constitutes one of the weaknesses of the cycle-of-erosion hypothesis. However, many events, such as renewed uplift, can occur to prevent the complete development of a peneplain.

Perhaps the main objection to the Davis cycle is that it requires

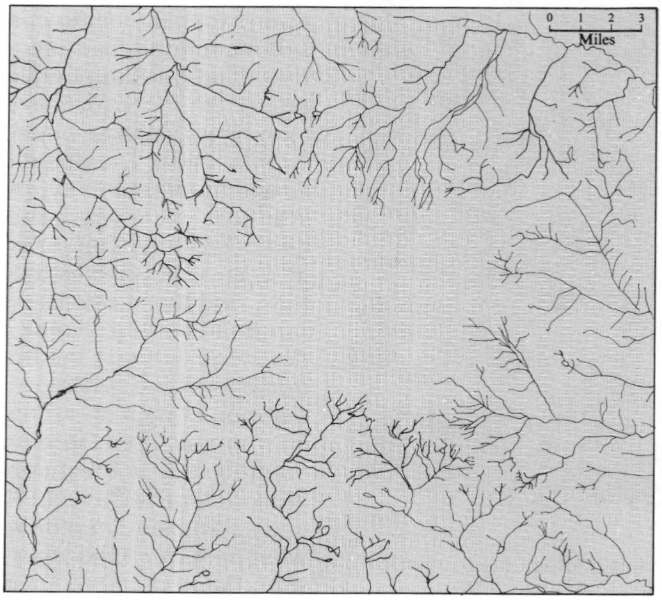

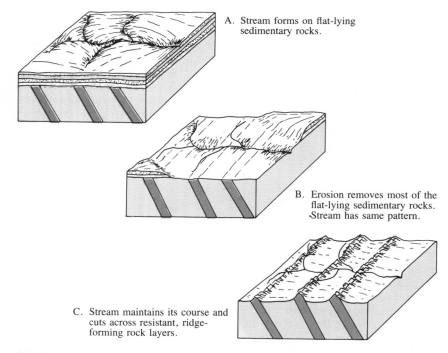

FIGURE 8–39 Radial stream pattern. From U.S. Geological Survey, Mount Rainier quadrangle, Washington.

A. Stream forms on flat-lying sedimentary rocks.

B. Erosion removes most of the flat-lying sedimentary rocks. Stream has same pattern.

C. Stream maintains its course and cuts across resistant, ridge-forming rock layers.

FIGURE 8–40 Development of a superposed river.

FIGURE 8–41 A typical scene in the arid Southwest. The lack of vegetation and the fact that the flat-lying sedimentary rocks can easily be traced for many miles make recognition of the immense amount of erosion that has occurred easy.

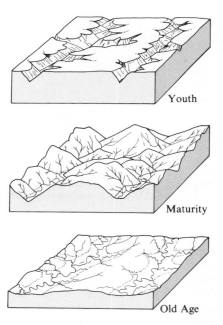

Youth

Maturity

Old Age

FIGURE 8–42 The cycle of erosion as envisioned by Davis.

FIGURE 8–43 An example of youthful topography in eastern Utah. The flat areas are being attacked by headward erosion of the streams. Oblique aerial photograph from U.S. Geological Survey.

FIGURE 8–44 An example of mature topography in the Ozark Plateau, Missouri. No flat areas remain. Oblique aerial photograph from U.S. Geological Survey.

rapid uplift and then a stillstand during which erosion occurs. If the uplift of an area is slow enough, perhaps erosion keeps pace with the uplift and no change in form occurs. However, any type of uplift from rapid to slow, or from accelerating to slowing, is geologically possible, so at best the Davis cycle describes a special case.

Another objection is that the hills might not be reduced in the manner envisioned by Davis. Instead, a balance might result when the hill slopes and stream gradients become adjusted. Such an equilibrium can, in a way, be thought of as an extension of the graded-stream concept discussed earlier. After this stage is reached, the shape of the area remains substantially the same as erosion progresses. Thus there is no great change in shape, as there is in the Davis cycle of erosion, because once an equilibrium shape develops, no new shape can form unless the balance is upset by uplift or some other change. (See Fig. 8–45.)

In an effort to resolve this latter objection, a number of recent studies have been made of hill slopes. The studies have been made in areas where the climate, vegetation, relief, and rock type are uniform. The slope angle of such hillsides is related to both stream gradient and density of streams. Where direct measurement of slope erosion was made, it was found that during retreat of slopes the slope angle remains constant if the debris is removed from the foot of the slope. (See Fig. 8–46.) If the erosional debris is not removed from the foot of the hill, the slope angle is reduced. The latter case is probably more common in humid areas where weathering may be deep and creep may be active. All of this suggests that the Davis cycle is not only a special case, but an oversimplification as well.

The length of time and the size of the area considered in the study of landscape may also influence one's ideas. A single hillside or

169

stretch of river may be in equilibrium during the time of a short study of the type just described; however, if the whole drainage area is considered over a geologically long time interval, erosion must remove appreciable amounts of mass and so changes the system. Thus, in the short term, landforms may not seem to change with time, but over a long interval they might actually do so. The dynamic earth may not remain still long enough, however, for these changes to occur, for the earth's surface is the result of the constant struggle between erosion and uplift.

ENVIRONMENTAL HAZARDS OF RUNNING WATER

"Floods are as much a part of the phenomena of the landscape as are hills and valleys; they are natural features to be lived with, features which require certain adjustments on our part."[1]

Some primitive peoples adapted reasonably well to rivers and accepted seasonal or occasional floods as part of life. As humans developed and built bigger and better structures for their comfort, they became less willing to accept the natural behavior of rivers.

A typical river valley consists of a channel, a flat area (called a floodplain), and sloping valley sides. (See Fig. 8–47.) Most of the time the flow is confined to the channel; but at times of high flow, when the channel cannot carry the water, it spills onto the aptly named floodplain. The floodplain is the channel of the river during times of high water. This is the normal, expected behavior of a river. People who live or build structures on floodplains must expect these floods (See Fig. 8–48.) Flood control is never one hundred percent effective and can

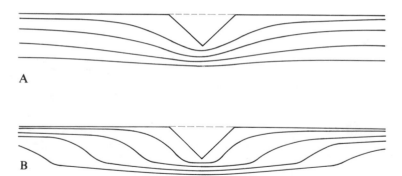

FIGURE 8–45 Successive profiles of the erosion of a river valley contrasting two theories of the development of slopes. A. The Davis concept of downwearing of hillsides. B. Backwearing of hillsides, preserving approximately the same slope.

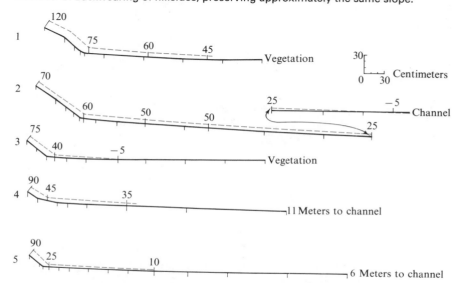

FIGURE 8–46 Measurement of erosion in the Badlands of South Dakota. The dashed profiles are July 1953, and the solid lines are May 1961. The numbers are the erosion depth in millimeters. From S. A. Schumm, *Bulletin,* Geological Society of America. Vol. 73, 1962.

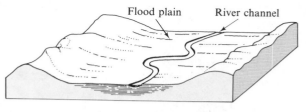

FIGURE 8–47 A typical river valley, showing the channel and the floodplain.

increase the actual damage as we will see later. Floods are not entirely bad events, however, because the silt deposited on the floodplains renews the soil of these generally very productive agricultural areas.

Rivers erode on the outsides of curves and deposit on the insides of curves. (See Fig. 8–14.) This means that the location of a river is always changing. Rivers are commonly used as property boundaries as well as state and in-

[1] W.G. Hoyt and W.B. Langbein, *Floods.* Princeton University Press, 1955.

FIGURE 8-48   Flooding of the Mississippi River at West Alton, Missouri. The river is generally contained behind the row of trees in the middle of the photograph. Wide World Photos.

ternational boundaries. Such boundaries are not fixed and lead to many problems. In some cases the boundary is fixed at the position of the river on a certain date. Along the Mississippi, a farmer with land inside a meander loop may wake up some morning and find that his land is separated by the river from the rest of the state and county to which he pays taxes. He may have trouble getting police and fire protection. On the other hand, he now may have an ideal location for moonshining. Recently the United States and Mexico concluded a treaty fixing the international boundary near El Paso. (See Fig. 8-49.)

Storms and mudflows are geologically rapid but normal events (See Fig. 8-50.) Intense storms that cause floods, landslides, or mudflows can occur almost anywhere. In the United States, such storms are common in the desert areas of the Southwest where little note is taken of them because few people are affected. A few cities have expanded into hazardous parts of this region in Arizona

and especially in southern California; the resulting problems will be described later. The hurricanes of the Gulf and southeastern coasts of the United States cause similar problems, although flooding of the coastal areas is the usual result. Some of these storms cause problems as far north as New England. Hurricanes are generally accompanied by heavy rainfall that chokes the rivers, and strong onshore winds drive ocean waters onto the low coastal plain. The low barometric pressure associated with such storms increases

the flooding by allowing much higher than normal tides. Hurricane Camille, which hit the Gulf Coast in August 1969, had winds up to 322 kilometers (200 miles) per hour, a low pressure of 905 millibars, and dumped up to 76 centimeters (30 inches) of rain in a day. The East Pakistan (now Bangladesh) flood of December 1970, killed between 250,000 and 300,000 people. This flood, erroneously described in the newspapers as a tidal wave, was caused by a hurricane (called a cyclone in that area) that had winds up to about 193 kilometers (120 miles) per hour and a low pressure of about 950 millibars. The wind and low pressure caused a storm surge about 5.5 meters (18 feet) high [normal tides are about 1.8 meters (6 feet)], which overran the low, flat area in the Bay of Bengal. In June 1972, 38 centimeters (15 inches) of rain fell on Rapid City, South Dakota, causing a flood that killed 237 people and left thousands homeless.

Mudflows caused by relatively rare but intense thunderstorms are not confined to desert areas. They can occur on any steep slope with enough overburden or loose material and not too much underbrush. Forested slopes, especially those with big trees whose shade prevents growth of underbrush, are susceptible to this type of erosion. The following example is from the Sierra Nevada mountains near Lake Tahoe on the California-Nevada boundary. The Lake

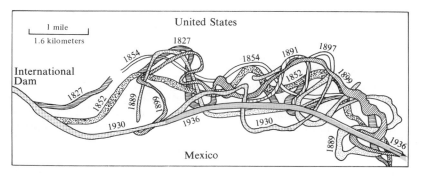

FIGURE 8-49   Changes in the Rio Grande River near El Paso from 1827 to 1936. After Boggs, 1940.

171

FIGURE 8–50 Mudflow near Big Sur, California. A forest fire destroyed the vegetative cover, and subsequent heavy rains caused the mudflow. Photo from California Department of Transportation.

Tahoe area, because of its beauty, its nearness to San Francisco and Sacramento, and its climate that makes it an excellent skiing as well as summer resort area, is currently undergoing urbanization. In the late afternoon of August 25, 1967, more than an inch of rain fell in the drainage basin of Second Creek. This caused a mudflow of more than 38,300 cubic meters (50,000 cubic yards) in the 3.9 square-kilometer (1.5 square-mile) basin. The mudflow caused extensive damage to homes, townhouses, apartment houses, and roads. A 2-meter- (7-ft.-) diameter boulder moved in this flow, and at some places, the flow deposited up to 6 meters (20 feet) of debris. Similar events have been observed elsewhere in the Sierra Nevada mountains, and the results of this type of erosion can be seen in many drainage basins. This particular event was estimated to have reduced the basin by about 0.6 centimeter (0.02 foot) and to have a recurrence interval of less than 50 years. This mudflow was a natural event and was not caused by the urbanization, although some conservationists blamed the re-

cent building in the area. This event did affect the building programs in the area because it showed that structures should not be built in the path of potential mudflows and that much larger culverts and other drainage structures had to be installed on structures such as roads that must cross potential mudflow paths. Mudflow paths must be considered when purchasing or developing mountain areas. That no mudflow or flash flood has occurred in the memory of residents does not mean that one will never occur.

Although not directly related to running water, snow avalanches are another geologic hazard in mountain resort areas. Avalanche chutes are also common in the Lake Tahoe area as well as any mountain area with moderate to heavy winter snows. Avalanche danger is controlled at ski resorts as a result of studies and regulations developed by the Forest Service and the National Ski Patrol System. In lift-served ski areas, the common method of controlling avalanches is to cause them to occur by blasting or other means at times when no one is

skiing. Thus many small avalanches may fall harmlessly in a given chute. In uncontrolled areas an avalanche may not fall until a great depth of snow has accumulated. The author has seen three- to five-foot-diameter trees broken by large avalanches in the Cascade Mountains of Washington. In general, avalanche chutes can be recognized by either a lack of vegetation where the slides occur at grass root level, or by underbrush-covered slopes (slide alder and the like) where sliding occurs after the brush is covered by snow. Slopes that avalanche frequently rarely allow trees to root on them, but it must also be remembered that at many places avalanches occur on forested slopes without disturbing fairly widely spaced trees of any size. Like mudflows and flash floods, avalanches must be considered in buying or developing mountain property. The writer has been offered attractive resort property with all of these hazards clearly shown in the landscape to anyone with an elementary understanding of geologic processes. Unfortunately, land sellers and developers rarely notice or understand these processes.

FLOODS—DAMMED RIVERS AND DAMNED RIVERS

Any process that puts more water into a river than the river can carry will cause a flood. Intense precipitation and rapid melting of snow, especially in mountainous headwaters, are common causes. Other natural processes such as dams caused by landslides can cause upstream floods. In this latter case, commonly when the lake behind the slide overtops the slide-dam, the rapidly moving water quickly erodes the slide, emptying the lake rapidly and causing downstream flooding. When the ice breaks up each spring on northern rivers, the floating ice blocks themselves can cause damage to bridges and the like; and ice jams cause floods first upstream and then downstream in the same way as the slide dams just described. Except for landslides, the natural processes described are seasonal and are controlled by climatic processes.

Flood prediction requires much data and cooperation between hydrologists and meteorologists and even at best is not an exact science. The hydrologist must know such characteristics of the drainage basin as how long it takes for a rainfall of a given intensity to cause an increase in discharge of the river. This will occur after a low-intensity rain has saturated the ground, and almost at once in a high-intensity storm, but will also depend on how much moisture was in the soil at the start of the storm. He or she must also know the characteristics of the river. How much discharge can it carry before it floods? How much area will be flooded as the height of water rises above bank-full? How rapidly will the wave of high water move downstream? The meteorologist must predict the length and intensity of the storm. If melting snow is involved, he or she must consider the many factors involved in melting snow, such as temperature, humidity, and wind, as well as the amount and condition of the snow. A number of stream gauging stations that report stream height and discharge, and weather stations that report precipitation and other weather elements, are necessary to gather the needed data in a timely fashion.

Data on historical floods have been published by the U.S. Geological Survey in a series of publications called "Hydrologic Investigations Atlases." These atlases show areas flooded and stream gauge data and other data. Flood frequency data show the expected recurrence of floods of different height—that is, the size flood that can be expected every 25 and 50 years, and so forth. If one of these atlases is available for your area, it should be consulted.

Accurate flood warnings can reduce damage by 5 to 15 percent and save many lives. The forecasts must be accurate because if too conservative, people will be caught unprepared, and if unnecessary, will be disregarded. Flood preparations are costly and time consuming. Predictions are difficult because a small error in estimating the height of water as flood level or levee height is approached can make the prediction inaccurate. On the other hand, the more time there is between warning and flood, the better the people are prepared.

A flood consists of several phases, each of which can be at least to some extent controlled. The first phase is the land phase and is characterized by flow of water on the surface. This occurs when the ground surface is either impermeable or saturated so the water must flow overland to the streams. Typically this occurs either during very intense storms or after the ground has been saturated by a long-duration storm. Typical damage caused by this phase is gullying. Control measures are revegetation and terracing. A gross example of how this phase leads to floods occurs when forest or brush fires destroy vegetation. In the following rainy season the water that was previously used by the plants runs off and floods occur downstream. This happens almost every year in southern California.

The second phase of a flood is the channel phase, and this, of course, occurs downstream from the land phase. The objective now is to keep the water within the channel. Dikes, dams, and channel improvements are the main methods of control. Dikes or levees may be temporary or permanent and will prevent flooding only if they are higher than the water level. They are brittle defenses in the sense that where overtopped and broken, the flooding is locally worse than if the water, in the absence of dikes, had spread over a larger area upstream. Channel improvements consist of dredging and straightening the channel to permit more efficient movement of the water. Dams hold back the water and release it at a rate that will not permit flooding downstream. Again, if the dam is too small, or the reservoir is allowed to fill at the wrong time, dams are brittle defenses, and once overtopped, the resulting flood is more destructive.

Dams are expensive and controversial. Many dams are multipurpose and produce electricity and irrigation water as well as flood control. Recreational use is also important. These sometimes conflicting uses lead to controversy. Large dams may affect several states or other countries which, of course, leads to other problems of a nonscientific nature. The scientific controversy that bears on these social and political problems is: Which is more efficient, big dams or little dams in flood control? Should floods be controlled by small dams in the small drainage basins in which they form, or is one large dam nearer to the place to be protected more efficient? There seems to be no clearcut answer to this question. Probably each river system needs a carefully tailored approach using both methods of flood control. A larger question is: What other benefits can or should be gained from dams—power, irrigation, and recreation? This question has social, political, and scientific aspects.

URBANIZATION, FLOODS, AND EROSION

The main effect of urbanization is to increase the runoff, which, in turn, causes increased erosion and, of course, the eroded material is subsequently deposited. The resulting flooding, erosion, and deposition, as we have seen, may also occur in urbanized areas,

FLOOD-HAZARD MAPPING[1]

Flood-hazard mapping is a means of providing floodplain information for planning and management programs. Such information should be designed to assist officials and private interests in making decisions and alternative plans concerning the development of specific lands subject to flooding. Proper use of flood-hazard mapping will help to:

1. Prevent improper land development in floodplain areas.
2. Restrict uses that would be hazardous to health and welfare and which would lead to undue claims upon public agencies for remedy.
3. Encourage adequate stream channel cross-section maintenance.
4. Protect prospective home buyers from locating in flood-prone areas.
5. Preserve potential for natural ground-water recharge during flood events.
6. Guide the purchase of public open space.
7. Avoid water pollution resulting from the flooding of sewage treatment plants and solid waste disposal sites that were located on floodplains.

WHAT IS A FLOOD-HAZARD MAP?

A flood-hazard map uses as its base a standard U.S. Geological Survey topographic quadrangle which includes contours that define the ground elevation at stated intervals. (See Appendix B.) The area inundated by a particular "flood of record" is superimposed in light blue on the map to designate the "flood-hazard area." Also marked on the flood-hazard map are distances (at half-mile intervals) along and above the mouth of each stream and the locations of gauging stations, crest-stage gauges, and drainage divides. Figure 1 shows part of the Elmhurst quadrangle, a typical flood map.

PROFILES AND PROBABILITIES

Accompanying the flood-hazard map are explanatory texts, tables, and graphs, which facilitate their use. One set of graphs shows the linear flood profiles (see Fig. 2) of the major streams in the quadrangle; from them, the user

can tell how high the water rose at any given point during one or more floods.

Another valuable tool (Fig. 8–53) is a set of graphs showing probable frequency of flooding at selected gauging stations. These charts indicate the average interval (in years) between floods that are expected to exceed a given elevation. Frequencies can also be expressed as probabilities which make it possible to express the flood risk or "flood hazard" for a particular property; for example, a given area may have a five-percent chance of being inundated by flood-waters in each year.

HOW TO USE A FLOOD-HAZARD MAP

To illustrate the use of a flood-hazard map, assume that you own property along Salt Creek, near Elmhurst and about half a mile south of Lake Street. Perhaps you plan to build there, and you want to know the risk of being flooded. You examine the Elmhurst quadrangle flood-hazard map (Fig. 1) and note that your property is located at a point 23.5 miles above the mouth of Salt Creek. (The river miles are shown on the map.)

One of the graphs accompanying the map is a flood-frequency curve for Salt Creek (Fig. 8–53). This curve, however, is for a particular point on Salt Creek—the Lake Street Bridge. To apply the flood-frequency relationship to your own property will require an adjustment for the water-surface slope between the two points. So you consult another graph, the one which shows profiles or high-water elevations, of floods along Salt Creek (Fig. 2). There, you find that at the Lake Street Bridge (river mile 24) the 1954 flood crested at 671.5 feet, while at the point you are interested in (river mile 23.5) the crest was at 671 feet.

Returning now to the flood-frequency curve, you find that the 1954 flood has an eight-year "recurrence interval," meaning that, over a long period of time, floods can be expected to reach or exceed that level on an average of once every eight years. That level, you have already found, is 671.5 feet at Lake Street and 671 feet at your property. Another way of thinking

[1]This material is very slightly edited from J. R. Sheaffer, D. W. Ellis, and A. M. Spieker, *Flood-Hazard Mapping in Metropolitan Chicago*. U.S. Geological Survey Circular 601-C, Washington, D.C.: U.S. Government Printing Office, 1970, 14 pp.

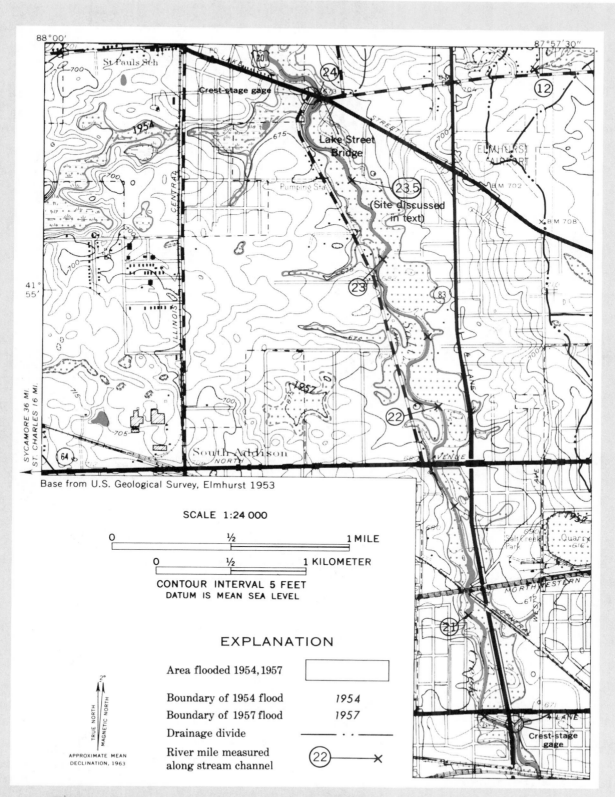

Base from U.S. Geological Survey, Elmhurst 1953

SCALE 1:24 000

0   ½   1 MILE

0   ½   1 KILOMETER

CONTOUR INTERVAL 5 FEET
DATUM IS MEAN SEA LEVEL

TRUE NORTH
MAGNETIC NORTH
2°
APPROXIMATE MEAN
DECLINATION, 1963

EXPLANATION

Area flooded 1954, 1957

Boundary of 1954 flood   *1954*

Boundary of 1957 flood   *1957*

Drainage divide   — · · —

River mile measured
along stream channel   (22) —×

FIGURE 1 Flood-hazard map of part of the Elmhurst quadrangle. Adapted from Ellis, Allen, and Noehre (1963).

175

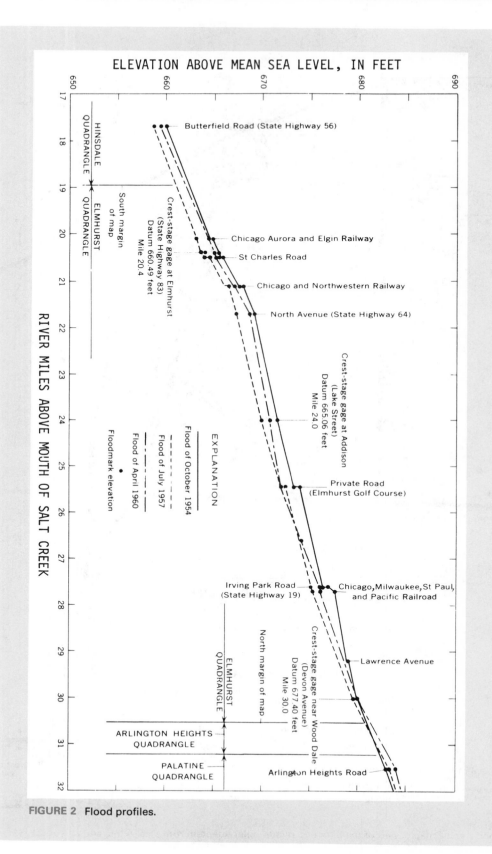

FIGURE 2 Flood profiles.

of it is this: if you were to erect a building on your property at 671-foot elevation, the chances of a flood reaching the structure in any given year would be approximately one in eight. These are only odds—probabilities—and the actuality may be better or worse. But the odds are poorer than most property owners are willing to accept, so you will probably want to seek better odds at higher ground.

Suppose you were willing to accept a flood risk of 1 every 25 years: What is the ground elevation at which a building should be situated to enjoy that much security?

The flood-frequency curve indicates that, at the Lake Street Bridge, an elevation of 672.3 feet corresponds to the 25-year recurrence interval. You now plot this elevation at river mile 24 on the flood-profile chart and draw a straight line through the point you have plotted and parallel to the 1954 flood profile. You now have the profile for a flood with a 25-year recurrence interval, and it shows that the elevation reached by such a flood at your property would be 671.8 feet. Using the line you have drawn, you can determine corresponding elevations (for the same recurrence interval) at other points along Salt Creek. And, of course, you can use the method outlined to approximate the elevation at your property for other recurrence intervals—up to 50 years.

---

compounding the problems (see Fig. 8–51). In a given drainage basin, a rainfall will, after some time period, cause an increase in the flow of the main stream. The length of the time lag and the amount of the increase in the flow will depend on the amount and the intensity of the rainfall, as well as the characteristics of the river. Fig. 8–52 shows an example. Urbanization of a drainage basin will increase both the peak runoff and the total runoff because buildings and paving reduce the infiltration.

The increase in discharge causes increased erosion. The first effect is a deepening of the channel and removal of accumulated debris. This increases the capacity of the river so that it can carry more water without flooding. As the main channel deepens, the tributary streams also deepen because their base level, which is the elevation at which they join the main stream, is lowered. Gullies may appear at many places in the drainage basin.

The increased runoff means less infiltration, so less recharge of ground water supplies. The water table will fall, and water wells must be drilled deeper.

A dam causes changes in a stream. Dams have finite lives because the sediment carried by the stream eventually fills the lake created by the dam. This deposition upstream from the dam occurs not only in the lake but also all along the stream and its tributaries because the rising lake is the new base level of the river. This upstream deposition causes problems, especially for farmers whose irrigation ditches are silted up. Downstream the opposite problem, erosion, occurs. The clear water released from the dam no longer has the sediment load that it carried in the natural state, so it uses its excess energy to erode and deepen its channel. The tributaries also deepen to keep pace with the main stream, and gullies form where they can. These events have all happened to a greater or lesser extent at some dams and have resulted in many lawsuits. Climatic changes can also cause these effects, so some experts deny that dams are the whole cause of these troubles.

Other effects of dams are not always considered until after they are built. If the dam was built for water supply or irrigation, the amount of evaporation from the large surface area of a lake may reduce the amount of useful water. Upstream, the rising water table, due to the lake, may waterlog the soil, causing problems for the farmer. This may also cause a change in the natural vegetation that may or may not be good.

Another unexpected human-caused change occurred in Lake Michigan. The navigation channel built to bypass Niagara Falls allowed the alewives to reach Lake Michigan with unfortunate results for the native fish. The dams in the Northwest, especially on the Columbia River, prevented salmon from swimming upstream to their spawning grounds and so reduced the salmon in the Pacific Ocean. Fish ladders were added or built into new dams, reducing this problem.

The increased erosion caused by either urbanization or dams may make the water muddy, a type of pollution. To this may be added domestic sewage and industrial waste. The river is then no longer useful as a water source or for its esthetic beauty; and the long political, social, and economic tasks of cleaning up the river must be faced. The maximum amount of erosion occurs during the construction phase of urbanization when the natural slope of the ground is altered. More erosion occurs after urbanization than be-

177

A

B

FIGURE 8–51 A. Urbanization at the top of the hill caused mudslides that threatened the lower homes in this Los Angeles subdivision. B. Installation of retaining walls and drains protects the lower homes. Photos from Los Angeles Dept. of Building and Safety.

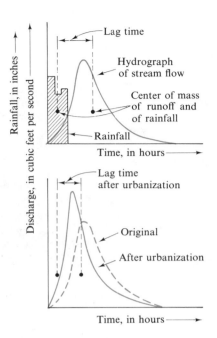

FIGURE 8–52 The effect of urbanization on stream flow. After urbanization, the peak flow is higher and occurs sooner. From U.S. Geological Survey Circular 554, 1968.

fore, in spite of the reduced area where erosion can occur because of building and paving. This is because the increased runoff effectively erodes the uncovered areas.

Southern California, a semiarid area that recently has been urbanized, has many flood and erosion problems. The streams head in mountains, flow across alluvial

fans, then across flat valleys into the Pacific Ocean. As described earlier, storms cause intense flows of water or mudflows. The debris is largely deposited on the alluvial fans at the foot of the mountains. As urbanization moved inland, first the valleys and then the fans were built upon. The rivers in many cases only rarely have water in them, so were disregarded in the building. Alluvial fans are built by the rivers changing their position on the fan from time to time. A mud or debris flow is deposited in the channel on the fan, choking it, and the next flow moves on the fan at a different, lower area. In this way, over a long time, the semicircular fan is built. The river valley and the fan are not distinctive features on the ground, and so the tendency is to disregard them. They are, however, clear on topographic maps and air photos.

Thus flooding, debris flows, or mudflows cause damage on the fans. In the natural state, much of the water seeps into the ground on the fan. Where building prevents this, the water moves down the valley, flooding the urbanized areas there. Much work has been done to minimize these problems. Debris basins have been constructed behind dams to stop the sediment from reaching the fans. Flood conveyance channels have been constructed to carry the water to the ocean. This work was begun as long ago as 1915, or twenty years before the federal government began any flood control work. These measures have saved countless lives and probably billions of dollars damage, but even so, lives and property are still lost. Zoning laws to control the use of the hazardous areas could prevent some of this loss. This simple measure is difficult to achieve because of political and economic complications. One facet of the problem is the large number of municipalities involved. In Los Angeles County alone, the number of incorporated areas has increased from 44 to 78 between 1935 and 1969.

Many of the ways of preventing flood damage have already been mentioned such as dams, debris basins, and zoning laws. Flood-proofing structures by making the lower floors waterproof, either permanently or during flood dan-

ger, can reduce damage. Accurate flood forecasting can make emergency floodproofing effective. Flood-hazard maps, which are available for many areas, can be used for zoning and for predicting how often a given site will be flooded. (See Fig. 8–53.)

Floodplain zoning is perhaps the best answer in spite of the relocation problems for those now using the floodplains. The resulting green belts could be parks that would improve urban life. The fi-

nancial institutions that provide the money for buildings could protect themselves and their borrowers by consulting flood-hazard maps before lending money. Municipalities should not build streets or sewers in these areas, and utility companies should not provide service. All of these agencies would save themselves and their clients money by these simple measures.

Other environmental aspects of rivers are discussed in Chapter 12.

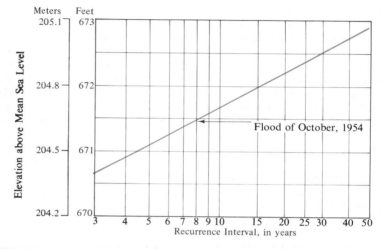

FIGURE 8–53 Frequency of floods on Salt Creek at Addison (Lake Street) near Chicago. This example shows how often a flood of a given height can be expected on the average. From U.S. Geological Survey Circular 601-C, 1970.

SUMMARY

Rivers are agents of erosion, transportation, and deposition. They are the most important agents of transportation. Weathering, downslope movements, and rivers shape the landscape.

Average velocity of river flow is less than 8 kilometers (5 miles) per hour.

Discharge or *flow* is the amount of water passing a given cross-section of the river in a time unit (area of cross-section times the velocity, usually given in cubic feet per second, cfs). The factors controlling the flow vary from place to place on a river, and vary with season.

The main geologic work of rivers is transportation. *Dissolved load* comes from chemical weathering and solution by the river. *Suspended load* is material carried in suspension—clay, and, in turbulent water, silt. *Bed load* is material that rolls or slides along the river bottom, in some cases only during floods.

River erosion is mainly by solution and abrasion. Maximum erosion occurs in areas with about 25 centimeters (10 inches) of rain per year.

River erosion deepens, lengthens, and widens the valley. *Headward*

erosion is downcutting at the head of a stream; it lengthens the valley. The valley widens by creep and landsliding, and by lateral cutting by the stream.

Distinctive curves called *meanders* form which distribute the river's loss of energy most uniformly. Developing meanders may intersect, cutting off a stretch of river, forming *oxbow lakes.*

A graded stream is one that has just the right gradient to be able to transport the debris delivered to it by its own erosion, tributaries, downslope movements, and rain wash. The gradient of a given

179

stretch is maintained by the river's eroding or building up its bed in response to climate or other conditions. The concave slope of the long profile of a river provides for the most uniform rate of work along the river, balancing slope with discharge and velocity.

Base level is the limit to which a river can erode, the elevation of its mouth.

Fluvial (river) deposits are *deltas* (formed where rivers enter standing water), *alluvial fans* (formed where rivers flow out of mountains into broad, flat valleys), and *levees* (formed along river banks when floodwaters spill over, reducing velocity). Rivers also build up their beds by deposition. Deposition occurs when velocity is slowed.

Terraces form from changes in river regimen or conditions; they can form by either deposition or erosion.

Braided streams form when load is large and coarse for the slope and discharge. The channel divides and rejoins many times.

Narrow, *youthful* valleys have V-shaped cross-sections and are generally formed by downcutting, non-graded rivers near their headwaters. They commonly have falls, rapids, and lakes, which are temporary features.

Valley widening begins after a stream has reached grade, usually first near the mouth. Meanders are common features. In *mature* valleys the meander belt occupies most of the width of the valley. *Old age* valleys are much wider than the meander belt.

A V-shaped valley being cut in a wider valley indicates uplift or some change in river's regimen. Slow uplift can cause meanders to be entrenched in narrow V-shaped valleys.

A river valley consists of a channel, a floodplain, and sloping valley sides. The floodplain is the channel of the river during times of high water, and flooding is the normal behavior of the river.

Drainage patterns are affected by climate, runoff, vegetation, and underlying bedrock structure. The rill with the most water downcuts fastest, and by headward erosion captures other rills *(stream piracy),* accelerating its development into a master stream.

Uniformitarianism is the doctrine that the present processes at work, given enough time, caused the present landforms.

William Morris Davis' *cycle of erosion* is a deductive sequence of landforms from initial uplift to complete leveling by weathering and erosion. *Youth* is characterized by deeply incised streams cutting a relatively flat, newly uplifted area. *Maturity,* the stage of maximum relief, exists when no flat interstream areas remain. *Old age* is the stage of rounded, lowered interstream areas and wide valleys. *Peneplain* is the term given to the ultimate flat eroded surface, a rare feature on the earth. Davis' cycle has objections to it, and at best is a special case or an oversimplification.

A balance might result when hill slopes and stream gradients reach equilibrium, and no new shape can develop unless the balance is upset.

The flat part of a river valley is called a *floodplain;* a potential user of such an area should be aware that it may be flooded at times.

Storms and mudflows are also normal, expectable hazards.

Floods occur when the river has more water than can move in its channel. Flood prediction is difficult, but accurate flood warning can reduce damage.

Dams and levees are brittle flood defenses because once overtopped, the amount of flooding is greater.

Urbanization increases the peak and the total runoff, causing more erosion and deposition.

Dams cause deposition upstream and erosion downstream.

Alluvial fans may be subjected to mudflows.

Zoning is probably the only effective way to limit development of hazardous areas.

QUESTIONS

1 How is most of the energy of a river used?
2 How do you account for the fact that the velocity of a river increases downstream?
3 What are the three ways by which a river transports material?
4 What are the processes of river erosion?
5 Describe the sequence of events in the formation and migration of a curve in a river course. (Use diagrams.)
6 A placer deposit is a place where a river has deposited fragments of heavy minerals, such as gold. Where would you look in a river for such deposits?
7 Can you think of some reasons why doubling the velocity of a river increases the amount of material transported four to eight times but only increases the maximum size of material transported four times?
8 Show with a series of sketches the erosion of a river valley.
9 Describe the erosion of an upland area. How does the topography change as erosion proceeds? Your answer depends on several assumptions you must make.

10 What is the final surface produced by long erosion called?
11 What are some characteristics of narrow river valleys?
12 Discuss whether building should be allowed on the floodplains of rivers.
13 Under what conditions are mudflows most apt to occur?
14 How can flood danger be reduced?
15 How does urbanization affect the runoff of streams?
16 Describe the evidence showing rapid uplift of an area.
17 To what was Twain referring in the quotation at the beginning of this chapter?

SUPPLEMENTARY READINGS

General

Carter, L. J., "Soil Erosion: The Problem Persists Despite the Billions Spent on It," *Science* (April 22, 1977), Vol. 196, No. 4288, pp. 409–411.

Curtis, W. F., J. K. Culbertson, and E. B. Chase, *Fluvial-Sediment Discharge to the Oceans from the Conterminous United States.* U.S. Geological Survey Circular 670, Washington, D.C.: U.S. Government Printing Office, 1973, 17 pp.

Davis, W. M., *Geographical Essays,* especially Part II, "Physiographic Essays." New York: Dover Publications, Inc., (1909) 1954, 777 pp. (paperback).

Flemal, R. C., "The Attack on the Davisian System of Geomorphology: A Synopsis," *Journal of Geological Education* (January 1971), Vol. 19, No. 1, pp. 3–13.

Hunt, C. B., *Physiography of the United States.* San Francisco: W. H. Freeman and Co., 1967, 480 pp.

Judson, Sheldon, "Erosion of the Land," *American Scientist* (Winter 1968), Vol. 56, No. 4, pp. 356–374.

Leopold, L. B., and W. B. Langbein, "River Meanders," *Scientific American* (June 1966), Vol. 214, No. 6, pp. 60–70.

Leopold, L. B., and Thomas Maddock, Jr., *The Hydraulic Geometry of Stream Channels and Some Physiographic Implications.* U.S. Geological Survey Professional Paper 252, 1953, 57 pp.

Leopold, L. B., M. G. Wolman, and J. P. Miller, *Fluvial Processes in Geomorphology.* San Francisco: W. H. Freeman and Co., 1964, 504 pp.

Morgan, J. P., "Deltas—A Résumé," *Journal of Geological Education* (May 1970), Vol. 18, No. 3, pp. 107–117.

Morisawa, Marie, *Streams: Their Dynamics and Morphology.* New York: McGraw-Hill Book Co., 1968, 175 pp. (paperback).

Peixoto, J. P., and M. A. Kettani, "The Control of the Water Cycle," *Scientific American* (April 1973), Vol. 228, No. 4, pp. 46–61.

Pillsbury, A. F., "The Salinity of Rivers," *Scientific American* (July 1981), Vol. 245, No. 1, pp. 54–65.

Schumm, S. A., "The Development and Evolution of Hillslopes," *Journal of Geological Education* (June 1966), Vol. 14, No. 3, pp. 98–104.

Thornbury, W. D., *Principles of Geomorphology,* 2nd ed., Chapters 5, 6, 7, 8, pp. 99–208. New York: John Wiley & Sons, Inc., 1969, 594 pp.

Environmental Hazards

Bue, C. D., *Flood Information for Flood-Plain Planning,* U.S. Geological Survey Circular 539, Washington, D.C.: U.S. Government Printing Office, 1967, 10 pp.

Chorley, R. J., *Water, Earth, and Man.* London: Methuen and Co., Ltd., 1969, 588 pp.

Glancy, P. A., *A Mudflow in the Second Creek Drainage, Lake Tahoe Basin, Nevada, and Its Relation to Sedimentation and Urbanization.* U.S. Geological Survey Professional Paper 650C, Washington, D.C.: U.S. Government Printing Office, 1969, pp. C195-C200.

Guy, H. P., *Sediment Problems in Urban Areas.* U.S. Geological Survey Circular 601-E, Washington, D.C.: U.S. Government Printing Office, 1970, pp. E1–E8.

Leopold, L. B., *Hydrology for Urban Land Planning—A Guidebook on the Hydrologic Effects of Urban Land Use.* U.S. Geological Survey Circular 554, Washington, D.C.: U.S. Government Printing Office, 1968, 18 pp.

Rantz, S. E., *Urban Sprawl and Flooding in Southern California.* U.S. Geological Survey Circular 601-B, Washington, D.C.: U.S. Government Printing Office, 1970, pp. B1–B11.

Sheaffer, J. R., and others, *Flood-Hazard Mapping in Metropolitan Chicago.* U.S. Geological Survey Circular 601-C, Washington, D.C.: U.S. Government Printing Office, 1970, pp. C1–C14.

Thomas, H. E., and W. J. Schneider, *Water as an Urban Resource and Nuisance.* U.S. Geological Survey Circular 601-D, Washington, D.C.: U.S. Government Printing Office, 1970, pp. D1–D9.

U.S. Geological Survey, Hydrologic Atlases Portraying Areas Struck by Hurricane Camille on the Mississippi Gulf Coast. HA 395 to HA 408, 1969.

9

DESERTS AND THE GEOLOGIC WORK OF WIND

As outlined in the introduction to the preceding chapter, some of the classical studies on erosion by running water were made in the arid American southwest. Other aspects of those same studies will be considered in this chapter. Thus it should be no surprise that the prevailing opinion is that running water, although rare, is the main agent of erosion there. Wind erosion is much less effective. However, in drier areas, such as parts of north Africa, wind is an important geologic agent. Recognition of this has come in part from study of photographs made by the earth-orbiting satellites. (See Fig. 9–1.) Much of our knowledge of wind erosion comes from wind tunnel experiments published in 1941 by British Brigadier R. A. Bagnold, who became interested in deserts when posted in Egypt.

GEOLOGIC PROCESSES IN ARID REGIONS

Dry regions form over one-quarter of the total land surface of the earth. Most of these arid areas are hot deserts or near-deserts, but some of the dry areas are cold. The main desert areas are in the subtropics near thirty degrees north and south latitude. They are caused by the primary atmospheric circulation of the earth that causes dry air to descend at these latitudes, and the resulting compression heats the already dry air. Other arid areas occur in the rain shadows of high mountain ranges. Southwestern United States is an example of such an area. The moist winds from the Pacific Ocean are cooled and precipitate their moisture when they are forced to rise over the Sierra Nevada and other mountain ranges in California. After passing over these ranges, here, too, the dry descending air is heated by compression.

In most arid areas, the lack of moisture changes the emphasis

but not the types of geologic processes. Rainfall, although infrequent and irregular, is the main agent of erosion. Rain may fall only once in many years at any one place, but is generally in the form of an intense cloudburst or thunderstorm. The lack of vegetation enables such storms to erode very efficiently by splash and runoff. Flash floods in normally dry stream channels many miles from the storm may trap unwary travelers. The runoff carries much loose surface material to the stream channels, and the resulting flow may be a thick mudflow rather than a normal stream. Such mudflows can carry huge boulders and can do great damage to anything in their path. Weathering is slow, and because mechanical weathering is much more important in arid regions than chemical weathering, little soil is formed. Limestone that is rapidly dissolved in humid climates forms bold cliffs in the desert.

DESERT LANDSCAPES

Infrequent rain and the resulting lack of moisture cause desert landscapes to develop slowly. Because of the lack of rain, throughgoing rivers rarely develop. Exceptions are the Colorado and Nile Rivers that begin in high, well-watered mountains and then cross deserts. Most deserts, however, are areas of internal drainage, and this profoundly affects their erosional development.

The development of drainage patterns in deserts is well shown by the Great Basin of southwestern United States. This is an area of many mountain ranges formed by relatively recent faulting. The Basin and Range country is another name sometimes applied to much of this region because basins separate the ranges. The basins are low areas, and lakes may form in them during times of heavy rainfall, such as the ice age described in the next chapter.

Lake-deposited sediments are fairly common in these basins, and dry lake beds form the surfaces of many. The lakes generally contain salt and other dissolved materials that come from the rainwater and chemical weathering. Thus many of them are what are called alkali lakes, and the dry lake beds are alkali flats. Great Salt Lake and Bonneville Salt Flats are examples. Because these lakes are not connected to the ocean, they develop in much the same way as do the oceans. They are, in effect, small oceans.

The usual development of drainage patterns in a desert begins with erosion in the mountains and deposition in the basins. The deposition may occur well out in the basin or as alluvial fans near the mountain front. These processes continue until the basin is filled to the level that the ephemeral streams can flow over the lowest divide separating it from an adjoining basin. (See Fig. 9–2.) The processes now continue in the new basin until it, too, is filled. While the second basin is being filled, the divide between the two basins is generally lowered by erosion; this initiates a stage of erosion in the first basin because of a lowered base level. (See Fig. 9–3.) This complicated series of events will continue until an outlet to the ocean is established; however, a connection to the ocean is rarely developed because lack of rain makes these processes very slow.

The erosional reduction of desert mountains is somewhat different from that of more humid climates. The differences may be basic, but are more likely the result of different emphasis of the various erosional processes. Also,

Motorcycles and dune buggies have destroyed the vegetative cover in the area of the Mojave Desert in California. Dust plumes from this area have been photographed from space. Photo from John Nakata, U.S. Geological Survey.

FIGURE 9-1 Satellite photographs of sandy deserts or sand seas. A. Rows of star dunes. Algerian desert. B. Star dunes similar to A but not in rows. Algerian desert. ERTS photographs.

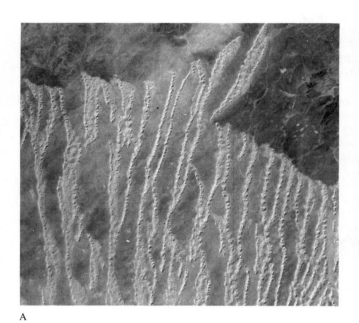

A

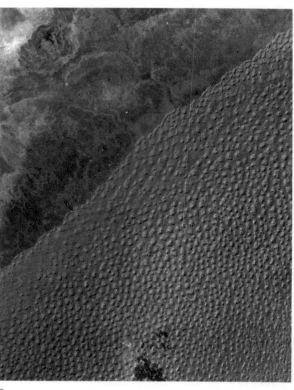

B

as just noted, local base levels and small drainage areas are involved, rather than large watersheds, so that each desert range or part of a range develops at its own rate, making the developmental sequence easy to see.

The first step is generally the development of *fans* at the mountain front. After this, the wearing back of mountains begins. *The initial steep mountain front retreats, and a fairly smooth surface in the bedrock is eroded. This surface is called a* pediment *and is generally continuous with the fan. The pediment is commonly covered by a thin layer of gravel, making it very difficult to distinguish from a fan unless the generally present gullies cut through the gravel. The method by which pediments are formed is not clear, but streams issuing from the mountains may change their courses back and forth across the pediment in a process similar to the way alluvial fans are formed. Runoff is probably also important in pediment

formation. Pediments have concave-upward slopes, much like river profiles. The slope of most pediments is between one-half and seven degrees and is related to the particle size of the gravel

that covers it, and probably the amount of runoff. The steep mountain front behind the pediment seems to maintain its slope angle during retreat. The various stages in the erosion of a desert

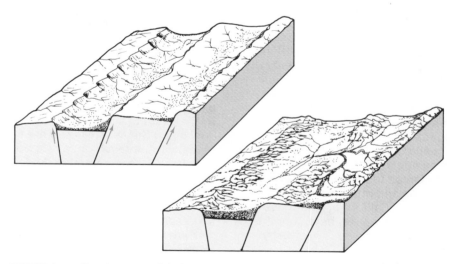

FIGURE 9-2 Development of drainage systems in the Basin and Range area. As each basin is filled, the drainage spills over the lowest divide into the next basin. This process continues, slowly integrating the drainage of many basins until the ephemeral streams reaches the ocean—a goal rarely reached.

184

FIGURE 9–3   A complex drainage system that resulted from integration of basins. The low range in the foreground and middleground is being vigorously eroded, probably because its base level was lowered by the integration of basin drainage. Some possible future captures can be seen. In the far middleground, several basins that have integrated are visible. Note the lack of water in the stream courses and the almost complete lack of vegetation in this view near the southern border of California. Photo from U.S. Geological Survey.

range are shown in Figs. 9–4, 9–5, and 9–6. The examples shown are faulted mountains similar to those in the arid areas of southwestern United States. Other types of mountains are found in other arid regions. The formation of faulted mountains is described in Chapter 15.

GEOLOGIC WORK OF WIND

EROSION

Only in dry areas is wind an active agent of erosion, transportation,

and deposition. The rapid heating and cooling of deserts leads to strong winds. The behavior of wind is similar to that of water in some ways, but it is much less effective. Wind erosion occurs by two processes, deflation and sandblasting (or abrasion). *Deflation is simply the removal by wind of sand- and dust-sized particles.* Typical places where deflation may take place are dry, unvegetated areas such as deserts, dry lake or stream beds, and actively forming, glacier-outwash plains in dry seasons. Deflation may, in some areas, produce hollows,

sometimes called blowouts, that are recognized by their concave shape, which could not have been produced by water erosion. (The only other common way that closed depressions form is by solution of soluble rocks, such as limestone, by ground water.) At other places deflation removes the fine material from the surface, leaving behind pebbles, to produce a surface armored by a pebble layer against further deflation. Such surfaces are called *desert armor* or *pavement.* (See Fig. 9–7.) Alternate wetting and drying can also cause pebbles to rise by a

185

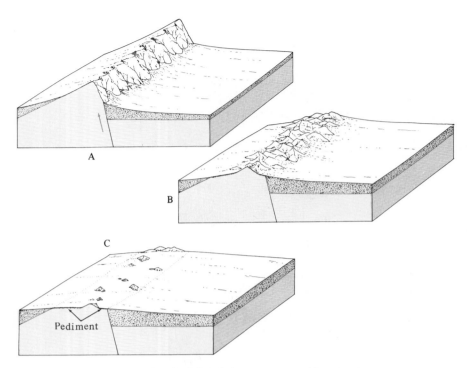

FIGURE 9–4 Formation of pediments. A. Shortly after uplift. B. Pediments have begun to form. C. Late stage. The mountain range is mainly pediment.

process somewhat like frost heaving discussed in Chapter 7. *Sandblasting by wind-driven sand grains can cause some erosion near the base of a cliff or on a boulder* (see Fig. 9–8), but probably is much less effective than weathering. (See Fig. 9–9). Sandblasting does produce some interestingly shaped pebbles called *ventifacts,* or *dreikanter* (German for "three edges"). *Sand hitting a pebble on the side facing the prevailing wind wears a rounded surface that may be polished and slightly pitted.* Most ventifacts have several such surfaces, three being common. The other surfaces may be produced by other common wind directions or, probably more commonly, by the turning of the pebble, perhaps due to the undermining of the pebble by deflation of its supporting sand. (See Fig. 9–10.)

Wind transportation is similar to water transportation in that both a suspended load and a bed load are involved. The *suspended load produces dust storms,* and the *bed load forms sandstorms.* Most windblown sand is less than 1 millimeter (mm) in diameter, and the average size is about $\frac{1}{4}$ mm. Sand begins to move when the wind reaches 17.7 kilometers (11 miles) per hour. All material finer than $\frac{1}{16}$ mm (silt) is considered dust, and most windblown dust is much smaller than this. The separation into these two size ranges is very sharp, much more so than in the case of water transportation, so that wind-deposited sediments are generally much better sorted by size than water-laid deposits. The reason for this better sorting is that wind can raise sand particles only a few feet above the ground, although it can raise dust thousands of feet. At one place where measurements were made, 50 percent, by weight, of the sand was raised less than 13 centimeters (5 inches) above the ground

FIGURE 9–5 The faulted eastern face of the Sierra Nevada is only slightly modified by erosion. Photo from U.S. Geological Survey.

FIGURE 9–6   The main range shown here has been embayed by erosion with the development of pediments and fans. In the right foreground, the mountains have been reduced to low pediments. Photo from U.S. Geological Survey.

FIGURE 9–7   Desert armor or pavement, Greenwater Range, eastern California. Photo by C. S. Denny, U.S. Geological Survey

FIGURE 9-8   Granite outcrop that has been undercut by windblown sand. Atacama Province, Chile. Photo by K. Segerstrom, U.S. Geological Survey.

FIGURE 9-9   Double Arch in Arches National Monument, Utah. The arches are believed to be partly the result of wind erosion of thick sandstone beds. Water from infrequent rains probably loosens the sand near the surface, and wind removes the loose sand grains and causes some abrasion. Photo from National Park Service.

the action of water. The process produces a bouncing motion of the sand grains. It begins when a grain rolls into or over another grain. When the first grain hits, its impact transfers its kinetic energy to one or more grains which, in turn, also bounce up and, on falling, continue the process by hitting other grains (Fig. 9–12). Except in large downdrafts, dust cannot be lifted by wind alone because at ground level there is a thin zone of no wind about one-thirtieth of the diameter of the average particle. The impact of falling sand grains can put the dust into motion, and the turbulence of the wind will tend to keep it in motion.

The amount of windblown material raised by a storm can be very high. For example, during some of the intense dust storms in the United States in the 1930s (Fig. 9–13), the load in the lower mile of the atmosphere in some areas was estimated at over 32,675 metric tons per cubic kilometer (150,000 tons per cubic mile) of air.

Since the 1930s when the photographs shown in Figures 9–11 and 9–13 were taken, wind erosion has been controlled to some extent by planting rows of trees, called shelter belts. The rows of trees across the prevailing wind direction slow the wind near the surface, and this plus other methods of conserving soil moisture have largely eliminated the "dust bowl" of the 1930s.

DEPOSITION

Wind deposits are of two types— *loess composed of dust,* and drifts and dunes composed of sand. *Loess forms thick, sheetlike deposits that are relatively unstratified because of their rather uniform grain size, generally in the silt size range.* (See Fig. 9–14.) Thick loess deposits in China are believed to have originated from the deserts of Asia. Many American loess deposits are associated with glacial deposits and are be-

and 90 percent was raised less than 64 centimeters (25 inches), although some large fragments rose over 3 meters (10 feet). At this place, maximum abrasion occurred 23 centimeters (9 inches) above the ground. Such abrasive effects a few inches above ground are common on fence posts and

power poles in areas of sandstorms. As might be imagined, sandstorms are associated with dust storms, but dust storms may occur without sandstorms; so, in general, dust is moved much farther than sand. (See Fig. 9–11.)

The mechanism for moving sand by wind is similar, in part, to

A

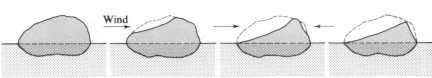

B

FIGURE 9–10 A. Ventifacts. Photo by M. R. Campbell, U.S. Geological Survey.
B. Formation of ventifacts by wind erosion. The main facet faces the prevailing wind
direction. Other faces are formed by other common wind directions, or the pebble is
turned, exposing another side to the wind.

FIGURE 9–11 A dust storm caused by a cold front at Manteer, Kansas, in April 1935.
The dust is in the cold air mass and rises to great heights. Photo from NOAA.

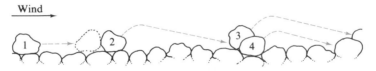

FIGURE 9–12 Movement of bed load. Sand grain 1 rolls over and hits grain 2 which
bounces up and hits grains 3 and 4, both of which also bounce up and hit more sand
grains.

189

FIGURE 9–13   Abandoned farm in Oklahoma in the 1930s. Destruction of the natural vegetation by plowing and drought resulted in dust storms that removed the topsoil. Photo from U.S. Department of Agriculture.

lieved to have formed when glacial retreat left large unvegetated areas on which winds could act. Fertile soils have formed on loess deposits that cover much of the central plains of the United States.

Windblown sand may accumulate as *sheets* or in long, narrow *stringers.* Such deposits are probably similar to loess deposits. The variety and extent of windblown sand deposits was realized when satellite images of the entire earth became available. Sand dunes attract our attention because of their variety. A *dune is a mound or ridge of windblown sand with one or more slipfaces* (Fig. 9–15). *The wind moves the sand grains up the gentle slope of the dune, and the grains fall off the top, forming the slipface.* This process causes many types of sand dunes to move as indicated in Fig. 9–15.

FIGURE 9–14   Loess near Vicksburg, Mississippi. The upper layer of loess is somewhat different from the lower. Although loess can stand in vertical cliffs, it is very easily eroded where a little water spills over the cliffs. Photo by E. W. Shaw, U.S. Geological Survey.

Another result of this motion of the sand grains in a dune is the development of *cross-bedding* in sand dunes. One way that cross-bedding forms in water-deposited sedimentary rocks is shown in Fig. 4–27. Very well developed cross-bedding in ancient sandstones with uniform grain size (well sorted) suggests that the sandstone may have been deposited in an arid region (Fig. 4–28B). On present-day dunes, ripple-marks

and footprints of desert animals are common; these features, too, are found in ancient dune deposits.

Several distinct types of dunes can be recognized. Many dunes probably start in the lee of an obstacle (Fig. 9–16). The most common dunes are *barchans* (Figs. 9–17 and 9–18A). Barchans have a single slipface and form where the winds come from a single direction. If there is more sand avail-

able, the bachans grow into each other, forming wavy rows called *barchanoid ridges* (Fig. 9–18B). If more sand is available, *transverse ridges* develop (Fig. 9–18C).

The shapes of some dunes are controlled by moisture or vegetation that restricts the movement of the sand grains. *Blowouts* (Fig. 9–19) and *parabolic dunes* (Fig. 9–20) form in this way. Like barchans, they form in areas with prevailing winds from a single di-

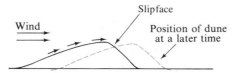

FIGURE 9–15   The movement of a sand dune. The wind moves individual sand grains along the surface of the dune until they fall off the steep face.

FIGURE 9–16   Accumulation of sand in lee of fence post.

FIGURE 9–17   Barchan dunes in the Columbia River Valley near Biggs, Oregon. Photo by G. K. Gilbert, U.S. Geological Survey.

A. Barchans   B. Barchanoid ridge   C. Transverse dune

FIGURE 9–18   A. Barchans. B. Barchanoid ridge. C. Transverse ridge. These dunes have one slipface. As sand supply increases, barchans grade into transverse ridges. After E. D. McKee, U.S. Geological Survey Professional Paper 1052.

191

rection, but slipfaces develop only where the sand grains are free to move.

Other dune types form where there is more than one common wind direction. *Linear dunes,* also called *seifs* and *longitudinal dunes,* form where there are two or more winds from the same general direction (Fig. 9–21). These dunes are ridges with two slipfaces. *Reversing dunes* are also ridges with two slipfaces; but in this case the two winds are from opposite directions, and the dune forms perpendicular to the winds. With three or more wind directions, *star dunes* (Fig. 9–22) form.

Domes are dunes with no obvious slipfaces, but when they are excavated, cross-bedding is revealed. This suggests that they once had slipfaces that have been obscured.

The classification of sand dunes is summarized in Fig. 9–23. As might be expected, many dunes and dune areas are not easily classified because the dunes have intergrown.

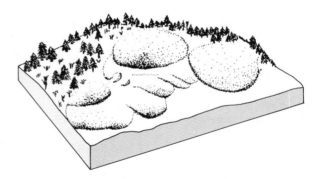

FIGURE 9–19 Blowout dunes. After E. D. McKee, U.S. Geological Survey Professional Paper 1052.

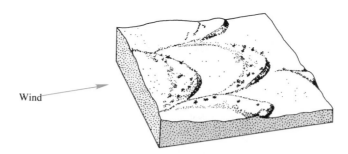

FIGURE 9–20 Parabolic dunes. After E. D. McKee, U.S. Geological Survey Professional Paper 1052.

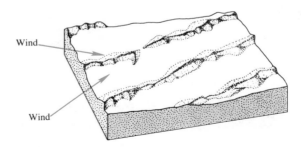

FIGURE 9–21 Linear dunes, also called *seifs* and *longitudinal dunes.* Arrows show wind direction. After E. D. McKee, U.S. Geological Survey Professional Paper 1052.

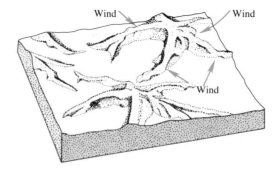

FIGURE 9–22 Star dunes form with three or more wind directions. After E. D. McKee, U.S. Geological Survey Professional Paper 1052.

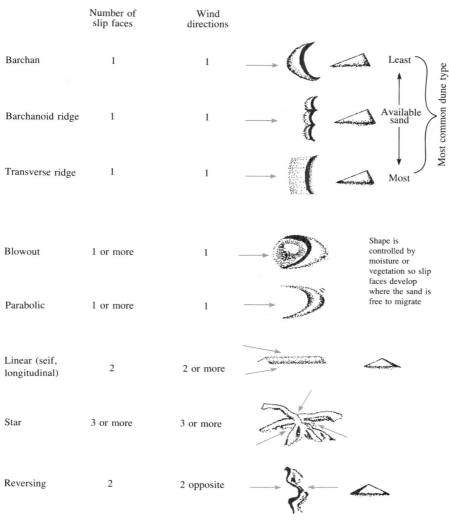

	Number of slip faces	Wind directions		
Barchan	1	1		Least
Barchanoid ridge	1	1		Available sand
Transverse ridge	1	1		Most
Blowout	1 or more	1		
Parabolic	1 or more	1		
Linear (seif, longitudinal)	2	2 or more		
Star	3 or more	3 or more		
Reversing	2	2 opposite		

Most common dune type

Shape is controlled by moisture or vegetation so slip faces develop where the sand is free to migrate

FIGURE 9–23 Classification of sand dunes.

SUMMARY

Intermittent rainfall and running water are the main agents of erosion in most arid areas, but wind may be important in very arid areas.

Rainfall erodes very efficiently by splash and runoff. Runoff carries much loose surface material to stream channels, and the resultant flow may be a thick, destructive mudflow.

Mechanical weathering is more important than chemical weathering in arid regions, so little soil is formed.

Limestone may form cliffs rather than be dissolved as in more humid climates.

Throughgoing rivers rarely develop because of lack of rainfall.

Lakes may form in basins during heavy rainfall; they generally contain salt and other dissolved material from rain or chemical weathering. They are called *alkali lakes,* or, when dry, *alkali flats.*

Drainage patterns in deserts develop as the result of erosion in mountains and deposition in basins or on alluvial fans. When a basin is filled and the mountains are reduced sufficiently, the stream flows over the lowest divide into the adjacent basin. This lowers the base level and so initiates new erosion in the first basin and deposition in the adjacent basin.

Pediments are smooth, gravel-covered, concave-upward erosional surfaces cut in the bedrock of desert mountains. They are continuous with the alluvial fans deposited at the mountain fronts.

Wind erosion occurs by two processes. *Deflation* is the removal of sand- and dust-sized particles. It may produce closed depressions called *blowouts*. The removal of fine material may leave a surface armored by pebbles. *Sandblasting* is abrasion by wind-driven sand grains.

Ventifacts, or *dreikanter,* are pebbles with polished facets or surfaces produced by sandblasting.

Wind transportation involves *suspended load* (dust storms) and *bed load* (sandstorms). All material finer than $\frac{1}{16}$ mm is considered dust; most windblown dust is smaller. Most windblown sand is less than 1 mm, average about $\frac{1}{4}$ mm. The size separation is very sharp, and wind-deposited sediments are well-sorted by size because dust can be raised much higher and transported longer than sand.

Loess is a wind deposit composed of silt; it forms thick, sheet-like, unstratified deposits.

Dunes are naturally formed accumulations of windblown sand. Unless fixed by vegetation, dunes move slowly across the desert.

A *dune* is a mound or ridge of windblown sand with one or more *slipfaces.* The wind moves the sand grains up the gentle slope of the dune, and the grains fall off the top, forming the *slipface.* Dune types are summarized in Fig. 9–23.

QUESTIONS

1 What is the role of climate in erosion and the development of landforms?
2 Describe pediments.
3 Briefly describe wind erosion (two kinds).
4 Wind-deposited silt is called _____.
5 Deposits of windblown sand are called _____.
6 Describe the sorting of wind deposits.
7 How might wind-deposited sediments differ from water-deposited sediments?
8 Describe the development of desert pavement.
9 Why is wind not an important agent of erosion in humid, windy climates?
10 Explain how a slipface forms.
11 What types of dunes have only one slipface?
12 With two or more prevailing wind directions, which types of dunes may form?

SUPPLEMENTARY READINGS

Bagnold, R. A., *The Physics of Blown Sand and Desert Dunes.* London: Methuen & Co., Ltd., 1941, 265 pp.

Carter, L. J., ''Soil Erosion: The Problem Persists Despite the Billions Spent on It,'' *Science* (April 22, 1977), Vol. 196, No. 4288, pp. 409–411.

McKee, E. D., *A Study of Global Sand Seas.* U.S. Geological Survey Professional Paper 1052, Washington, D.C.: U.S. Government Printing Office, 1979, 429 pp.

Péwé, T. L. ed., *Desert Dust: Origin, Characteristics, and Effect on Man.* GSA Special Paper 186. Boulder, Colorado: Geological Society of America, 1982, 303 pp.

Sharp, R. P., ''Wind-driven Sand in Coachella Valley, California,'' *Bulletin,* Geological Society of America (September 1964), Vol. 75, No. 9, pp. 785–804.

Thornbury, W. D., *Principles of Geomorphology,* 2nd ed., Chapter 12, pp. 288–302. New York: John Wiley & Sons, Inc., 1969, 594 pp.

10 GEOLOGIC WORK OF ICE— GLACIERS

Some say the world will end in fire,
Some say in ice.

> Robert Frost
> "Fire and Ice"

The recognition that glaciers in the recent geologic past had covered much of the northern hemisphere was made in 1836 in the Alps. From observations that deposits far down valleys were similar to the materials currently being deposited by glaciers, Charpentier and others had concluded that glaciers were at one time much more extensive. In 1836 Charpentier showed the field evidence to Jean Louis Rodolphe Agassiz (1807-1873). Agassiz, at first skeptical, was quickly convinced. He quickly grasped that glaciation could not occur only in Switzerland, but that a lower temperature would be worldwide, and this realization led to the recognition of the Ice Age. He announced his conclusions in 1837, and in a few years the evidence for past continental glaciation was recognized all over Europe and North America. Agassiz traveled extensively in his study of glaciation and spent his last 27 years at Harvard. Prior to Agassiz, glacial deposits were thought to be remnants of the Flood or to have been carried on icebergs.

At the time of Agassiz' discovery, uniformitarianism was being adopted by many as a result of the publication of Lyell's book in 1830. Neptunism had died out, and the other theory, catastrophism, was on the wane. Catastrophism is the idea that geological changes come about by rapid changes, such as earthquakes creating great valleys. The climatic changes necessary to bring about an ice age seemed rapid, so Agassiz' discovery slowed the acceptance of uniformitarianism for a short time.

Although glaciers are much less important than rivers in overall erosion, they have shaped many of the landforms of northern North America and Eurasia. In addition, most of the high mountain ranges of the world have been greatly modified by mountain glaciers.

FORMATION OF GLACIERS

A *glacier is a mass of moving ice.* Glaciers form as a result of accumulation of snow in areas where more snow falls than melts in most years. This can occur at sea level at the poles to about 6000 meters (20,000 feet) near the equator. The accumulation of snow must, however, become thick enough so that it recrystallizes to ice. The recrystallization process depends on the pressure of the overlying snow, which transforms the light, loosely packed snow into small ice crystals; with increased pressure, the small crystals become larger. (See Fig. 10-1). The process is quite similar to the metamorphism of sandstone into quartzite and of limestone into marble. The tendency of snow to recrystallize and form larger crystals is well known to anyone who skis or who has lived in snowy areas. Falling snowflakes are generally small and light, but after a few sunny days, or in the spring, they are much larger at the surface; digging will reveal even larger crystals. When the amount of ice formed by this recrystallization becomes large enough that it flows under its own weight, a glacier is born. Ice, as we see it, is a brittle substance, so the fact that a glacier flows under its own weight, much like tar, may seem strange; however, a mass of ice a few hundred feet thick behaves as a very viscous liquid.

Glaciers are of two main types: the alpine or mountain glacier, and the much larger continental or icecap glacier. Present-day continental glaciers occur only near the poles, but mountain glaciers occur at all latitudes where high enough mountains exist. (See Table 10-1.)

MOUNTAIN GLACIERS

BUDGET AND MOVEMENT

Mountain glaciers (see Fig. 10-2) *develop in previously formed stream valleys, which, because they are lower than the surrounding country, become accumulation sites for snow.* When enough ice has formed, the glacier begins to move down the valley. How far down the valley it extends will depend on how much new ice is formed and how much melting occurs each year. As the glacier extends further down the valley, the amount of melting each year will increase until a point is reached at which it equals the amount of new ice added. Such a glacier is in equilibrium. If the amount of snowfall increases or the summers become cooler, it will advance further; and if the snowfall decreases or the summers become warmer, it will melt back. However, it generally takes a number of years for the more- or less-than-normal snowfall of a single year, or group of years, to become ice and to reach the snout of the glacier. This means that although a glacier is sensitive to climate changes, it tends to average snowfall and temperature changes over a number of years. Hence, the advance or the retreat of gla-

Icebergs forming where a glacier meets the ocean. Storströmmen, east Greenland. Reproduced with the permission (A.477/73) of the Geodetic Institute, Denmark.

197

ciers gives some information on long-term climatic changes. Study of old photographs and maps shows, for instance, that the gla-

ciers in Glacier National Park—and elsewhere—have retreated in the last few decades. Glaciers reached maximum advances about 1825,

1855, and 1895. In general, they have been receding since the turn of the century although with some minor advances. An advance appears to be occurring at present. One point needs emphasis: *the glacial ice is always moving down-valley, even if the snout of the glacier is retreating because of melting.*

TABLE 10–1 Glaciers in the United States. Data from U.S. Geological Survey.

State	Approximate Number of Glaciers	Total Glacier Area	
		Square Miles	Square Kilometers
Alaska	?	About 17,000	About 44,030
Washington	800	160	414
Wyoming	80	18	47
Montana	106	10	26
Oregon	38	8	21
California	80	7	18
Colorado	10?	1	2.6
Idaho	11?	1	2.6
Nevada	1	0.1	0.3
Utah	1?	0.1?	0.3?

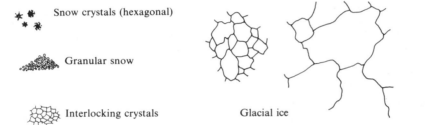

Snow crystals (hexagonal)

Granular snow

Interlocking crystals

Glacial ice

FIGURE 10–1 Some of the steps in the transformation from snow to glacial ice. Crystals in glacial ice can be several inches in diameter.

The movement of glacial ice can be measured by driving a series of accurately located stakes into the glacier and surveying them periodically. As would be expected, because of friction, the sides of the glacier move more slowly than the center. The upper 30 to 60 meters (100 to 200 feet) of the glacial ice behaves brittlely so that *the different rates of flow at the surface open large cracks, called crevasses,* in this brittle zone. (See Figs. 10–3 and 10–4.) The movement of mountain glaciers varies from less than an inch a day to over 15 meters (50 feet) a day.

EROSION

Erosion by mountain glaciers produces such spectacular mountain scenery as Yosemite Valley in the Sierra Nevada of California. Glaciers erode by plucking large

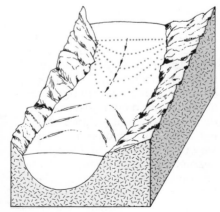

FIGURE 10-3 Movement of glaciers and the development of crevasses. The movement is shown by the successive locations of stakes driven into the ice in the upper part of the figure. The opening of crevasses in the brittle ice as a result of the greater movement of the center of the glacier is shown in the lower part of the figure.

FIGURE 10-2 Mountain glaciers. Several mountain glaciers join in this valley. Note the crevasses on the surface of the ice and the moraines both on the valley sides and separating the ice tongues from the different glaciers. Photo from Swissair Photo Ltd.

FIGURE 10-4 Oblique aerial view of Mount Hood, Oregon, showing crevasses on the glaciers. Note the U-shaped valleys. Photo from U.S. Geological Survey.

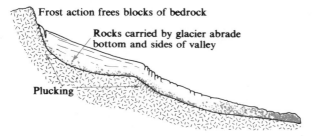

FIGURE 10-5 Longitudinal section through a mountain glacier showing development of a cirque by plucking, and reshaping of the valley by plucking and abrasion. This sketch shows an early stage in the development of a glacial valley.

gradient of the valley. This deepening, together with the steepening of the sides and the head of the valley, produces a *large amphitheaterlike form that closely resembles a teacup cut vertically in half. Cirque is the name applied to this form.*(See Fig. 10–6.)

A glacially eroded valley is recognized by the U-shape, the cirque at the head, and the grooves, scratches, and polishing due to abrasion. (See Fig. 10–9.) In addition, any small hills or knobs in the valley have been overridden by the glacier, and they are rounded and smoothed by abrasion on the struck side and steepened by plucking on the lee side. (See Fig. 10–10.) Many glaciated valleys are nearly flat with rises or steps where more resistant rocks crop out. (See Figs. 10–11 and 10–12.) A glacier moving down a valley also tends to straighten the valley, because the glacier cannot turn as abruptly as the original river could. This has the effect of removing the spurs or ridges on the insides of curves of the stream valley. The amount of ice in a main-stream valley, whose

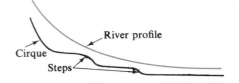

FIGURE 10-6 Glacial erosion of river valleys. Steps are commonly developed and may contain lakes.

blocks of bedrock and by abrasion. *Plucking is accomplished by meltwater that flows into joints in the bedrock and later freezes to the main mass of ice which, on advancing, pulls out the blocks of loosened bedrock.* These blocks arm the moving ice with rasplike teeth that grind the sides and bottom of the glacial valley (Fig. 10–5). Glacial erosion greatly modifies the shape of the stream valley occupied by the glacier. Most of the erosion is probably done by plucking, and abrasion generally smooths and even polishes the resulting form. A glaciated valley differs from a stream valley in that it is deepened, especially near its head, and the sides are steepened so that its cross-section is

changed from V-shaped to U-shaped (Figs. 10–6, 10–7, and 10–8). Glacial erosion is most active near the head of the glacier, and the deepening there flattens the

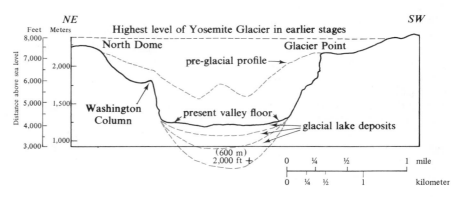

FIGURE 10-7 Glacial modification of Yosemite Valley, California. From Clyde Wahrhaftig, California Division of Mines and Geology Bulletin 182, 1962.

199

FIGURE 10-8 Oblique aerial photograph of a fiord. A fiord is a U-shaped, glacially eroded valley that is an arm of the sea. Valley glaciers in this fiord are fed by a continental glacier. Evighedfjord, west Greenland. Reproduced with the permission (A.477/73) of the Geodetic Institute, Denmark.

source is near the crest of the mountain range, is generally much greater than the amount of ice in a tributary-stream valley; hence, the main valley is eroded deeper by the ice than is the tributary. After the ice melts, *the tributary stream, in a valley called a hanging valley, is higher than the main stream into which it flows via a waterfall.* (See Figs. 10–13 and 10–14.)

Even the high part of a mountain range that is above the level of the glaciers develops characteristic forms. Increased frost action here produces narrow, jagged ridges extending up to the peaks. *The peaks develop pointed, pyramidal shapes, called horns,* that are largely due to headward erosion by cirque development. (See Figs. 10–2 and 10–12.)

DEPOSITION

Glaciers also deposit the debris that they carry. *Drift is the name used for all types of glacial deposits, and till is used for ice-deposited sediments. The material in the glacier is carried by the ice to the snout, where it is deposited to form the moraine.* (See Fig. 10–15.) The *terminal moraine marks the furthest advance of the glacier down the valley. If the glacier retreats from this position it may deposit a series of end moraines* above the terminal moraine. Because the glacier carries a great variety of sizes of fragments, moraines are unsorted; huge boulders are mixed with all sizes of fragments down to silt and finer. (See Fig. 10–16.) The pebbles are generally faceted and striated by abrasion. Because the snout of a glacier is curved, the terminal moraines are characteristically curved. They continue along the sides of a glacier as *lateral moraines.* The lateral moraines *develop, in much the same manner as terminal moraines, as a result of melting of the slow-moving ice at the valley sides.* (See Figs. 10–2 and 10–15.) Many of the rocks deposited by glaciers are scratched or striated from abrasion during transport. The meltwater leaving the snout of a glacier also transports and deposits glacial material, forming outwash deposits that in some cases form thick sheets. The meltwater from an active glacier is colored white due to *fine material, called rock flour, from the abrasive erosion of the glacier.* This material resembles clay, but microscopic examination shows that it is composed of fine mineral fragments, not clay. It is the source of the *varved "clays"* to be discussed later. *When a glacier melts, its transported load is dropped wherever it happens to be and forms ground moraine. Upon melting, blocks of ice buried in ground moraine leave depressions, which may become small ponds called kettles.* (See Fig. 10–17.)

CONTINENTAL GLACIERS

The geologic work of continental glaciers is very similar to that of mountain glaciers. Their erosive effects are much less and are generally limited to rounding the topography and scraping off much of the soil and overburden. (See Figs. 10–19, 10–20, and 10–21.) Their deposits are like those of mountain glaciers, but are more extensive. One type of deposit that helped lead to the recognition of continental glaciation is the *erratic boulder.* (See Fig. 10–22.) *This term is applied to boulders, some of which weigh up to 18,000 tons, that are foreign to their present locations.* In some areas, occurrences of distinctive erratic boulders can be traced back to their origin, thus indicating the path of the glacier.

Continental glaciers occupied parts of North America and Europe in relatively recent geologic time, and they occupy Greenland and Antarctica now. (See Fig. 10–23.) During the recent glacial pe-

A

B

C

FIGURE 10-9 Evidence of glacial abrasion. A. Polished and grooved granite caused by the rock-studded base of a moving glacier. Photo by Mary Hill, California Division of Mines and Geology. B. Glacial grooves partly destroyed by mechanical weathering. Photo by C.W. Chesterman, courtesy California Division of Mines and Geology. C. Glacial scratches from continental glacier. Clinton, Massachusetts. Photo by W.C. Alden, U.S. Geological Survey.

FIGURE 10-10   Small hills of resistant rock are smoothed by abrasion on the struck side and steepened by plucking on the lee side. The arrows show the direction of the main force of the glacier that moved from right to left. The jointing in the rock is indicated. From F.E. Matthes, U.S. Geological Survey Professional Paper 160, 1930.

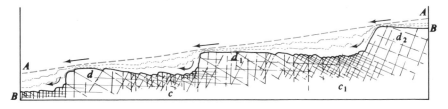

FIGURE 10-11   Origin of steps in a glacial valley. *AA* is the profile of the preglacial valley and *BB* the post-glacial profile. Closely jointed rocks at c and c_1 are readily plucked by the glacier, but sparsely jointed rocks as at d, d_1, d_2 are reduced by abrasion. From F.E. Matthes, U.S. Geological Survey Professional Paper 160, 1930.

FIGURE 10-12   A glacial valley showing rock steps with lakes on the steps. Photo by S.A. Davis, courtesy California Division of Mines and Geology.

FIGURE 10-13 The development of Yosemite Valley illustrates both river and glacial erosion. A. The broad valley stage prior to the uplift of the Sierra Nevada. B. After uplift, the rivers downcut about 214 meters (700 feet). C. After the second uplift, the Merced River cut a canyon about 366 meters (1300 feet) below A. D. Yosemite Valley just after the glacier retreated. The lake was dammed by glacial moraine, and the filled lake is the present valley floor. B, Mount Broderick; BV, Bridalveil Creek; C, Clouds Rest; CC, Cascade Cliffs; CR, Cathedral Rocks; EC, El Capitan; EP, Eagle Peak; HD, Half Dome; LC, Liberty Cap; LY, Little Yosemite Valley; MR, Merced River; MW, Mount Watkins; ND, North Dome; R, Royal Arches; SD, Sentinel Dome; TC, Tenaya Creek; W, Washington Column; YC, Yosemite Creek. From F.E. Matthes, U.S. Geological Survey Professional Paper 160, 1930.

FIGURE 10-14   Yosemite Valley. Figure 10-13 shows the steps in the development. Note that the present river is very small for the size of the valley. Photo by S.A. Davis, courtesy California Division of Mines and Geology.

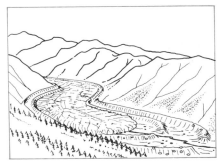

FIGURE 10-15   Idealized sketch of a glacier leaving successive moraines as it melts back. Both end moraines and lateral moraines are shown. From F.E. Matthes, U.S. Geological Survey Professional Paper 160, 1930.

FIGURE 10-16   Glacial moraine containing many kinds of rock in all sizes from boulders to silt. Photo by Mary Hill, courtesy California Division of Mines and Geology.

FIGURE 10-17　Pond created by the melting of a block of ice buried in moraine. Note the variety of sizes of rocks in the moraine. Baird Glacier, southeastern Alaska. Photo by A.F. Buddington, U.S. Geological Survey.

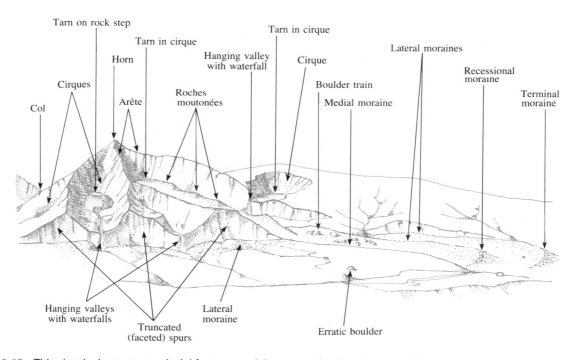

FIGURE 10-18　This sketch shows many glacial features and the names that have been applied to them.

FIGURE 10-19 Aerial view of the 1918-meter (6288-foot) summit of Mount Washington, New Hampshire, the highest point in the northeast. The continental glaciers overrode and smoothed all of the rounded mountains in the view. Photo by Dick Smith, New Hampshire Division of Parks.

FIGURE 10-20 Glacially eroded area near Hidden Lake, Northwest Territories, Canada. Continental glaciers have removed the soil and caused many small depressions occupied by lakes. The bedrock structures can be seen. The light-colored areas are granite bodies, and the white lines are pegmatite dikes; both features are discordant to the foliation of the bedrock. Oblique aerial photo from Department of Energy, Mines and Resources, Ottawa, Canadian Government Copyright (A5032-62.R).

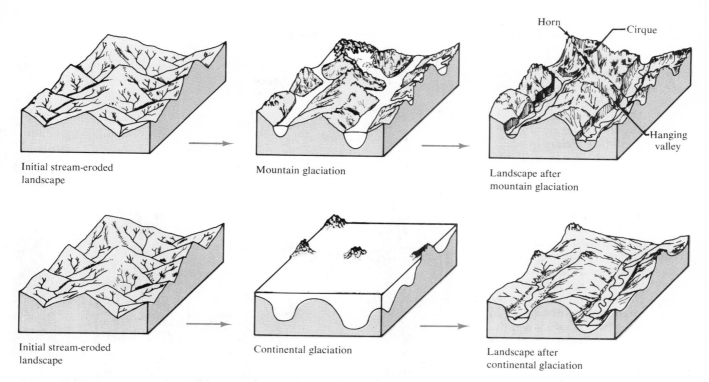

Initial stream-eroded
landscape

Mountain glaciation

Landscape after
mountain glaciation

Horn

Cirque

Hanging
valley

Initial stream-eroded
landscape

Continental glaciation

Landscape after
continental glaciation

FIGURE 10-21   Mountain and continental glacial erosion of a hypothetical landscape are shown. Continental glaciers create smooth, rounded topography, and frost action above mountain glaciers forms jagged peaks.

FIGURE 10-22   Erratic boulders dropped by a glacier when it melted. Photo by Mary Hill, courtesy California Division of Mines and Geology.

riod, there were four main times of glacier development in North America and at least five in Europe. In North America, there were several centers of ice accumulation in Canada. Glaciers moved out from these centers in all directions. Thus, although the ice moved generally southward in the United States, it moved northward in northern Canada. In no sense did the last glacial age result because ice from near the north pole moved south. Recognition of four advances depends largely on field observations. In some places fresh, unweathered moraines overlie weathered moraines which in turn overlie more deeply weathered moraines. Tracing such relationships led to the recognition of four main stages with the extents shown on the map in Fig. 10–24.

Moraines, however, are not the only types of deposits left by continental glaciers. In some places, *blocks of ice were buried in the moraine and when the ice melted, depressions formed.* These depressions are called *kettles,* and lakes form in many of them (Fig. 10–17).

Eskers are long, sinuous ridges, generally 3–15 meters (10–50 feet) high, that run for fractions of a mile to many miles. They are believed to be the filling of meltwater channels that ran under the glacial ice (Fig. 10–25).

Drumlins are streamlined hills 7.5–60 meters (25–200 feet) high and several hundred meters long. Their origin is not known, but the streamlined shape suggests that they formed under the moving glacial ice. They are not symmetrical, and the blunt end is in the direction of the origin of the ice. (See Fig. 10–26.)

PLEISTOCENE GLACIATION

The most recent continental glaciation of Pleistocene age consisted of four major advances and

FIGURE 10-23 The edge of a continental glacier where it meets the ocean. Inglefieldland, northwest Greenland. Reproduced with the permission (A.477/73) of the Geodetic Institute, Denmark.

began about 2.5 million years ago. (See Table 10–2) This date is estimated from the degree of weathering of deposits, from the amount of soil produced between glacial advances, from the study of fossils, from radioactive dating, and from other information obtained by a variety of methods. The last glacial advance began about 30,000 years ago, reached its maximum advance about 19,000 years ago, and retreated about 10,000 years ago. Another minor advance occurred between 1550 and 1850 and is sometimes called the Little Ice Age. These dates are fairly accurately known from carbon-isotope studies. This method is based on the fact that organic carbon has a fixed ratio of carbon isotopes when formed, but this ratio changes after the death of the organism due to radioactive decay of one of the carbon isotopes (carbon 14). The method is

not suitable for material more than about 40,000 years old. Radioactive dating methods are discussed elsewhere.

Before radiocarbon was used, other methods, particularly the study of varved clays, were used. This method is much like the study of tree rings, as the varves are assumed to be yearly deposits in glacial lakes. The dark part of each varve is deposited during the summer, and the light part in the winter. (See Fig. 10–27.) Thus the thickness of varves records the climatic conditions, and the sequence of relative thicknesses is correlated from lake to lake until the life span of the glacier is covered; then the total number of varves is counted. Apparently due to error in correlation, this method gave too high an age estimate.

The climatic conditions that cause glaciation also had far-reaching effects, even far from the

208

areas of actual glacial accumulation. It is estimated that sea level was lowered about 152 meters (500 feet) during the glacial advances, and that if the present-day glaciers melt, sea level will rise between 30 and 46 meters (100 and 150 feet).

CAUSE OF GLACIATION

GLACIAL CLIMATE AND GLACIAL LAKES

To discover the cause of glacial periods, we must first see under what conditions glaciers form. Average temperatures must be lower than at present, but not greatly lower; and precipitation must be high. These two conditions will cause the accumulation of snow, which will form glaciers. The temperature must be low enough to insure that precipitation will be in the form of snow, but if it is too low, it will inhibit precipitation. This latter point is illustrated by the lack of glacier formation, because of low precipitation, in the many very cold regions today. An average annual temperature decrease of about 6°C will probably cause a new onset of glaciation.

Heavy precipitation during at least parts of the glacial age is suggested by the huge lakes that formed in areas south of the glaciers, where it was too warm for snow to accumulate. On the other hand, a humid climate without much evaporation may also have been involved in the formation of these lakes. These lakes formed in the basin areas of western United States, and the present Great Salt Lake is but a small remnant of Lake Bonneville, which covered much of northwestern Utah. Other lakes covered large areas in Nevada, and some of the present saline lakes, such as Carson Sink, are remnants. (See Fig. 10–28.) These lakes are saline because,

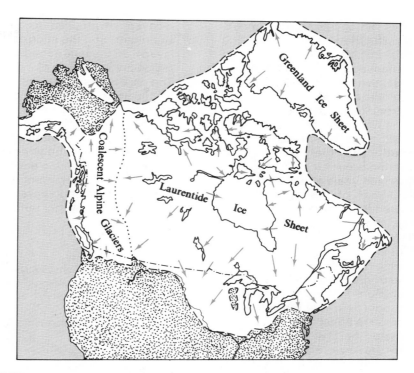

FIGURE 10-24 Maximum extent of Pleistocene ice sheets. Glaciers also formed in the mountains in both eastern and western North America. The Laurentide ice sheet is now known to have consisted of many smaller centers of accumulation with radial flow outward from each.

FIGURE 10-25 Esker. The long, somewhat curving ridge is an esker and probably originated as a meltwater channel under the continental glacier. Oblique aerial photo from Department of Energy, Mines and Resources, Ottawa, Canadian Government Copyright (A2711-94).

without outlets to the sea, dissolved material builds up in them while evaporation keeps their levels fairly constant.

THEORIES

A number of theories of glacier origin have been proposed, but none is completely satisfactory. A successful theory must explain the change in climate just discussed. It also must account for the multiglacial advances that occur during an ice age, such as the four separate advances of the recent ice age, as well as for other periods of glaciation at different places on the earth. The other periods of glaciation occurred about 275 and 600 million years ago. (See Table 10–3.) Many of these glaciations occurred far from the present poles. The main theories are as follows:

1. Changes in the amount of energy received from the sun. Astronomers detect very slight changes in the sun's energy production, very much too small to affect the earth's climate. The 11-year sunspot cycle, however, shows that the sun's processes do change, and many other stars are variable stars.

    A variation of this theory is that as the sun moves through space in its trip around the center of the Milky Way galaxy, it may encounter regions of space containing dust. Such dust would limit the amount of the sun's energy reaching us. Dust clouds of this type have not been detected near the sun. Both of these ways of limiting the amount of the sun's energy reaching us are not testable, and no evidence for them exists.

2. Changes in the amount of the sun's energy reaching the earth's surface, resulting from changes in the earth's atmosphere. Dust from volcanic eruptions could cause cooling of the surface. On the other hand, dark particles in the atmosphere might absorb the sun's energy and so heat the atmosphere. Volcanic dust does eventually settle to the earth's surface where it forms thin layers of volcanic ash. The number of these layers can be counted in sedimentary cores

recovered from deep-ocean sediments, and the ages of the layers can be determined by radioactive methods of dating. The results of such studies are inconclusive, and few people believe that volcanic dust is the main cause of ice ages.

Decrease in the amount of carbon dioxide in the atmosphere would cool the earth, as this gas allows radiant energy from the sun to pass through, but prevents radiation from the earth from leaving. Thus the carbon dioxide in the atmosphere acts much like a greenhouse; however, it is doubtful if the necessary changes in the amount of carbon dioxide are possible.

3. Continental drift caused by plate tectonics can move continents nearer to the poles and so help to start an ice age. At high latitudes temperatures are lowered, and this would aid, but not cause, develop-

TABLE 10–2 North American glacial stages.

Age in Millions of Years before Present	Glacial Advance	Interglacial Times
0	Wisconsinan	
		Sangamonian
.5	Illinoian	
1		Yarmouthian
1.5	Kansan	
		Aftonian
2		
2.5	Nebraskan	
3		

ment of glaciers. Continents near the poles would tend to obstruct poleward movement of warm tropical ocean water because the continents would be in the way of such water movements. Mountain building and uplift in parts of the continents would also cause locally cooler temperatures and would affect atmospheric circulation. Probably continents must be near the poles for a glacial age to occur, but some other cause must also be in operation.

4. Periodic changes in the earth's motion around the sun affect the amount of the sun's energy received at any point on the earth. This is an old theory recently revived. In the 1930s Milutin Milankovich, a Serbian mathematician, suggested that the effects of the earth's motions could cause glaciers to form. His calculations were based on ideas that were nearly 100 years old then.

Milankovich recognized three periodicities. The earth's orbit around the sun is not quite circular, and the average distance between sun and earth changes slightly with about a 105,000-year period. The calculated effect of this period is very small because the total energy input from the sun only varies about 0.1 percent. The seasons are caused by the tilt of the earth's rotational axis relative to the plane of the earth's orbit. The tilt is currently 23.5 degrees, but it varies between about 22 degrees and 24.5 degrees, with a period of 41,000 years. The changing tilt angle changes the contrast between the seasons. The third periodicity is in the direction in which the earth's axis points. This motion is known as the *precession of the equinoxes,* and the period of this motion is about 26,000 years. The precession determines where on the orbit the seasons occur, that is, whether winter in the northern hemisphere occurs when the earth is closest or farthest or some other distance from the sun. This motion also accen-

FIGURE 10-26　Aerial view of drumlins, Stefansson Island, Northwest Territories. The ice flowed toward the upper right. Photo from Geological Survey of Canada.

FIGURE 10-27　Varves, or yearly layers. Photo shows varves in the Green River Formation, Colorado. Diameter of area photographed is 5 centimeters (2 inches). Photo by W.H. Bradley, U.S. Geological Survey.

211

tuates the contrast between seasons.

The effects of these cycles can be tested in several ways. Cores of sedimentary rocks that range in age from the present back through much of the last ice age can be obtained in some lakes and in the deep ocean. The pollen in the cores obtained on the continents can be studied, and climatic changes can be inferred from the types of trees and plants. In the marine cores tiny fossils can also be studied, and the temperature of the seawater can be determined from study of the isotopes of oxygen present.

The cores that have been studied revealed climatic cycles of 23,000 years, 41,000 years, and 100,000 years in remarkably close agreement with the theoretical predictions. This suggests that these orbital motions do affect climate; however, in detail the agreement is not so good. The 100,000-year cycle is dominant in the cores, and the theory predicts that it should have the smallest effect. There is also a lag of 7000 to 8000 years between the predicted and actual time of maximum extent of glaciation. Finally, the glacial advances in the last ice age lasted between 15,000 and 30,000 years, not the 10,000 predicted.

5. Changes in the circulation of the oceans. This is a promising hypothesis that seems adequate for the recent glacial periods. The hypothesis begins with an ice-free Arctic Ocean. With the present pole positions and our present atmospheric circulation, the ice-free Arctic Ocean would cause the heavy precipitation necessary to initiate a glacial advance. By the middle of the glacial advance the Arctic Ocean would freeze, cutting off much of the supply of moisture. Precipitation would continue until the North Atlantic became too cold to provide enough moisture. Melting would then begin and continue until the Arctic Ocean was again ice-free. The stage is now set for the cycle to begin again.

This hypothesis can account for glacial periods only when the poles are located as they are now. If both poles were over open ocean, then the atmospheric circulation could not develop glacial periods. Thus, to initiate a glacial period, the poles would have to shift. Thus, this theory accounts for the four glacial ad-

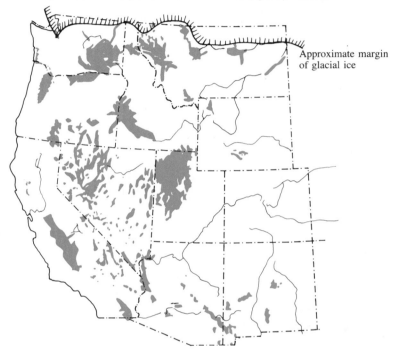

FIGURE 10-28 Map showing the known lakes of glacial times in western United States. From J.H. Feth, U.S. Geological Survey Professional Paper 425B, 1961.

TABLE 10-3 Glacial periods in the geologic past. The geologic periods are discussed in Chapter 16. See Fig. 16-2. Data from Steiner and Grillmair, 1973.

Age		Location
Million Years	Geologic Period	
0–3	Pleistocene	Northern Hemisphere
235–320	Permian–Carboniferous	Southern Hemisphere (See Chapter 14)
410–470	Ordovician–Silurian	Africa, South America, Europe, and northern North America
570—680	Late Precambrian	Australia, Europe, Greenland, Newfoundland, Brazil, Algeria, China
740—825	Precambrian	Australia, Europe, Siberia, China, North America, Africa
950	Precambrian	Africa, Asia
2,300	Precambrian	North America

vances of the recent glacial period; it can also explain the irregular periods of older glaciation. Recent studies, however, suggest that the present Arctic ice may be much older than the last glacial advance, a finding which casts doubt on this theory.

All of the causes of glaciation discussed have both advocates and dissenters because none of them is entirely satisfactory.

SUMMARY

A glacier is a mass of moving ice, which forms as a result of accumulation of more snow than can melt. The thick snow recrystallizes under pressure to ice. When enough ice is formed, it flows downhill as though it were a very viscous fluid.

Mountain or *alpine glaciers* develop in previously formed stream valleys, whose relative deepness makes them accumulation sites for snow.

The snout of the glacier advances or retreats in response to long-term climate changes in snowfall and temperature. The glacial ice is always moving down-valley, even though the snout may be retreating because of melting.

Crevasses, or large cracks, are formed by different rates of flow in the brittle upper portion of the glacier.

Glaciers erode by two methods. *Plucking* is the process in which meltwater flows into joints in bedrock, later freezes to the main ice, which, on advancing, pulls out the loosened block of bedrock. *Abrasion* is performed by the transported blocks of rock grinding, scratching, and polishing the sides and bottom of the valley.

Glaciated valleys have U-shaped cross-sections rather than the stream valley's V shape. The sides are steepened; the valley is flattened and straightened.

A *cirque* is the large amphitheater-shaped form cut at the head of a glaciated valley.

A *hanging valley* is the glaciated valley of a tributary stream. The glacier erodes the main valley deeper, steepening its sides, and the water flows from the hanging valley into the main stream via a waterfall.

A *horn* is the pointed pyramidal peak formed above glacier level by headward development of adjacent cirques.

Drift is the name for all types of glacial deposits.

Till is the term used for ice-deposited sediments.

Moraines are unsorted deposits of rock material carried by the glacier. *Terminal moraines* are characteristically curved deposits at the snout. *Lateral moraines* are deposits along the glacier's sides. *Ground moraine* is the deposit left when a glacier melts and drops the load it is transporting.

Rock flour, composed of fine mineral fragments, colors meltwater from an active glacier white; it forms varved clay deposits.

Continental or *icecap glaciers* cover large areas. In relatively recent geologic time there were four advances in parts of North America and at least five in Europe. Continental glaciers occupy Greenland and Antarctica now.

Erratic boulders are boulders foreign to their present locations that were glacially transported.

Kettles are depressions formed by the melting of blocks of ice buried in moraines.

Eskers are long, sinuous ridges, believed to be the filling of meltwater channels that ran under glaciers.

Drumlins are blunt-ended, streamlined hills composed of till.

The Pleistocene glaciation began about three million years ago. The last glacial advance began about 30,000 years ago; maximum, 19,000 years ago; retreat, 10,000 years ago. The Little Ice Age was a minor advance between 1550 and 1850.

Dating of glaciation is by several methods. More recent than 40,000 years is dated by radioactive decay of carbon 14. Before radiocarbon was used, varved clays (yearly layers deposited in glacial lakes, dark in summer and light in winter) were counted and correlated. Other methods are study of weathering of deposits, soil produced between advances, fossil study, and other means.

Glaciers form under special conditions. The temperature must be low enough to insure that the precipitation is snow, but not so low as to inhibit precipitation. The precipitation must also be heavy. Where it is too warm for snow to accumulate, the heavy precipitation may form huge lakes such as Lake Bonneville in Utah, whose remnant is Great Salt Lake, and lakes in Nevada, one of whose remnants is Carson Sink.

Older glaciation occurred 275 million years ago in Africa, India, and Australia, and also 600 million years ago.

Theories of causes of glacier formation are:

Changes in amount of energy received from the sun, due to changes either in output by the sun or in the amount that penetrates our atmosphere.
Continental drift.
Periodic changes in the earth's orbit.
Changes in circulation of the oceans.

Melting and forming of glaciers causes changes in sea level.

QUESTIONS

1 What is a glacier?
2 The two main types of glaciers are _____ and _____.
3 What changes does a mountain glacier produce in its valley?
4 How does a present-day glacier show climatic changes?
5 Why do crevasses form on the surface of glaciers?
6 Describe the processes of glacial erosion.
7 What are the erosional effects of continental glaciers?
8 What is a glacial erratic?
9 How many glacial stages are recognized in the Pleistocene?
10 Which way did the ice move in northern Canada? What is the evidence?
11 What type of climate is most favorable for the development of glaciers?
12 What evidence do we have as to the climate during glacial times in southwestern United States?
13 List some suggested causes of continental glaciation.
14 Outline the theory of glacier origin that depends on changes in ocean currents between the Arctic and the Atlantic oceans.
15 Which of the three agents of transportation — wind, water, or ice — is able to transport the largest boulders; which is most limited as to the largest size it can transport?
16 How do mechanical weathering and glacial erosion work together?
17 What is the evidence for continental glaciation?
18 How do glacially deposited rocks differ from those found in river deposits?
19 How does the fine material in a river with active glaciers at its head differ from the material in a river in a nonglacial area?
20 Describe several types of glacial deposits.

SUPPLEMENTARY READINGS

Beaty, C. B., "The Causes of Glaciation," *American Scientist* (July-August 1978), Vol. 66, No. 4, pp. 452–459.

Davis, W.M., *Geographical Essays,* especially "The Sculpture of Mountains by Glaciers," "Glacial Erosion in France, Switzerland, and Norway," and "The Outline of Cape Cod." New York: Dover Publications, Inc., (1909) 1954, 777 pp. (paperback).

Damuth, J.E., and R.W. Fairbridge, "Equatorial Atlantic Deep-sea Arkosic Sands and Ice-age Aridity in Tropical South America," *Bulletin,* Geological Society of America (January 1970), Vol. 81, No. 1, pp. 189–206. Describes climatic changes far from the areas of glaciation.

Denton, G.H., and S.C. Potter, "Neoglaciation," *Scientific American* (June 1970), Vol. 222, No. 6, pp. 101–110.

Donn, W.L., "Causes of the Ice Ages," *Sky and Telescope* (April 1967), Vol. 33, No. 4, pp. 221–225.

Flint, R.F., *Glacial and Quaternary Geology.* New York: John Wiley & Sons, Inc., 1971, 892 pp.

Hays, J.D., John Imbrie, and N.J. Shackleton, "Variations in the Earth's Orbit: Pacemaker of the Ice Ages," *Science* (December 10, 1976), Vol. 194, No. 4270, pp. 1122–1132.

Hough, J.L., *Geology of the Great Lakes.* Urbana, Ill.: University of Illinois Press, 1958, 313 pp.

Hutter, Kolumban, "Glacier Flow," *American Scientist* (January-February 1982), Vol. 70, No. 1, pp. 26–34.

Imbrie, John, and K.P. Imbrie, *Ice Ages. Solving the Mystery.* Short Hills, N.J.: Enslow, 1979, 224 pp.

Kerr, R.A., "Milankovich Climate Cycles: Old and Unsteady," *Science* (September 4, 1981), Vol. 213, No. 4512, pp. 1095–1096.

Koteff, Carl, and Fred Pessl, Jr., *Systematic Ice Retreat in New England.* U.S. Geological Survey Professional Paper 1179, Washington, D.C.: U.S. Government Printing Office, 1981, 20 pp.

11 SHORELINES

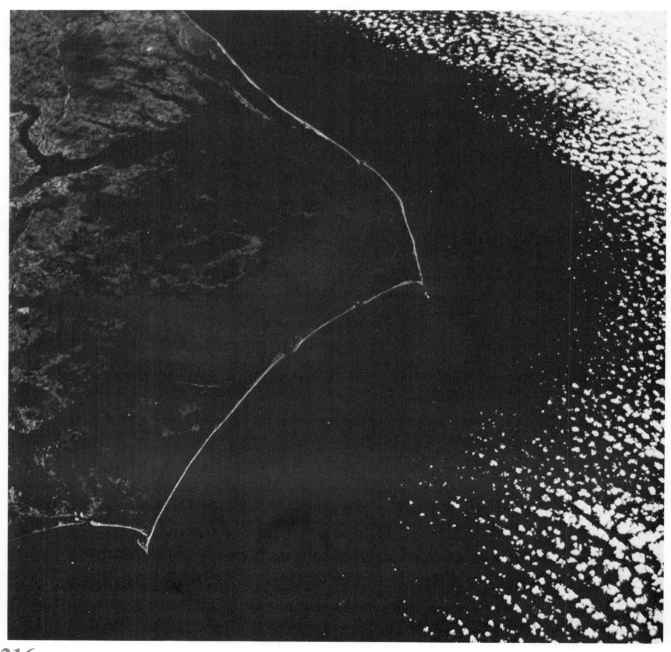

Waves breaking on the shore are also agents of erosion, transportation, and deposition. Breaking waves have a great deal of energy, and anyone who watches them cannot fail to be impressed by this. (See Fig. 11–1.) It is so obvious that they are agents of erosion that for many years they were considered more important than the much less obvious, less spectacular rivers. Before discussing shoreline processes, we will consider the origin of the various movements of ocean waters.

CIRCULATION OF THE OCEANS

The movement of the surface waters of the oceans is caused by wind and the earth's rotation. The wind blowing over the water exerts a drag and sets the water in motion. In this process, first ripples and then waves are generated. The circulation developed in this manner is shallow, at the most only a few hundred feet deep; and the water moves slowly, generally between a fraction of a mile per hour and a few miles per hour. The resulting circulation is rotational because the earth's rotation and the continents deflect the moving water. The earth's rotation causes a deflection of the water's forward motion to the right in the northern hemisphere, and to the left in the southern hemisphere. In the southern part of the southern hemisphere, where no continents intervene, the circulation is latitudinal. (See Fig. 11–2.)

The deep circulation is caused largely by density differences in the ocean water. The cold, heavy Arctic, and especially the Antarctic, waters flow near the bottom toward the equator and in some cases to points beyond. In contrast, the waters of the warm Mediterranean, which have a high salinity because of evaporation, flow into the Atlantic at middle depths.

Much more study is needed before the circulation of the oceans can be described in detail, but the circulation is much less important geologically than waves. Oceanic circulation is, however, a very important means of heat transfer, greatly affecting the climates of the world and so affecting geologic processes.

TIDES

Tides are associated with gravitational attraction, mainly due to the moon. Newton's law of gravity shows that there is a mutual attractive force between two bodies. The oceans are free to move and so are deformed by this force. Thus, the water on the side of the earth toward the moon moves a little closer to the moon, causing a high tide at that point. It is not quite so apparent that a high tide forms at the point on the opposite side of the earth at the same time. The reason for the two tidal bulges is that centrifugal forces caused by the earth's rotation are also involved in producing tides. In an equilibrium system such as the earth-moon system, the gravitational force between the two bodies is exactly balanced by the centrifugal force. This balance is exact at the centers of the two bodies, but at the earth's surface the two forces are not equal. On the side near the moon, the gravitational force toward the moon is greater; and on the opposite side, the centrifugal force acting in the opposite direction is greater. (See Fig. 11–3.) Because the moon orbits the earth, it rises about 51 minutes later each day. Thus, successive high tides occur about 12 hours and 25 minutes apart. The sun also affects tides because of its gravitational attraction, and all of these effects change somewhat with the season. This is the origin of tides, but this theoretical approach cannot be used to predict either the height or the time of actual tides at a coastline because tides are greatly affected by the shape of the coastline and the configuration of the ocean bottom.

Tidal currents may move sediments at some places; but, except at these places, tides probably are not important agents of transportation or erosion. A notable exception occurs where tides move through narrow inlets, constantly scouring these entrances to many good harbors that would otherwise be blocked. (See Fig. 11–4.) The vertical movement of the tides is in the range of a few feet to a few tens of feet, and this constantly changing datum complicates the near-shore processes.

Do not confuse tidal currents with the misnamed *tidal waves* (seismic sea waves) that are caused by earthquakes in the oceans, discussed in Chapter 13.

The solid rock earth is, like the oceans, also deformed by the tidal forces. Although difficult to measure, the movement is three to six inches.

WAVES

Waves can be described by their height and wavelength. These terms are defined in Fig. 11–5. The wavelength and *period* are related to the velocity at which the wave travels by the equation

Wavelength = period × velocity

The velocity, however, depends on the wavelength; and longer wavelength waves (longer period waves) travel at higher velocities. Period is much easier to measure than wavelength, so most relations are expressed in terms of period. In deep water, the velocity in kilometers per hour is about 5.6 times the period in seconds (3.5 times for miles per hour). The

Barrier beaches, North Carolina. Cape Hatteras juts farthest into the Atlantic Ocean. Cape Lookout is near the bottom of the picture. Photo from NASA.

height of a wave has almost no effect on its period or velocity.

Waves, like surface circulation, are caused by wind. Wind blowing over water causes ripples to form; and, once formed, the ripples provide surfaces on which the wind can act more efficiently. The buildup of ripples into waves is a complex process. Turbulence in the wind forms ripples of differing size and direction that combine to form other waves. The energy of the shorter wavelength waves is largely absorbed by longer waves, increasing the height of the longer waves. The height of the waves produced depends on the strength of the wind, how long the wind blows, and how large an area the wind blows over. Most large waves originate in storms; but, once formed, the waves can travel great distances with little loss of energy. Thus, a storm's energy may be transmitted by waves to a distant shore.

Typical waves in a storm area have periods near 10 seconds. As the waves move out of the storm area, their periods (and wavelengths) increase. At distances of a few thousand miles from the area of their formation, waves have periods between 15 and 20 seconds and may travel faster than the winds that generated them.

In deep water, waves are surface shapes; and though the shape moves forward with the wind, the water essentially does not. An individual water particle in a wave moves in a circular path and returns to almost its original position. (See Fig. 11–6.) Thus, in deep water a floating object bobs up and down. The wind does drag the water slowly forward, causing the slow surface circulation (see Fig. 11–7); but in the case of waves, it is the form, not the water, that moves. The motion is much like a wind wave through tall grass or wheat.

In shallow water the motion is very different. (See Fig. 11–8.) Because waves form at the surface, their energy decreases downward. At a depth of one-half the deep water wavelength, the water motion is negligible. As long as the water is this deep, a wave is unaffected by the depth of the water. If a wave travels into water shallower than this, however, it feels the bottom and changes. The

FIGURE 11–1 Breaking waves are agents of erosion. Photo from California Division of Beaches and Parks.

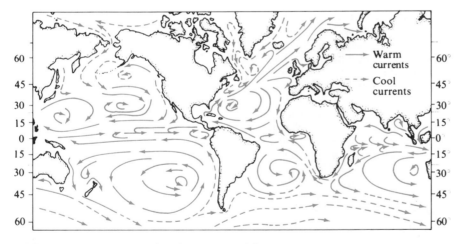

FIGURE 11–2 Circulation of surface waters of the oceans.

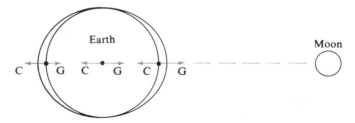

FIGURE 11–3 Origin of tides. On the side of the earth toward the moon, the gravitational attraction (G) between the moon and the earth is greater than the centrifugal (C) force. On the other side of the earth, the centrifugal force is greater. At the earth's center, the two forces balance. The tidal bulge is greatly exaggerated, and the earth–moon distance is greatly shortened.

FIGURE 11-4 Vertical aerial photograph of inlet to Apalachicola Bay, Florida. The outgoing tide has carried out into the ocean the sediment brought into the bay by the Apalachicola River. At the time the photo was taken, clear water was moving through the inlet and the cloudy water forms a turbidity current in the ocean. In the upper right of the photo, clouds of the fine sediment are being moved by wave action.

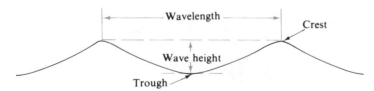

FIGURE 11-5 Characteristics of a wave.

wave travels slower, and some of its energy is used in moving small grains on the bottom back and forth. The height of a wave increases as the water becomes shallower; and on a sloping shore, the front of a wave is slowed more than the rear. These effects cause waves to break, creating surf. In the *surf zone,* the breaking waves

run up on the shore and are active agents of erosion and transportation. (See Fig. 11-9.) Their return to the ocean creates the "undertow" feared by many swimmers. On the landward side of breaking waves, the surface water moves on shore and the bottom water moves outward, thus tending to upset a bather. *At times the returning wa-*

ter is concentrated at a few places, creating what are called *rips,* or *rip currents.* (See Fig. 11-10.)

Along many coasts, after breaking, the water moving onshore forms new waves that in turn break closer to the shore. This process may be repeated several times, especially if the waves are large and the bottom slopes gently.

Wave erosion, then, is concentrated in shallow water. Waves generally break at depths of about one and one-half times their height. Waves over 6 meters (20 feet) high are rare, although waves up to 30 meters (100 feet) high have been reported. At depths less than 9 meters (30 feet), therefore, the majority of wave energy is expended and most of the geologic work of waves, or surf, is done. Because waves first begin to act on the bottom at about one-half their wavelengths, however, some wave erosion can occur as deep as 305 meters (1000 feet). Waves with 10-second periods have wavelengths of about 156 meters (500 feet), and 20-second waves have about 610-meter (2000-foot) wavelengths.

Another effect of shallow water on waves is to cause their bending, or refraction. When a wave feels the bottom, it slows. If parts of the wave do not feel the bottom, these parts are not slowed. In this case, because different parts of the wave are traveling at different speeds, the wave bends or refracts. The effect of refraction is to concentrate the waves on headlands, as shown in Fig. 11-11.

WAVE EROSION

Waves erode mainly by hydraulic action and abrasion, and to a much smaller extent by solution. *Hydraulic action is the impact and pressure caused by waves.* Storm waves, in particular, are able to move huge blocks of bedrock. Breaking waves can cause great pressure in cracks by the

219

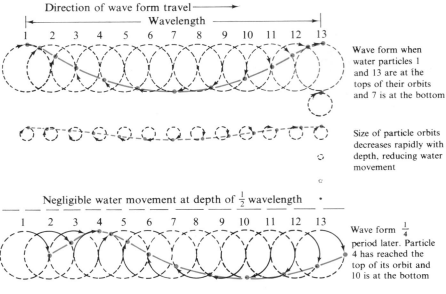

Direction of wave form travel⟶
⟵ Wavelength ⟶

1 2 3 4 5 6 7 8 9 10 11 12 13

Wave form when water particles 1 and 13 are at the tops of their orbits and 7 is at the bottom

Size of particle orbits decreases rapidly with depth, reducing water movement

Negligible water movement at depth of ½ wavelength

1 2 3 4 5 6 7 8 9 10 11 12 13

Wave form ¼ period later. Particle 4 has reached the top of its orbit and 10 is at the bottom

FIGURE 11–6 Water particles move through a circular orbit as a wave passes. Almost no water motion occurs below a depth of one-half wavelength.

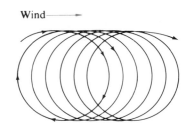

Wind⟶

FIGURE 11–7 The wind moves the water slowly forward, causing the surface circulation of the oceans. The circles show the actual movement of water particles.

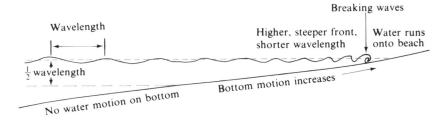

Wavelength

½ wavelength

No water motion on bottom

Bottom motion increases

Higher, steeper front, shorter wavelength

Breaking waves

Water runs onto beach

FIGURE 11–8 The changes that occur when a wave moves onto the shore.

FIGURE 11–9 Aerial view of breaking waves and surf zone. Photo from California Division of Beaches and Parks.

water's force alone or by compressing air in cracks. A breakwater weighing 2600 tons has been moved in this manner. Windows in lighthouses hundreds of feet above sea level have been broken by rocks thrown up by storm waves. The effects of storm waves in moving large fragments may be likened to river floods, during which the largest fragments are moved. Newspapers occasionally show pictures of large and small boats thrown long distances inland in this way. Waves breaking on a coastline very commonly develop cliffs by undercutting the bedrock. (See Fig. 11–12.)

Abrasion is the grinding done by material moved by the waves. It is especially effective in the surf zone, but can occur, though only to a much smaller extent, in the entire area where the waves feel bottom and where some of their energy is dissipated on the bottom. This action rounds and reduces the size of the pebbles. Waves running up on the shore carry pebbles and sand grains up with them and then back as they return to the sea. This rolling, churning action is very similar to the movement of the bed load of a river. Acting in the surf zone, such *abrasion cuts a smooth surface, generally at a depth of 9 meters (30 feet) or less. The surface is called a surf-cut bench.* (See Fig. 11–12.) The presence of such a bench may cause the waves to break far from the shore and so limit the effect of the waves on the actual shore.

Wave erosion is less effective than generally thought. It is estimated that only about one percent of the sediments in the oceans came directly from wave erosion. Most beaches are formed of material moving onshore. At some places wave erosion is very active, and rates as high as 9 meters (30 feet) per year of recession are known. (See Fig. 11–13.) Most wave erosion occurs during large storms.

WAVE TRANSPORTATION

The movement of sediment by waves is a selective process that effectively sorts the sediment by size, especially in the near-shore surf zone. Only sand and coarser material are found in the surf zone, and the main movement of

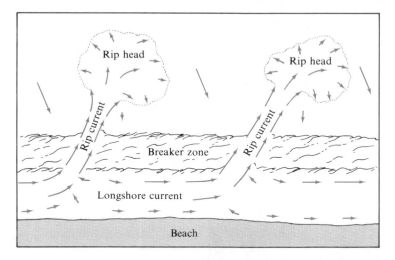

FIGURE 11–10   Rip currents are formed by the return of the water that is moved on-shore by the waves. The location of the rip currents is variable. They can carry an unwary swimmer into deep water.

FIGURE 11–11   Wave refraction along the California coast. Note the changes in wavelength that occur in the bay on the right. Photo from U.S. Department of Agriculture.

221

FIGURE 11-12   Surf-cut terrace and undercut cliff near Arecibo, Puerto Rico. Photo by C. A. Kaye, U.S. Geological Survey.

FIGURE 11-13   Photo taken in 1970 of a highway south of San Francisco that was built in 1936. Photo by Raymond Sullivan, California Division of Mines and Geology.

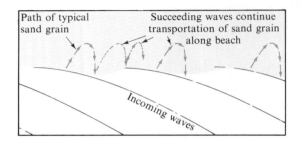

FIGURE 11-14   Movement of sand along a beach. The incoming waves are not parallel to the shoreline and move sand grains forward perpendicular to the wave. On retreat, the water flows perpendicular to the shore. Thus each wave moves material a little further along the beach.

this material is along the shore. In this active zone the waves run up on the beach, carrying sand and coarser material as both bed load and suspended load. The slope of this area is such that the receding wave, aided by gravity, carries this material back with it, even though it loses some of its water by percolation into the beach. Any finer material in the receding wave is carried to deeper water by the "undertow." Thus the sand and coarser materials move back and forth in this zone. This material moves parallel to the beach because refraction of the waves generally causes them to move diagonally toward the shore. The waves moving onshore carry the sand forward in this diagonal direction. The receding water, however, moves the sand back down the slope under the influence of gravity. Thus, in each back and forth movement the sand is carried further down the beach. (See Fig. 11-14.) The magnitude of this longshore transportation has been measured with marked sand grains and at places amounts to a few thousand tons (dry weight) per day under normal conditions and several times this much during storms. Another measure of the longshore movement is the effect of barriers sometimes constructed to protect beaches or provide harbors for small boats, as shown in Fig. 11-15.

In the zone where the waves feel bottom, sand moves shoreward when the bottom is disturbed by the waves. The returning water flows offshore with a more constant, slower movement that can only move finer material. Thus, in this zone sand moves shoreward and finer materials may tend to move offshore. (See. Figs. 11-4 and 11-16.)

In deeper water, however, only the long wavelength waves disturb the bottom materials, and the movement is symmetrically back and forth as the wave passes. The bottom generally slopes seaward so gravity may help to move some

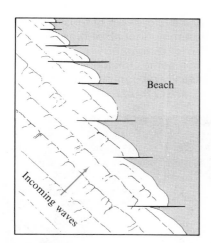

FIGURE 11–15 The effect of barriers on longshore movement of sand is shown in this oblique view of a beach.

of the disturbed sediments seaward.

During storms beaches lose sand. This typically occurs during winters. The sand returns to the beach soon after the waves become normal again. On some shores the high storm waves move some of this sand inland beyond the beach. Wind may also move sand inland. In general, storm waves remove sand from the beach and deposit it as a bar at the edge of the surf zone, probably because the higher wave crests of the storm waves need a shallower slope to maintain the equilibrium slope on which they move the sand back and forth. (See Fig. 11–16.) On some coasts some of the sand is carried to deeper water, and it continues to move seaward down the gentle slope of the shelf.

Rip currents transport fine sand offshore, especially during times when onshore winds as well as waves move surface water onshore.

Sand also moves offshore in submarine canyons. This occurs where sand moving along the shore meets a canyon. For this reason beaches may end abruptly near submarine canyons.

Storm waves or storm winds may stir up the bottom sediments, and the resulting turbid water may move offshore. Turbid water a few tens of miles offshore has been noted from satellites.

WAVE DEPOSITION

Wave deposits are of two main types: beaches and sand bars, either offshore or attached to the shore.

Beaches are the more common wave-deposited features. Some of the material forming a beach comes from wave erosion, and some of it comes from other agents of erosion, especially rivers. In recent years many rivers have been dammed so they deliver little sand to the beaches. As a result many beaches are becoming smaller. The size of the sediment in a general way controls the slope of the beach, with pebble beaches being steeper than sand beaches. As noted before, beaches are in delicate balance and change drastically when wave and current conditions change. (See Fig. 11–17.)

The sediment carried along the shore is deposited as sand or gravel bars at headlands and across bays, as shown in Fig. 11–18. These *bars* and *offshore bars* are important in the development of shorelines.

DEVELOPMENT OF SHORELINES

The great variety of present-day coastlines suggests that shorelines are complex areas; many factors must be considered. Only a few of the better understood cases will be considered here. Any ac-

tual shoreline must be studied individually, taking into account such factors as rock type, rock structure, size of waves, direction of waves, number of storms, tidal range, and submarine profile. All coastlines, like all other parts of the crust, are constantly changing, and the type of changes depends partly on the recent changes in regimen. As with rivers, major changes are initiated mainly by changes in elevation.

Two types of sea level changes are possible. The water level itself may rise or fall, or the land may be uplifted or depressed. Sea level may change as a result of uplift or depression of the ocean basins. Evidence that such changes have taken place will be presented in later chapters. Another way that sea level is changed is by the formation and melting of glaciers. This has happened in the recent past, and it is very difficult to separate the changes caused by the recent withdrawal and return of the sea from the normal shoreline processes. Uplifts and depressions of the continents are also common and will also be described later. Changes in sea level from this second cause are more localized, but cannot always be distinguished from rise or fall of the ocean itself. Many coastlines show evidence of repeated relative changes of land and sea. The melting of the last glacial advance caused a rise in sea level, but the effects of this are not apparent on many coasts.

If relative changes in sea level are the main factors in coastal development, then coasts could be classified as either *emergent* or *submergent.* This was one of the

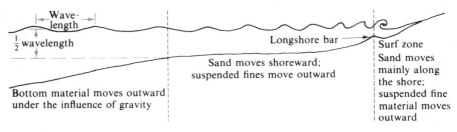

FIGURE 11–16 Transportation in the near-shore area.

early classifications used, but such a classification does not take into account many other important factors. No entirely satisfactory genetic classification of shorelines have yet been proposed. On some coasts, erosion is the dominant process, and on others it is deposition. Adding these factors to sea level changes, and taking into account rock types, wave energy, and time, can probably describe the development of most coasts, assuming that all of these factors do not change.

On some coastlines *emergence* is easily recognized. An *emergent coastline* forms when either sea

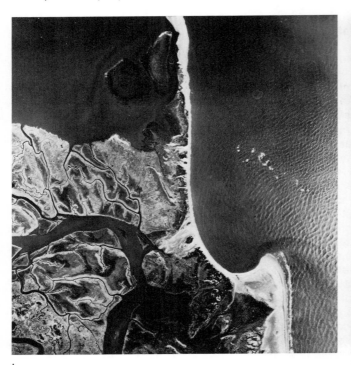

A

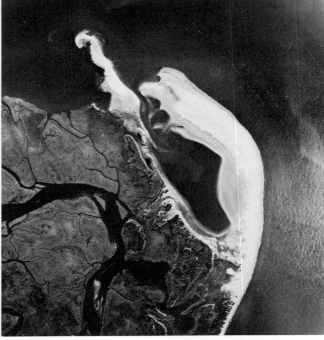

B

FIGURE 11-17 Shoreline changes at Little Egg Inlet, New Jersey. A. 1940. B. 1957. C. 1963. These three photographs taken in 1940, 1957, and 1963 show changes in Island Beach south of the inlet. The tidal channels in the salt marsh serve as reference marks for comparison of shoreline changes from year to year. In 1940, a long, narrow, north-trending beach and spit separated the marsh from the ocean. By 1957, the north-trending beach and spit had been replaced by a northwest-trending beach and spit, shorter and slightly west of those of 1940. In addition a new northwest-curving spit about a quarter of a mile wide had been built east of the position of the 1940 spit and extending northward for about a mile. By 1963, the curving spit had grown northwestward an additional 305 meters (1000 feet) and changed shape to some extent. Photos and caption from C.S. Denny and others, U.S. Geological Survey Professional Paper 590, 1968.

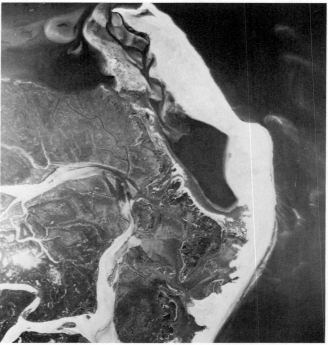

C

FIGURE 11-18   Oblique aerial photo of Morro Bay, California, showing a sand bar enclosing a bay. Photo from California Division of Beaches and Parks.

level has dropped or the continent has risen relative to sea level. Rising land or falling water level will bring *beaches* and *surf-cut benches* above sea level. In some cases, several benches are visible above a present beach. This is common at many places along the California coast where recent faulting and uplift are responsible. (See Figs. 11–19 and 11–20.) At some places warping of the wave-cut benches reveals differential uplift.

In the same way simple submergence can be imagined, but in this case the results are not so obvious. A *submergent coastline would develop if either the land went down or the ocean rose.* A submerged coastline should have *deep embayments* and *estuaries* caused by the filling of coastal valleys with seawater. (See Figs.

11–19 and 11–21.) This is, of course, not the only way that such a coast can develop. Few coastlines are of this type, although the recent melting of the glaciers should have made almost all coasts submergent.

Assuming stability, it is possible to describe the development of some types of shorelines. An initially irregular coastline with headlands and embayments will be greatly modified by wave action. Refraction of the waves will concentrate the wave erosion on the headlands (see Fig. 11–22), and a wave-cut bench and sea cliffs will develop there. The waves will carry some of the eroded material into the embayments, which, together with river-eroded material, will tend to fill them. Sand bars of various types will also develop across the embayments. Ultimately, a smooth coast will develop. A coastline of this type tends to retreat landward.

Barrier island coasts have developed on many shallow smooth coasts. These coasts have offshore bars or barrier islands with shallow lagoons between the islands and the mainland. Their origin presents some problems. These coasts form in areas where the seaward slope is gentle and there is an abundant supply of sediment. At some places the barrier islands are formed mainly by longshore movement of sediments, much as sand bars and spits are formed (Fig. 11–23). In most cases the sediment appears to be moving shoreward, and the islands may have originated from earlier beaches above which onshore winds had caused the development of sand dunes. These beaches are believed to have formed during glaciation when sea level was lower and winds may have been more active. Melting of the glaciers would cause sea level to rise and flood the area behind the beach-dune complex. The resulting *lagoon* may be partially filled by material carried into it by rivers. (See Fig. 11–23.)

A. Neutral coastline

B. Submergent coastline. Wave action may develop cliffs on the headlands, and the eroded material may form sand bars across the bays

D. Profile of a coastline that has been uplifted (or the sea level fallen) several times, producing a series of wave-cut terraces and cliffs

C. Emergent coastline

FIGURE 11-19   Possible types of shorelines, A. A neutral coastline. B. The same coastline after the land has sunk or sea level has risen. Note the drowning of the river valleys, producing bays and estuaries. Such irregular coastlines are typical of submergence. C. The same coastline after the land has risen or sea level has sunk. Note the wave-cut cliff produced by breaking waves on the newly exposed land. D. The profile of a coastline that has emerged in several stages similar to the coastline in C.

FIGURE 11-20 Sea cliffs undergoing active erosion by waves and uplifted surf-cut benches on the skyline. Photo from Oregon State Highway Department.

FIGURE 11-21 A deeply embayed submergent or drowned coastline. Marlborough Sound, New Zealand. Photo from New Zealand National Publicity.

226

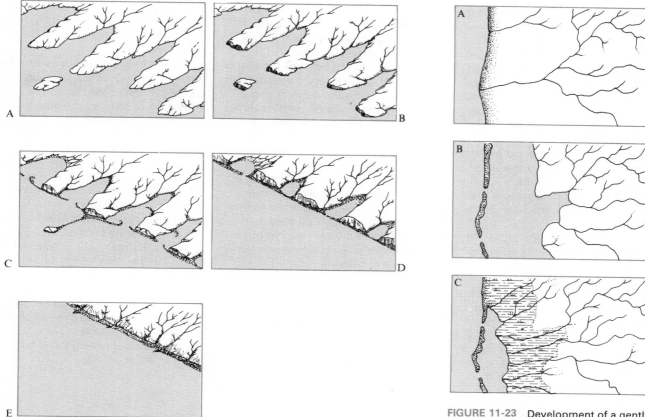

FIGURE 11-22   Development of an initially irregular coastline.

FIGURE 11-23   Development of a gently sloping, smooth coastline. A. Beach-dune complex forms at a lower sea level during glaciation. B. Melting of the glaciers floods the area behind the beach-dune complex, forming a lagoon. C. River deposition begins to fill the lagoon.

SUMMARY

The circulation of the surface waters of the earth is caused by winds and the earth's rotation. The deep circulation is caused by density differences in ocean water.

Tides on the opposite sides of the earth are caused by the gravitational and centrifugal forces of the earth-moon system. Tides are not important agents of transportation or erosion except where tidal currents move through and scour narrow inlets.

For waves, wavelength = period × velocity. In deep water, the velocity in kilometers per hour is about

5.6 times the period in seconds. Wave height has almost no effect on period or velocity.

Waves are caused by wind. In deep water the wave shape moves forward with the wind, but the individual water particles move in circular paths almost returning to their original positions. At a depth of $\frac{1}{2}$ wavelength, water motion is negligible.

In water shallower than $\frac{1}{2}$ wavelength the wave is slowed. The height of the wave increases and the front of the wave is slowed more than the rear, causing the waves to

break, creating surf. Breaking waves are active agents of erosion and transportation.

Rips or *rip currents* are caused when the water returning to the ocean from waves breaking on the beach is concentrated at a few places.

Waves generally break at depths of $1\frac{1}{2}$ times their height. The majority of the geologic work of waves is done at depths less than 9 meters (30 feet).

The refraction or bending of waves is caused if parts of the wave do not feel bottom and are not

227

slowed; refraction concentrates the waves on headlands.

Waves erode by hydraulic action (impact and pressure of waves) and by abrasion (grinding by material moved by waves). Only about one percent of ocean sediments comes directly from wave erosion; most wave erosion occurs during large storms.

Surf-cut benches are cut by abrasion in the surf zone, generally at a depth of 9 meters or less. The presence of the bench may cause the waves to break far from the actual shore.

Sand and coarser materials are transported back and forth in the surf zone. Wave refraction generally causes the waves to move diagonally toward shore, so each onshore wave moves the sand along the beach, and the returning water moves it back down the slope. In this way sand moves along the beach.

Between the surf zone and the depth of $\frac{1}{2}$ wavelength, in general, sand moves shoreward and finer material moves outward.

During storms beaches lose sand. In general, the storm waves deposit the removed sand as a bar at the edge of the surf zone, and it returns to the beach when the waves are normal again.

Wave deposits are of two types: *beaches* and *sand bars*.

Some of the material forming a beach comes from wave erosion, some from rivers and other agents. Pebble beaches are steeper than sand beaches. Beaches change drastically with changing wave and current conditions.

Sand or gravel bars, at headlands and across bays, are formed of sediment carried along the shore and are important in the development of shorelines.

Two types of sea level changes are possible. Water level may rise or

fall, or the land be uplifted or depressed. Formation and melting of glaciers affects the rise and fall of sea level.

No entirely satisfactory classification of shorelines has been proposed. Erosion, deposition, rock type, wave energy, and time are all factors in the development of shorelines. Emergent and submergent coasts can be imagined.

A *submergent* coastline would result if land went down or ocean rose; it would have deep embayments and estuaries.

An *emergent* coastline would result from rising land or falling water level, and may result in beaches or surf-cut benches being above sea level.

If the shore is stable, an initially irregular coast will tend to become straighter in time as a result of erosion and deposition.

Gently sloping coastlines tend to have barrier islands.

QUESTIONS

1 What causes the surface circulation of the oceans?
2 Explain why high tides occur simultaneously on both sides of the earth.
3 What are the relationships among wavelength, period, and velocity of water waves?
4 Describe the motion of water in waves in both deep and shallow water.
5 Describe the processes and the results of wave erosion.
6 Discuss the transportation of both sand and finer material by waves.
7 Do sea cliffs always mean uplift of land (or lowering of sea level)?
8 How might one distinguish between uplift of the continent and depression of sea level at a given place?
9 Discuss whether sea level is a horizon to which absolute movements of the continents can be referred?
10 Describe the changes that will occur on an irregular stable coastline.

SUPPLEMENTARY READINGS

Bascom, Willard, "Ocean Waves," *Scientific American* (August 1959), Vol. 201, No. 2, pp. 74–84. Reprint 828, W. H. Freeman and Co., San Francisco.

Bascom, Willard, "Beaches," *Scientific American* (August 1960), Vol. 203, No. 2, pp. 80–94.

Bascom, Willard, *Waves and Beaches.* Garden City, N.Y.: Doubleday and Co., Inc., 1964, 267 pp. (paperback).

Carson, Rachel, *The Sea Around Us,* revised edition. New York: Oxford University Press, 1961, 230 pp. (also available in paperback).

Dolan, Robert, et al., "Shoreline Erosion Rates along the Middle Atlantic Coast of the United States," *Geology* (December 1979), Vol. 7, No. 12, pp. 602–606.

Dolan, Robert, Bruce Hayden, and Harry Lins, "Barrier Islands," *American Scientist* (January-February 1980), Vol. 68, No. 1, pp. 16–25.

Gross, M. G., *Oceanography,* 3rd ed. Co-

lumbus, Ohio: Charles E. Merrill Publishing Co., Inc., 1976, 138 pp. (paperback).

Higgins, C. G., "Causes of Relative Sea-level Changes," *American Scientist* (December 1965), Vol. 53, No. 4, pp. 464–476.

Hoyt, J. H., *Field Guide to Beaches.* ESCP Pamphlet Series, PS-7. Boston: Houghton Mifflin Co., 1971, 45 pp.

Hoyt, J. H., "Shoreline Processes," *Journal of Geological Education* (January

1972), Vol. 20, No. 1, pp. 16–22.

Inman, D. L., and B. M. Brush, "The Coastal Challenge," *Science* (July 6, 1973), Vol. 180, No. 4094, pp. 20–32.

King, C. A. M., *Beaches and Coasts.* London: Edward Arnold, 1959, 403 pp.

Kort, V. G., "The Antarctic Ocean," *Scientific American* (September 1962), Vol. 207, No. 3, pp. 113–128. Reprint 860, W. H. Freeman and Co., San Francisco.

Munk, Walter, "The Circulation of the Oceans," *Scientific American* (September 1955), Vol. 193, No. 3, pp. 96–104. Reprint 813, W.H. Freeman and Co., San Francisco.

Schubel, J.R., and D. W. Pritchard, "The Estuarine Environment," *Journal of Geological Education,* Part I (March 1972), Vol. 20, No. 2, pp. 60–68. Part II (September 1972), Vol. 20, No. 4, pp. 179–188.

12

GROUND WATER AND WATER RESOURCES

All the rivers run into the sea;
Yet the sea is not full:
Unto the place from whence the rivers come,
Thither they return again.

Ecclesiastes 1:7

GROUND WATER

Ground water is an important source of water, especially in dry areas. In many of the countries of north Africa and the Near East, ancient peoples, hundreds of years before Christ, dug tunnels many miles long to collect and transport ground water. In Europe in medieval times, a technique of well drilling was invented, and in a few areas the wells flowed naturally.

It was not until the latter part of the seventeenth century that it was shown that the origin of river water is rainfall. When this measurement was undertaken, however, it was discovered that only a fraction of the amount of water that fell as rain on a drainage basin flowed out as the river. The fate of the rest of the precipitation is not hard to imagine. Some of it evaporates, some is used by plants, and some seeps into the ground. The latter is the origin of ground water.

In the middle 1850s, Henry Darcy, a French engineer, worked out the law of fluid flow through a permeable medium. In the last half of the nineteenth century, a parallel development of ground water geology occurred in Europe and the United States, and the same discoveries were made in some cases in several countries. In the United States in 1877, T.C. Chamberlin wrote the first report on artesian wells (to be described). In 1888, J. W. Powell, who was then the Director of the United States Geological Survey, stimulated much work on ground water. He was deeply concerned that the Great Plains might be opened to homesteading on the same basis as the better-watered parts of the country were. He felt that much larger homesteads would be necessary because of the dryness of that area. In the

early 1900s, the first quantitative studies of ground water movement were made possible by the development of new instrumentation. A classic study of ground water in the United States was made in 1923 by Oscar Meinzer, who directed the ground water branch of the United States Geological Survey for many years.

Ground water is a very important part of the water resources of many areas. Since World War II the problem of water resources has received much attention, and all aspects, including ground water, are being studied vigorously. Water studies have important legal, social, and economic aspects as well as the mainly scientific aspects considered here.

MOVEMENT OF GROUND WATER

The surface below which rocks are saturated with water is called the water table. (See Fig. 12–1.) Some water is retained above the water table by the surface tension of water. (See Fig. 12–2.) In general, the water table is a reflection of the surface topography, but is more subdued, that is, has less relief, than the surface topography. (See Fig. 12–3.) Lakes and swamps are areas where the land surface is either below or at the water table. Springs occur where the water table is exposed as on a valley side. The position of the water table changes seasonally, which explains why some springs are dry in summer. *Springs may form where the water table is perched. This situation results where the downward percolation of rainwater is stopped by a relatively impermeable rock, such as a shale.* Springs of this type may flow only during the wet season. (See Fig. 12–4.)

The movement of ground water is controlled by the physical properties of the rocks. The amount of water that can be stored in the rocks is determined by the *porosity, the amount of pore, or open, space.* The availability of the water is determined by *the interconnections of the pore space, which is, of course, the permeability* (Fig. 12–5). The most common reservoir rock is sandstone, although fractured granite or limestone, as well as many other rock types, can serve equally well. (See Fig. 12–6.)

A very important type of well is the *artesian well. The reservoir of an artesian well is confined above and below by impermeable rocks. The reservoir is charged at a place where it is exposed, and the water is forced to move through the reservoir. A well drilled into an artesian reservoir may flow without pumping because the released water seeks its own level— that of the recharge area. However, the resistance to flowage of the water through the reservoir prevents the water from flowing to the same elevation as its recharging area; therefore, flowing wells must be lower in elevation than the recharge area.* (See Fig. 12–7.)

Ground water is a very valuable economic commodity. It is recovered from wells for domestic, industrial, and agricultural use. (See Fig. 12–8.) In areas where ground water is used extensively, care must be taken that, on the average, no more water is withdrawn in a year than is replaced by natural processes or artificial infiltration. This can be determined by seeing that the water table is not lowered. The conservation of ground water is very important because it moves very slowly, and many years may be required to replace hastily pumped water. The

Dry valleys and sinkholes near Timaru, New Zealand. Photo by S.N. Beatus, New Zealand Geological Survey.

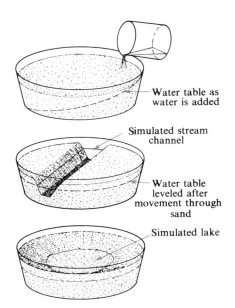

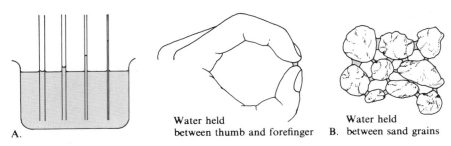

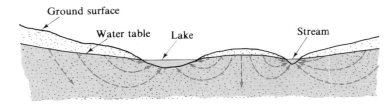

FIGURE 12–1 An illustration of ground water. The dish is filled with sand, and the water poured in percolates into the sand, filling the pore spaces. A hole in the sand can represent a well, or a slot a stream channel. This experiment works best in a transparent dish with water colored by a few drops of ink.

FIGURE 12–2 Capillary action is caused by surface tension and can cause water to rise above the water table. A. In a very narrow tube, water rises well above the water level. B. In the ground, the narrow spaces between particles take the place of the tubes. For this reason, the top of the water table is a thin zone rather than a sharp line. This is also the way that soil retains its moisture.

FIGURE 12–3 The water table has less relief than the topography. The arrows show the movement of the ground water.

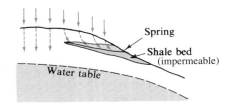

FIGURE 12–4 Spring formed by a perched water table created by impervious shale bed. Such springs tend to dry up in late summer.

Well-sorted sandstone. Large amount of pore space gives high porosity and permeability.

Poorly sorted sandstone. Has much lower porosity and permeability.

FIGURE 12–5 The sorting of sandstone affects its porosity and permeability.

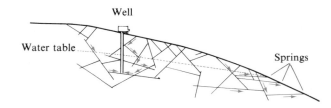

FIGURE 12–6 Springs and wells can also occur in fractured rocks such as granite.

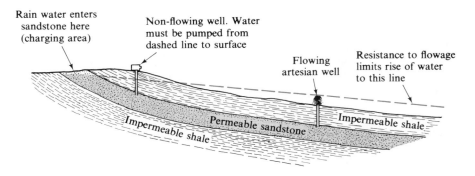

FIGURE 12–7 Artesian water results when the flow of ground water is restricted by impermeable layers.

WATER WITCHING

Water witching, or dowsing, or divining, has been with us a long time. Some say the origins are found in the Bible (Numbers XX, 9–11):

> And Moses took the rod from before the Lord, as he commanded him.
>
> And Moses and Aaron gathered the congregation together before the rock, and he said unto them, Hear now, ye rebels; must we fetch you water out of this rock?
>
> And Moses lifted up his hand, and with his rod he smote the rock twice: and the water came out abundantly, and the congregation drank, and their beasts also.

Accounts of the use of the divining rod in the ancient civilizations are found in the records of the Greeks and Romans. More modern usage can be traced to Germany, where it was described in 1556 by Johannes Agricola. From there it spread all over Europe and then to the New World. Throughout this whole period, it was denounced by most scientists and by many clerics. Luther tried to stop the practice in 1518, and it was forbidden by the Inquisition in 1701.

Most water witches use a forked stick or rod. The most favored types are peach, willow, hazel, and witch hazel, although any type may be used; some use metal rods and bent coat hangers. Most water witches hold the rod with their palms up and the butt end of the rod pointing up. When the rod passes over water, the butt of the rod turns down. Some water witches will predict the depth of the water and/or the quantity from the number of times the rod bobs or twists. Most water witches walk on the ground, but some move their rods over maps. Almost all dowsers talk about veins of water, as well as underground rivers and domes of water, so their descriptions are quite different from those in this chapter.

Divining rods are also used by some to locate ore deposits, oil, buried or hidden treasure, lost land surveys, and even to locate criminals. When used to locate ore, some dowsers put a small cap of the metal sought on the rod. Not all diviners use rods; some use pendulums or even keys on a key chain.

Water witching seems to be a mixture of witchcraft and folklore, so why has it persisted, and why is it routinely used by otherwise sensible people? Some would say there must be something to it because it has lasted so long. Of course, so have astrology, fortune telling, and palm reading. Most evidence in favor of water witching comes from uncontrolled tests, that is, anecdotes and similar reports. In many areas, most wells do produce water, so it is unknown whether another site would have been better. In controlled tests in more difficult areas, witched wells seem to be no better than random drilling, whereas geological locations are somewhat better, depending on the amount of geologic information available.

The cost of drilling a water well for just domestic water can be in the range of a few thousand to several tens of thousands of dollars. Vogt and Hyman suggest that this cost, together with the uncertainty that the well will be successful, causes anxiety. Calling in a water witch adds very little to the cost of the well and relieves the landowner of the responsibility of locating the well. Water witches are always positive in their locations, and this reduces the uncertainty and anxiety of the landowner. If the well is dry, the landowner is not at fault, and this is a great relief.

In the areas where geological knowledge of the underground is good (that is, places where there are many wells), geologists can do a better job of well location, and they, too, can relieve the landowner of the responsibility of well location. Even in such places, witches are used because geologists are not nearly so positive in their statements and locations. The geologist trying to give an honest evaluation of the problems in well location does not relieve anxiety the way a water witch does.

REFERENCES

A. J. Ellis. *The Divining Rod, A History of Water Witching,* U.S. Geological Survey Water-Supply Paper 416. U.S. Government Printing Office, Washington, D.C. (1917, reprinted 1957).

Ray Hyman and E. Z. Vogt. "Some Facts and Theories on Waterwitching in the United States," *Geotimes* (March 1958), Vol. II, No. 9, pp. 6–7, 15.

Kenneth Roberts. *Henry Gross and His Divining Rod.* Garden City, N. Y.: Doubleday, 1951, 310 pp.

Kenneth Roberts. *The Seventh Sense.* Garden City, N. Y.: Doubleday, c1953, 337 pp.

Kenneth Roberts. *Water Unlimited.* Garden City, N. Y.: Doubleday, c1957, 285 pp.

U.S. Geological Survey. *Water Dowsing,* 1980-341-618-8. U.S. Government Printing Office, Washington, D.C., 15 pp.

E. Z. Vogt and Ray Hyman. *Water Witching.* University of Chicago Press, 1959.

FIGURE 12–8 Ground-water areas in the United States. Color shows areas underlain by reservoirs capable of yielding to a single well 190 liters (50 gallons) per minute (g.p.m.) of water containing less than 0.2 percent dissolved solids (includes some areas where more mineralized water is actually used). From U.S. Department of Agriculture, 1955 Yearbook, *Water.*

average rate of movement of ground water through rocks is only about 15 meters (50 feet) per year, although at some places the movement is much faster.

This slow movement of the water requires another caution. *If water is withdrawn from a well by pumping too rapidly, the water is removed from the immediate vicinity of the well faster than it can be replaced. The result is called a cone of depression* (Fig. 12–9). If cones of depression are developed around several closely spaced wells, the water table can be drastically lowered. Thus, the spacing and the rate of pumping of wells need to be regulated in order to insure the most efficient use of the water. Figure 12–10 shows the changes that have occurred as ground water use has increased on Long Island. Note that as fresh water was removed faster than it was replaced, the heavier salt water from the ocean encroached. At some places the water has been withdrawn so rapidly that hundreds of years may be required to replace it. Such usage is similar to mining, in the sense that a natural resource is used up.

Another effect of the withdrawal of ground water at a more rapid rate than it is replaced is the gradual *subsidence* of the surface of the ground. Subsidence has also occurred at some places because of the withdrawal of oil from the ground. (See Fig. 12–11.) Up to 6 meters (20 feet) of subsidence caused by withdrawal of ground water has occurred in parts of the Central Valley of California.

Compaction and subsidence generally do not occur when fluids are withdrawn from sand or gravel. The water is in open pore space, and the weight of the overlying sediments is borne by the pebbles or sand grains. Clay layers, or lenses, behave much differently. Although clay may hold

large amounts of water, only a little water flows from clay into a well that penetrates it. If water is withdrawn from the coarser sediments that surround the clay, the buoyant effect of the water is removed, and the pressure of weight of the overlying sediments must be borne by the clay. This pressure squeezes water out of the clay, reducing the volume of the clay. Even if the surrounding rocks are again saturated, the clay will not be able to absorb more water.

Pollution is a problem in some areas where well water is used. Drainage from cesspools, toxic waste dumps, feedlots, and so forth can contaminate nearby wells. This danger is always present but is most acute where the ground water flows freely as in wide joints or cavernous limestones. If the polluted water seeps slowly enough through sand or gravel, especially above the water table, it is commonly purified in only a few tens of feet of travel. The purification is accomplished by filtering and oxidation. (See Fig. 12-12.) Advantage of this means of purifying water has been taken lately by spraying sewage and industrial waste on the ground surface, especially in wooded areas. In this way the ground water is recharged and the contaminants removed. However, surface disposal or burial both have the danger of contaminating water, especially underground water. In times of flood or heavy precipitation, surface waters can also be polluted. Decomposing garbage in a landfill produces meth-

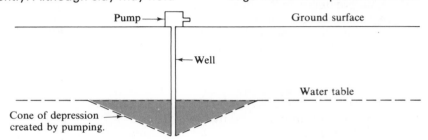

FIGURE 12–9 A cone of depression results when water is pumped too rapidly from a well.

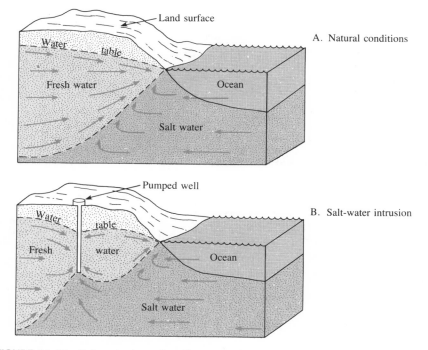

FIGURE 12–10 Salt water intrusion. Overpumping ground water can cause salt water intrusion near a coastline.

ane, carbon dioxide, ammonia, and hydrogen sulfide. The carbon dioxide combines with water, forming carbonic acid that increases the amount of many materials that can be dissolved in the water. Such water generally pollutes surface or ground water into which it seeps. It is possible to avoid pollution from disposal sites by paying attention to the ground water conditions as shown in Fig. 12-13.

One special type of underground disposal is the pumping of fluids into deep rocks. This is a common way to dispose of oilfield brines. The earthquake problems at the Denver Arsenal, described in the next chapter, suggest caution.

GEOLOGIC WORK OF GROUND WATER

Much of the geologic work of ground water was covered in the

FIGURE 12–11 Flooding at Long Beach, California, caused by subsidence resulting from withdrawal of oil. Photo courtesy of City of Long Beach.

235

discussion of weathering in Chapter 4. Water is a very good solvent, especially if it contains carbon dioxide; and the role of ground water in solution and cementation has been described. It is not surprising that most ground water contains dissolved material and so is what is termed hard water. This is the origin of the taste of most spring and well water. In some cases ground water contains so much dissolved material that it is not suitable for domestic or industrial use. Some ground water used for irrigation deposits dissolved salts in the soil and so can ruin soil unless enough water is used to flush the salts continually from the soil. Some of the water encountered in drilling is salty, and some of these waters were once seawater that was buried with the sediments when the bed was deposited in the ocean.

Ground water forms caves, but exactly how is not known. They are formed mainly in limestone, but can form in any soluble rock. (See Fig. 12-14.) Calcite is soluble in water, especially in water that has carbon dioxide dissolved in it, as noted in the discussion of chemical weathering. Probably caves form in the zone of seasonal water table fluctuation by solution of limestone along joint planes. Lowering of the water table may make the cave accessible. In times of flood caves may be enlarged by erosion by running water. A process going on in caves (Fig. 12-15) is the deposition of calcite in the form of *stalactites, stalagmites,* and other features, all of which are caused by the evaporation of carbonate-charged water.

In limestone areas, caves or other channels may carry most of the water. *The surface water may sink underground and flow through caves. In such an area there are only a few short surface streams, and they end in closed depressions called sinks.* (See Fig. 12-16.) This, as might be expected, produces an uncommon type of

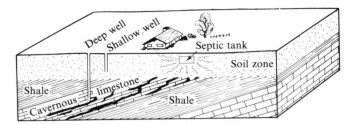

FIGURE 12–12 Pollution of wells. In this example, because of the cavernous limestone, the shallow well is more likely to produce pure water than is the deep well that is further away from the septic tank.

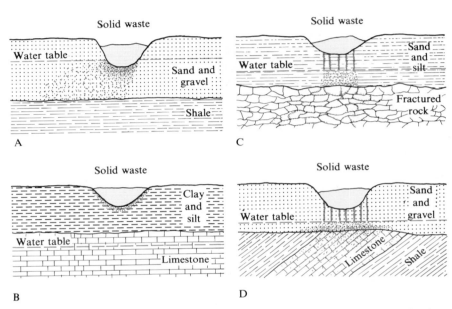

FIGURE 12–13 Effects of waste disposal on ground water. Only in B is contamination prevented by impermeable rocks. In all the other examples, contaminants from the waste move through permeable rock to the water table. From U.S. Geological Survey Circular 601F, 1970.

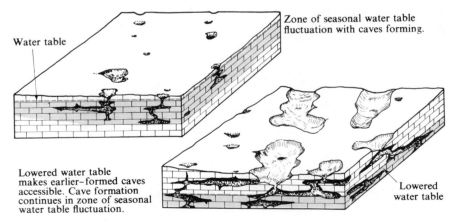

FIGURE 12–14 Development of caves. The caves begin by solution of limestone in the zone of seasonal water table fluctuation. Lowering of the water table makes the caves accessible. An area of cavernous limestone generally has few, if any, surface streams.

236

FIGURE 12–15 Stalactites and stalagmites in Carlsbad Caverns, New Mexico. The stalactites form by dripping water that evaporates, depositing its dissolved calcite. Stalagmites build up when drops of water with dissolved calcite fall on the floor of the cave and evaporate. Photo by National Park Service.

FIGURE 12–16 This home in Florida fell into a sinkhole 158 meters (520 feet) long, 38 meters (125 feet) wide, and 18 meters (60 feet) deep. Solution of limestone by ground water caused this sink. Photo from U.S. Geological Survey.

surface topography. It is worth noting that only in the case where caves are developed is anything like an underground stream developed; many people have the erroneous impression that all ground water flows in underground streams similar to surface streams. In the same sense, there are no underground lakes, but only reservoir rocks whose pore space is filled with water.

HOT SPRINGS AND GEYSERS

Geysers are periodic discharges of hot ground water, in contrast to *hot springs which flow more or less continuously.* (See Fig. 12-17.) Studies of the isotopes in hot spring and geyser waters have shown that most, if not all, of it is ordinary ground water that has been heated. This heating is due to deep circulation of the water or to contact with hot igneous rock bodies. (See Fig. 12-18.) Geysers result when water accumulates in underground chambers where it is heated. The pressure of the overlying water causes the boiling point near the bottom to rise. The heating also causes the column of water to expand and spill over near the top. This reduces the pressure, and the superheated water at the bottom flashes into steam and causes the geyser to erupt. (See Fig. 12-19.) This process is repeated more or less periodically in a geyser. All ground water contains dissolved material, and the hot waters contain, in general, more. Thus, hot springs and geysers commonly deposit calcite and other minerals. (See Fig. 12-20.)

WATER RESOURCES

Water is one of the cheapest and most useful and important commodities that we have. Most people take water for granted, but

FIGURE 12–17 Old Faithful at Yellowstone National Park in eruption. Photo by National Park Service.

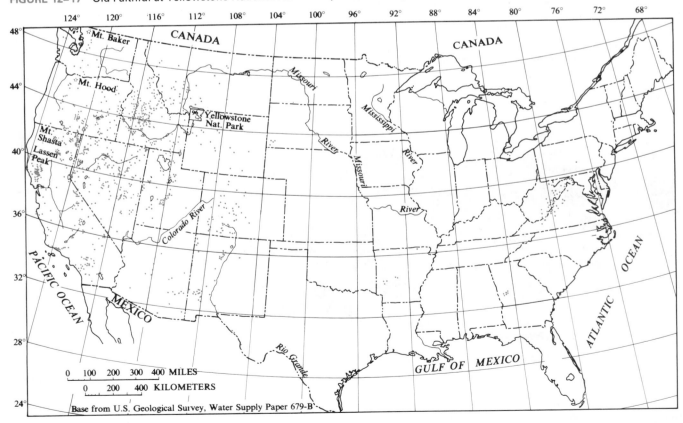

FIGURE 12–18 Locations of hot springs and geysers in the United States. They are mainly found in the geologically young, active areas. From G.A. Waring, U.S. Geological Survey Professional Paper 492, 1965.

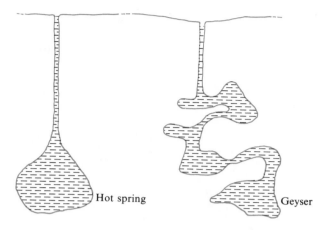

FIGURE 12–19 Idealized diagrams showing hot spring and geyser. A hot spring in which the water temperature no place exceeds the boiling point may have almost any configuration. If, however, high temperatures are involved, the configuration must be such that the heat is easily distributed by convection as in the figure shown. A geyser can form if the heat is not distributed by convection. In the figure above, the water near the bottom is heated to near its boiling point. The boiling point is higher there than at the surface because the weight of the water above increases the pressure. The water higher in the geyser system is also heated and so expands and flows out at the top. This reduces the weight of the water on the bottom and, at the reduced pressure on the bottom, boiling occurs. The bottom water flashes into steam, and the expanding steam causes an eruption.

FIGURE 12–20 Terraces built by hot spring deposits at Mammoth Hot Springs, Yellowstone National Park. Photo by National Park Service.

more and more often water shortages in different areas are reported in the newspapers. Are these shortages ominous signs that some regions have reached their limit of development, or are we mismanaging our water resources? Water is a worldwide problem, but only the United States, excluding Alaska and Hawaii, will be discussed here. The same natural problems occur ev-erywhere, and they are aggravated to a greater or lesser extent by the stage of industrial and technological development of the country.

The United States receives an average of 76 centimeters (30 inches) of rain per year over the entire country. Of this 30 inches, about 21.5 are lost by evaporation or used by plants. The remaining 8.5 inches become the runoff of rivers and the ground water; this is the water available for use by people. (See Fig. 12–21.) This available water amounts to about 4,536,000 million liters (1,200,000 million gallons) per day, and we withdraw about 1,398,600 million liters (370,000 million gallons) per day. It thus appears that we have more than three times as much available water as we need. The true situation is not this simple, unfortunately, and several other factors must be considered. They are the natural distribution of water, the need for runoff to carry away refuse, and the water used by plants that are the base of the food chain on which all life depends.

NATURAL DISTRIBUTION OF WATER

Precipitation is not uniform, and neither is the distribution of population. Precipitation ranges from about 216 centimeters (85 inches) per year in the northwestern United States to about 10 centimeters (4 inches) in the southwest. (See Fig. 12–22.) If individual stations are considered, the range is even greater. These are average figures and may vary from year to year by large amounts, especially in the drier regions. Climate, too, varies; and as shown in Fig. 12–23, the potential or possible evaporation and consumption of water by plants exceeds the rainfall in most of western United States. These are the arid regions where runoff occurs only during infrequent heavy storms. The season at which rain occurs also affects the use of the water. In most of eastern United States, much of the rain falls during the growing season, but in the west, most of the precipitation comes during the winter months. For all of these reasons, some places have sufficient water, a few places have a surplus, and many areas have a deficiency, even though the overall available water is more than three times the amount withdrawn.

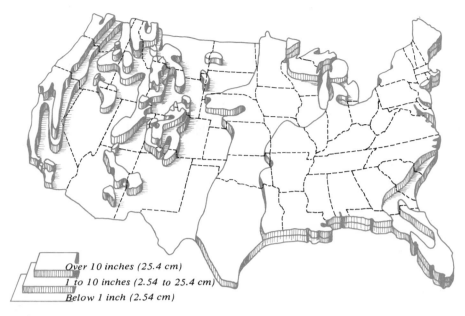

FIGURE 12-21   Average annual runoff of rivers in the United States. From U.S. Department of Agriculture, 1955 Yearbook, *Water.*

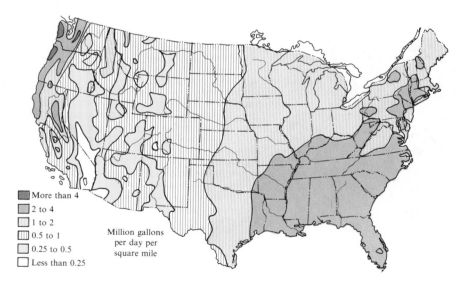

FIGURE 12-22   Average annual precipitation in the United States. Units are millions of gallons per day per square mile. From A.M. Piper, U.S. Geological Survey Water Supply Paper 1797, 1965.

In an effort to overcome these problems of natural distribution, dams and other structures have been built to control and divert rivers. Many of these projects have multiple purposes and may provide such benefits as municipal water supplies, flood control, irrigation, electricity, and recreation areas. Deciding which rivers to control, and for which benefit, is always a problem because of the competition for use of the water and use of reservoir sites. Also, any reservoir or lake that is formed increases overall evapora-

tion, and thus some water is lost. In arid regions this loss can be high, but recent studies suggest that much of it can be eliminated by placing a film of fatty alcohol on the water surface.

Like precipitation, river flow is seasonal; and during times of heaviest flow, floods can occur and much runoff is lost to the ocean. The maximum flow of a river is, in general, five to ten times the minimum flow. The most regular major river in the United States is the St. Lawrence, whose average maximum flow is less than twice its minimum flow. This is because its main flow comes from the Great Lakes. (See Fig. 12-24 and Table 12-1.) In the same way, dams that create lakes can make the flow of rivers more uniform and so increase the use of the water. At places such dams may eliminate most floods; but if, in an exceptional year, a flood does occur, much damage may result because the previously avoided flood area is now in use.

USE AND CONSUMPTION OF WATER

Distinction must be made between the water *used* and the water *consumed. Only the water lost to the atmosphere by evaporation or plants is consumed. The rest of the water withdrawn for use is returned either to the ground or to the runoff, and thus is available for other use.* In some cases the water returned is unaffected by the use, but in most cases its quality is impaired. The water withdrawn is not the only water used by people; the runoff itself is used for some purposes in most rivers. Much of the water used to grow crops comes from soil moisture and so is not withdrawn. All of these complications must be taken into account to understand water needs.

Only about one-quarter of the withdrawn water is actually consumed, and this is a small percentage of the total available water.

240

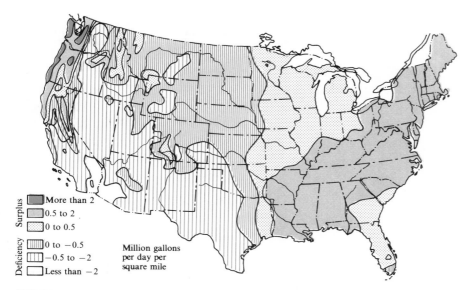

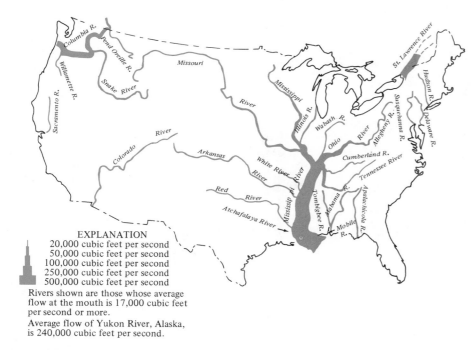

FIGURE 12–23 Water surplus and deficiency in the United States. Average precipitation less potential evaporation and consumption by plants is plotted on this map. From A.M. Piper, U.S. Geological Survey Water Supply Paper 1797, 1965.

EXPLANATION
20,000 cubic feet per second
50,000 cubic feet per second
100,000 cubic feet per second
250,000 cubic feet per second
500,000 cubic feet per second
Rivers shown are those whose average flow at the mouth is 17,000 cubic feet per second or more.
Average flow of Yukon River, Alaska, is 240,000 cubic feet per second.

FIGURE 12–24 Large rivers in the United States. From U.S. Geological Survey Circular 686, 1974.

The reuse of water in public water supplies is well illustrated by communities along a river. An upstream community withdraws its water supply from the river and discharges its wastes at a lower point on the river. The downstream community also draws its water from the river and so reuses some of the water. It is estimated that 60 percent of the public water supplies withdraw some used water.

This reuse works well only if upstream communities adequately treat their sewage before dis-

charging it into the river. Detergent types have recently been changed so that sewage treatment plants can prevent foaming in downstream water supplies. Unfortunately, only about two-thirds of the population of the United States have sewers, and only about two-thirds of these sewer systems treat the sewage before discharging it. Much of the water withdrawn for industrial use is also returned. In some cases this industrial return, like the domestic return, must be purified either naturally or by treatment before it can be reused. In other cases the industrial return is only warmer, but even this can drastically alter the ecology of a stream.

The runoff is not wasted water. It is used to produce hydroelectric power, for recreation, for transportation, and, as described above, to carry away wastes. The widespread pollution of rivers and streams suggests that we are nearing maximum usage for waste disposal in many areas, although many improvements in treating waste before discharge into rivers are possible. The magnitude of the waste disposal problem is illustrated by the fact that the city of St. Louis, on the Mississippi River, produces about 756,000 liters (200,000 gallons) of urine and 400 tons of solid body waste every day, in addition to industrial wastes.

Rivers are used to carry away the excess heat produced in many industrial processes and especially at power plants. (See Fig. 12–25.) It may seem strange that heat energy must be disposed of but the reason is that in any engine using heat energy, not all of the energy is available to do work. This is a well-known law of thermodynamics and is the reason why a perpetual-motion machine cannot be invented. Steam boilers are less than 40 percent efficient and nuclear burner reactors are about 30 percent efficient. The rest of the energy must go to the environment as heat. The current nuclear

241

TABLE 12–1 Large rivers of the world. Data from U.S. Geological Survey, Water Resources Division, 1961.

Rank	River	Country	Drainage Area (thousands of square miles)	Average Discharge at Mouth (thousands of cubic feet per second)
1	Amazon	Brazil	2,231	3,000 to 4,000
2	Congo	Zaire	1,550	1,400
3	Yangtze	China (mainland)	750	770
4	Bramaputra	Bangladesh	361	700
5	Ganges	India	409	660
6	Yenisei	USSR	1,000	614
7	Mississippi	USA	1,244	611
8	Lena	USSR	936	547
9	Paraná	Argentina	890	526
10	St. Lawrence	USA & Canada	498	500
26	Danube	Romania	315	218 (Largest river in Europe)

A

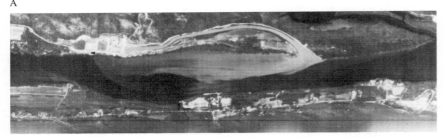

B

FIGURE 12–25 Infrared images showing water temperature at Connecticut Yankee nuclear power plant on the Connecticut River near Haddam. A. Plant is not operating. B. Plant in operation. Light-colored areas are warm water from the plant. Both images were made at flood tide. Photos from Environmental Analysis Department, HRB-Singer, Inc.

reactors must use water cooling. In other cases, cooling towers can be used so that the heat escapes into the atmosphere. The amount of water needed for cooling will increase. Raising the temperature of a river, lake, or the shallow nearshore ocean can have a deleterious effect on the aquatic life. Heating the atmosphere can change the climate, and the climate near cities is changing.

A current problem is that the runoff water from fertilized fields carries some of the fertilizer to rivers. In rivers and lakes the fertilizer provides nutrients that increase the growth of algae. The algae use the oxygen dissolved in the water, and the lack of oxygen causes the death of fish and other aquatic life. Phosphates in laundry detergents have the same effect. Pesticides used on crops get into rivers in this way, too. Long-lived insecticides such as DDT have gotten into most types of life in this way, and from animals such as birds eating poisoned insects. Thus, keeping soil productive without causing harm to the environment is a problem that requires much study.

The main consumption of withdrawn water is by irrigation. It was noted earlier that over 70 percent of total precipitation either evaporates or is used by plants. Plants, then, consume most of our water and must be considered in any study of water resources.

All life on the earth ultimately depends on vegetation. Plants are the base on which the food chain, or pyramid, is built. The amount of water needed to grow plants is very large. An acre of corn gives off about 15,120 liters (4000 gallons) of water per day to the atmosphere. Two hundred and twenty-seven kilograms (500 pounds) of water are required to grow 454 grams (1 pound) of wheat (dry weight). Most of this water is lost to the atmosphere through the leaves. About one-half of the wheat is discarded in milling, so about 454 kilograms (1000 pounds) of water are required to produce 454 grams (1 pound) (dry weight) of bread. Thus the production of the 1.1 kilograms (2.5 pounds) of bread that could be a daily diet requires 1135 kilograms (2500 pounds) of water, or 1134 liters (300 gallons). Meat requires even more water. About 13.6 kilograms (30 pounds) of alfalfa and 45 liters (12 gallons) of drinking water are required to produce 454 grams (1 pound) of beef. The 30 pounds of alfalfa require 8694 liters (2300 gallons) of water to grow. Thus the minimum water

needed to maintain life, on bread alone, is 1134 liters (300 gallons) per day, and if meat is included, is about 9450 liters (2500 gallons) per day. Water is also required to produce clothing, and in the same way, cotton requires much less water than wool.

With an appreciation that he may be oversimplifying, the writer ventures that the United States can be assured of sufficient water of acceptable quality for essential needs within the early foreseeable future, provided that it (1) informs itself,

much more searchingly than it has thus far, in preparation for the decisions that can lead to prudent and rational management of all its natural water supplies; (2) is not deluded into expecting a simple panacea for water supply stringencies that are emerging; (3) finds courage for compromise among potentially competitive uses for water; and (4) accepts and can absorb a considerable cost for new water-management works, of which a substantial part will need be bold in scale and novel in purpose.[1]

SUMMARY

The origin of ground water is precipitation that seeps into the ground.

The *water table* is the surface below which rocks are saturated with water. The water table is a reflection of the surface topography with less relief. Lakes and swamps are areas where the land surface is below or at the water table. The position of the water table changes seasonally.

Springs occur where the water table is exposed, as on a valley side.

A *perched* water table is caused by a relatively impermeable rock (such as shale) stopping the downward percolation of rainwater. It may result in a spring.

The *porosity* of a rock is a measure of its open or pore space.

The *permeability* of a rock is a measure of the interconnections of the pore space, affecting the ability of water or petroleum to move through the rock. Poorly sorted or fine-grained sedimentary rocks have low permeability, although they may have high porosity.

The most common reservoir rock is sandstone, although fractured granite or limestone or other rocks can serve as well.

An *artesian well* has its reservoir confined above and below by impermeable rocks. A well drilled into such a confined reservoir will flow nearly to the height of the charging area (the area where the water enters the reservoir rock) because the released water seeks its own level.

The conservation of ground water is important as its average rate of movement through rocks is only about 50 feet per year, and many years may be required to replace hastily pumped water.

A *cone of depression* is the result of pumping water from a well faster than it can be replaced. Cones of depression around a number of closely spaced wells drastically lower the water table. Rapid removal of ground water or petroleum may also cause subsidence of the ground surface.

Pollution is a major problem in use of ground water and in waste disposal.

Water is a good solvent and ground water plays an important role in weathering. Most ground water contains dissolved material and is, therefore, hard or occasionally unsuitable for use.

Caves probably form in the zone of seasonal water table fluctuation by solution of limestone by ground water.

Sinks are closed depressions in areas of cavernous limestone.

Geysers are periodic discharges of hot ground water; *hot springs* flow more or less continuously.

The United States receives an average of 76 centimeters (30 inches) of rain per year. Of this, 54.6 centimeters (21.5 inches) are lost by evaporation or used by plants, leaving 21.6 centimeters (8.5 inches) available for use in the form of runoff and ground water. More than one-third of the available water is being used. However, distribution of population and of precipitation is not uniform, so many areas have a deficiency of available water.

Water that is *consumed* is the water lost to the atmosphere by evaporation or plants; the main consumption of withdrawn water is by irrigation.

Water that is *used* is returned to the ground or runoff and is available for reuse. Its quality may not be impaired by its use, although it commonly is polluted by heat, fertilizer,

---

[1]A.M. Piper, U.S. Geological Survey Water Supply Paper No. 1797, "Has the United States Enough Water?"

phosphates, insecticides, and so forth.

All life on earth ultimately depends on vegetation, as plants are the base of the food chain or pyramid. A diet of vegetable matter and clothing of cotton requires much less water than the use of meat and wool.

Our water resources will have to be very carefully managed so that we will have enough water in the future.

QUESTIONS

1 What is the general relationship between the water table and the topography?
2 What is porosity?
3 What is permeability?
4 What happens near a well that is pumped too fast?
5 The average rate of movement of ground water is _____.
6 Sketch an artesian well. How long ago did the water in such a well fall as rain?
7 Describe capillary action as it applies to geology.
8 Sketch several types of springs.
9 Sketch a perched water table.
10 In what type of rock is pollution most likely to become a problem?
11 Describe one way a geyser eruption may occur.
12 How are caves formed?
13 Describe how the topography is different in an area underlain by soluble limestone compared to an area of sandstone. Assume humid climate.
14 What is the nature of the United States' water problem?
15 What is the main use of runoff at present?

SUPPLEMENTARY READINGS

General

Baldwin, H.L., and C.L. McGuiness, *A Primer on Ground Water.* Washington, D.C.: U.S. Government Printing Office, 1963, 26 pp.

Bradley, C.C., "Human Water Needs and Water Use in America," *Science* (October 26, 1962), Vol. 138, No. 3539, pp. 489–491.

Dunne, Thomas, and L.B. Leopold, *Water in Environmental Planning.* San Francisco: W.H. Freeman, 1978, 818 pp.

Hackett, O.M., *Ground-Water Research in the United States.* U.S. Geological Survey Circular 527, Washington, D.C.: U.S. Government Printing Office, 1966, 8 pp.

Keefer, W.R., *The Geologic Story of Yellowstone National Park.* U.S. Geological Survey Bulletin 1347, Washington, D.C.: U.S. Government Printing Office, 1971, 92 pp.

Keller, W.D., "Drinking Water: A Geochemical Factor in Human Health," *Bulletin,* Geological Society of America (March 1978), Vol. 89, No. 3, pp. 334–336.

Leopold, L.B., and W.B. Langbein, *A Primer on Water.* Washington, D.C.: U.S. Government Printing Office, 1960, 50 pp.

Swenson, H.A., and H.L. Baldwin, *A Primer on Water Quality.* Washington, D.C.: U.S. Government Printing Office, 1965, 27 pp.

Thomas, H.E., and L.B. Leopold, "Groundwater in North America," *Science* (March 5, 1964), Vol. 143, No. 3610, pp. 1001–1006.

U.S. Geological Survey, *Before the Well Runs Dry. A Handbook for Designing a Local Water Conservation Plan.* Washington, D.C.: U.S. Government Printing Office, 1980, 95 pp.

Vogt, E.Z., and Ray Hyman, *Water Witching, U.S.A.* Chicago: University of Chicago Press, 1959.

Caves

Ford, T.D., and C.H.D. Cullingford, *The Science of Speleology.* New York: Academic Press, 1976, 593 pp.

Moore, G.W., "Origin of Limestone Caves: A Symposium," *Bulletin,* National Speleological Society, 1960, Vol. 22, Part 1, 84 pp.

Moore, G.W., and G.N. Sullivan, *Speleology: The Study of Caves.* 2nd ed. Teaneck, N. J.: Zephyrus Press, 1978, 150 pp.

Waste Disposal

Clark, J.R., "Thermal Pollution and Aquatic Life," *Scientific American* (March 1969), Vol. 220, No. 3, pp. 18–27.

Merriman, Daniel, "The Calefaction [warming] of a River," *Scientific American* (May 1970), Vol. 222, No. 5, pp. 42–52.

Schneider, W.J., *Hydrologic Implications of Solid-Waste Disposal.* U.S. Geological Survey Circular 601-F, Washington, D.C.: U.S. Government Printing Office, 1970, 10 pp.

13

EARTHQUAKES AND THE EARTH'S INTERIOR

From the scourge of the earthquake, O Lord, deliver us.
—From the Litany of the Roman Catholic Church

Knowledge of the earth's interior cannot be obtained by direct observation and, as a result, comes from very diverse kinds of data. Astronomical observations reveal the size and shape of the earth and indicate that the interior is much different from the surface. Study of earthquake vibrations reveals much of the internal structure, and meteorites provide clues to the possible internal composition. The use of data from such diverse fields shows why the geologist's interest and training must be so broad.

SHAPE, SIZE, AND WEIGHT OF THE EARTH

SHAPE AND SIZE OF THE EARTH

The Greeks and other ancient peoples were excellent observers. They observed the earth's round shadow during eclipses of the moon. From this, and from noting that ships sailing away appear to sail downhill, they concluded that the earth is a sphere. They felt that this inference must be correct because they considered the sphere to be the most perfect of shapes.

About 235 B.C., in Egypt, Eratosthenes went a step further and measured the diameter of the earth. His method was simple (Fig. 13–1) and involved only two assumptions:

1. The earth is a sphere.
2. The sun's rays are all parallel.

His method required the measurement, at the same time at two places, of the angle between the sun's rays and a plumb bob in the plane defined by the sun's rays and the plumb bob. The two places had to be widely separated to obtain a measurable angle, and he had to know the distance be-

tween the two cities. He chose noon, the time when the shadow is shortest and points due north (he had no other way to synchronize the time), and two cities that he believed to be on a north-south line so that the line between the two cities would be in the same plane as the sun's rays and the shadows. He chose to make the measurement on the day that the sun was directly overhead at noon at Aswan; he measured 7½ degrees at Alexandria. He then used simple geometry to compute the diameter.

He knew that for a circle,

$$C = \pi D$$

and for any arc of a circle,

$$C' = \frac{\theta}{360} \pi D \text{ or } D = \frac{C'}{\pi} \frac{360}{\theta}$$

His measurement was about 15 percent too high (we are not sure about the length of his linear unit), which was remarkably accurate for the time. Later, less accurate measurements indicated a much smaller earth and resulted in Columbus' thinking that he had reached Asia.

More precise repetition of this measurement at different places in the last few hundred years has shown that the diameter of the earth is less if the measurement is made nearer the poles. From this we reach the conclusion that the earth is ellipsoidal, not truly spherical. The equatorial bulge is apparently due to the rotation of the earth. The difference is not great:

Equatorial radius:
  6,378,388 meters = 3963.5 miles

Polar radius:
  6,356,912 meters = 3950.2 miles

Difference:
  21,476 meters = 13.3 miles

The most precise knowledge we have on the shape of the earth

comes from observations made on satellites. The earth's gravitational pull on a satellite determines its orbit; therefore, the shape of the earth's mass controls the exact orbit. The orbit can be measured by radar, and the shape of the earth can be computed. As a result, we now know that the cross-section at the equator is not an exact circle and that the southern hemisphere is slightly larger than the northern. These discrepancies are, however, very small.

WEIGHT OF THE EARTH

To weigh the earth is a more difficult problem than to determine its general shape. Work on this problem was not possible until after Newton, in 1687, deduced the behavior of bodies at rest and in motion. One of the results of the three laws of motion that bear his name was his law of universal gravitation. With this law, which states that the gravitational attraction between two bodies is proportional to their masses and inversely proportional to the square of their distance apart, he was able to explain the motion of the

The San Andreas fault south of San Francisco. A number of the homes shown in this 1965 photo have been damaged or removed because of landslides and seismic damage. The fault zone shows clearly in this photo (arrows), but is now obscured by subsequent building. Photo by R.E. Wallace, U.S. Geological Survey.

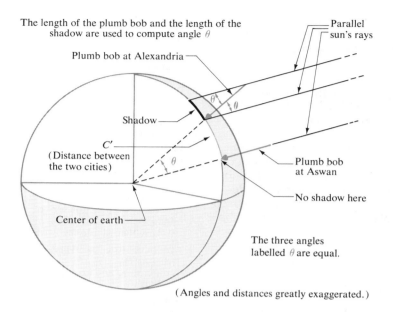

The length of the plumb bob and the length of the shadow are used to compute angle θ

Plumb bob at Alexandria

Parallel sun's rays

Shadow

C'
(Distance between the two cities)

Center of earth

θ

Plumb bob at Aswan

No shadow here

The three angles labelled θ are equal.

(Angles and distances greatly exaggerated.)

FIGURE 13–1 Measurement of the earth's diameter.

solar system. Stated mathematically,

$$F_{\text{gravitational}} = G\frac{m_1 m_2}{R^2}$$

where G is the universal gravitational constant.

The universal gravitational constant G should not be confused with the acceleration of gravity [generally called g and equal to about 9.8 m/sec^2(32 ft/sec^2)]. One of Newton's important contributions was to differentiate between the *weight,* or force acting on an object, and its *mass,* an inherent, unvarying property. The background to this is that Aristotle stated that a heavy object falls faster than a light one; Galileo showed experimentally that both fall at the same speed; and Newton showed why this is true. Newton's second law states that the acceleration (rate at which speed changes) of an object is proportional to the force acting on it and inversely proportional to its mass:

$$a = \frac{F}{m}$$

(more commonly written $F = ma$)

The force F acting on a falling object is its weight. Because weight and mass are related, the quantity F/m is a constant for all falling

bodies at any one place; hence, they all accelerate at the same rate. In the case of a falling body, a is the acceleration due to gravity, g. Note that because the earth is not a sphere, all points on the surface are not equidistant from the center of mass of the earth, and the weight F of an object, which is due to its gravitational attraction, is not a constant. Thus g, the acceleration of gravity, varies with latitude and elevation. Perhaps this should be taken into account in determining world's records in athletic events.

Note that although it is relatively easy to measure g at any point, this is not the universal constant of gravitation G, and there is

no relationship between them that will enable calculation of G unless the mass of the earth is known.

In the case of any object on the earth, its weight is equal to its mass times the acceleration of gravity, or

$$F_{\text{weight}} = mg$$

This weight is also the gravitational force between the earth and the object:

$$F = G\frac{m_{\text{object}} m_{\text{earth}}}{(R_{\text{earth}})^2}$$

Combining, we get

$$F = mg = G\frac{m_o m_e}{(R_e)^2}$$

Solving algebraically for m_e, we obtain

$$m_e = \frac{g(R_e)^2}{G}$$

This equation cannot be solved until either m_e (mass of earth) or G (gravitational constant) is known.

The earth was accurately weighed in 1798 by directly measuring the force between two masses so that G could be calculated. (See Fig. 13–2.)

The mass of the earth is computed using the relation derived above:

$$m_e = \frac{g(R_e)^2}{G}$$

All quantities on the right side of the equation are known, and substitution leads to the result that

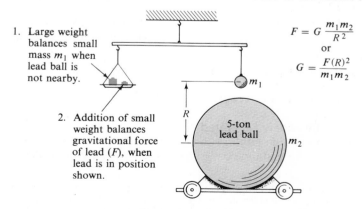

1. Large weight balances small mass m_1 when lead ball is not nearby.

2. Addition of small weight balances gravitational force of lead (F), when lead is in position shown.

$$F = G\frac{m_1 m_2}{R^2}$$

or

$$G = \frac{F(R)^2}{m_1 m_2}$$

m_1

R

5-ton lead ball

m_2

FIGURE 13–2 Determination of the gravitational constant. The figure is diagrammatic, and the balance used was of a different, more sensitive design.

the earth has a mass of about 7 times 10^{21} tons (or 7 followed by 21 zeros). Note that in English units the pound is considered by some a force, and by others, a mass unit. It is considered a unit of mass here.

DISTRIBUTION OF MASS IN THE EARTH

A surprising result (predicted by Newton on the basis of planetary motion) is that this mass requires the average specific gravity of the earth to be about 5.5; that is, on the average, a given volume of it weighs 5.5 times as much as does the same volume of water. This is surprising because the rocks at the surface have average specific gravities less than 3. Thus the rocks at depth must have much greater specific gravities than those at the surface. This suggests that the composition of the deeper parts of the earth is very different from that of the crust.

The distribution of mass in the earth determines its motion in space. One can imagine two spheres of equal size and weight; one has a uniform composition, but the second has its weight concentrated near the center with the rest hollow, except for spokes that separate the shell from the core. These two spheres will have different mechanical behavior. If we roll them down an inclined plane, they will travel at different rates, the uniform sphere moving more slowly.

The earth's distribution of mass is determined by study of the earth-moon system. The moon's gravitational attraction on the earth's equatorial bulge causes the earth's axis to wobble very slowly. This motion is called the *precession of the equinoxes* and is quite similar to the wobbling motion of a top as it slows down. *Precession* is a term used to describe this type of wobble, and *equinox* is a term that designates a relationship between the earth's position and the "fixed" stars on certain days, occurring twice a

year when day and night are of equal duration. This wobble requires 25,735 years to return to the same point, and accounts for the fact that the Egyptian astronomers noted a different star in the position of Polaris, which is our present north star.

Because the amount of wobble (precession) caused by the moon's attraction depends on the mass distribution in the earth, it can be used to calculate the distribution of mass in the earth. This method also reveals that the earth is much denser at deep, than at shallow, levels.

EARTHQUAKES AND SEISMOLOGY

Although earthquakes cause great damage and are greatly feared, they do provide much useful information about the earth's interior. Study of waves generated by earthquakes deep in the earth, for example, reveals that the earth has a layered structure. Combining this information with the distri-

bution of mass deduced from astronomical observations leads to a model of the earth. First, the cause of earthquakes will be examined, and then the types of vibrations produced by earthquakes. The effects of these vibrations at the surface, which include the damage they cause, will be considered next; and finally, how their travel times through the earth reveal the earth's deep structures will be described.

CAUSE OF EARTHQUAKES—FAULTS

Earthquakes are vibrations caused by movement of rocks. The surface on which such movement occurs is called a *fault*. Faults are easily recognized at places where the movements have brought diverse rock types into contact. Some shallow faults cause surface breaks called *fault scarps*. The amount of displacement of surface features is not generally an accurate indicator of the amount of movement at depth, as shown in Fig. 13–3. (See also Figs. 13–4

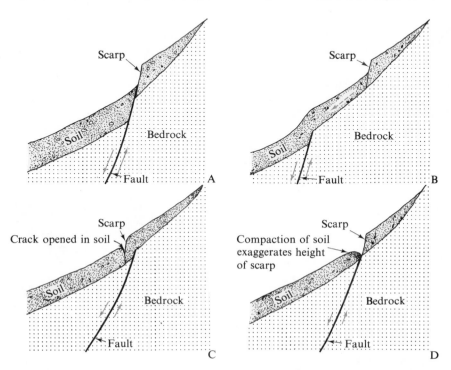

FIGURE 13–3   Types of fault scarps that develop with mostly vertical movement. Parts A and B after I. J. Witkind, U.S. Geological Survey Professional Paper 435, 1964.

FIGURE 13-4 Scarp produced by the Hebgen Lake, Montana, earthquake in 1959. Note the subsidence below the scarp. Photo by J. R. Stacy, U.S. Geological Survey.

and 13-5.) These movements cause vibrations; *seismology is the study of these vibrations.* The speed at which such vibrations, caused by either earthquakes or explosions, travel through the earth provides much information about rock types and structure. Fault types will be described more completely in Chapter 15.

Faulting is one of the few geologic processes that is rapid or catastrophic. Earthquakes result from sudden movement by faulting; but apparently the actual movements within the earth are slow, and rupture occurs only when the stress has built up to the point that it exceeds the strength of the rocks and a new fault is born. Most earthquakes occur on existing faults, and the movement occurs when the forces exceed the friction on the fault plane. This is suggested by the slow move-

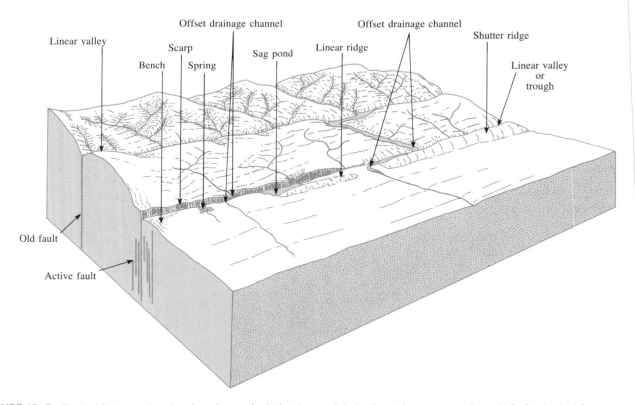

FIGURE 13-5 Typical features that develop along a fault that has mainly horizontal movement. From U.S. Geological Survey Professional Paper 941-A.

ments that have been detected on some active faults. This was first noted after the 1906 earthquake in San Francisco on the San Andreas fault. Accurate surveys, which had been made on both sides of the fault before the earthquake, showed that a small amount of movement had occurred between the time of the surveys. Resurvey after the earthquake showed that about 6 meters (20 feet) of absolute horizontal movement occurred in the earthquake, in agreement with the 6 meters of relative movement measured by surface features. This led to the theory that most earthquakes occur when the energy stored by elastic deformation in the rocks on both sides of the fault is enough to rupture the rocks or to overcome the friction on an existing fault plane. This *elastic rebound* process is shown in Fig. 13–6. This theory explains the surface movements of most earthquakes. Friction on the fault plane may cause sticking after some movement has occurred, and so the total strain may not be relieved in a single earthquake.

The actual mechanism of earthquakes is probably more complicated because some sort of lubri-

cation appears necessary to reduce the friction on the fault plane before movement occurs. In any case, although rocks do not seem to be elastic, they do deform elastically and so can store great amounts of energy that are released suddenly in earthquakes. The process is similar to that of a bow and arrow.

Slow creep has been detected along several faults in California in agreement with the elastic rebound theory. (See Fig. 13–7.) At several places buildings, roads, pipelines, and railroads cross active faults, and this slow movement has caused damage. (See Fig. 13–8.) Recording instruments have been installed at some of these places. Some examples of this movement are:

Warehouse at Irvington on Hayward fault: 7.6 cm (6 in.) between 1921 and 1966 (Fig. 13–9)
Railroad tracks at Niles on Hayward fault: 20.3 cm (8 in.) between 1910 and 1966
University of California stadium at Berkeley on Haward fault: 3.2 cm ($1\frac{1}{4}$ in.) between 1954 and 1966
Winery near Hollister on San Andreas fault: 8.3 cm ($3\frac{1}{4}$ in.) in $7\frac{1}{2}$ years

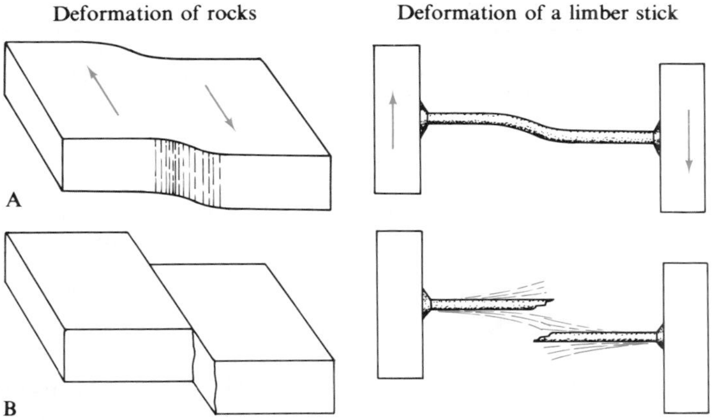

The movement appears to occur in spasms of a small fraction of an inch followed by weeks or months of quiet.

WAVE TYPES

Earthquakes are recorded by seismographs. (See Figs. 13–10 and 13–11.) Three types of waves are sent out by an earthquake. (See Fig. 13–12.) They are

1. *P waves*—primary or push (compression) waves. "Primary" because they travel fastest and, therefore, arrive first at a seismic station. "Push" because they vibrate as compressions and rarefaction.
2. *S waves*—secondary or shake (shear) waves. "Secondary" because they are the second arrivals at a seismic station. "Shake" because they vibrate from side to side.
3. *Surface waves*—slow-moving waves with motion similar to the waves caused by a pebble tossed into a pond.

The propagation, or movement, of these waves will show more of the differences among them. The energy of an earthquake travels in all directions away from the source. We are interested only in the particular rays that are received at one point. The travel of the P waves is easily demonstrated. If a part of a spring is compressed and then quickly released, the resulting wave moves along the spring by successively compressing and spreading the spring. The S-wave movement is shown by displacing, and then releasing, one end of the spring. The wave moves along the spring by successive displacements to one side and the other. (See Fig. 13–13.) This same motion can be demonstrated by shaking a rope. (See Fig. 13–14.) Surface waves are ground motion caused by P and S waves reaching the surface.

Deformation of rocks   Deformation of a limber stick

FIGURE 13–6 Origin of earthquakes. A portion of the earth's crust is shown on the left and a limber stick on the right. A. Slow deformation of the crust is caused by internal forces. B. When the strength of the rocks is exceeded, they rupture or fault, producing earthquake vibrations. Earthquakes on old faults result when the friction along the plane of the old break is exceeded.

The transmission of energy by the surface wave is more complex, with two types of waves involved, and cannot be discussed here beyond the previous mention of the water-wave analogy. Typical seismic records show more than three waves because these main waves are commonly reflected by the layers within the earth.

DAMAGE

Earthquakes cause damage in a number of ways. They set off landslides, they produce seismic sea waves, and they cause uplift or subsidence of large areas. However, the principal damage done by earthquakes is due to differential movements of buildings. These movements are caused mainly by the surface waves. The P and S waves vibrate fast but with small movement and, therefore, cause very little damage to structures. The surface waves, on the other hand, have much larger *amplitude (movement)* in lower frequencies and are the cause of most direct damage done by earthquakes. This is easy to understand, for if one part of a building is moved a few inches in one direction, while another part moves in another direction, the building will be damaged. Most earthquake-resistant buildings are designed to be flexible, so that such movements will not damage them. The building site will also affect the amount of movement that a structure experiences. In general, buildings on bedrock will be damaged less than those built on less consolidated, easily deformed material such as natural or artificial fills, because the softer material slows the waves, causing an increase in the amplitude just as occurs with the ocean waves described in Chapter 11. These statements apply only to earthquake-wave damage and do not include the sometimes total destruction that occurs at places

FIGURE 13-7 Recent horizontal movements in the San Francisco Bay area revealed by surveying. After U.S. Geological Survey.

where there is an actual break or displacement at the surface.

In urban areas, another major cause of property damage due to earthquakes is the fires started by crossed electric wires and broken gas lines. These fires cannot be controlled because of broken water mains and disrupted communications. In general, private homes have not been too badly damaged in most recent earthquakes be-

cause the usual wood-frame house is a reasonably flexible structure. In some cases, the home may be too badly damaged to be salvageable, but it rarely collapses, killing the occupants. Particularly in many older homes the damage may be in the form of the building being pushed off its foundation. Most modern building codes require a tight bond between the foundation and the

FIGURE 13-8 The Baldwin Hills Dam in Los Angeles was built across a creeping fault. When it failed, 5 lives were lost, and 15 million dollars property damage resulted. The dam is on the left, and in the foreground are a swimming pool and the foundations of damaged homes. Photo from U.S. Army.

FIGURE 13-9 Damage to warehouse built across the Hayward fault at Irvington, California. The originally straight wall is bent several inches near the far end of the second window.

building. Typically the part of the average home that is damaged in an earthquake is the chimney. Because it is made of brittle bricks, a chimney tends to be cracked or to fall down.

In certain parts of the world these statements are not true, par-

ticularly in the Near East where most dwellings are built of adobe brick. In an earthquake the brick, being a strong, rigid material, withstands the earthquake until its strength is overcome, at which time the building collapses, commonly killing some of the occu-

pants. In the Near East, after such an experience, a city will be completely rebuilt right on top of the old ruins. This is one cause of the layered cities that archeologists excavate in these areas. In high-rise buildings the damage pattern is quite different. Most modern high-rise buildings are built of steel and concrete and are designed to be flexible. Much of the major structural damage to such buildings occurs where horizontal members are pulled away from vertical members. The tying of such joints is a critical part of such buildings. In general, an attempt is made in high-rise buildings to tie the foundation to bedrock. Generally this means a very deep foundation. In some areas this is not possible and other types of foundations are designed.

Some earthquakes, particularly the January 1968 earthquake in Caracas, Venezuela, the earth-

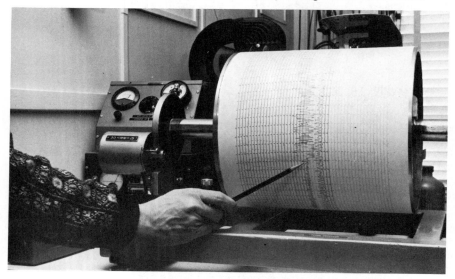

FIGURE 13–10 Long-period seismograph showing record of earthquake near Taiwan on March 12, 1966. Photo from NOAA.

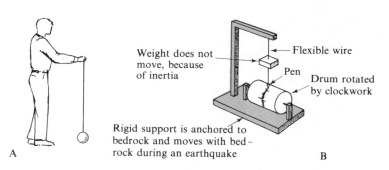

FIGURE 13–11 The operation of a seismograph. A. A weight on a string illustrates the principle of the seismograph. If the arm moves rapidly, the weight remains still. B. In this diagrammatic model of a seismograph, the support is attached to bedrock and moves when the earth quakes. The weight tends to remain still, so the pen records the relative movement between the chart that moves with the bedrock and the weight that does not move. Actual seismographs are designed to respond to movement in a single direction. Instead of using a pendulum, the weights are suspended like doors and so respond only to horizontal movements at a right angle to the plane of the weight.

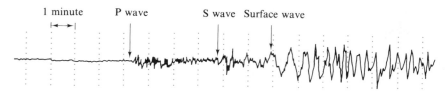

FIGURE 13–12 A seismograph record showing the three waves.

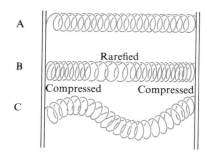

FIGURE 13–13 Types of waves. A. A spring stretched between two supports. B. A push wave showing compression and rarefaction. C. A shake wave.

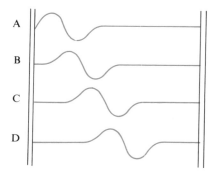

FIGURE 13–14 Propagation of an S wave shown by displacement of a rope. Part A shows initial displacement; B, C, and D show displacements at later times.

quake of March 1970, in Santa Rosa, California, and the February 1971 earthquake in San Fernando, California, caused much damage to well-designed buildings. Engineers and engineering geologists are closely studying the damage that occurred at these places in hopes of finding ways to design buildings that are more earthquake resistant. (See Fig. 13–15.) Part of the problem of this unexpected damage is apparently related to the vibration frequency that the building experiences. If the building is shaken in such a way that it begins to resonate at its natural frequency, much damage can result. The amount of movement is, in this way, magnified. The vibration frequencies of most high-rise buildings are quite low. If such buildings are built on relatively soft foundation material, even with well-designed foundations, apparently the high-frequency earthquake waves are filtered out by the soft foundation material, and the low frequencies move the foundation. Thus, in some cases, well-designed buildings on soft foundation materials are shaken at the natural frequency of the building. The length of time that the earthquake vibrations last is probably also very important in determining the amount of damage. The longer the earthquake, the more opportunity for the high-rise to begin moving at its natural frequency. Although eyewitness observers are not too reliable, earthquake vibrations in major earthquakes seem to last anywhere from about a minute to three or four minutes. In general the amount of damage caused by a nearby earthquake is more related to the local geology, that is, the underlying rock, than it is to distance from the fault or from the epicenter of the earthquake. This applies to those places within a few miles to perhaps a few tens of miles from the actual site of the earthquake.

Many people have a great fear of falling into a fissure created by

FIGURE 13–15   Damage in downtown Anchorage, Alaska, caused by the Good Friday Earthquake of 1964. The differences in amount of damage in the various buildings suggest that changes in building codes or building practices are necessary to protect lives. Photo from NOAA.

an earthquake. Fig. 13–3 shows how small fissures can form in earthquakes, but this figure also shows that such cracks are generally shallow. The only recorded cases of loss of life in this way are a woman in a Japanese earthquake and a cow in the 1906 San Francisco earthquake. In both cases the walls of the cracks failed, and the victims were buried before they could climb out.

The likelihood of being injured or killed in a major earthquake in the United States is exceedingly small, but when a major earthquake does strike, the results can be severe, as witnessed by the 116 deaths in 1964 in Alaska; 28 in 1959 at Hebgen Lake, Montana; 115 in 1933 at Long Beach, California; and 64 in 1971 at San Fernando, California. In 1972 over 10,000 were killed at Managua, Nicaragua. If you feel the earth shaking under your feet and realize that you are experiencing an earthquake, you should dive under a table or desk, get in a doorway, or, in short, get into some position where you will be protected

against material falling from the ceiling. The worst possible thing to do is to run out into the street, where you may be struck by a falling cornice, chimney, or other object. The elimination of buttresses, cornices, and other masonry structures that may fall in an earthquake is one of the major considerations in designing earthquake-resistant buildings.

Most earthquakes are followed by aftershocks, some of which can be as severe as the original earthquake itself. For this reason, it is quite dangerous to reoccupy a building after an earthquake, even though the building has withstood the earthquake, because it may now, in its weakened condition, fail in the less severe aftershocks.

Seismic Sea Waves cause much earthquake damage. They are commonly called tidal waves, but this is a very poor term because tides have nothing to do with this type of wave. For this reason the Japanese word for these waves, *tsunami,* is in common use for them; but the translation of tsu-

255

nami means tidal wave, so it, too, is not really suitable. *Seismic sea wave* is a much better, more descriptive term.

Seismic sea waves are generated by faults or earthquakes under the sea. These undersea movements can cause waves that travel across oceans as small waves imperceptible to ships at sea. In shallow water at shorelines, these waves become huge and cause great destruction. The initial movement of the water is commonly a recession followed in a few minutes by huge waves, some of which have been more than 61 meters (200 feet) high, although they are generally a few tens of feet high. Because they are so destructive to life and property, warnings are now sent by radio when seismographs detect an earthquake under the sea. These waves travel at velocities of 644 to 725 kilometers (400 to 450 miles) per hour and can cause damage on the opposite side of an ocean. (See Fig. 13–16.)

SIZE OF EARTHQUAKES— INTENSITY AND MAGNITUDE

The size of an earthquake can be measured in many different ways. The amount of surface displacement is one measure; however, as shown in Fig. 13–3, the surface displacement may be different from the displacement at depth. Surveying shows the absolute displacement, which may differ from the relative displacement shown at the surface. Moreover, most earthquakes have no surface displacement, although in some it is appreciable. The maximum surface displacement so far recorded is 14.4 meters (47 feet 4 inches) at Yakutat Bay, Alaska, in 1899, and in the Good Friday earthquake of 1964 parts of the sea floor off Alaska were raised over 15.25 meters (50 feet).

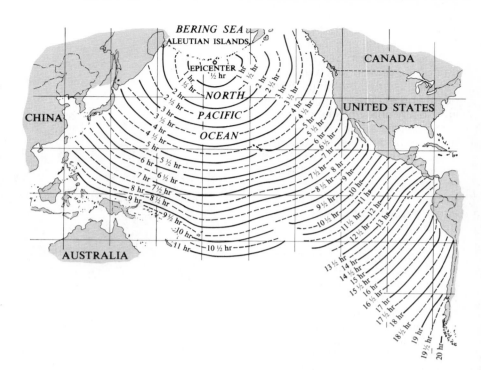

FIGURE 13–16 This map shows the travel times in hours for seismic sea waves to reach most of the Pacific Ocean for an earthquake off the Aleutian Islands. Courtesy of NOAA.

Another measure of the size of an earthquake is the size of the area in which the displacement occurs. In Alaska's Good Friday earthquake of 1964, the observable crustal deformation covered an area of between 171,100 and 199,600 square kilometers (66,000 and 77,000 square miles). (See Fig. 13–17.) This is the largest of any one historical earthquake, although the 1906 San Francisco earthquake had the longest length of surface movement in a single earthquake. A 330-kilometer (205-mile) stretch of the nearly 966-kilometer- (600-mile-) long San Andreas fault moved in that earthquake.

The duration of shaking is another measure of an earthquake.

None of these ways of measuring an earthquake enables direct comparison of earthquakes, and none is useful for the many earthquakes without surface displacement. Seismologists have devised *intensity* scales to overcome these

difficulties. *Intensity scales rate the damage that occurs to substantial buildings.* (See Table 13–1.) The estimation of intensity on the basis of damage and eyewitness reports is somewhat subjective, but works reasonably well. The main problem is that the damage depends somewhat on the local surface conditions, and so the intensity grades of many earthquakes follow the surface geology. (See Fig. 13–18.) Intensity maps are made by sending postcards to numerous people near the earthquake, asking them to describe the surface effects. In this way the epicenter of an earthquake can be located reasonably well. Notice that intensity enables direct comparison of surface effects only and that a deep earthquake will, in general, have a lower intensity than a shallow earthquake, even though both release the same amount of energy.

The *magnitude of an earthquake is a measure of the amount*

TABLE 13-1 Earthquake intensity scale. This scale indicates the amount of surface damage caused by an earthquake. (Modified Mercalli Intensity Scale of 1931)

I. Not felt except by a very few under especially favorable circumstances.

II. Felt only by a few persons at rest, especially on upper floors of buildings. Delicately suspended objects may swing.

III. Felt quite noticeably indoors, especially on upper floors of buildings, but many people do not recognize it as an earthquake. Standing motorcars may rock slightly. Vibration like passing of truck.

IV. During the day, felt indoors by many, outdoors by few. At night, some awakened. Dishes, windows, doors disturbed; walls make creaking sound. Sensation like heavy truck striking building. Standing motorcars rocked noticeably.

V. Felt by nearly everyone, many awakened. Some dishes, windows, etc., broken; a few instances of cracked plaster; unstable objects overturned. Disturbances of trees, poles, and other tall objects sometimes noticed. Pendulum clocks may stop.

VI. Felt by all, many frightened and run outdoors. Some heavy furniture moved; a few instances of fallen plaster or damaged chimneys. Damage slight.

VII. Everybody runs outdoors. Damage negligible in buildings of good design and construction; slight to moderate in well-built ordinary structures; considerable in poorly built or badly designed structures; some chimneys broken. Noticed by persons driving motorcars.

VIII. Damage slight in specially designed structures; considerable in ordinary substantial buildings with partial collapse; great in poorly built structures. Panel walls thrown out of frame structures. Fall of chimneys, factory stacks, columns, monuments, walls. Heavy furniture overturned. Sand and mud ejected in small amounts. Changes in well water. Persons driving motorcars disturbed.

IX. Damage considerable in specially designed structures; well-designed frame structures thrown out of plumb; great in substantial buildings, with partial collapse. Buildings shifted off foundations. Ground cracked conspicuously.

X. Some well-built wooden structures destroyed; most masonry and frame structures destroyed with foundations; ground badly cracked. Rails bent. Landslides considerable from riverbanks and steep slopes. Shifted sand and mud. Water splashed (slopped) over banks.

XI. Few, if any, (masonry) structures remain standing. Bridges destroyed. Broad fissures in ground. Underground pipelines completely out of service. Earth slumps and land slips in soft ground. Rails bent greatly.

XII. Damage total. Waves seen on ground surfaces. Lines of sight and level distorted. Objects thrown upward into air.

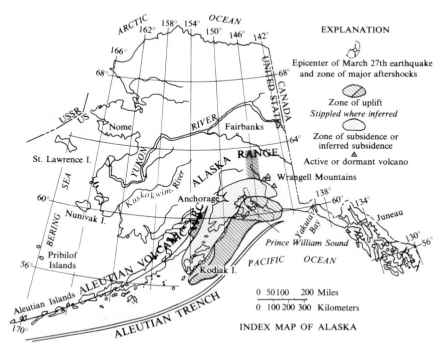

FIGURE 13-17 Map of Alaska showing the area where elevations changed in the Good Friday earthquake of 1964. From George Plafker, *Science* (June 25, 1965), Vol. 148, p. 1676.

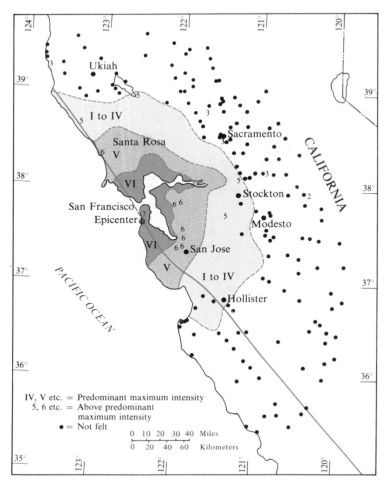

FIGURE 13-18 Map of the reported intensity of the March 22, 1957, earthquake. Movement apparently occurred on the San Andreas fault that is also shown. After W. K. Cloud, California Division of Mines and Geology Special Report 57, 1959.

The magnitude scale in common use is the Richter Scale. The largest magnitude yet recorded on this scale is 8.9, and the smallest felt by humans is 2. The San Francisco earthquake of 1906 was 8.3. For each increase of one unit on the scale, for example, from 6.5 to 7.5, the energy released increases about 31 times and the recorded vibrations at a distant station increase about 10 times. Table 13-2 shows the relationship between intensity and magnitude.

Many more earthquakes occur than cause damage great enough to be reported in the newspapers. Although the average annual number of earthquakes strong enough to be felt at least locally is estimated at tens of thousands, most of the earth's seismic energy is released by the few large earthquakes. It is estimated that about 16 earthquakes of magnitude 7, 5 of magnitude 7.5, and 0.1 to 0.2 of magnitude 8 occur each year.

To these 20 or so largest earthquakes, we must add about 500 magnitude 5 to 6, and 100 magnitude 6 to 7 earthquakes to include all the damaging earthquakes. The energy released in these 600 magnitude 5 to 7 earthquakes is about the same as that released in one magnitude 7.5 earthquake, so many small earthquakes are not a safety valve preventing a large earthquake.

EXAMPLES

Alaska's Good Friday earthquake occurred at 5:36 P.M. local time on March 27, 1964. It was centered in south-central Alaska and had a magnitude of 8.4 to 8.6. It thus re-

of energy released. This figure is what is reported in newspapers and on the radio soon after an earthquake occurs. The magnitude of an earthquake is determined from the amount of movement on a standard seismometer. This amount of movement is corrected for the distance between the instrument and the earthquake. The early reports of an earthquake commonly give slightly different magnitudes from different seismic stations. This occurs because the movement of the standard seismometers is somewhat affected by the foundations on which they rest. Another reason for this slight uncertainty is that magnitude determinations are based on the

wave with the largest amplitude. For most earthquakes this is the surface wave, but the very deep earthquakes have no surface waves. Therefore, a magnitude scale based on P and S waves had to be devised for deep earthquakes. The relationship between these two scales has slight discrepancies.

TABLE 13-2 The relationship between the Richter Magnitude and the Mercalli Intensity Scales at the epicenter of an earthquake.

Modified Mercalli Intensity (at the epicenter)	I	II	III	IV	V	VI	VII	VIII	IX	X	XI	XII
Richter Magnitude		—2—	3		4		—5— 6			—7— 8		

leased about twice as much energy as the San Francisco earthquake of 1906 that had a magnitude of 8.3. Figure 13–17 shows the large area in which the land level changed. Figure 13–17 shows the epicenter of the main shock and also the zone of aftershocks that presumably outlines the area where fault movement occurred at depth. Aftershocks are common in most big earthquakes and are believed to be caused by readjustments made necessary by the main movement. In many earthquakes, the aftershocks cause almost as much damage as the main earthquake. The seismic sea waves caused by this earthquake caused much damage in Alaska and as far away as California, and were recorded in Japan. The vibrations caused many landslides, which were especially damaging to coastal installations built on unconsolidated sediments. Examples are shown in Figs. 7–17, 7–18, and 7–19. In this case, a previous geologic report had suggested that the particular clay of the area was susceptible to landsliding.

The Hebgen Lake, Montana, earthquake of August 17, 1959, was a shallow earthquake of magnitude 7.1. It produced two more or less parallel fault scarps 11.3 kilometers (7 miles) long and up to 6 meters (20 feet) high. (See Fig. 13–4.) Tilting movements are clearly shown by the displaced shorelines of Hebgen Lake. This area is close to Yellowstone National Park, and the shocks affected some of the geysers there, mainly by increasing their activity at least temporarily. The main damage in this remote area was the triggering of a huge rockslide. Unfortunately, the slide fell on a campground, burying a number of people. (See Fig. 7–14.) It also dammed the Madison River, creating a problem. It was feared that when the lake created by the dam overtopped the dam, the flowing water might rapidly erode the

rockslide debris and cause a flood. A channel was later bulldozed through the rockslide to control the water flow.

EARTHQUAKE PREDICTION AND CONTROL

The goal of seismologists is the prediction of the time, place, and magnitude of earthquakes so that property damage and casualties can be reduced. Recently and inadvertently human actions have triggered some earthquakes, suggesting the possibility that earthquakes can be controlled. Such control would enable us to cause several small, harmless quakes rather than allow one natural destructive quake.

"Earthquake weather" is believed by many to be a precursor of earthquakes. Studies of many destructive earthquakes over the last few hundred years have revealed no correlation between weather and earthquakes, but the folklore persists, probably because it is so longstanding.

Giant earthquakes, those with magnitudes over 8.5, seem to occur in periods. Since 1800, most have occurred in three periods: 1835–1847, 1896–1911, and 1933–1942. These periods were at times when the earth's rate of rotation was slower than normal. The amount of slowing is only a few one-thousandths of a second per year. It is not clear whether the slowing causes the earthquakes, or the earthquakes cause the slowing. In any case, it seems reasonable that a shifting of some of the earth's mass, as occurs in a giant earthquake, should affect the earth's rotation.

The places where earthquakes are most apt to occur are easily predicted on the basis of past experience. (See Fig. 13–19.) The earthquake epicenters in Fig. 13–19 outline the tectonic plates discussed in the next chapter. As the

maps show (Figs. 13–20A and B), most earthquakes in the United States occur in California and in a belt through part of the Rocky Mountains. One obvious way to reduce earthquake damage would be to avoid these areas. Such a solution is not practical because it would mean avoiding most of California, the most populous state; however, anyone moving into an earthquake area should be aware of the potential danger.

Within these broad earthquake areas, the epicenters are located mainly along known faults so it is possible to avoid these high-danger areas. The faults to avoid are those that are active. Such faults are those that are creeping or are the sites of previous earthquakes. Some other faults, generally much older, are no longer active and so are much less dangerous. Just because no historical earthquake has occurred on a fault and there is no evidence of recent movement, the fault is not proven inactive. Faults are generally weak areas, and movement on one could be caused by a nearby large earthquake on an apparently unrelated fault. Also, as noted earlier, the amount of damage is controlled by the nature of the building and the rocks on which it is built.

The San Andreas fault in California is one of the longest faults, and because it has been the site of several large earthquakes, it has been studied closely. It extends about 966 kilometers (600 miles) through California from the northern Coast Range to the Mexican border. It is a fracture zone a few hundred meters to about 2 kilometers wide, composed of many more or less parallel faults. The total amount of movement is difficult to measure, and may be up to 564 kilometers (350 miles).

The behavior of this fault varies along its length. Figure 13–21 shows that there are three active areas characterized by creep and/or many small, frequent earthquakes. In marked contrast are the

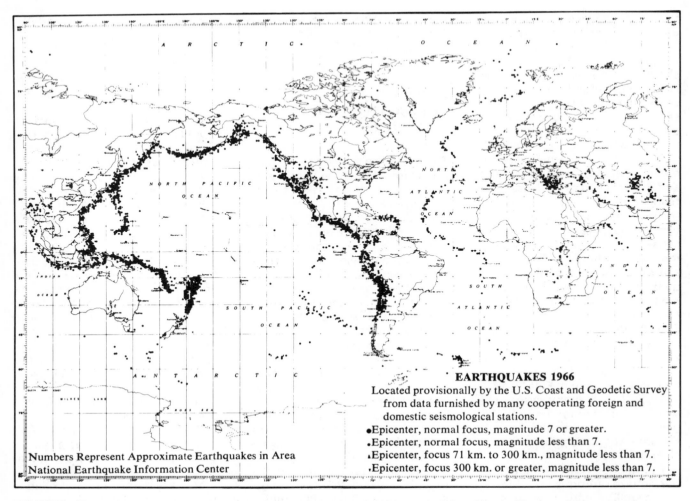

EARTHQUAKES 1966
Located provisionally by the U.S. Coast and Geodetic Survey
from data furnished by many cooperating foreign and
domestic seismological stations.
- Epicenter, normal focus, magnitude 7 or greater.
- Epicenter, normal focus, magnitude less than 7.
- Epicenter, focus 71 km. to 300 km., magnitude less than 7.
- Epicenter, focus 300 km. or greater, magnitude less than 7.

Numbers Represent Approximate Earthquakes in Area
National Earthquake Information Center

FIGURE 13-19 Earthquake zones of the earth shown by the locations of 1966 earthquakes. The significance of these zones is described in Chapter 14. Courtesy of NOAA.

areas where large movements [up to 6 meters (20 feet)] occurred in recent large earthquakes. These parts of the fault are apparently locked, and few earthquakes occur at those places. These large segments of the earth's crust are capable of storing large amounts of strain and may sometime release their strain in a large earthquake. Thus, these are the areas where large destructive earthquakes may occur. These differences in behavior seem to be related to both rock type and composition of the ground water. Most quakes on the San Andreas fault are shallow and so are capable of causing much damage.

The detailed prediction of earthquakes is based on observation of events that occur just before an earthquake, such as creep, tilting, increase in microearthquakes, changes in strain, and changes in such physical properties of rocks as electrical conductivity and magnetic susceptibility. Instruments set up near faults are necessary to make such observations, and such data gathering is one of the main types of current research into earthquake prediction. The Japanese have been issuing general earthquake predictions for some areas since 1966. Their predictions are based on slight tilting and swarms of microearthquakes that

precede potentially damaging earthquakes by a number of weeks.

The first prediction of a major earthquake was made by the Chinese in early 1975 after a false alarm the preceding December. This prediction was based on overall seismicity, tilting, microearthquakes, and changes in water level in wells. Changes in animal behavior occurred just prior to the earthquake.

Another precursor of some earthquakes is the release of radon gas. Small amounts of radon are present in most rocks because radon is a product of the radioactive decay of uranium and thorium, both of which are present in very small amounts in crustal rocks. The radioactive radon is generally detected in drill holes that are instrumented to detect it. Other gases may also be released at the time of, or just before, an earthquake. Many eyewitnesses to earthquakes have reported explosionlike noises, similar to thunder. The noise may come from the escape of gas under high pressure or the explosion of a combustible gas. A less likely explanation is direct audible detection of the seismic energy.

The San Andreas fault in California is our best instrumented fault. There, deformation of deep rocks changes their magnetic properties and so causes slight changes in the local magnetic field a few tens of hours before creep of a few millimeters occurs. So far these magnetic changes have occurred only before such movements. In a few cases earthquakes have followed these movements by a few days. In other places along the San Andreas, changes in rate of creep or even reversals in direction of creep have preceded earthquakes.

Since 1974, some earthquakes in the United States have been predicted using a method discovered by the Russians. The velocity of P waves slows locally before an earthquake and returns to normal velocity just before the earthquake occurs. The time required for these velocity changes is proportional to the magnitude of the eventual earthquake. The local velocity change is, of course, mea-

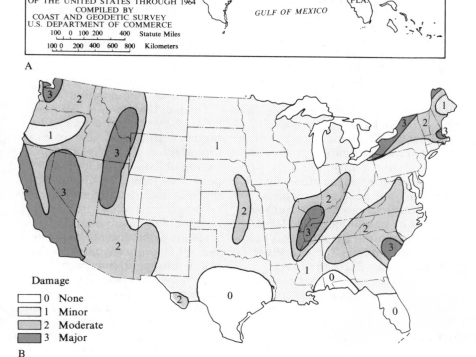

A

FIGURE 13–21 Earthquake activity on the San Andreas fault in California. The three stippled regions are areas of creep and many small to severe earthquakes. Few earthquakes occur in the areas of the 1857 and 1906 breaks, but because the strain energy may build up to high levels in these "locked" areas, a few very severe earthquakes may be expected. In the area south of San Francisco, the San Andreas fault may be locked because the activity there occurs along other faults. After C. R. Allen, 1968, and L. C. Pakiser and others, 1969.

FIGURE 13–20 A. Destructive and near-destructive earthquakes in the United States through 1964. The smallest circles are intensity VII to VIII, the large solid circles are intensity VIII to IX, the smaller ringed circles are intensity IX to X, and the large ringed circles are intensity X to XII. The numbers are the number of earthquakes. B. Seismic risk map for conterminous United States. The map divides the United States into four zones: Zone 0, areas with no reasonable expectancy of earthquake damage; Zone 1, expected minor damage; Zone 2, expected moderate damage; and Zone 3, where major destructive earthquakes may occur. Both A and B courtesy of Environmental Science Services Administration, U.S. Coast and Geodetic Survey, 1969.

sured on waves from distant earthquakes. The velocity change apparently is caused by movement of ground water into the many microcracks that develop in a rock when it is stressed to near rupture.

Flashing and glowing lights have also been observed during many earthquakes, and their cause is not known. This suggests that we still have much to learn about earthquakes and their prediction.

In spite of some of these early successes in earthquake prediction, many problems still exist. Parts of the San Andreas fault in California are closely monitored by instruments in the search for earthquake precursors. A few moderate earthquakes, such as one of magnitude 5.7 in August 1979 about 100 kilometers (60 miles) south of San Francisco, have occurred with no detectable precursors. A possible reason is that most fault planes where exposed at the earth's surface have irregular surfaces. Similar irregularities at depth may concentrate the stress in a small volume, causing slippage there, triggering an earthquake. Whether this is the reason or not, some earthquakes have no precursors, and prediction of such earthquakes is not possible now.

An approach to earthquake safety may be control of quakes. Our knowledge of this came from accidentally triggering earthquakes. The best publicized of these events were the many small earthquakes near Denver that were caused by pumping waste fluids into a disposal well at the Rocky Mountain Arsenal. These earthquakes, some of which had magnitudes up to 5.5 and caused some damage, began in 1962. Until that time there had been few earthquakes in the Denver area, and it was noted that the earthquakes followed periods of pumping waste fluids into the well. A series of seismic stations were set up to monitor the well, and they

revealed that the earthquakes occurred in a zone about 8 kilometers (5 miles) long and up to 1.6 kilometers (1 mile) deeper than the 3660-meter (12,000-foot) well. Analysis of the data suggested that the movements resulted from a reduction in friction caused by the increased pore pressure due to pumping. Soon after this discovery, similar microearthquakes were observed at the Rangely oil field in northwestern Colorado. Here, water was being pumped into the ground to increase the recovery of oil.

These discoveries stimulated the search for other artificial earthquakes. In 1945 a paper had appeared that documented hundreds of small earthquakes in the area of Lake Mead since the building of Hoover Dam. One of these quakes had a magnitude of almost 5, and two had magnitudes of about 4. It was suggested that the weight of the water had caused movements on old faults. More recently in Koyna, India, a lake created in 1962 and 1963 by construction of a dam apparently caused a number of earthquakes that culminated in 1967 with a magnitude 6.5 earthquake that caused about 200 deaths. The geologic environment at Koyna is similar to that at Grand Coulee, Washington, where no earthquakes associated with the lake have been observed. Other occurrences of earthquakes caused by dam-impounded lakes are a magnitude 5.8 earthquake at Lake Kariba on the Zambezi River, Zambia, and a 6.3 magnitude quake at Kremasta Dam, Greece. In most cases a reservoir depth of over about 150 meters (500 feet) seems necessary to trigger an earthquake.

The exact cause of any of these earthquakes is not clear. At the Denver well, they could be due to reduced friction, as suggested above, or to temperature changes caused by the injection of cool fluids. Withdrawal of fluids at Denver suggests that very little mixing of waste fluids and ground water

occurred. At the dams, increased weight could be the trigger, or the lakes could cause inseeping of fluids and so the situation could be much like that at the injection wells. No earthquakes are associated with many high dams. Much more study is necessary, and it is hoped that such studies will be included in the construction costs of new large dams. The subsurface geology of oil fields is generally well studied, making them good places for earthquake study if fluid injection is used for secondary recovery late in the production history of the field.

Earthquakes also have been triggered by underground nuclear blasts. The earthquakes produced in this way have all been within about 24 kilometers (15 miles) of the actual blast and of lower magnitude than the nuclear blast. At the Benham test of 1968, the nuclear blast had a magnitude of 6.3, and the largest associated aftershock had a magnitude of 5.

In the deep gold mines of South Africa, mining at depths of 3.2 kilometers (2 miles) has caused earthquakes of magnitude 4.

Thus, at the moment, we may be on the verge of being able to predict earthquakes, and we can trigger them. This combination may enable us to reduce the dangers of earthquakes, but there are many serious problems that must be overcome before this goal is reached. Prediction alone will reduce the likelihood of casualties but not all property damage. Control by triggering many small earthquakes instead of allowing a large natural one could reduce both casualties and property damage, especially if earthquake-resistant construction is practiced more extensively. Fluid injection wells currently appear best for such control. For instance, a series of such wells along the San Andreas fault could be used to dissipate the strain in a number of small harmless quakes or in creeping movement. The problem is that the San Andreas fault, which is our best-studied fault because it causes so many earthquakes, crosses some densely populated areas. Any experimental program triggering earthquakes along it is likely to be blamed for all earthquake damage that occurs. The resulting lawsuits would become a serious problem because the results of such a program are difficult to predict and it is almost impossible to distinguish between natural earthquakes unassociated with fluid injection and those caused by our interference. In spite of such hypothetical problems, the benefits appear so great that a program of earthquake control should be started as soon as practical. Until such time, the protection of people and structures must depend on zoning laws, building codes, and foundation and structure designs.

LOCATION OF EPICENTERS

The point at which the vibrations of an earthquake originate is called the focus, and the point on the surface directly above is called the epicenter. The different speeds of the P and S waves can be used to locate the *epicenter.* A P wave can travel across the earth in about a half an hour, an S wave travels at about one-half this speed, and the surface wave would require several hours to make the trip. Because the S wave travels more slowly than the P wave, *the difference in arrival time between the P and S waves is proportional to the distance away from the station that the earthquake occurred.* As shown in Fig. 13–22, reports from three stations are needed to locate an epicenter.

To use this simple principle, it was first necessary to determine the velocities of the P and S waves. This was done by finding the approximate location of an earthquake from intensity reports. Then the P and S times at many seismic stations were compared with the distance between the sta-

tions and the earthquake. The simplest way to do this is to plot a graph of distance *versus* arrival time for each wave. Generally, it was necessary to refine such a determination by adjusting the location of the earthquake, which was known only approximately from the intensity, until the best velocity determination was found. This was repeated for many earthquakes until the travel-time curves for the whole earth were determined. In this process the internal structure of the earth was determined.

It is possible to determine the depth of an earthquake from the arrival times of vibrations at nearby stations. Most earthquakes are shallow, less than 56 kilometers (35 miles), but some are as deep as 700 kilometers (435 miles). Deep earthquakes were first suspected from the lack of surface waves from some earthquakes.

INTERNAL STRUCTURE OF THE EARTH

Earthquake waves have revealed the layered structure of the earth. The details of the analysis are best seen in Fig. 13–23. Three main layers were soon distinguished: the *crust,* a very thin shell between 4.8 and 56 kilometers (3 and 35 miles) thick; the *mantle,* the next layer, about 2898 kilometers (1800 miles) thick; and the

core, with a radius of about 3418 kilometers (2123 miles). The velocity of earthquake waves changes abruptly at the contacts between these layers, but, as we will soon see, other boundaries more difficult to detect are also important.

Seismic study of the earth's interior began in earnest near the beginning of the twentieth century. Earthquake detectors had been built as early as 132 A.D. in China, but it was not until late in the nineteenth century that effective seismographs became available. It was suggested as early as 1906 that the earth had a core, and shortly thereafter its existence was confirmed.

If the earth were homogeneous, seismic waves would travel in straight lines; and if the layers were homogeneous, the seismic waves would travel in straight lines in each layer. The density of the layers increases with depth, and so the seismic velocity also increases and this makes the seismic-wave paths curve (Fig. 13–23).

Seismic waves behave much as does light. At each interface where the velocity changes, the wave is refracted, or bent. At these layer boundaries, some of the wave is also reflected; and though these reflected waves are detected by seismographs, we will not consider them here.

CORE—SHADOW ZONES

The evidence for the core is that at distances between about 11,431

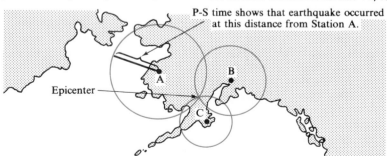

FIGURE 13–22   A, B, and C are seismic stations. The P-S times at each station indicate that the earthquake occurred somewhere on the circle drawn around each station. The point where all three circles intersect is the epicenter.

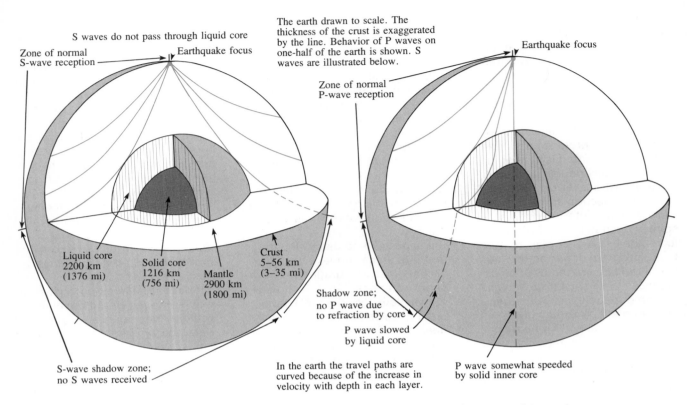

FIGURE 13-23 Behavior of P and S waves in the mantle and the core reveals the layered structure of the earth.

and 15,939 kilometers (7100 and 9900 miles) from an earthquake, no direct P or S waves are received. Beyond 9900 miles, P waves are received, but no direct S waves are received past 7100 miles. (See Fig. 13-23.) This last observation shows that the core is liquid because shake or shear waves cannot pass through a liquid although push waves can. The P waves that go through the core are slowed, and those that pass through the inner part of the core travel somewhat faster than those that go through the outer core. This suggests that the inner part of the core is solid. The P-wave shadow zone is caused by the bending or refraction of the waves as they go into layers of different velocity.

At the boundaries of each of these layers, and even at the lesser velocity changes within the layers, the waves are also reflected. These reflected waves

complicate the simple explanation given above. It was not until the mid-1930s that the depths and velocities of each of the layers were calculated. Since about 1960 it has been possible to refine some of this work because extremely sensitive seismograph systems have been built to detect atomic bomb explosions.

CRUST AND MANTLE— THE MOHO

The crust is the thin shell that overlies the mantle. *A very marked change in seismic velocity separates the crust from the mantle. This change is the Mohorovičić discontinuity,* commonly called the M-discontinuity or Moho. In 1909, Andrija Mohorovičić, a Yugoslavian seismologist, noted that the velocity of P waves changed sharply at distances more than about 201 kilometers

(125 miles) from the earthquake. (See Fig. 13-24.) He interpreted this to mean that at a depth of about 32 kilometers (20 miles), an abrupt change in velocity occurs. Later work shows that this change is worldwide and occurs at depths of about 32 kilometers (20 miles) under the continents and about 11 kilometers (7 miles) below the oceans. (See Figure 13-25.)

Other smaller differences in travel times (velocity) reveal a layering within the crust itself, as shown in Fig. 13-26. The continental crust is believed to be granitic because intrusive and metamorphic rocks of this composition underlie the sedimentary rocks that cover much of the surface, and because granitic rocks have the observed seismic velocities. In the same way, the oceans are believed to be underlain by basalt, and the lower layer under the continents is believed to be basalt. Although this layer has the composition of basalt, seismic data suggest that it is in the form of gabbro or amphibolite, both of which have the same chemical composition as basalt.

The Mohorovičić discontinuity is a marked change in seismic velocity within a thickness of 0.16 to 3.2 kilometers (0.1 to 2 miles). It is believed to be caused by a change in composition from peridotite or dunite in the mantle to basalt in the lower layer of the crust. The depth of the Moho and the seismic velocity in the upper mantle both vary (Fig. 13–27), suggesting that the Moho may be a complex transition.

Although the Moho is seismically the most obvious discontinuity in the upper part of the earth, it is probably not the most important zone. The *low-velocity zone* in the upper mantle is probably more involved in the formation of surface structures.

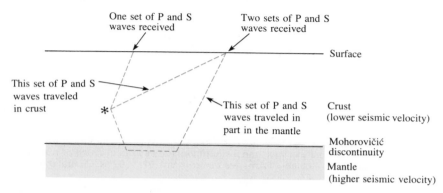

FIGURE 13–24 Determination of the depth of the Mohorovičić discontinuity. At distances over 100–150 kilometers (60–90 miles), the waves traveling through the mantle arrive first because the seismic velocity is higher in the mantle.

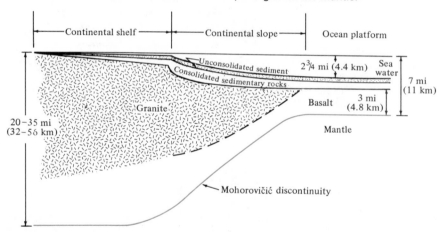

FIGURE 13–25 Generalized cross-section of continent-ocean boundary at a geologically stable area. The sedimentary rocks on the continental shelf and slope generally have more complex structures than shown. From seismic data.

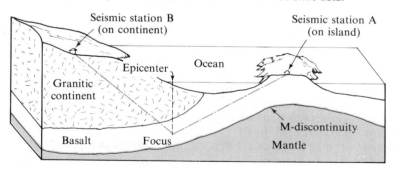

FIGURE 13–26 Block diagram showing paths of seismic waves in oceanic basalt and continental granite. Both stations A and B are equidistant from the focus of the earthquake, but station A receives waves first because they travel faster in basalt than in granitic rock.

LITHOSPHERE AND ASTHENOSPHERE—THE LOW-VELOCITY ZONE

The upper mantle is also layered, but the layers here could not be detected until sensitive seismographs were available. *At depths between about 70 kilometers (40 miles) and 250 kilometers (150 miles), seismic shear waves (S waves) are both slowed and attenuated.* This is called the *low-velocity zone,* and the top of this zone varies in depth between 50 kilometers (30 miles) and 125 kilometers (75 miles). The low-velocity zone is in the upper mantle. The slowing has not been detected everywhere under the continents, but it is generally present under the oceans. *The rocks above the low-velocity zone are brittle and are called the litho-sphere* (See Fig. 13–28). The lithosphere is the crust and the upper mantle. The plates that move on the earth's surface are composed of lithosphere.

The layer below the lithosphere is called the *asthenosphere;* this is the layer in which the lithosphere's plates move. The asthenosphere is the low-velocity zone and is a weak or soft zone in the upper mantle. At the temperature and pressure that exist at the depth of the asthenosphere, the upper mantle rock, believed to be peri-

265

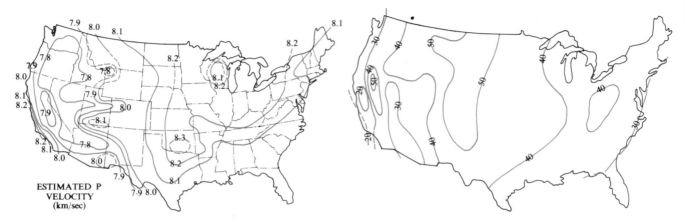

FIGURE 13–27 Lateral variation in the structure of the United States. Note the rough correspondence with the geologic provinces shown in Fig. 15–37. A. Seismic velocity in the upper mantle. The velocity is less than 8 kilometers per second in much of western United States. After Eugene Herrin and James Taggart, Seismological Society of America *Bulletin,* Vol. 52, 1962. B. Crustal thickness. The contours are in kilometers. After L. C. Pakiser and J. S. Steinhart, 1964.

dotite, is very close to its melting point. It is estimated that between 1 and 10 percent of the asthenosphere is melted. This partial melting would cause the slowing and attenuation of shear waves that is observed in the low-velocity zone. Partial melting of the upper mantle rocks has also been suggested as the origin of the basalt that rises at the mid-ocean ridges.

COMPOSITION OF THE LAYERS OF THE EARTH

The velocity of waves is related to the physical properties of the medium through which they pass; hence, it is possible to make some inferences about the material that composes the deep parts of the earth. Remember also the density distribution of the deeper parts of the earth, which was learned from the earlier discussion of the weight of the earth. Both methods of investigation lead to the specific gravity of the mantle, varying from about 3 near the top to about 6 at its base. Similarly, the core has a specific gravity of about 9 at its top and about 12 at the center.

Because the earth's interior is not accessible to direct sampling, all knowledge of its composition is

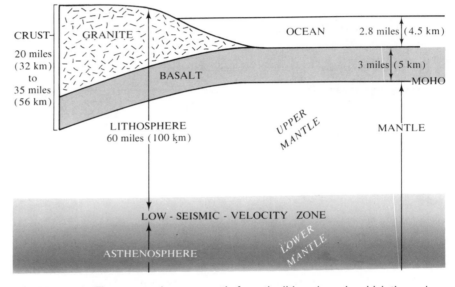

FIGURE 13–28 The crust and upper mantle form the lithosphere, in which the rocks behave as solids. In the asthenosphere, below the low-velocity zone, the rocks flow.

based on indirect evidence. Meteorites provide some clues about the earth's composition. The earth and the meteorites are believed to have formed at the same time and from the same material. *Meteorites are the objects that fall into the earth from space.* Many of them are burned by frictional heat in the atmosphere to form what are called "shooting stars." Some meteorites are quite large and form impact craters, such as Meteor Crater in Arizona. The craters on the moon, which is not pro-

tected by an atmosphere, are believed to have formed in this way.

Meteorites are of three types: *stony meteorites, composed mainly of olivine and pyroxene* (peridotite is the rock name for this composition); *iron meteorites, composed mainly of iron and nickel* (see Fig. 13–29); and *stony-iron meteorites, which are mixtures.*

The mantle is believed to resemble stony meteorites in composition. The upper mantle is probably peridotite, and the physi-

FIGURE 13-29 Cross-section of iron-nickel meteorite from Henbury, Australia. The crystal pattern has been shown by acid etching after cutting. Photo from Ward's Natural Science Establishment, Rochester, N.Y.

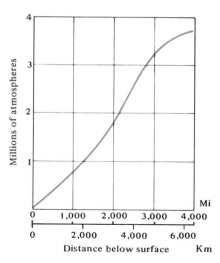

FIGURE 13-30 Pressures within the earth. Adapted from John Verhoogen, *American Scientist,* Vol. 48, 1960.

cal properties of peridotite fit those inferred from seismic and other indirect methods of study of the mantle. Below the low-velocity zone, at depths below the surface between 360 and 400 kilometers (220 and 240 miles), the seismic velocity increases abruptly. At the temperature and pressure at that depth, olivine, the main mineral in peridotite, changes into a different, more compact crystal structure. At depths between 650 and 700 kilometers (400 and 430 miles), another velocity increase occurs. There the temperature and pressure are such that olivine breaks down into oxides of iron, magnesium, and silicon. Both of these velocity changes suggest that the upper mantle is composed of peridotite.

The core is generally believed to be mainly molten iron with some nickel. On both seismic and chemical evidence, it has been suggested that iron and sulfur are the main elements in the core. In reaching these assumptions, allowance must be made for the increased temperature and pressure deep within the earth, for these will greatly modify the properties of rocks within the earth.

TEMPERATURE AND PRESSURE WITHIN THE EARTH

The pressure deep in the earth is reasonably well known, but the temperature is less well known. The pressures are due to the weight of the overlying material, and, with our knowledge of specific gravity gained from seismic studies, can be calculated from the law of gravity. (See Fig. 13-30.) These pressures are hydrostatic; that is, they are equal in all directions because the rocks in the mantle and crust are not, as we

will see, completely rigid, even though they are solid. A simple analogy will illustrate hydrostatic pressure. Bricks piled one on the other exert a pressure only on their base, because they are rigid. A pile of sand, however, unless confined by a wall, will assume a cone shape because the sand is not rigid; hence, the pressure due to its weight is exerted in all directions.

Data from drill holes and deep mines reveal that the temperature rises with depth in the earth at the approximate rate of 1°C per 27 meters (1°F per 50 feet). If this rise continues to the center of the earth, the temperature there would be 222,000°C (400,000°F),

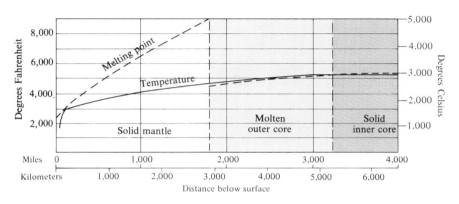

FIGURE 13-31 Estimated temperatures in the earth. Adapted from John Verhoogen, *American Scientist,* Vol. 48, 1960.

an unreasonably high temperature—hotter than the sun. The temperature in the earth is estimated from knowledge of the melting points of the materials assumed to form each layer, taking into account the rise in melting point at high pressures. The temperature distribution is shown in Fig. 13–31.

SUMMARY

The diameter of the earth is easily measured using the method shown in Fig. 13–1.

The polar radius of the earth is 6356.912 kilometers (3950.2 miles), and the equatorial radius is 6378.388 kilometers (3963.5 miles).

The mass of the earth was calculated from Newton's law of gravity as soon as the constants in the equation were determined by experiment.

The average specific gravity of the earth is 5.5, but the surface rocks are less than 3, indicating that the deep rocks have high specific gravities.

Earthquakes are vibrations caused by rapid movement on a fault.

A few faults move by slow *creep.*

P *waves* are primary (first to arrive) or push (compression) waves.

S *waves* are secondary (second to arrive) or shake (shear) waves.

Surface waves are slow moving, low frequency, large amplitude waves that cause most earthquake damage.

Earthquake damage can be reduced by constructing flexible buildings with good foundations in bedrock.

Seismic sea waves or *tsunami* are destructive waves caused by undersea earthquakes.

Earthquake *intensity* is a measure of the damage to buildings.

Earthquake *magnitude* is a measure of the energy released by an earthquake.

Earthquake prediction may be possible based on creep, microearthquakes, and especially local changes in the velocity of seismic waves.

Earthquakes can be triggered by pumping fluid into deep wells, formation of large lakes by dam construction, and underground nuclear blasts.

The *focus* is the place where the earthquake occurs.

The *epicenter* is the place on the surface above the focus.

The difference in arrival time between the P and S waves is proportional to the distance to the epicenter. Such measurements at three stations are necessary to locate the epicenter.

The *crust* is 4.8–56 kilometers (3–35 miles) thick, and is separated from the *mantle* by the *Mohorovičić (M or Moho) discontinuity.*

The mantle is about 2898 kilometers (1800 miles) thick.

The *core* has a radius of 3418 kilometers (2123 miles), and S waves cannot pass through the outer core, indicating that it is liquid.

The continental crust is 32–56 kilometers (20–35 miles) thick and is composed of granitic rocks.

The oceanic crust is about 4.8 kilometers (3 miles) thick and is composed of basalt.

The Mohorovičić discontinuity is probably caused by a difference in composition of the layers.

The *low-velocity zone,* about 100 kilometers (60 miles) down, is believed to be the level at which floating occurs.

In the *lithosphere,* as the rocks above the low-velocity zone are called, the rocks behave as solids.

In the *asthenosphere,* below the low-velocity zone, the rocks can flow slowly.

Meteorites are our best clues as to the composition of the deep interior of the earth.

Stony meteorites are composed of olivine and pyroxene (peridotite), and the mantle may have a similar composition.

Iron meteorites are composed of iron and some nickel, and this may be the composition of the core.

QUESTIONS

1 How was the size of the earth first measured? (Use a diagram.)
2 How was the weight of the earth determined?
3 What is the average specific gravity of the earth, and what problem does this pose?
4 What is the immediate cause of an earthquake? What is the ultimate cause of an earthquake?
5 How does a seismograph work?
6 Briefly describe each of the three seismic waves.
7 Which of the earthquake waves causes the most damage? How is most of the property damage due to earthquakes caused? How might some of this damage be avoided?

8 What does the magnitude of an earthquake mean?

9 How is the epicenter of an earthquake determined? (Diagram might help.)

10 Describe the behavior of S waves passing through the earth.

11 How can earthquakes be predicted?

12 How can people trigger earthquakes?

13 Describe the behavior of P waves passing through the earth.

14 How is the focus of an earthquake determined?

15 The deepest earthquakes are _____ kilometers.

16 The core has a radius of _____ kilometers and is believed to be composed of _____. What is the evidence for this composition?

17 The mantle of the earth is _____ kilometers thick and is believed to be composed of _____. What is the evidence for this composition?

18 The crust is separated from the mantle by the _____.

19 The lithosphere is separated from the asthenosphere by the _____.

SUPPLEMENTARY READINGS

Anderson, D. L., "The San Andreas Fault," *Scientific American* (November 1971), Vol. 225, No. 5, pp. 52–68.

Bolt, B. A., "The Fine Structure of the Earth's Interior," *Scientific American* (March 1973), Vol. 228, No. 3, pp. 24–33.

Bolt, B. A., *Earthquakes: A Primer.* San Francisco: Freeman, 1978, 272 pp. (paperback).

Boore, D. M., "The Motion of the Ground in Earthquakes," *Scientific American* (December 1977), Vol. 237, No. 6, pp. 68–78.

Cox, K. G., "Kimberlite Pipes," *Scientific American* (April 1978), Vol. 238, No. 4, pp. 120–132.

Grantz, Arthur, and others, *Alaska's Good Friday Earthquake, March 27, 1964.* U.S. Geological Survey Circular 491, 1964, 35 pp.

Hammond, A. L., "Dilatancy: Growing Acceptance as an Earthquake Mechanism," *Science* (May 3, 1974), Vol. 184, No. 4136, pp. 551–552.

Hansen, W. R., and others, *The Alaska Earthquake, March 27, 1964: Field Investigations and Reconstruction Effort.* U.S. Geological Survey Profes-sional Paper 541, 1966, 111 pp.

Iacopi, Robert, *Earthquake Country.* Menlo Park, Calif.: Lane Book Co., 1964, 192 pp. (paperback).

Kate, R. W., and others, "Human Impact of the Managua Earthquake," *Science* (December 7, 1973), Vol. 182, No. 4116, pp. 981–990.

Kerr, R. A., "Prospects for Earthquake Prediction Wane," *Science* (November 2, 1979), Vol. 206, No. 4418, pp. 542–545.

Kerr, R. A., "Quake Prediction by Animals Gaining Respect," *Science* (May 16, 1980), Vol. 208, No. 4445, pp. 695–696.

Kerr, R. A., "Assessing the Risk of Eastern U.S. Earthquakes," *Science* (October 9, 1981), Vol. 214, No. 4517, pp. 169–171.

National Science Foundation and U.S. Geological Survey, *Earthquake Prediction and Hazard Mitigation Options for USGS and NSF Programs.* Washington, D.C.: U.S. Government Printing Office, 1977, 76 pp.

O'Nions, R. K., P. J. Hamilton, and N. M. Evensen, "The Chemical Evolution of the Earth's Mantle," *Scientific Ameri-can* (May 1980), Vol. 242, No. 5, pp. 120–133.

Press, Frank, "Earthquake Prediction," *Scientific American* (May 1975), Vol. 232, No. 5, pp. 14–23.

Smylie, D. E., and L. Mansinha, "The Rotation of the Earth," *Scientific American* (December 1971), Vol. 225, No. 6, pp. 80–88.

Stevenson, D. J., "Models of the Earth's Core," *Science* (November 6, 1981), Vol. 214, No. 4521, pp. 611–619.

Walker, Jearl, "How to Build a Simple Seismograph to Record Earthquake Waves at Home," *Scientific American* (July 1979), Vol. 241, No. 1, pp. 152–161.

Wallace, R. E., *Goals, Strategy, and Tasks of the Earthquake Hazard Reduction Program.* U.S. Geological Survey Circular 701, 1974, 27 pp.

Witkind, I. J., and others, *The Hebgen Lake, Montana, Earthquake of August 17, 1959.* U.S. Geological Survey Professional Paper 435, 1964, 242 pp.

Wylie, P. J., "The Earth's Mantle," *Scientific American* (March 1975), Vol. 232, No. 3, pp. 50–63.

14 PLATE TECTONICS– OCEANS AND CONTINENTAL DRIFT

The birth of the oceans is a matter of conjecture, the subsequent history is obscure, and the present structure is just beginning to be understood. Fascinating speculation on these subjects has been plentiful, but not much of it predating the last decade holds water. . . .

H. H. Hess, "History of Ocean Basins," p. 599, in Petrologic Studies: A Volume to Honor A. F. Buddington, *Geological Society of America, November 1962.*

The oceans are geologically young features, constantly in the process of formation. As new ocean floor forms, it moves, carrying continents with it. This chapter about the oceans is present-day geology—the study of plate tectonics in action. In the next chapter, the generally much older continents are the subject; their formation and deformation are the consequences of the processes described here.

OCEANS

Until quite recently very little was known about the more than 70 percent of the earth's crust that is covered by ocean water. The continents, except for the continental shelves that are parts of the continents covered by ocean, can be studied directly. The oceanic crust must be studied by such indirect means as depth soundings and seismic studies and by bottom samples and cores. This difference in the kinds of data available results in knowing different things about these two major parts of the earth's crust. One very obvious aspect of this is that, with a few exceptions, only large structures are known in the oceans. Hopefully, future, more detailed studies and new methods will provide as much information about the oceanic crust as is known of the continental crust.

This chapter describes the features of the ocean and their relationship to crustal structures.

Oceanography began in the 1870s with the voyage of the British ship *Challenger,* the first oceanographic research vessel; however, very little was learned about the shape of the ocean bottoms until the invention of the echo sounder in the 1920s (Fig. 14–1.) Prior to that time, soundings had to be taken with a weight on a line. It was not until World War II that enough soundings were made to discover that some submarine mountains have flat tops. During the late 1950s, in connection with the International Geophysical Year, enough soundings were made to discover the unity of the mid-ocean ridges. These discoveries were made possible by the development of the continuous echo sounder, which can operate while a ship is under way. Using sound waves, it is possible to obtain a topographic profile of the bottom. By using other energy sources, such as dynamite or electric sparks, it is possible to penetrate to various depths into the bottom rocks and so to learn more about the structures. These techniques have been used mainly on the continental shelves in exploration for potential oil-bearing structures. (See Fig. 14–2.)

Beginning in 1970, deep-sea drilling became a reality. The drilling ship *Glomar Challenger* can drill deep holes in deep water and recover cores from the holes. This provided our first samples of rocks below the surface of the ocean floors and helped to confirm many of our ideas about continental drift.

OCEAN BASIN FLOORS

The *ocean basin floors* cover about 40 percent of the earth's surface. (See Fig. 14–3.) Seismic studies have revealed that they have remarkably uniform structure. (See Fig. 14–4.) Under a depth of 4.4 kilometers (2.75 miles) of water lies a thin layer of unconsolidated sediments averaging about 305 meters (1000 feet) thick. This first layer is composed of red clay and oozes. Below this is the second layer of irregular thickness, averaging about 2 kilometers (1.25 miles) thick. This layer is believed to be composed of basalt and consolidated sediments. The third layer extends to the Moho and is everywhere very close to 4.8 kilometers (3 miles) thick. The composition of this layer is not known but is traditionally thought to be basalt.

The oceanic floors are generally fairly smooth surfaces. They are interrupted at places by *low hills, submarine mountains, mid-ocean ridges,* and *fracture zones,* all of which will be described below. The submarine mountains indicate some volcanic activity. The low hills are probably also volcanic, but may in part be caused by faulting or other processes. Erosion is very slow on the oceanic floors, so minor features tend to be preserved. Sedimentation, too, is slow for the most part and in areas far from land is limited to dust settling on the ocean and biologic material, especially shells, sinking from the near surface. Perhaps the smoothest surfaces on

Atolls in Tuamotu Archipelago, South Pacific. Photo from NASA.

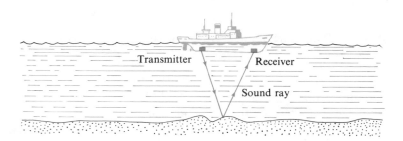

FIGURE 14–1 An echo sounder. Depth = ½ speed of sound × echo time.

the earth are parts of the oceanic floors near the continental slopes. The smoothness is probably the result of burial of the surface irregularities by sediments moving down the continental slopes, perhaps as density currents. (See Fig. 14–5.)

The age of the ocean floors is remarkably young in geologic terms—less than 160 million

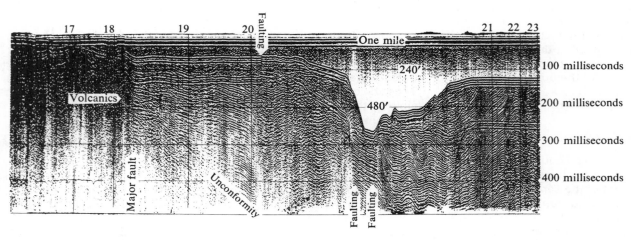

FIGURE 14–2 Sparker profile near Santa Cruz Island, California. An electric arc created waves that penetrated the bottom sediments and were reflected back to the receiver. The sea floor is shown by the strong reflection at a depth of about 92 meters (300 feet) on the right side. The sedimentary rocks that form most of the cross-section are in fault contact with the volcanic rocks on the left (major fault). Three other faults in the sedimentary rocks are also labeled. Note that the sea floor is displaced about 2.4 meters (8 feet) by the fault labeled on the top. Courtesy of R. F. Herron, who made the profile using EG&G International, Inc. equipment.

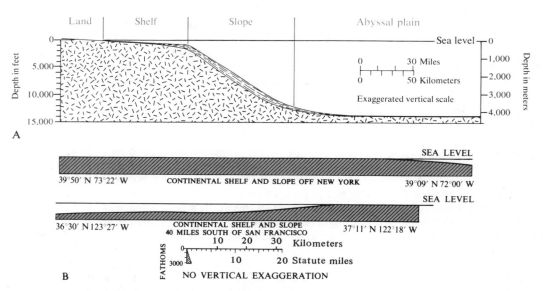

FIGURE 14–3 A. Continental margin showing typical continental shelf, continental slope, and abyssal plain. B. Actual profiles of continental margins with no vertical exaggeration. Part B from K. O. Emery and others, U.S. Geological Survey Professional Paper 260A, 1954.

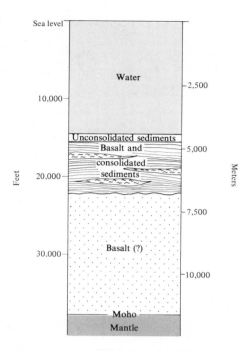

FIGURE 14–4 Typical section of rocks on the ocean floors.

years. *The youngest parts are near the mid-ocean ridges, and the oldest near continents.*

SUBMARINE MOUNTAINS

The *submarine mountains* are mountains, *rising thousands of feet above their bases, which are apparently basaltic volcanoes.* Some of these, such as the Hawaiian Islands, extend above sea level.

In some areas these submarine mountains are *flat topped;* they are called *guyots.* (See Fig. 14–6.) These flat tops are now at depths up to a mile below the surface of the ocean. They are especially interesting because their *flat tops were probably produced by wave erosion at sea level,* although some types of submarine volcanism may produce initially flat-topped mountains. Wave-cut benches are being formed now on the present islands, and if this process continues, such islands will become flat topped. Lowering sea level or uplifting an island can pre-

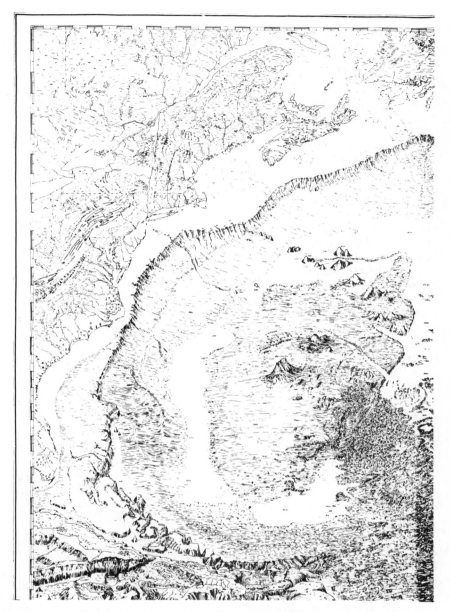

FIGURE 14–5 The continental margin and ocean floor off eastern United States. The continental shelf and slope are cut by submarine canyons. The smooth surfaces in the deep ocean below the continental slope are probably depositional. The rough topography to the southeast and the scattered hills are from volcanic activity. This is a portion of the Physiographic Diagram of the North Atlantic Ocean published by the Geological Society of America © Copyright by Bruce C. Heezen and Marie Tharp. Reproduced by permission.

FIGURE 14–6 Flat-topped seamount. Note the small wave-cut bench.

273

serve such benches, and this can be seen on present-day islands and coastlines. (See Fig. 11–19.) Similar benches are revealed in profiles of some of the flat-topped seamounts (Fig. 14–6). Samples dredged from flat-topped seamounts contain basalt fragments and Cretaceous and Early Cenozoic shallow-water fossils, suggesting the age of the wave erosion.

The *coral atolls* of the Pacific may also reveal subsidence. The corals live in the near-surface zone and form an offshore reef. (See Fig. 14–7.) If the island sinks slowly enough, the corals will build up the reef so that they remain near the surface where their food supply is. This process may continue until the island is completely submerged, perhaps quite deeply, and only a coral reef remains. At Eniwetok the coral is 1372 meters (4500 feet) deep.

The submarine mountains reveal some of the history of the oceans. Because they are basalt volcanoes, they record the importance of volcanic activity in the oceans. The flat-topped seamounts and the atolls reveal large changes in elevation of the deep ocean floors. Figure 14–8 shows that the distribution of flat-topped seamounts and atolls is not random, but rather that a large elongate area in the southwest Pacific Ocean has subsided, perhaps indicating a former position of the mid-ocean ridge or rise. The Cretaceous and Early Cenozoic shallow-water fossils dredged from the flat tops suggest the time when this subsidence occurred. Note that the atolls are mainly on the south flank of this area, suggesting that climate may have controlled the growth of coral. Present areas of coral growth are restricted to about 30 degrees on each side of the equator. The present depths below sea level of the flat-topped seamounts are not regular; therefore, as suggested above, differing rates of subsidence may also have influenced coral growth.

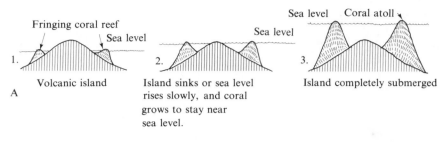

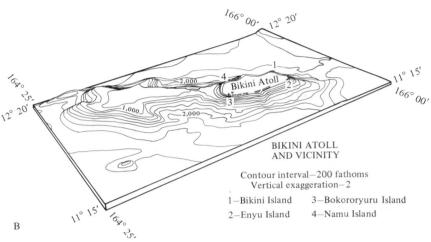

FIGURE 14–7 Coral atolls. A. Development of a coral atoll. B. Perspective diagram of Bikini Atoll and Sylvania flat-topped seamount adjoining it on the west. Bikini Atoll formed on a flat top, unlike the atoll shown in A. Most of Bikini Atoll is submerged, and only the small color areas in the diagram are islands. Part B from K. O. Emery and others, U.S. Geological Survey Professional Paper 260A, 1954.

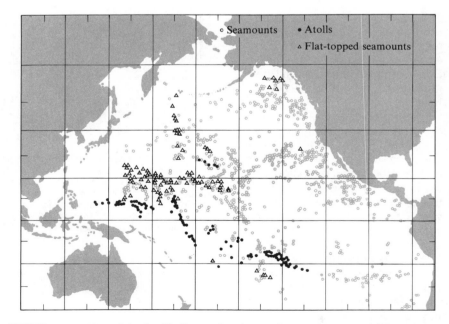

FIGURE 14–8 Map of the Pacific Ocean showing atolls, seamounts, and flat-topped seamounts. The coral atolls are almost entirely confined to the warm waters within 30 degrees of the equator. The flat-topped seamounts also form distinct clusters. The data on atolls, seamounts, and flat-topped seamounts are from H. W. Menard, *Marine Geology of the Pacific,* McGraw-Hill, 1964.

274

MID-OCEAN RIDGES

The *mid-ocean ridges form the longest continuous topographic mountain range on the earth.* Their extent can be seen on the map in Fig. 14–35. Their length is about 64,000 kilometers (40,000 miles); their width is 483 to 4830 kilometers (300 to 3000 miles); they rise up to about 3050 meters (10,000 feet) above their base; and at places, the highest points form islands. Two parts of this ridge are fairly well known: the Mid-Atlantic Ridge and the East Pacific Rise.

The Mid-Atlantic Ridge has been known for many years because islands such as Iceland, the Azores, St. Paul Rocks, Ascension, St. Helena, and Tristan da Cunha are parts of it. That it is part of a much larger feature was not known until the 1950s; at that time, many soundings greatly increased our knowledge of its nature. (See Fig. 14–9.) It consists of numerous ridges and lines of peaks paralleling the main ridge that extend outward for hundreds of miles on the flanks (Fig. 14–10). The ridges and peaks probably are caused mainly by faulting. The central crestal zone generally has one or more discontinuous *median rift valleys* that are up to 48 kilometers (30 miles) wide. The seismic activity of the ridge is shown by the numerous earthquake epicenters located on it. The bedrock, revealed in dredged samples, is mainly basalt, some of which is pyroclastic with some metamorphic equivalents of basalt such as greenschist and amphibolite, and some serpentine.

The Mid-Atlantic Ridge has been studied directly on the few islands

FIGURE 14–9   The ocean floor on the Mid-Atlantic Ridge. Basalt rocks and some bottom-dwelling organisms can be seen. Courtesy R. M. Pratt.

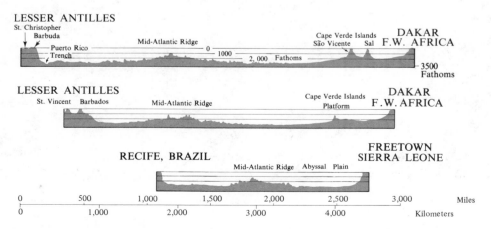

FIGURE 14–10   Transatlantic topographic profiles showing the Mid-Atlantic Ridge. Vertical exaggeration 40:1. From B. C. Heezen, Geological Society of America Special Paper 65, 1959.

275

that are part of the ridge. The islands are mainly basaltic, and recent volcanic activity on Tristan da Cunha and Iceland has also been basaltic. St. Peter and St. Paul Rocks are tiny islands on the ridge, and sheared peridotite (mylonite) is found there. Thus, these rocks may be mantle material somehow brought to the surface by the process that formed the ridge. Peridotite has also been dredged from the Indian Ocean. Elsewhere the rocks are mainly basaltic (including gabbro) with some serpentine. The central rift valley passes through Iceland; and though the evidence is not clear, dilation of up to 386 kilometers (240 miles) may have occurred. (See Fig. 14–11).

The East Pacific Rise is similar to the Mid-Atlantic Ridge. It has similar topography except that it lacks the rift valley at its crest. It is also an active seismic area. It rises up to three miles above its base and forms the Galápagos and Easter Islands. It meets North America just south of Baja California, leaves in northern California, and extends to British Columbia. Measurements on the East Pacific Rise show that the heat flow near the crest is up to five times greater than average, and on the flanks is lower than average (Fig. 14–12). *The average heat flow on the rise is about twice the normal oceanic heat flow.* This high heat flow extends into the Gulf of California. Seismic studies using underwater explosions have revealed that the crust under the East Pacific Rise is only about three-quarters of its thickness elsewhere in the Pacific.

The mid-ocean ridge appears to go inland in Africa where it may form the *African rift valleys* that, up until the discovery of the mid-ocean ridge, were thought to be unique. These African areas are now being studied closely in the hope of learning more about these ridges. The African rift valleys are unsymmetrical but similar to the rifts on the mid-ocean ridges, and are associated with high plateaus, indicating uplift like the mid-ocean ridges. The African rift valleys are not all the same age, and most show evidence of recent geologic activity, such as volcanoes and changes in elevation shown by drainage features. The volcanoes are not ordinary basalt, but are alkali-rich rocks, some of which are extremely rare types. (See Fig. 14–13.) Nearby in the Red Sea, unusual hot brines containing metals are found on the mid-ocean ridge.

The mid-ocean ridges are young, active parts of the crust, and they play an important role in the development of the crust.

FIGURE 14–11 Looking south along the Thingvellir valley, southwest Iceland. This valley is probably an extension of the Mid-Atlantic Ridge. The scarp is about 40 meters (130 feet) high. Photo by S. Thorarinsson.

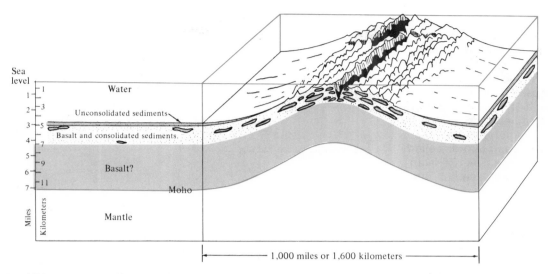

Sea level

Water

Unconsolidated sediments

Basalt and consolidated sediments

Basalt?

Moho

Mantle

Miles / Kilometers

1,000 miles or 1,600 kilometers

FIGURE 14-12 Mid-ocean ridge. A cross-section of typical ocean floor is shown on the left. The ocean crust thins at the ridge. Note the vertical exaggeration.

CONTINENT-OCEAN BORDERS

The change in elevation between the continents and the oceans is one of the main features on the elevation diagram, Fig. 15–1. Two types of continent-ocean borders can be easily distinguished. They are the *continental slopes that are geologically relatively stable and the active volcanic arcs.* Sedimentation on the shelf and slope was described in Chapter 4.

CONTINENTAL SLOPE

The continental slope appears to be the simpler of the border types, but this may be only apparent. The continental slope begins abruptly at the edge of the continental shelf. The shelves of the world average about 72 kilometers (45 miles) wide, but the range is from near zero to 1290 kilometers (800 miles). Their average depth is near 60 meters (200 feet), and their edge is generally at 120 to 150 meters (400 to 500 feet), although the range is from near sea level to 610 meters (2000 feet). The slope angle changes from about 1.9 meters per kilometer (10 feet per mile) on the shelf to be-

W   E

Late Miocene – early Pliocene

1. Volcanics erupted on crest of uplift
2. Faulting on west side of rift, flexuring on east side

W   E

Late Pliocene

3. Faulting of floor of rift; renewal of movement on main fractures; new fractures on rift shoulders
4. Volcanic activity in rift floor

W   E

Quaternary

5. Further uplift of rift shoulders; renewal of movement on faults in rift floor; new closely spaced fractures develop in median zone
6. Small plugs and larger volcanoes built in rift floor; some volcanoes on the rift shoulders

0   50 Miles

0   50   Kilometers

FIGURE 14-13 Development of the Central Rift in Kenya. From B. H. Baker, in *East African Rift System,* University College, Nairobi, 1965.

tween two to six degrees on the slope in a distance of under 32 kilometers (20 miles). Some conti-

nental slopes are as steep as 45 degrees. Compared to the average slope of most mountain ranges, 5 degrees is very steep. The average drop on the continental slope is about 3660 meters (12,000 feet), and the slope angle lessens with depth, making the distinction between the ocean floor and the slope difficult. Thus, the slopes are a few tens of miles wide on the average. The region between the slope and the ocean floor is called the *continental rise.* The rise is apparently composed of sediments. Much of the east coast of the United States is typical of continental slopes (Fig. 14–3).

Continental slopes are composed of sediments. Submarine canyons are cut into the continental slope at many places, as described in Chapter 4. This implies that, at least at these places, erosion is at work to lower the angle of the continental slopes. However, some deposition is taking place also; the sediments were described in Chapter 4. Some of the erosion may be caused by deep currents parallel to the slopes as well as by turbidity currents moving down the slopes. At places, the slope is bare of unconsolidated sediments. Thus, both erosion and deposition occur on the continental slopes. (See Fig. 14–14.)

Some continental slopes are in part formed by reefs, as shown in Fig. 14–14. The barrier reef of Australia is an example.

Seismic studies show that the continental crust ends at the continental slope, but the details of the structure in this area are not clear. Faulting has been suggested by some as the origin of continental slopes because many are seismically active.

VOLCANIC ARCS

At other places, particularly on the rim of the Pacific Ocean, the border is more active with *chains of volcanoes forming arcs.* (See Fig. 14–15.) The arcs are seen at places where the volcanoes are islands. At other places the volcanoes are at the edge of the conti-

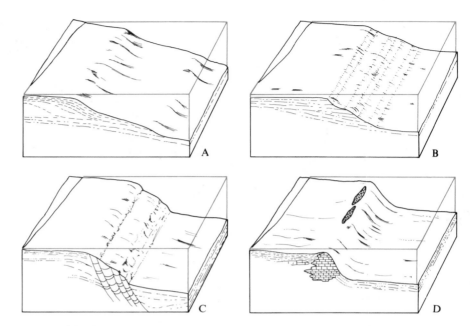

FIGURE 14–14   Continental slopes. A. Depositional. B. Erosional. C. Faulted. D. Reef.

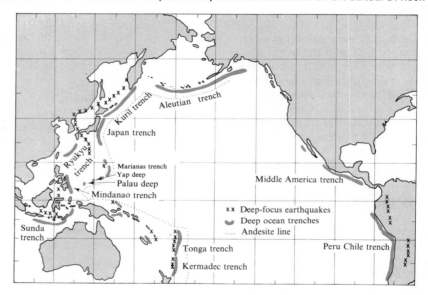

FIGURE 14–15   Map of the Pacific Ocean showing the relationships among island arcs, deep-focus earthquakes, and deep trenches. The volcanic rocks on the continent side of the andesite line are andesite, and basalt occurs on the ocean side. The epicenters of shallower earthquakes lie between the deep earthquakes and the island arcs.

nent. The island arcs were noted early in the history of geologic oceanography, but recognition of many of their features came much later. *Deep trenches exist on the ocean side of volcanic arcs.* During the 1920s and early 1930s, the Dutch geologist F. A. Vening-

Meinesz devised a method of measuring the value of gravitational attraction at sea. Using a method that measured the period of a pendulum, he had to work in Dutch and U.S. Navy submarines to avoid the effects of ocean waves. He made the extremely important discovery that the *acceleration of gravity is low (negative gravity anomaly) over these deep trenches.* (See Fig. 14–16.) *The volcanic arcs are also active seismically, and about the same time it was discovered that some of the earthquakes come from very deep in the mantle.* (See Figs. 14–15 and 14–17.) Such a concentration of features on the narrow border between the continent and the ocean has important geologic implications.

The volcanic arcs pose many problems, but also give much insight into earth processes. The volcanoes are andesitic in composition. Small batholiths of similar chemical composition are found on some, such as the Aleutian Islands. This is also the composition of the volcanic arcs in the Andes Mountains of South America and the Cascade Range in North America that, together with the island arcs, form what has been somewhat poetically called the "rim of fire" of the Pacific Ocean. The deep oceanic trough off South America and the deep-focus earthquakes inland show that they are similar to the volcanic island arcs of the western Pacific. On the oceanward side of this rim of andesite volcanoes, all of the volcanic islands are basaltic (see *andesite line* in Fig. 14–15).

The foci of the earthquakes associated with volcanic arcs are on a plane sloping from the trench under the volcanoes. The deepest earthquakes on the earth, up to 645 kilometers (400 miles) deep, occur here.

These oceanic features are involved in the shaping of the earth's surface, but some other aspects must be considered before we return to the role of the oceans.

CONTINENTAL DRIFT

HISTORY AND EVIDENCE

Continental drift was first seriously proposed in 1910 by Alfred Wegener, a German meteorologist who marshalled evidence from geology and meteorology. His theory was not accepted, however, because his mechanism was not adequate. Much of the geologic evidence for continental drift comes from the southern hemisphere, so geologists there tended to accept drift while European and American geologists did not. In 1937, A. L. Du Toit, a South African, published a book, *Our Wandering Continents,* with a better but still unconvincing mechanism. Thus it was, until after World War

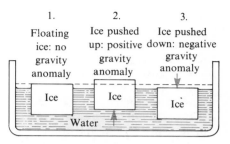

FIGURE 14–16 Gravity anomalies: Station 1 gravity is normal. At Station 2 the attraction of gravity is higher because more mass is present below the station. At Station 3 there is less mass below the station because the lighter ice displaces the heavier water, so the gravity force is less.

II when new evidence reopened the question. In the 1950s two discoveries were made. P. M. S. Blackett and S. K. Runcorn in England showed, as will be described below, that the magnetism in rocks strongly suggests that the continents have moved relative to each other. At the same time, the American oceanographers Bruce Heezen and Maurice Ewing showed the extent of the mid-ocean ridge.

In 1960 Harry Hess of Princeton University proposed sea-floor spreading, a theory somewhat similar to one proposed in 1928 by the Englishman Arthur Holmes. Almost at once, many geologists began to take continental drift se-

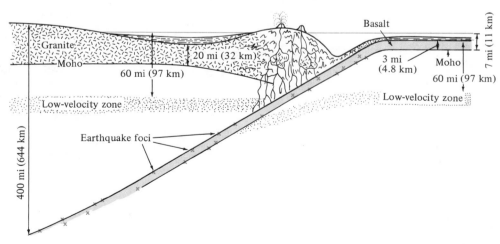

FIGURE 14–17 Volcanic island arc, showing location of earthquake foci. Because of the distances involved, this figure is not drawn to scale.

riously. The early and middle 1960s was a time of active study of the magnetism of the ocean floors. As results of these studies were reported, more and more geologists accepted continental drift and sea-floor spreading. In 1965, J. Tuzo Wilson, a Canadian, proposed transform faults, and in 1968, Bryan Isacks, Jack Oliver, and Lynn R. Sykes of Columbia University supported the theory with evidence from seismology. In 1970, the deep-sea drilling project of *Glomar Challenger* started returning more evidence for the theory.

Thus, in less than ten years, geology underwent a revolution in thinking. Perhaps the reason that sea-floor spreading and plate tectonics were so quickly adopted is that it is a unifying theory in the sense that it brings together observations from many fields in earth science into a single theory. Some of those who are not convinced say that Neptunism was also a theory that explained many observations and so simplified and unified geology.

The idea of continental drift has been with us since the first accurate maps showed that Africa and South America fit together. During the last part of the nineteenth century, European geologists were deciphering the thrust structures in the Alps and American geologists were mapping the West, which was just opening up. As a result, they took little notice of some startling discoveries made in India, Australia, and South Africa. At each of these places a peculiar flora, called the *Glossopteris* flora after its most abundant genera, was found in beds overlying glacial deposits. Since then the same sequence has been found in South America and Antarctica. The glacial deposits are from continental, not mountain, glaciers and rest on polished and striated basement rocks. (See Fig. 14–18.) The striations suggest that in many cases the ice moved from areas that are now occupied by

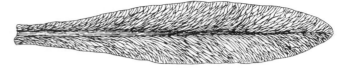

A

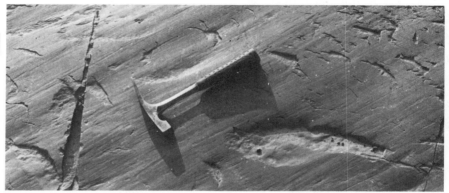

B

C

D

FIGURE 14–18 Evidence for continental drift. A. *Glossopteris.* This leaf has been found in beds overlying ancient glacial deposits in India, Australia, South Africa, and Antarctica. B. Glacial polish in Australia giving clear evidence of ancient glaciers there. Ice moved toward upper right. C. Striated boulder from ancient glacial moraine in Australia. D. Glacial polish from glaciers of the same age in Africa. Etching by Bushmen on left. Parts B, C, and D, photos by Warren Hamilton.

oceans. The moraines contain many boulders of rock types not found on that continent, and in some cases the foreign boulders match well with outcrops on the adjoining continent. The questions immediately arise: How could extensive glaciers form in these tropical areas, and how did the *Glossopteris* flora spread across the oceans now separating these areas? (See Fig. 14–19.) These areas are so widespread that shifting the rotational axis of the earth does not help because even in the most favorable location, some of the glacial areas are still within ten degrees of the equator. If all of the continents of the southern hemisphere had been in contact at one time, these occurrences would be explained; but the continent would have to break up and drift to explain the present geography. (See Fig. 14–20.) Thus began the idea of continental drift. Today the evidence is even more compelling because of the discovery in Antarctica of *Glossopteris* and other fossils, indicating a humid, temperate climate at one time.

The *Glossopteris* flora found just above the glacial deposits is believed to be evidence of previous connection among the southern continents. It is possible that seeds (spores) could have floated across the oceans to distribute these plants, but it seems rather unlikely. Today, a newly formed volcanic island soon develops plants, but the seeds are carried by birds. The first flying animals did not appear until tens of millions of years after the supposed separation and drift. Land bridges and island chains have been suggested in the past, but the known oceanic structure precludes this possibility.

The swimming reptile *Mesosaurus,* found in both South America and south Africa, is also strong evidence favoring connection of these two continents. Although a swimmer, this aquatic reptile lived in fresh water, and it is very unlikely that it could swim across an ocean. Other reptiles that lived during the time of the alleged connection have been found on the southern continents. The distribu-

tion of some of these reptiles requires a single connected continent.

Piecing together the continents is like putting together a jigsaw puzzle whose pieces do not quite match. Fitting together the various layers of sedimentary rocks is like putting together a multilayer jigsaw puzzle with many of the pieces missing. The generalized stratigraphy of the southern continents is shown in Fig. 14–21. The similarity among these continents in the Carboniferous Period is striking. This similarity persists into the Jurassic in all but Australia. This stratigraphy and fossils in Australia suggest that it became isolated from the other continents after the glaciation and that it has remained isolated to the present. The Australian fauna and flora of the present appear to have migrated from time to time very selectively from mainland Asia. The present flora and the marsupial fauna are very different from those of the other continents. The placental mammals that populate the other continents did not reach Australia.

Radioactive dates and deformation of the rocks both show a good match between the basement rocks of west Africa and northern Brazil. (See Fig. 14–22.)

The fit of the continents on a map seems to be too good to be accidental. The fit is even better if, instead of the shoreline, the true

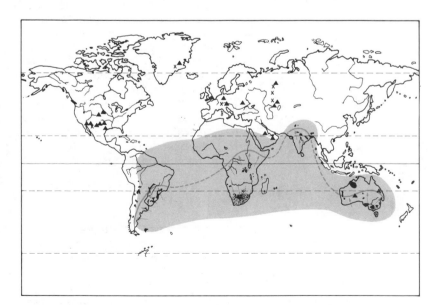

FIGURE 14–19 Map showing the present distribution of the ancient glaciation of the southern hemisphere. The area involved in the hypothetical southern continent existing before drifting is stippled. The areas of glacial deposits are indicated and extend from north of the equator to near the south pole. Reefs (crosses) and evaporites (triangles) far to the north indicate much warmer climates. Reefs and evaporites from Opdyke, 1962.

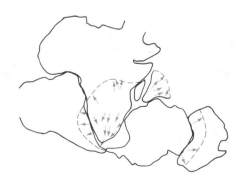

FIGURE 14–20 Reconstruction of the ancient continent of the southern hemisphere. The glaciers and their directions of movement are indicated.

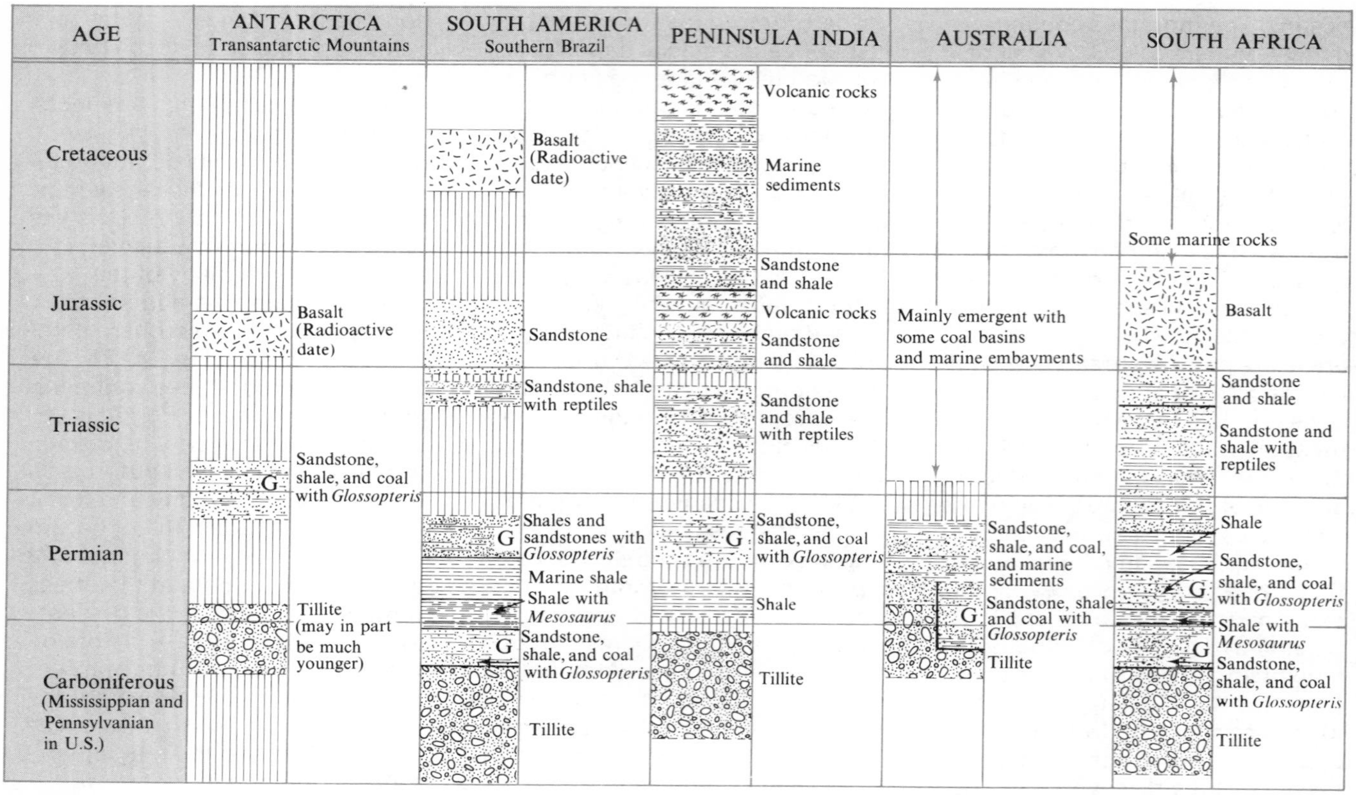

AGE	ANTARCTICA Transantarctic Mountains	SOUTH AMERICA Southern Brazil	PENINSULA INDIA	AUSTRALIA	SOUTH AFRICA
Cretaceous		Basalt (Radioactive date)	Volcanic rocks; Marine sediments		Some marine rocks
Jurassic	Basalt (Radioactive date)	Sandstone	Sandstone and shale; Volcanic rocks; Sandstone and shale	Mainly emergent with some coal basins and marine embayments	Basalt
Triassic	Sandstone, shale, and coal with Glossopteris (G)	Sandstone, shale with reptiles	Sandstone and shale with reptiles		Sandstone and shale; Sandstone and shale with reptiles
Permian	Tillite (may in part be much younger)	Shales and sandstones with Glossopteris (G); Marine shale; Shale with Mesosaurus; Sandstone, shale, and coal with Glossopteris (G)	Sandstone, shale, and coal with Glossopteris (G); Shale	Sandstone, shale, and coal, and marine sediments; Sandstone, shale and coal with Glossopteris (G)	Shale; Sandstone, shale, and coal with Glossopteris (G); Shale with Mesosaurus (G)
Carboniferous (Mississippian and Pennsylvanian in U.S.)		Tillite	Tillite	Tillite	Sandstone, shale, and coal with Glossopteris; Tillite

FIGURE 14–21 Simplified stratigraphic columns in the southern hemisphere. The age terms are described in Chapter 16. Vertical lines indicate no rocks of that age.

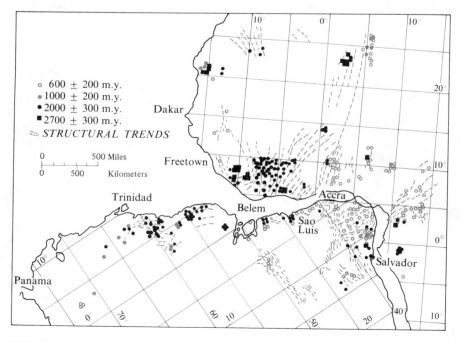

FIGURE 14–22 West Africa and South America fitted together, showing the coincidence in both structural trends and radioactive ages of the rocks (m. y. is million years). From P. M. Hurley and others, *Science* (August 4, 1967), Vol. 157, p. 496.

edge of the continent, the continental shelf, is used. Later deformation or erosion does spoil the match at some places. There are, however, problems. In most reconstructions, Spain must be rotated, and there is no room for much of Central America; but these problems seem small in the overall view. Most reconstructions are based on magnetic data, so they will be discussed next.

EARTH'S MAGNETIC FIELD

It is the earth's magnetic field that aligns a compass needle and makes it point toward magnetic north. The earth behaves much as if it had a giant bar magnet in its core. (See Fig. 14–23.) However, even though the core is made of iron and nickel, no such magnet can exist at the temperatures there. Iron loses its magnetism

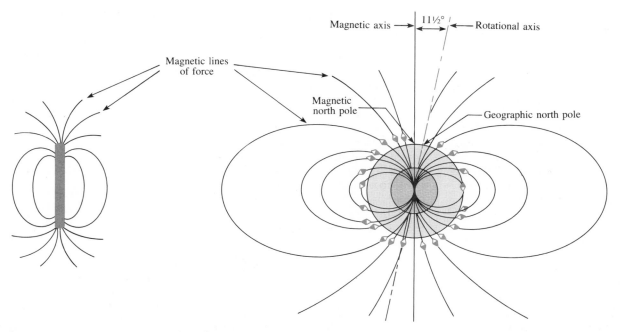

FIGURE 14–23 The earth's magnetic field is much like that of a bar magnet. The magnetic poles do not coincide with the geographic poles that are determined by the rotation axis. A compass needle that is free to move in the vertical plane will point in the directions shown.

when heated above a certain temperature—well below its melting point. Also the earth's magnetic field is not that of a simple *dipole, which is the field of a bar magnet.* (See Fig. 14–24A.) The earth's field consists of a *dipole field* and a *nondipole field.* One effect of the nondipole field is that the south magnetic pole, at the edge of Antarctica, is 2093 kilometers (1300 miles) from the point that is antipodal to the north magnetic pole, which lies north of Canada. However, for the most part, the earth's field can be considered nearly a dipole field.

The earth's magnetic field causes a compass to point toward the magnetic pole, but if the compass is pivoted to move in the vertical direction also, it will indicate inclination as well. (See Fig. 14–23.) At the magnetic equator, the field is horizontal; at the poles, vertical; and in between, it varies. The intensity of the field at the poles is about twice that at the equator. At some places records of the direction and inclination of the earth's field have been kept for

hundreds of years, and they reveal changes in direction and inclination. (See Fig. 14–24.) More recent studies also show variation in intensity. Study of the *remanent magnetism* of rocks (to be discussed) indicates that several times in the geologic past the di-

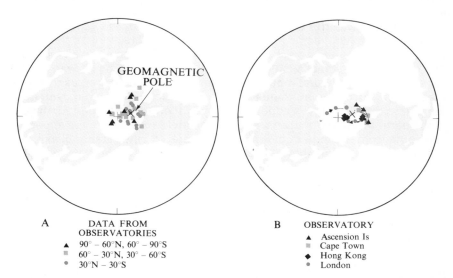

FIGURE 14–24 Variations in the earth's magnetic field. A. Variations in position of magnetic north pole computed in 1945 from direction and inclination of the earth's magnetic field at various places. These calculations were made assuming a dipole field and show the effect of the nondipole part of the earth's field. B. Changes in the position of the magnetic north pole with time. Time between points is 40-50 years. In both figures the cross shows the position of the north geographic pole, and the X, the average 1945 position of the north geomagnetic pole. From Allan Cox and R. R. Doell, Geological Society of America, *Bulletin,* Vol. 71, 1960.

rection of the earth's field has reversed, i.e., the north pole of a compass needle would point toward Antarctica. The timing of these reversals in the last 150 million years is well enough known that they can be used to date rocks. (See Fig. 14–25.) Any theory for the origin of the earth's magnetic field must explain these observations.

The earth's magnetic field is believed to originate from electric currents in the core. It is a well-known law in physics that a moving current creates a magnetic field, and conversely, a current is induced into a conductor moving in a magnetic field. Electrical currents in the core could start in a number of ways, such as through a weak battery action caused by compositional differences. Once started, these currents could be amplified by a dynamo action in the earth's core. A dynamo produces electric current by moving a conductor in a magnetic field. In the earth the core is the conductor, and because it is fluid, it can move. The mechanism envisioned is similar to a dynamo which produces an electric current, and this current is fed to an electric motor that drives the dynamo. Friction and electrical resistance would prevent such a perpetual motion machine from continuing to operate unless more energy is added. In the earth this additional energy can come from movement of the fluid core. Calculation shows that very little additional energy needs to be added in the case of the earth; and the necessary energy for core motion, probably in the

form of convection, could come from many sources.

Thus, a current is somehow started in the earth's core, and this current produces a magnetic field. The conducting core moves through the magnetic field, and a current is induced in this part of the core, starting the whole process over. (See Fig. 14–26.) In a stationary earth, the movements of the core would probably be more or less random and the magnetic fields produced would probably cancel each other. The rotation of the earth tends to orient the motions of the core so that the motions and, therefore, the currents are in planes perpendicular to the rotational axis. Viewed from outside the earth, the overall current produced would be parallel to latitudes. Such currents produce a dipole magnetic field centered on the rotational axis.

A field produced in this way would have all of the characteristics of the earth's magnetic field. The main field would be a dipole field; superimposed on this would be other fields caused by variations in the core's motion. This can account for the variations in intensity and location of the earth's field. The earth's dipole field does not coincide with its rotational axis, however, as this theory predicts, although if averaged over a long enough period, the dipole field would perhaps appear centered on the rotational axis. The positions of the magnetic pole deduced from remanent magnetism of Pleistocene and younger rocks cluster around the rotational axis, suggesting that this is true.

The present dipole field is 11.5 degrees from the rotational axis and lies northwest of Greenland, some distance from the magnetic north pole north of Canada. Jupiter is the only other planet, as far as we know, with an appreciable magnetic field, and its field is also about 11.5 degrees from its rotational axis.

REMANENT MAGNETISM

Much of the recent data on continental drift have come from measuring the remanent magnetism of rocks. *Remanent magnetism is the magnetism frozen into a rock at the time the rock forms.* For instance, in the cooling of basalt, when the temperature reaches a certain point, which is well below the crystallization temperature, the magnetic minerals, mainly magnetite, are magnetized by the earth's field. On cooling below this temperature, their magnetism is not greatly affected except by very strong fields, many times stronger than any in the earth. Thus it is possible to measure the direction

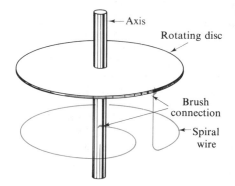

FIGURE 14–26 The disc dynamo illustrates in a general way how the earth's magnetic field is believed to originate. The disc is attached to the axis and rotates with the axis. The spiral wire is in electrical contact with the axis and the disc by means of brushes. To start this self-sustaining dynamo, suppose that a current is flowing in the spiral wire. This current produces a magnetic field; the disc is a conductor moving through the field, and so a current is induced in the disc. This current flows to the spiral wire by means of the brush and so continues the processes.

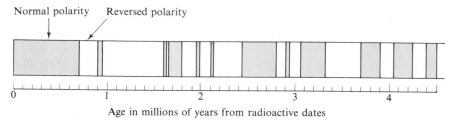

FIGURE 14–25 A few of the many known reversals of the earth's magnetic field. From data by Cox.

and inclination of the earth's field at the time the basalt cooled. The inclination can be used to calculate the distance to the magnetic pole.

Remanent magnetic data for rocks of different ages have been collected on each of the continents. The study of magnetic data from old rocks is called *paleomagnetism* (old magnetism). The magnetic poles for each age (geologic period) on each continent group in a fairly small area. However, in general, for any given continent, each geologic period has a different location for the magnetic pole, and the poles for any period fall in different locations for each continent. (See Fig. 14–27.) Here, then, is evidence of continental drift. If the magnetic pole moved, it should be at the same place during each period for each continent; but instead, each continent gives a different location for each period. Thus, if the interpretation of the paleomagnetic data is correct, the continents must have moved relative to each other. (See Fig. 14–28.) The magnetic pole may also have moved. Note in Fig. 14–27 that the magnetic data indicate no drift of North America relative to Europe since Eocene.

Paleomagnetic data have certain limitations. They tell the direction and distance to the magnetic pole. Thus they show the rotation of a continent and its latitude. (See Fig. 14–29.) They do not indicate the longitude, and the continent may be anywhere on the indicated latitude. This gives the imagination rather free rein in proposing reconstructions, particularly since many of the rocks are incompletely mapped and dated. (See Fig. 14–30.) In spite of the close grouping of poles for a given period, the possibility remains that the magnetic pole may not have been coincident with the geographic pole. At present, they differ by 11.5 degrees.

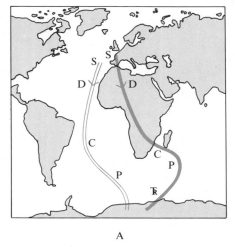

(1) Europe
(2) North America
(3) Australia
(4) India
(5) Japan

Pl Pliocene	*About 6 million years ago*
M Miocene	*About 18 million years ago*
E Eocene	*About 50 million years ago*
K Cretaceous	*About 100 million years ago*
J Jurassic	*About 160 million years ago*
Ŕ Triassic	*About 200 million years ago*
P Permian	*About 250 million years ago*
C Carboniferous	*About 310 million years ago*
D Devonian	*About 375 million years ago*
S Silurian	*About 420 million years ago*
Ĉ Cambrian	*About 550 million years ago*

FIGURE 14–27 On this diagram, for each continent the position of the magnetic pole relative to the present position of each continent for each geologic period is indicated by the letters. The lines drawn through these locations show the apparent drift of the magnetic pole for each continent. The diagram indicates that movement of the continents relative to each other must have occurred but does not show the route of drifting because of the limitations of paleomagnetic data shown in Fig. 14–29. From Allan Cox and R. R. Doell, Geological Society of America, *Bulletin,* Vol. 71, 1960.

SEA-FLOOR SPREADING— MOVEMENT OF THE SEA FLOOR

Sea-floor spreading is the theory that oceanic crust is created at mid-ocean ridges and then moves

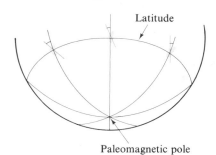

FIGURE 14–29 Paleomagnetic data can only show the latitude and orientation of a continent. The horizontal component of the paleomagnetism points the direction toward the pole, and the vertical inclination gives the latitude. The continent could have been in that orientation anywhere on that latitude.

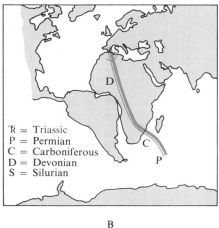

A   B

FIGURE 14–28 A. The curves show the position of the magnetic pole for each geologic period relative to the present positions of the continents. B. Shows that if South America and its polar curve are moved so that the two polar curves match, the two continents join. The position at which they join is geographically and geologically the most likely.

toward the volcanic arcs where it is consumed. (See Fig. 14–31.) The volcanic activity at mid-ocean ridges suggests that new crust could be forming there, and the age of the ocean floors strengthens this idea. The age of the sea floors is determined from geologic study of sediments and the magnetic properties of the oceanic crust. Magnetic evidence was crucial in the establishment of the theory of sea-floor spreading.

AGE OF SEA FLOORS

The evidence that new sea floor is created at the mid-ocean ridges is

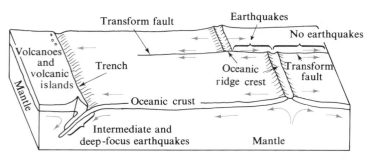

FIGURE 14–31 Schematic representation of the oceanic crust. Note the formation of crust at ridges and destruction in trenches.

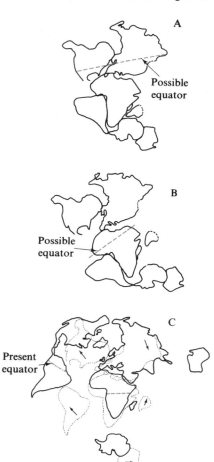

FIGURE 14–30 Movement of continents. A. Late Paleozoic time, 250 million years ago. B. End of Jurassic Period, 136 million years ago. C. End of Cretaceous Period, 65 million years ago, is shown dotted and present positions are shown by the solid lines. Based on R. S. Deitz and J. C. Holden, 1970.

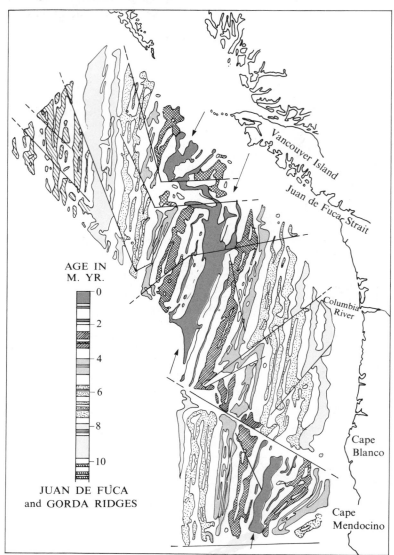

FIGURE 14–32 Magnetic anomalies in part of the ocean along the west coast of North America. Positive magnetic anomalies are indicated. They are believed to be formed where the remanent magnetism of the underlying rocks reinforces the earth's field. The intervening white areas have lower than normal magnetism and may be places where the underlying rocks have reversed magnetism. The straight lines are faults that displace the pattern. The arrows indicate short lengths of oceanic ridge, and the pattern is parallel and symmetrical to the ridges. After F. J. Vine, *Science* (December 16, 1966), Vol. 154, p. 1406.

abundant. Radioactive dating of basalt samples shows a progressively older age away from the ridges. The sediments on the sea floor also thicken away from the ridges, and examination of drill cores reveals that the bottom sediments are older further away from the ridges. The sea floors are young geologically, and only a few samples older than 130 million years have been found. The oldest sediments so far found are sandstones about 200 million years old off South America and in the western Pacific.

MAGNETIC STRIPES

The earth's magnetic field is impressed on a rock at the time the rock is formed. Basalt, because it contains small amounts of magnetite, an especially magnetic mineral, faithfully records the earth's magnetic field at the time the basalt cools through a certain temperature (*Curie point*) that is well below its melting point.

Magnetic surveys near the mid-ocean ridge show bands of alternating high and low magnetic intensity that parallel the ridge as shown in Fig. 14–32. The meaning of these magnetic bands was not clear until magnetic studies were made on the continents. It was soon discovered that many rocks recorded a magnetic field opposite or reversed to the present magnetic field. Magnetic studies of each flow in areas of thick accu-

mulations of basalt flows showed that these reversed magnetic field rocks did not occur in a random way but, rather, that if radioactive dates of the rocks were also obtained, the magnetic reversals occurred simultaneously all over the earth. Starting with the youngest rocks, it has been possible to determine the times of reversals accurately enough to use the magnetic direction to date some young rocks. (See Fig. 14–25.) These reversals suggest that the bands of alternating high and low magnetic intensities paralleling the mid-ocean ridge could be caused by bands of ocean floor basalt that have normal and reversed magnetic fields. The normally magnetized basalt would reinforce the present magnetic field, giving a high magnetic intensity; and the reverse magnetized rocks would partially cancel the present field, giving a lower than normal magnetic intensity. Careful comparison of the magnetic intensity on each side of the mid-ocean ridge shows that the magnetic intensity on one side is nearly the mirror image of that on the other side. (See Fig. 14–33.) The final argument is to construct a model of the ocean floor by calculating the positions of bands of normal and reversed magnetic basalt and then to calculate the theoretical magnetic field produced by such a model. As shown in Fig. 14–34, this model produces a field almost identical to that found in nature.

PLATE TECTONICS

Plate tectonics is the theory that brings together all of the preceding observations into a unifying theory. All of the elements of plate tectonics have already been stated. The plates are created at mid-ocean ridges, consumed at volcanic arcs, and carry continents with them as they move. In this section, the plates themselves and how they move will be considered.

PLATES

The active lithospheric plates are outlined by earthquakes. Figures 14–35 and 13–19 show where earthquakes occur; their concentration is at the deep-sea trench—volcanic arc areas, to a lesser extent on the mid-ocean ridge, and at many places on the faults or fracture zones that offset the ridge. Near the volcanic arcs the earthquake foci tend to be deep, and elsewhere they are generally less than 64 kilometers (40 miles) deep. These earthquakes suggest

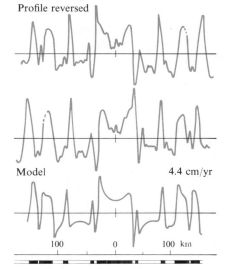

Profile reversed

Model   4.4 cm/yr

100   0   100 km

FIGURE 14–34 Magnetic profiles across the East Pacific Rise. The two upper curves are an actual magnetic profile, shown both as recorded and reversed. The lower curve was computed from a model, assuming a spreading rate of 4.4 centimeters per year. After F. J. Vine, *Science* (December 16, 1966), Vol. 154, p. 1409.

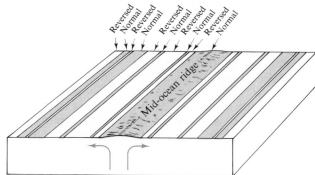

FIGURE 14–33 Formation of magnetic anomalies at a mid-ocean ridge. In this hypothesis the lavas that form the ocean floor cool at the ridge and are magnetized by the earth's field at the time they cool. Convection currents carry the lavas away from the ridge and so create magnetic anomaly patterns parallel and symmetrical to the ridge.

that the earth can be thought of as consisting of seven main plates and a few smaller plates, each one moving from the mid-ocean ridge to a trench or volcanic arc. These plates are indicated in Fig. 14–35.

The map shows that plates have three types of borders: the mid-ocean ridges (spreading centers, divergent boundaries) where they are created; the subduction zones (convergent zones) where they are consumed; and transform faults where they move past each other. If the plates are active, earthquake maps show these boundaries clearly; but if the plate is inactive, its margin may be difficult to find. An example of the latter case is the border between the Eurasian and American plates. Because this border is inactive, it is shown as an arbitrary line in Fig. 14–35.

A plate may be composed of ocean only, or it may carry a continent as well. Even the small plates are very much wider than they are thick. The plates are the lithosphere, so called because the rocks are rigid. Below the plates is the much weaker asthenosphere. The plates are rigid, meaning that the distances between points on a plate do not change as the plate moves. Distances between points on different plates do, of course, change.

TRANSFORM FAULTS AND FRACTURE ZONES

Although the mid-ocean ridge can be followed continuously through all of the oceans, its center is clearly offset at many places. (See Figs. 14–35 and 14–36.) The map pattern suggests that movement along faults has displaced the ridge, and this idea was strengthened by *fracture zones* (to be described) extending many hundred kilometers from the offsets. This simple interpretation is *not* correct, and the offsets of the mid-ocean ridge or spreading center are more complex. The features that offset the ridge are called *transform faults.*

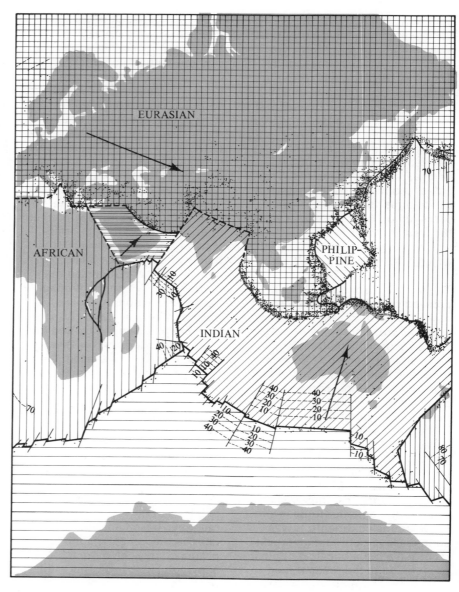

FIGURE 14–35 The crustal plates. In general, each plate begins at a ridge (heavy line) and moves toward a trench (hatching). Earthquake epicenters (dots) tend to outline the plates. Transform faults are shown by thin lines. The dashed lines show the ages (in

Mid-ocean ridges probably originate as discontinuous features; that is, the apparent offsets are original features. In any case, the movement on transform faults caused by formation of new ocean crust at segments of mid-ocean ridges is in the opposite direction to the apparent offset (Fig. 14–37). Detailed seismic studies reveal the nature of the movements on the faults that offset the mid-ocean ridge. It is possible to determine

the relative direction of the movement of the plates on each side of the fault from the direction of the first movement on seismograms. Such studies show that the movement on each block is away from the mid-ocean ridge, as expected if new crust is created there. Note, however, as shown in Fig. 14–37, that the relative movement is opposite to that expected by ordinary fault movement. This special kind of fault is termed a *transform*

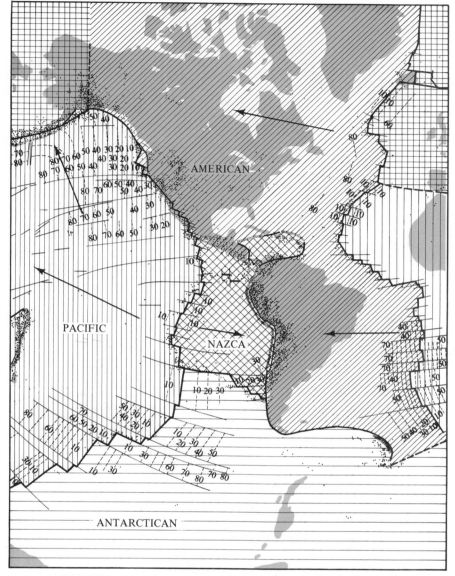

millions of years), based on magnetic data. The arrows show the general direction of movement of the plates. The information and interpretation are from many sources, mainly workers at Lamont-Doherty Geological Observatory.

has a south-facing scarp up to 2623 meters (8600 feet) high. The Murray fracture zone is about 4025 kilometers (2500 miles) in length and is a north-facing scarp up to 2135 meters (7000 feet) high. The Molokai fracture zone is less regular than the others. The Clarion fracture zone is a trough up to a mile deep for much of its length. The Clipperton fracture zone is over 9660 kilometers (6000 miles) long, about one-quarter of the earth's circumference. It is a zone about 80 kilometers (50 miles) wide of irregular relief of 305 to 610 meters (1000 to 2000 feet) and is less pronounced than the other fracture zones. Seismic studies show that the crustal thickness changes abruptly at some of these fracture zones.

Fracture zones are formed in the active parts of transform faults between the segments of mid-ocean ridge. In that region, the transform is probably a weak zone, and some of the basalt created at the ridge rises there. This intrusive activity, together with the movement, produces the irregular submarine topography. This topography, the fracture zone, is carried by the moving plate.

PLATE MOTIONS

The direction of movement can be determined from a map of the plate. Plate motion is, in general, from a mid-ocean ridge or spreading center toward a subduction zone. Transform faults reveal the actual direction because motion must be parallel to a transform fault. If the motion is not parallel to a transform, lithosphere would have to be either created or consumed at the transform. Most transform faults are at right angles to the spreading center, but plate motion can be at any angle from a spreading center, and a plate can move at any angle into a subduction zone.

In geometric terms, the movement of a plate can be described as a rotation about a pole. The

fault, and it occurs at mid-ocean ridges. The San Andreas fault and the oceanic fracture zones are believed to be transform faults. Some of the transcurrent faults described in the next chapter may be transform faults.

Note that earthquakes occur only on the parts of transform faults between segments of mid-ocean ridges. The scars of the transform faults extend long distances into both plates (Fig. 14–

37). These scars were discovered before transform faults were recognized and were called *fracture zones.*

Some of the most-studied fracture zones are in the Pacific Ocean. There the fracture zones are from 4025 to 9660 kilometers (2500 to 6000 miles) long (Fig. 14–36). The Mendocino fracture zone is the most northerly of the long fracture zones and is over 5635 kilometers (3500 miles) long and

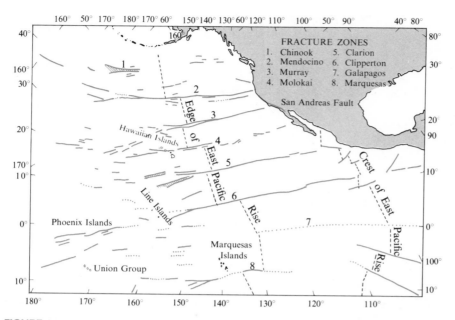

FIGURE 14–36 Principal fracture zones of the Pacific plotted on a great-circle projection. A straight line on this projection is part of a true great circle. Cross-hatching indicates areas of submarine volcanoes. Apparent displacements on the fracture zones are indicated by the displacements of the edge of the East Pacific Rise. From H. W. Menard, *Science* (January 6, 1967), Vol. 155, p. 73.

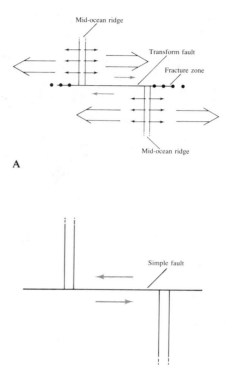

FIGURE 14–37 Transform fault compared with simple fault. A. Transform fault. The relative movement on the transform fault is caused by the new sea floor created at the two segments of the ridge. B. Two segments of once-continuous linear feature separated by a simple fault. Note that the relative movement, shown by the arrows, is opposite to that of the transform fault shown in A.

pole of rotation of a plate has no relationship to the geographic pole except that it is possible that the rotational pole of a plate could be at a geographic pole. A simple experiment can demonstrate. Cut a piece out of an old tennis ball and place this "plate" on another tennis ball. Notice that as you move the "plate," a point on the "plate" moves in a circular path on the tennis ball. The pole of rotation is the center of the circle (Fig. 14–38). You can tie a string to a thumbtack and attach it to the tennis ball at the pole of rotation and to the "plate." Further manipulation will convince you that movement of the "plate" can be described by the location of its pole and rate of rotation about that pole. Note also that the rate of movement of the plate is slower near its pole than farther away. The rate of plate movement is maximum 90 degrees from its pole (at the "equator" of its pole).

In our tennis ball experiment, it is easy to see the actual or absolute movement of the plate because we saw the "plate" move relative to the tennis ball. On the real earth we can only see relative motion; that is, we measure movement of one plate relative to another plate. We must measure movement in this way because we have no fixed points on the earth. All that we see on the earth's surface are the moving plates. If some of the hot spots or mantle plumes (discussed next section) are in fact at fixed positions in the mantle, they may enable us to determine the actual motion of the plates.

The rotational poles of relative motion of two plates can be found easily. The transform faults show the direction of motion, so any circle on the globe drawn at a right angle to a transform passes through the rotational pole. Such circles drawn at two or more places on two or more transforms will intersect at the rotational pole (Fig. 14–39).

Rate of movement of one plate relative to another is determined from magnetic stripes. The times when the magnetic field reversed are known, so the width of the stripes divided by the time over which they formed will give the rate of movement of the plate at that point. Absolute rates of movement can be determined from the ages of islands and seamounts that formed over fixed hot spots or mantle plumes.

An obvious goal of geologists is to determine the current relative motions of the plates. The map of the plates (Fig. 14–35) shows that the African and the Antarctic plates are almost completely bordered by spreading centers. For this reason, either or both of these plates are assumed to be stationary, and the motions of other plates are determined relative to them. The resulting relative motions are shown by the big arrows on the map. As might be expected, those plates with approximately equal amounts of spreading center and subduction zone show the most rapid movement. The difficulties involved in determining current motion of plates indicate the problems encountered in deciphering plate movements in the geologic past.

Current relative motion can be measured directly by the use of satellites and radio-emitting stars such as pulsars. A laser beam from earth is reflected by the satellite, and the round-trip time reveals the distance between the laser source and the satellite. If the orbit of the satellite is known, the laser source can be located within a few centimeters. From such locations, movements of one plate relative to another can be determined. The distance between two stations can also be determined by measuring the time difference in the arrival times of radio pulses from a pulsar. These methods are used to measure creep on faults as well as plate motions. The African plate is moving about 1 centimeter per year, and the Nazca plate off western South America is moving at about 15 centimeters per year.

If, as in the preceding discussion, a plate is bordered by spreading centers, either new sea floor is created on only one side of the spreading center, or the plate is expanding. If the plate is expanding, then the location of the spreading center must be moving. The existence of symmetrical magnetic stripes on both sides of mid-ocean ridges bordering both the African and Antarctic plates suggests strongly that those plates are expanding. The increase in surface area of those plates must be balanced by subduction of the surrounding plates. Thus, mid-ocean ridges or spreading centers and their associated transform faults can move; and if they can move, they, too, can be consumed at subduction zones.

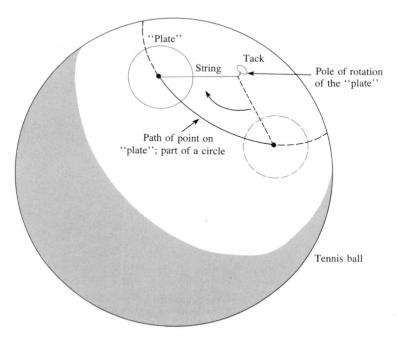

FIGURE 14–38   The movement of a plate can be described by its rate of rotation about its pole.

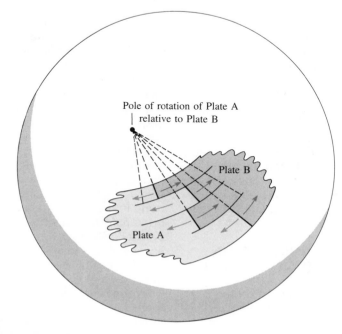

FIGURE 14–39   Lines at right angles to transform faults locate the rotation pole of Plate A relative to Plate B.

HOT SPOTS AND MANTLE PLUMES

Many volcanic areas are not near mid-ocean ridges or subduction zones. This suggests that hot material rises through the mantle at such places. These areas are called *hot spots,* and the hypothe-

291

sized column of rising hot material is termed a *mantle plume.* The underlying idea is that at mid-ocean ridges, magma rises as sheets, and that the mantle plumes are similar columns. Some hot spots occur on continents (Yellowstone National Park is an example), and some, such as the Hawaiian Islands, form in oceans.

The mantle plumes associated with some hot spots appear to be very long-lived and to be fixed in location. A lithospheric plate moving over such a place would have a line of volcanoes. The volcanoes should be progressively younger as the hot spot is approached, with active volcanoes at the hot spot. Such a situation is more easily recognized in an ocean than on a continent. The Pacific Ocean has a number of volcanic islands and submarine mountains that form in lines or chains. Accurate dating of the islands and seamounts is necessary to prove that they were formed when the Pacific plate moved over a mantle plume. The Hawaiian Islands appear to have formed in this way.

The active volcanoes in the Hawaiian Islands are at the southeastern end of the island group. From there, going westward toward Midway Island, the basalt on the island volcanoes gets progressively older (Fig. 14–40). Beyond Midway Island is the line of the north-trending Emperor Seamounts, and they continue the trend of older rocks to the north. These volcanic islands and seamounts are interpreted as having formed as the Pacific plate moved over the hot spot where the Hawaiian Islands are today. The change in direction of the volcanoes would, in this interpretation, be caused by a change in direction of the movement of the Pacific plate.

Near the Mid-Atlantic Ridge are other features that may be related to hot spots. Tristan da Cunha is an active volcanic area east of the Mid-Atlantic Ridge that may be caused by a mantle plume. Sub-

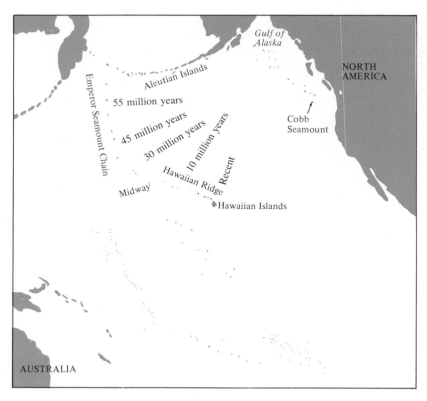

FIGURE 14–40 The line of seamounts and islands may have been formed as the Pacific plate passed over a hot spot. Approximate ages of the basalt are indicated.

marine ridges run northeast and northwest from Tristan da Cunha. These ridges are somewhat similar to mid-ocean ridges but have no seismic activity. They have been interpreted as having been formed as the African and South American plates moved over the hot spot that is now Tristan da Cunha.

The importance of hot spots is that they are our only means of measuring absolute motion of lithospheric plates, assuming that they form as just outlined. All of our other estimates of rate of plate movement are relative; that is, the movement is of one plate with respect to another.

RIFTING—FORMATION OF SPREADING CENTERS

When a new spreading center forms, the lithospheric plate is broken apart, or rifted. The process can occur on either continent or ocean, but because oceans are

created by this process, most spreading centers, mid-ocean ridges, are found in oceans. (See Figs. 14–41 and 14–42.) At a few places (Africa is an example) young spreading centers appear to be forming on the continent.

In map view, the mid-ocean ridges generally trend in the same direction for thousands of kilometers. The segments of mid-ocean ridge are much shorter and are separated by many transform faults that are at nearly right angles to the ridge segments. The young spreading centers in east Africa, the rift valleys, do not all trend in the same direction, and no transform faults have been recognized. Whether these differences are due to age or to some peculiarities of the African plate is not known.

Rifting requires the establishment of a spreading center by material rising from the mantle. We do not understand the details of how the present active spreading

FIGURE 14–41   View northeast across the Sinai Peninsula taken from *Gemini XI*. The southeast end of the Mediterranean Sea is in the upper left, and the north end of the Red Sea is in the lower right. The Sinai Peninsula is bordered by the Gulf of Suez and the Gulf of Aqaba. The Gulf of Aqaba is part of a linear depression on which the Dead Sea can be seen. The Red Sea may be a rift that originated when Africa and Arabia separated, and the two gulfs shown here may be related to that fracture. Photo from NASA.

FIGURE 14–42   Looking northeast at the south end of the Red Sea and the Gulf of Aden, taken from *Gemini XI* at an elevation of 872 kilometers (470 nautical miles) above Africa that forms the foreground. The Indian Ocean is to the right, and the southwest end of the Arabian Peninsula forms the top of the picture. The Red Sea and the Gulf of Aden may be fractures that formed when Africa and Arabia separated. The part of Africa where the Red Sea narrows (Afar Triangle) is recently uplifted ocean floor. Photo from NASA.

293

centers operate, so we do not know why rifting occurs and new spreading centers are formed. We also believe that the present mid-ocean ridges formed as a series of straight segments offset from each other. Transform faults are the result, not the cause, of the offset of the ridge segments.

Rifting begins with the rise of material from the mantle. One hypothesis is that a number of mantle plumes or hot spots form, and their heat causes domes to develop. The domes, when raised enough, would fracture; most commonly three fractures would form (Fig. 14–43). At each dome, one or two of the arms would become spreading centers, and transform faults would form between adjacent spreading centers. This process would result in a zig-zag spreading center unlike that of the present oceans. Another objection is that three-armed domes have not developed at the present-day hot spots. On the other hand, the rift valleys of east Africa have a map pattern consistent with such doming.

Another possibility is that the material rising from the mantle does so more or less as a sheet. The Line Islands of the Pacific Ocean may have formed in this way. The Line Islands, as their name implies, form a line. They were thought to have formed in the same way as the Hawaiian Islands, that is by passage of the Pacific plate over an active hot spot. However, when the rocks of the Line Islands were dated, it was discovered that the islands on more than half of the length of the island group all formed at the same time. A line of volcanic islands and seamounts about 2500 kilometers (1500 miles) long formed at the same time, suggesting that linear mantle plumes can form.

TRIPLE JUNCTIONS

The margins of a plate where it abuts another plate are of only three types: mid-ocean ridge or spreading area, volcanic arc or converging area, and transform fault. The mosaic of plates that cover the earth requires that junctions of more than two plates must occur. Study of Fig. 14–35, which shows the plates, reveals that at many places three plates meet at a point, forming a *triple junction*. Further study reveals that no more than three plates meet at a point; that is, there are no quadruple junctions. Most triple junctions are T-shaped; the others are Y-shaped.

Triple junctions may move with the plates, and they may be subducted. Such events may have been the origin of the bends shown by the magnetic stripes in parts of the Pacific Ocean. (See Fig. 14–44.)

THE CAUSE OF PLATE TECTONICS

To understand plate tectonics, we need to know both the mechanism or process that creates and moves plates, and the energy source that drives the process. Convection currents within the mantle have long been suggested as the driving force of most internal processes, but as we will see, other mechanisms may also be at work.

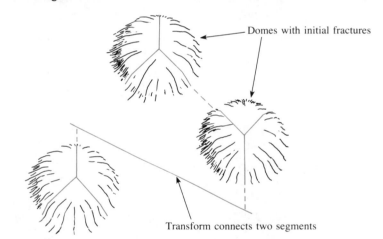

FIGURE 14–43   Rifting. Doming causes fractures that become spreading centers.

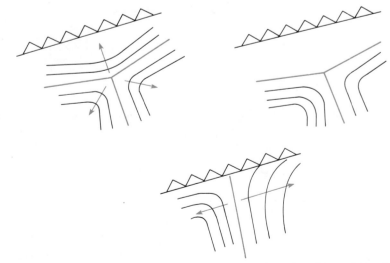

FIGURE 14–44   Movement of a triple junction into a subduction zone could be the origin of bends in magnetic stripes such as those in the Pacific Ocean.

The energy source is also in doubt, with radioactivity and deep, internal heat sources the leading contenders.

CONVECTION

Convection currents have been proposed as the cause of most geological features. *Convection cells develop when a fluid is heated at the bottom. As the warmed fluid expands, it becomes less dense and tends to rise; denser, cooler fluid at the top moves down to take the place of the rising, warm fluid,* as shown in Fig. 14–45. *In this process heat is transported upward by the movement of the warmed fluid.*

Although there is disagreement on the viscosity of the mantle, convection has been explored theoretically by a number of investigators who have established that it could exist in the mantle, even though the mantle is viscous enough to behave as a solid in transmission of earthquake waves. A substance may respond elastically to a rapid stress but may be deformed viscously by a stress applied over a long period of time. The latter case could occur in the mantle, especially in view of the temperatures and pressures there. Thus it is possible for a solid to flow at temperatures well below the melting point. A possible example of this behavior is the slow movement of glacial ice under the influence of gravity and its very different behavior when subjected to a rapid stress. Convection currents are believed to move at the rate of one to a few centimeters per year.

Many features of the mid-ocean ridges can be explained by convection currents. The high heat flow suggests upwelling convection currents in the mantle. The convection currents could cause the rise. The upwelling mantle material could melt partially as a result of the reduced pressure near the crust-mantle boundary, forming basalt. The basalt would

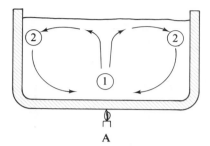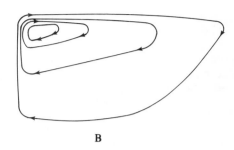

FIGURE 14–45 A. Two simple convection cells produced by heating at the bottom. Warm fluid at ① rises and cool fluid from ② falls to take the place of the rising warm fluid. B. Convection in a fluid whose viscosity depends on temperature and pressure. Heating is on the bottom.

become new ocean floor. (See Fig. 14–46.) Horizontal currents could cause sea-floor spreading and descending currents could cause trenches and volcanic island arcs.

The low-velocity zone in the upper mantle that was described earlier may be involved in sea-floor spreading. If the movement is due to convection, the convection could involve the whole mantle or only a part of it. Convection could be confined to the low-velocity zone, but such thin cells would not

be stable and would tend to break up into much smaller cells.

It has been suggested that the plates, extending down to the weak zone in the upper mantle, move as units, and the return flow is in the weak zone of the upper mantle. Such movement resembles convection (See Fig. 14–47).

GRAVITATION

The pull of gravity may be the force that moves the lithospheric

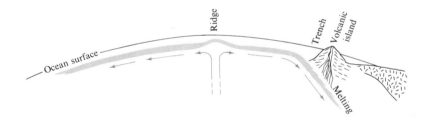

FIGURE 14–46 Formation of oceanic ridge, trench, and volcanic island arc by convection currents or sea-floor spreading.

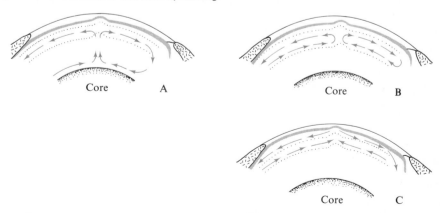

FIGURE 14–47 Convection currents. A. Deep convection involving much or all of the mantle. B. Thin convection confined to the low-velocity zone of the mantle. C. A convection-like flow in which the rocks above the low-velocity zone move and the return flow is in the low-velocity zone.

plates. At first glance this idea seems incredible, but it is supported by seismic data. At subduction zones, the descending slab of lithospheric plate causes many earthquakes. It is possible to determine whether an earthquake is caused by tension or compression by studying the first motions of the waves. These studies reveal that at subduction zones, the upper part of the descending slab is in tension, and the deeper part is in compression. This suggests that the weight of the plate pulls it down into the weak low-velocity zone, and compression occurs where the deeper parts of the plate come into contact with the stronger mantle.

The lithospheric plates are cold compared to the upper mantle, so they are denser than the layers through which they move. This difference in density is believed to be enough to make this mechanism work. At depth, the descending plate is heated by the surrounding hotter rocks, and the slab is assimilated into the mantle.

The height of the mid-ocean ridges may also provide a small gravity force. In this instance, the plate is simply sliding downhill. The ridge itself may be caused by expansion due to heating by the hot mantle material that rises there.

Gravitational pull is similar to convection in that both processes depend on differences in density. Indeed, they may both work together to cause plate tectonics. Gravitational pull may cause the plates to move, and convection in the mantle may help to propel the plates. If we knew the energy source, we might have help in deciding the relative importance of these possible mechanisms.

ENERGY SOURCES

The energy that runs the earth's internal engine probably comes from two sources: unstable radioactive elements that spontaneously decay, releasing energy; and internal energy stored within the earth. The energy is in the form of heat, and it is possible to measure the amount of heat coming from inside the earth.

The heat flow from the earth indicates the presence of sources of heat energy large enough to account for geologic processes. The total heat flow from the earth is about 1000 times more than the energy of earthquakes and 100 times more than volcanism, and is enough to cause geologic processes. The heat flow from within the earth is only about $\frac{1}{5000}$ of the amount of heat we receive from the sun. Heat flow is calculated from temperature measurements in drill holes and on the bottom of the ocean. Reliable heat-flow measurements were first made on the continents in the mid 1930s and at sea in 1950.

Heat-flow measurements show that although the amount of heat reaching the surface varies with the geologic location, on the average the heat flow is the same from both continents and oceans. As we will see, this is a surprising and, as yet, unexplained conclusion. The results are shown in Table 14–1.

The variations in heat flow from the oceans are easily explained by plate tectonics. At mid-ocean ridges, the heat flow is well above average, as might be expected in an area of active volcanism. The ocean basins are slightly below average, perhaps because the basalt crust cools as it moves away from the mid-ocean ridges. At the trenches, the heat from the descending slab of lithospheric plate is blanketed by the sediments in the trench, causing the low heat flow there. On the continents, similar variations are found, with low heat flow from old areas and high heat flow from areas of younger deformation. This high heat flow is in the volcanic and magmatic arc associated with subduction zones. Thus the heat flow data are compatible with plate tectonics.

Radioactivity is the source of at least some of the heat flow. Radioactive elements are concentrated in the crust and are not abundant in the mantle. The known radioactive content of crustal rocks of the continents is enough to account for the continental heat flow. The average heat flow from the oceans is about the same as the average from the continents. However, the thin layer of oceanic basalt does not contain enough radioactive elements to account for the oceanic heat flow. Therefore, some of the heat flow from the oceans must come from the man-

TABLE 14–1 Summary of heat flow measurements. The units are microcalories per square centimeter per second (10^{-6} cal/cm^2/sec). From W. H. K. Lee and Seiya Uyeda, "Review of Heat Flow Data," in *Terrestrial Heat Flow*, American Geophysical Union, Geophysical Monograph 8, 1965.

Average: 1.5 ± 0.15 (Same on continents and oceans)	
Continents	
Undeformed areas such as interior lowlands	1.54 ± 0.38
Areas of metamorphic rocks deformed more than 600 million years ago	0.92 ± 0.17
Areas deformed between 600 and 225 million years ago	1.23 ± 0.40
Areas deformed less than 225 million years ago	1.92 ± 0.49
Oceans	
Trenches	0.99 ± 0.61
Floors	1.28 ± 0.53
Ridges	1.82 ± 1.56

tle, and this is especially true of the mid-ocean ridges where the heat flow is very high.

Radioactive elements are fairly evenly distributed in the continental crust; geologically active areas, such as volcanoes and mountain belts, are no more radioactive than other areas. Radioactive energy is enough to cause all geologic processes, but the distribution of radioactivity suggests that it is not the energy source, so other sources must be considered.

Some other aspects of radioactivity should also be considered. As radioactive elements decay,

the total number of radioactive atoms is reduced; thus the amount of radioactive heating has decreased throughout geologic time, as was shown in Table 2–5. This implies that if radioactivity causes plate tectonics, plate movements should be slowing. Another point is that if convection in the mantle is involved in plate tectonics, a deep heat source is necessary because convection requires heating from below. Radioactivity is concentrated in the crust. Finally, analysis of heat-flow data strongly suggests that well over half of the heat flow, even on the continents,

comes from below the crust, and, of course, most of the heat flow in the oceans also comes from deep sources.

Some of the heat reaching the surface must come from deep in the earth. The deep source of heat is believed to be the liquid core. The heat could be released by crystallization of the core, and the inner core, believed to be solid, may be slowly enlarging. Internal heat could also come from recrystallization in the mantle, but we know very little about either the composition or the physical conditions there.

SUMMARY

The *ocean floors* cover about forty percent of the earth's surface. They are covered by about 4.4 kilometers (2.75 miles) of water, about 305 meters (1000 feet) of unconsolidated sediments, about 2 kilometers (1.25 miles) of basalt and consolidated sediments, and finally 4.8 kilometers (3 miles) of basalt that ends at the Moho.

The ocean floors are remarkably young geologically and are youngest near mid-ocean ridges.

Submarine mountains are basalt volcanoes.

Flat-topped seamounts record wave erosion followed by subsidence.

Coral atolls may form in tropical areas if the corals are able to grow as rapidly as the island subsides.

Mid-ocean ridges form the longest continuous mountain range on the earth. They are about 64,400 kilometers (40,000 miles) long, 483 to 4830 kilometers (300 to 3000 miles) wide, and rise up to 3050 meters (10,000 feet) above their base. They are composed of basalt and have a rugged topography, generally with a rift valley at the crest. Heat flow is high, and many earthquakes occur on mid-ocean ridges.

Continental slopes connect the continental shelves and the ocean floors.

Continental shelves have continental structure and average about 60 meters (200 feet) below sea level.

The *continental rise* is the region between the continental slope and the ocean floor.

Continental slopes are composed of sediments and are dissected by *submarine canyons.*

Volcanic arcs are composed of andesite volcanoes and have submarine trenches on their oceanic sides.

Negative gravity anomalies occur over the trenches, implying that the crust is thinner.

Deep-focus earthquakes, up to 644 kilometers (400 miles) deep, occur on a plane sloping beneath the continent at the trenches.

The evidence for continental drift includes paleomagnetic data, apparent fit of the continents, climatic data such as glaciation, apparent rifting, fossils such as *Glossopteris* and *Mesosaurus,* and age of the oceanic sediments.

The earth's magnetic field is nearly a dipole field, but the departures from a simple dipole are

enough so that the north magnetic pole is $11\frac{1}{2}°$ from the geographic pole, and the south magnetic pole is not antipodal to the north magnetic pole.

The dynamo theory explains the earth's magnetic field. In this theory, an electric current moves through the earth's conducting core, creating a magnetic field. The conducting core moves through this magnetic field, and a current is induced. This current starts the whole cycle over again.

Remanent magnetism is the magnetism frozen in a rock at the time the rock forms.

Paleomagnetic (old magnetic) data reveal the north magnetic direction and latitude of ancient rocks, but not where on that latitude the rock formed.

Sea-floor spreading is the theory that new oceanic crust is being created at the mid-ocean ridges, is then moving away from the ridges, and is being consumed at the submarine trenches.

The bottom sediments on the sea floor are younger and thinner near the mid-ocean ridges and become thicker and older away from the ridges. The oldest sediments so far found are latest Jurassic (about 160 million years old).

Basalt and other rocks are magnetized by the earth's magnetic field when they cool through a certain temperature (Curie point) that is well below their melting points.

The earth's magnetic field has reversed its polarity numerous times. These changes in magnetic direction are impressed on the new ocean floor created at the mid-ocean ridges and appear as symmetrical bands of alternating magnetic intensity parallel to the mid-ocean ridges.

Plate tectonics is the theory that new lithosphere is created at mid-ocean ridges and consumed at volcanic arcs.

Crustal plates are bounded by mid-ocean ridges, submarine trenches, and transform faults. They are the units of the earth's crust that move in the theory of sea-floor spreading. Plates can be composed of either continent or ocean or both.

Transform faults are the surfaces where one plate moves past another plate. The direction of movement revealed by seismic study is in the opposite sense to the apparent offset of the mid-ocean ridges associated with the two plates involved.

The movement of plates can be described as a rate of rotation about an axis.

Movement of one plate relative to another can be determined, but absolute motion is difficult to measure.

Hot spots and *mantle plumes* may cause lines of volcanoes on plates as they move over the hot spots. This may provide a reference for the measurement of absolute plate motion.

Rifting is the splitting of a plate by the formation of a new spreading center.

Triple junctions occur where three plates meet.

Convection currents may occur in the mantle and could account for sea-floor spreading.

Gravitational pull may also be a force that moves plates. The weight of a plate may pull it into a subduction zone.

The energy that causes deformation of the crust may come from radioactivity or may come from crystallization of the core.

QUESTIONS

1 What kinds of sediments are found on the ocean floors?
2 Describe the submarine mountains of the Pacific (composition, arrangement, etc.).
3 How do flat-topped seamounts form? What does this suggest about the history of the oceans?
4 How long are the mid-ocean ridges?
5 Describe a cross-section of the mid-ocean ridges. (A sketch might help.)
6 Sketch the ocean-continent boundary at a stable place, and label the parts.
7 What does a negative gravity anomaly mean?
8 Describe a typical island arc and sketch a cross-section.
9 What changes have occurred in the earth's magnetic field?
10 How is the earth's magnetic field believed to be caused?
11 How is remanent magnetism impressed on rocks?
12 What data on the previous position of a continent is obtained from remanent magnetism?
13 What evidence suggests that the southern continents were once joined?
14 What is the evidence for sea-floor spreading?
15 What are crustal plates? How many are there?
16 Describe transform faults.
17 Where does the energy come from that causes the deformation of rocks?
18 What are convection currents? What is the evidence for them in the mantle? What other force may move plates?
19 How is the location of the pole of relative motion of one plate relative to another plate determined?
20 What is the evidence for the existence of hot spots or mantle plumes?

SUPPLEMENTARY READINGS

Bullard, Sir Edward, "The Origin of the Oceans," *Scientific American* (September 1969), Vol. 221, No. 3, pp. 66–75.

Burke, K. C., and J. T. Wilson, "Hot Spots on the Earth's Surface," *Scientific American* (August 1976), Vol. 235, No. 2, pp. 46–57.

Carrigan, C. R., and David Gubbins, "The Source of the Earth's Magnetic Field," *Scientific American* (February 1979), Vol. 240, No. 2, pp. 118–130.

Dalrymple, G. B., and others, "Origin of the Hawaiian Islands," *American Scientist* (May-June 1973), Vol. 61, No. 3, pp. 294–308.

Dewey, J. F., "Plate Tectonics," *Scientific American* (May 1972), Vol. 226, No. 5, pp. 56–68.

Dickinson, W. R., "Plate Tectonics in Geologic History," *Science* (October 8, 1971), Vol. 174, No. 4005, pp. 107–113.

Dietz, R. S., "Geosynclines, Mountains and Continent-Building," *Scientific American* (March 1972), Vol. 226, No. 3, pp. 30–38.

Elders, W. A., and others, "Crustal Spreading in Southern California," *Science* (October 6, 1972), Vol. 178, No. 4056, pp. 15–24.

Emery, K. O., "The Continental Shelves," *Scientific American* (September 1969), Vol. 221, No. 3, pp. 107–122.

Hall, J. M., and P. T. Robinson, "Deep Crustal Drilling in the North Atlantic Ocean," *Science* (May 11, 1979), Vol. 204, No. 4393, pp. 573–586.

Hallam, A., "Continental Drift and the Fossil Record," *Scientific American* (November 1972), Vol. 227, No. 5, pp. 56–66.

Heezen, B. C., and I. D. MacGregor, "The Evolution of the Pacific," *Scientific American* (November 1973), Vol. 229, No. 5, pp. 102–112.

Heirtzler, J. R., and W. B. Bryan, "The Floor of the Mid-Atlantic Rift," *Scientific American* (August 1975), Vol. 233, No. 2, pp. 78–90.

James, David E., "The Evolution of the Andes," *Scientific American* (August 1973), Vol. 229, No. 2, pp. 60–69.

Kurtén, Björn, "Continental Drift and Evolution," *Scientific American* (March 1969), Vol. 220, No. 3, pp. 54–64.

Macdonald, K. C., and B. P. Luyendyk, "The Crest of the East Pacific Rise," *Scientific American* (May 1981), Vol. 244, No. 5, pp. 100–116.

Marsh, B. D., "Island-Arc Volcanism," *American Scientist* (March-April 1979), Vol. 67, No. 2, pp. 161–172.

McKenzie, D. P., and Frank Richter, "Convection Currents in the Earth's Mantle," *Scientific American* (November 1976), Vol. 235, No. 5, pp. 72–89.

McKenzie, D. P., and J. G. Sclater, "The Evolution of the Indian Ocean," *Scientific American* (May 1973), Vol. 228, No. 5, pp. 62–72.

Molnar, Peter, and Paul Tapponnier, "The Collision between India and Eurasia," *Scientific American* (April 1977), Vol. 236, No. 4, pp. 30–41.

Nierenberg, W. A., "The Deep Sea Drilling Project After Ten Years," *American Scientist* (January-February 1978), Vol. 66, No. 1, pp. 20–29.

Pollack, H. N., and D. S. Chapman, "The Flow of Heat From the Earth's Interior," *Scientific American* (August 1977), Vol. 237, No. 2, pp. 60–76.

Rona, P. A., "Plate Tectonics and Mineral Resources," *Scientific American* (July 1973), Vol. 229, No. 1, pp. 86–95.

Sclater, J. G., and C. Tapscott, "The History of the Atlantic," *Scientific American* (June 1979), Vol. 240, No. 6, pp. 156–175.

Smith, R. B., and R. L. Christiansen, "Yellowstone Park and a Window on the Earth's Interior," *Scientific American* (February 1980), Vol. 242, No. 2, pp. 104–117.

Tazieff, Haroun, "The Afar Triangle," *Scientific American* (February 1970), Vol. 222, No. 2, pp. 32–40.

Tokosöz, M. N., "The Subduction of the Lithosphere," *Scientific American* (November 1975), Vol. 233, No. 5, pp. 88–101.

Valentine, J. W., and E. M. Moores, "Plate Tectonics and the History of Life in the Oceans," *Scientific American* (April 1974), Vol. 230, No. 4, pp. 80–89.

Vine, F. J., "Sea-floor Spreading—New Evidence," *Journal of Geological Education* (February 1969), Vol. 17, No. 1, pp. 6–16.

15

Such changes in the superficial parts of the globe seemed to me unlikely to happen if the Earth were solid to the centre. I therefore imagined that the internal parts might be a fluid more dense, and of greater specific gravity than any of the solids we are acquainted with; which therefore might swim in or upon that fluid. Thus the surface of the globe would be a shell, capable of being broken and disordered by the violent movements of the fluid on which it rested.

Benjamin Franklin
Letter to Abbé Soulavie
September 22, 1782

The continents are both made and deformed by plate movements. Parts of the continents are very much older than the ocean crust, so it is from study of the continents that geologists have deciphered the earth's history. Thus the deformed rocks of the continents contain the record of past plate movements. The bent, broken, and metamorphosed rocks of the continents have fascinated geologists for over two centuries, and the older studies can now be reinterpreted in the light of plate tectonics. We will first consider why the crust has two levels—ocean and continent. Then we will proceed from small structures to huge structures, and we will relate them to plate tectonics.

ISOSTASY— DIFFERENCES BETWEEN CONTINENTS AND OCEANS

The elevations of the earth are summarized in Fig. 15–1. This diagram shows that the very high and very low portions of the earth are very small in area, and that two levels, the continents and the ocean floors, make up most of the surface. Note that the difference between the highest and the lowest points is about 19.3 kilometers (12 miles), which is very small compared with the radius of the earth ($19.3/6368 = 0.3$ percent), so that the earth is smoother than a billiard ball. This comparison shows that the features we observe on the surface are very slight compared with the earth as a whole.

Certainly the differences between ocean and continent are the most pronounced contrast on the earth's surface. One important contrast is in rock type: *The ocean basins are composed mainly of basalt; the continents are composed mainly of igneous and metamorphic rocks of granitic composition.* To understand how these compositional differences might account for the two main levels in the elevation diagram (Fig. 15–1), we must turn to a different type of study.

Surprisingly, an important clue came from a surveying error made about the middle of the nineteenth century when the English were mapping India. Their methods of surveying were much the same as our modern ways; only the instruments were different. Because it is easier and more accurate to measure angles than distances, most surveys use *triangulation.* To use this method, a base line is accurately measured, generally by actually taping it. From each end of this measured base line, the angle is measured to a third station that must be visible from both stations on the base line. The third station can now be located, either by calculating its position by trigonometry or by making a scale plot. The triangle is checked by measuring the angle at the third station. Now, two of these stations can be used to locate another station, and so forth, until all stations are located. (See Fig. 15–2.)

The location of each station in the Indian survey was determined also by astronomical observations in the same way that a navigator at sea locates his position. It was discovered that the northern sta-

tions near the Himalaya Mountains were located too far south by astronomical methods, according to the locations obtained by the method of triangulation. Although the differences in location by these two methods were small, they were greater than could be due to errors in measurements, especially as corrections for the curvature of the earth had been applied. Apparently, the great mass of the Himalaya Range attracted the plumb bob that was used to determine the zenith in the astronomical determination of location. A level bubble and a plumb bob are basically the same, and one or the other must be used to establish the zenith from which astronomical measurements are made. Knowing the size of the mountain range and using Newton's law of gravitation, it is easy to calculate the expected amount of attraction of the plumb bob by the range; in this case, the observed deflection was much less than the expected deflection (Fig. 15–3). This implies either that the mountains are composed of rock lighter in density than the plain, or, if both mountains and plains are composed of the same rock type, that the rock under the mountains is lighter in density than the deeper rocks. The latter case is supported by seismic data and geologic mapping and is called the *roots-of-mountains hypothesis.* (See Fig. 15–4.)

Grande Chartreuse, French Alps. Photo from Swissair.

301

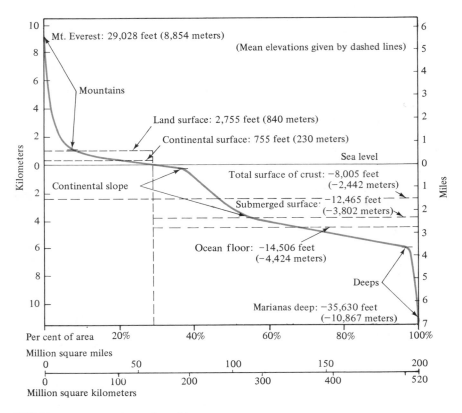

FIGURE 15-1 Elevations of earth's crust.

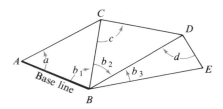

FIGURE 15-2 Location of points by triangulation. *AB* is base line. Angles *a* and b_1 are measured so that *C* is located relative to *A* and *B*. Angles *c* and b_2 are now used to locate *D*. Angles b_3 and *d* will now locate *E*, and the process can be continued.

$\therefore$1 cubic inch of granite weighs
   $2.7 \times 0.036 = 0.097$ pound
   1 mile = 5,280 feet = 63,360
   inches
$\therefore$A mile-high column of granite 1
   inch in cross-section weighs
   $63,360 \times 0.097 = 6,145.92$
   pounds

Therefore, the weight of a column of granite a mile high produces a pressure of 6146 pounds per square inch at its base. Thus, if the granite approaches four miles in thickness, its compressive strength is exceeded and the base will crumble under its own weight. The continental crust exceeds this thickness, and some mountain ranges are higher. How is this possible? A factor we have so far neglected, but which will not basically alter the problem, is the change in strength of granite due to the increase in temperature and pressure at depth. The specific gravity of rock will increase due to pressure, the pressure will tend to strengthen the rock, and the higher temperature will weaken it. The total of these effects is to increase greatly the height of the column of rock that is self-supporting, but clearly a limit will be reached. The continents

This suggests that the granitic continents, like the mountains, may stand higher because they are less dense and thicker than the basaltic oceans, granite being less dense than basalt. A possible layer in which the crustal rocks may float is now sought, and we will begin by considering the strength of crustal rocks.

Laboratory tests show that granite, the main rock composing the continents, has a compressive strength of about 22,000 pounds per square inch. The specific grav-ity of granite is about 2.7; that is, it weighs 2.7 times as much as an equal volume of water. From this we can calculate how high a column of granite can support itself; that is, at what height the granite will weigh more than its compressive strength.

1 cubic foot of water weighs 62.4 pounds
1 cubic inch of water weighs

$$\frac{1}{1,728} \times 62.4 = 0.036 \text{ pound}$$

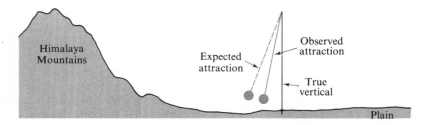

FIGURE 15-3 The gravitational attraction between the Himalaya Mountains and the plumb bob causes the plumb bob to be deflected from the true vertical. The deflection is less than expected, suggesting that mountains are underlain by lighter-density rocks. A plumb bob is simply a weight on a string.

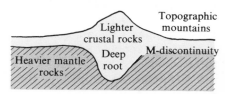

FIGURE 15-4 The root of a mountain range.

cannot be supported by their rigidity or strength.

Then what does support the continents? Mountain ranges stand high because they have light roots, much as a big iceberg stands higher above the water than a small iceberg because its root is deeper. In a similar way, the lithosphere of the earth floats on the asthenosphere, and the lighter rock of the continents floats higher than the thinner, denser rocks underlying the oceans. *The theory that the lithosphere floats on a viscous substratum (the low-velocity zone) is called isostasy.*

Further evidence for isostasy can be found in areas recently covered by continental glaciers, around some recent volcanoes, and by measurements of the gravitational attraction near most mountain ranges. Parts of the earth that were covered by thick accumulations of ice during the recent glacial age were apparently depressed by the weight, and now that the ice has melted, they are rising. This rising may be due in part to elastic or plastic compression of the crust, but calculation suggests that these effects are not enough to account for the postglacial uplifts and that isostasy was involved. The Hawaiian Islands are a group of young volcanoes, and their weight has apparently depressed the crust, creating a moat of deeper water surrounding them, as shown in Fig. 15–5. On the continent, erosional debris would mask a similar occurrence where a rapidly emplaced volcano depresses the crust. Measure-

ments of the gravity force—which is, of course, affected by the mass distribution under the point— when corrected for such effects as elevation and nearby mountain mass, show that most of the topographic features of the earth are compensated at depth, and thus tend to prove the hypothesis of isostasy.

DEFORMATION OF ROCKS

The continents display structures on many scales. Bent, broken, and otherwise deformed rocks are seen in small and large exposures. Study of these deformed rocks reveals much information on the kinds of forces that caused the deformation. The occurrence of these deformed rocks in distinct belts shows larger, more fundamental structures that occupy many thousands of square miles. These larger structures are best developed in mountain belts. Further study reveals many other unusual features of these large belts of deformed rocks and suggests that they may show how continents form. Thus, this chapter progresses from the small, easily studied features of deformed rocks to much larger, more conjectural structures.

When rocks are subjected to forces, they may either break or be deformed. Abundant evidence of both kinds of response is common as faults and folds, respectively. *Whether a rock breaks or is folded depends on its nature,*

which depends on the type of rock and the prevailing temperature and pressure. Thus ductile rocks tend to be folded and brittle rocks tend to be faulted. Folding is favored at depth, and faulting is more common near the surface.

FOLDS

Some general facts about the nature of the deforming stress can be learned from the study of folded and faulted rocks. *Folded rocks generally record compressive stress,* and folded rocks occupy less length than they did before they were folded (Figs. 15–6 and 15–7).

Folds may be upright or overturned, open or tight. *The basin-like folds are called synclines and the hill-like folds are called anticlines* (Fig. 15–8). A better definition, especially if the folds are overturned (as shown in Fig. 15–8), is that *in anticlines the oldest rocks are in the center and in synclines the youngest rocks are in the center.* Remember that the oldest rocks are at the bottom of horizontal sedimentary rocks. *A bed is overturned (upside-down) if it has been bent more than 90 degrees.*

Folds are not all horizontal but may *plunge,* as shown in Fig. 15–9. Notice the difference in the pattern of the ridges between the horizontal folds in Fig. 15–8 and the plunging folds shown in Fig. 15–9. A short anticline that plunges in both directions can be called a *dome.* (See Fig. 15–10.)

All rocks do not respond in the same way to the same stress. For

FIGURE 15–5 Depression of the crust into the viscous mantle by addition of a weight such as a volcano. As erosion removed the weight, the crust would return to its original position. The Hawaiian Islands and some other volcanic piles in the Pacific Ocean have apparently depressed the crust in this manner, because they are surrounded by a ring of deeper water.

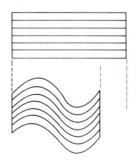

FIGURE 15–6 Folded rocks occupy less length than unfolded rocks.

303

example, if a series of sandstone and shale beds is folded, the weaker shales will flow into the crests of the anticlines and the bottoms of the synclines.

A special type of fold, caused in most cases by a vertical movement, is called a *monocline.* (See Fig. 15–11.) A fault is generally found below most monoclines. In the usual case, brittle basement rocks respond to the forces by faulting, and the more pliable overlying sediments respond by bending.

FAULTS

A fault is a break in rocks along which movement has taken place. Four types of faults can be distinguished, but there is much gradation among these types. Two of these fault types have predominantly vertical movement, and two have mainly horizontal movement.

The two types of faults with mainly vertical movement are shown in Figs. 15–12 and 15–13A. Note that *reverse faults* (see Fig. 15–14) *are caused by compression* and that the crust is shortened as a result of this type of faulting. *Normal faults* (see Fig. 15–15) *lengthen the crust and so are tensional features.* The term *normal* is poor because normal faults are not normal in the sense of being more common than others. At the time the term was first applied they were thought to be more common, so the term is an historical mistake. Compressive features such as folds and reverse faults are much more common, suggesting that most orogenic stresses are compressional. Normal faults are generally associated with the stretching that accompanies vertical movement, such as uplifts. Very few measurements of stress in rocks have been made; but where the vertical stress has been measured, it is generally more than the theoretical value, which is the weight of the overlying rocks. This suggests that vertical forces do exist in the crust.

Thrust faults are reverse faults that are nearly flat and so have mainly horizontal movement (Figs. 15–13B and 15–16). This definition may lead one to think they are merely a special case of reverse fault; in practice they are very different. This is not to imply

FIGURE 15–7   Folds in Rocky Mountains in southern British Columbia. Photo from Geological Survey of Canada.

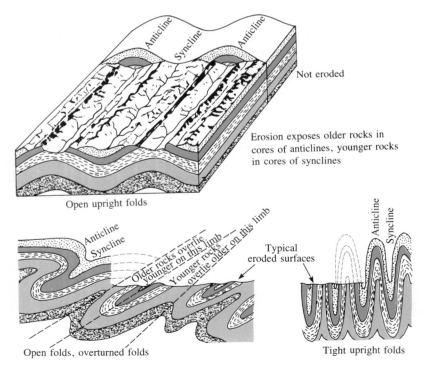

Anticline   Syncline   Anticline

Not eroded

Erosion exposes older rocks in cores of anticlines, younger rocks in cores of synclines

Open upright folds

Anticline   Syncline

Older rocks overlie younger rocks on this limb
Younger rocks overlie older on this limb

Typical eroded surfaces

Anticline   Syncline

Open folds, overturned folds

Tight upright folds

FIGURE 15–8   Types of folds.

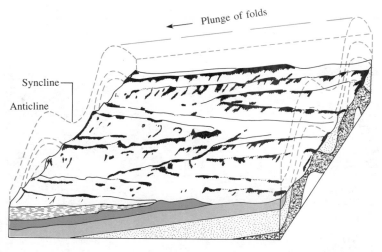

FIGURE 15–9 Plunging folds. In synclines erosion exposes younger rocks in the core and in anticlines, older rocks.

FIGURE 15–10 Oblique aerial photograph of a dome at Sinclair, Wyoming. Photo by J. R. Balsley, U.S. Geological Survey.

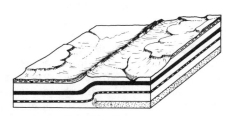

FIGURE 15–11 Monocline. This type of fold is commonly caused by faulting at depth.

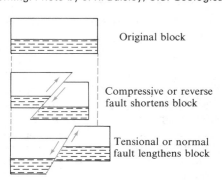

Original block

Compressive or reverse fault shortens block

Tensional or normal fault lengthens block

FIGURE 15–12 Normal and reverse faults.

305

A   B

FIGURE 15–13   A. A small normal fault. B. A thrust fault at Torquay, Devon, England. The flat upper beds have been thrust faulted above the folded limestones. A. by D. E. Winchester, U.S. Geological Survey; B. Crown Copyright, Geological Survey, reproduced by permission.

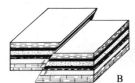
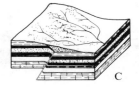
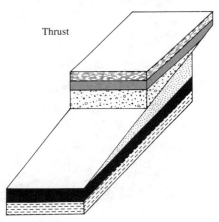

Thrust

FIGURE 15–14   Development of a reverse fault. Step B is never found in nature because erosion would remove the overhang. In C the bending of beds in the fault zone is shown.

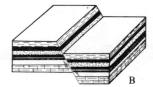

FIGURE 15–15   Development of a normal fault. In C the bending of beds in the fault zone is shown.

FIGURE 15–16   A thrust fault has a nearly horizontal fault plane.

that there are no borderline cases where either the term *reverse* or *thrust* can be used. Many thrust faults have relatively thin upper plates that have moved long distances. Such instances pose real problems because it can be calculated that the friction on the fault plane is so great that the stress necessary to overcome this friction and move the upper plate

would crumple the upper plate. The problem is discussed below. At other places thrust faults involve basement rocks and clearly reveal crustal compression and shortening. Some thrust faults are formed by extreme folding where the lower limb of the fold is pushed out as a thrust fault. In some cases this has occurred at such depth that the rocks were

plastic, and metamorphic rocks are in the cores of the folds. The most extreme example of thrust structure is in the Alps of Europe. Here, in the various ranges of the Alps, thrust sheets and great overfolds have moved many miles and have piled up, one on the other, as shown in Figs. 15–17 and 15–18.

The recognition of the great thrusts in the Alps began in 1841

306

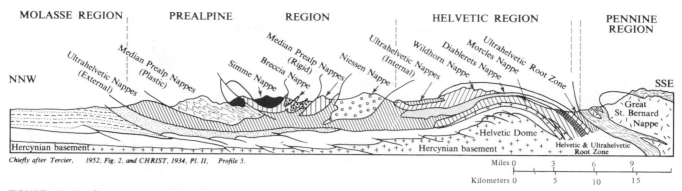

MOLASSE REGION | PREALPINE | REGION | HELVETIC REGION | PENNINE REGION

NNW

SSE

Chiefly after Tercier, 1952, Fig. 2, and CHRIST, 1934, Pl. II, Profile 5.

Miles 0   3   6   9
Kilometers 0   5   10   15

FIGURE 15–17 Cross-section through part of the Alps. Nappe is the term applied in the Alps to the overthrust rocks. Note the large distances that the overthrust rocks have moved. From J. C. Crowell, Geological Society of America, *Bulletin,* Vol. 66, 1955.

FIGURE 15–18 Mount Pilatus in the Alps south of Lucerne, Switzerland. A syncline apparently modified by faulting forms the summit of the mountain. The discordant beds on the left are probably thrust faulted against the syncline. Photo from Swissair Photo Ltd.

when Arnold Escher von der Linth, who was a professor at the University of Zürich, saw that at Segnes Pass in Glarus, older rocks overlay younger rocks with fault gouge or mylonite on the contact. He interpreted the relationship as thrust faulting, and he was able to convince other geologists; but such a structure was unheard of and he did not publish his conclusions. It was not until 1884 that Marcel Bertrand recognized and published the true thrust relation at Glarus, and at the same time he hinted that such thrusts might explain other problems in Alpine geology. In the years from 1865 to 1887, geologic maps of the Alps had been made. In the 1890s, the many, far-travelled thrusts, piled one on the other, were recognized.

In the early 1900s, thrusts were recognized in the Austrian Alps, then in southern France, the Pyrenees, Scotland, and in the northern Rocky Mountains.

The second type of fault with horizontal movement is the *transcurrent,* or *strike-slip, fault* (see Fig. 15–19). *Transcurrent faults are generally nearly vertical and may have considerable horizontal displacement.* Some of these transcurrent faults, such as the San An-dreas fault in California, are probably transform faults. In the Sangre de Cristo Mountains of New Mexico, a vertical transcurrent fault is believed to have a 37-kilometer (23-mile) horizontal displacement, based on the separation of a distinctive body of metamorphic rock. Most strike-slip faults are much smaller, and it is very difficult to prove long movements on any fault.

Most faults have neither all horizontal nor all vertical movement, but have diagonal movement. (See Figs. 15–20 and 15–21.) This is known from study of the grooves and scratches on fault planes that reveal the direction of the last movement. It should be clear from Chapter 13 that most faults have recurring movements; these scratches show only the last movement. It is not generally possible to determine the exact amount of movement on a fault. To determine the exact movement, a point that was once joined must be found on both sides of the fault. (See Fig. 15–22.) Generally, only planar features, such as dikes or sedimentary beds, are available to determine the movement, and such planar features do

FIGURE 15–19 Transcurrent, or strike-slip, fault.

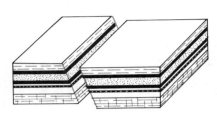

A. Normal fault with diagonal movement   B. Reverse fault with diagonal movement

FIGURE 15–20   Faults with oblique movements.

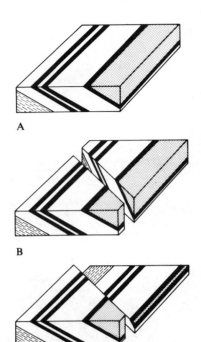

A

B

C

FIGURE 15–21   Map patterns in faulted rocks. Part A shows the original block. Part B shows the oblique-vertical movement on the fault. Part C shows the same area after erosion has leveled the uplifted block. Can the actual movement on the fault be determined from this map pattern alone?

not always permit unique solutions to the amount of actual movement. Also many fault planes are curved. The movement problem is further complicated on most faults by the deformation of the rocks near the fault. *On the fault plane the rocks are ground and milled by the movement, creating what is called* fault gouge. This is a type of metamorphism described in Chapter 5; on deep-seated faults some recrystallization may occur. The rocks near the fault are generally bent or dragged by the movement, making reconstruction of the prefaulting geometry very difficult.

Joints are cracks or breaks in rocks along which no movement has occurred. It is possible in some cases to show that the jointing was caused by the same forces that caused faulting or folding. Most rocks are jointed, and the origin of the joints is not always clear. Deep-seated rocks such as batholiths commonly have joints more or less parallel to the surface, suggesting that such joints form as the rock expands when the overlying rocks are eroded away. (See Fig. 4–4.)

INFERENCES FROM DEFORMED ROCKS

Some inferences about crustal forces can be made solely from study of deformed rocks. One of these is the importance of compression in production of

folded and thrust-faulted rocks. This conclusion seems clear, especially in areas where rigid metamorphic basement rocks have been folded or thrust along with the more plastic sedimentary rocks. At other areas, only the near-surface rocks have been compressed; the deeper rocks are not deformed. This is the case of some thrust faults where the upper plate is not strong enough to withstand the pressure necessary to push it. In such cases, perhaps uplift produces a slope down which the block slides under the influence of its own weight. Such a mechanism is mechanically possible on rather gentle slopes. Folding can also be caused by plastic beds slumping or sliding downslope. This can be seen, on a small scale, in many sedimentary beds. (See Figs. 15–23 and 15–24.) Another possible way to explain thrust faults is that the rocks may in some way be buoyed up or

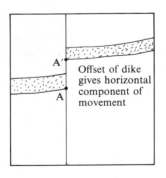

A and A′ locate the points of intersection of the near margin of the dike with the top of bed at the fault plane

Offset of dike gives horizontal component of movement

Displacement of horizontal bed gives vertical component of movement

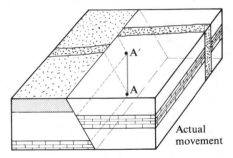

Actual movement

FIGURE 15–22   Determination of the actual movement on a fault. In most cases only one planar feature such as the displacement of the dike or of one of the sedimentary beds is available, and such features only reveal one component of the movement. In this figure the actual movement can be determined by finding the displacement of the intersection of the dike and one of the beds. Practically, in this case, the fault movement can be determined by finding the horizontal component by measuring the offset of the dike, and the vertical movement from the displacement of the horizontal sedimentary beds. This figure shows a highly idealized case. In most instances the rocks are more deformed, especially near the fault plane.

FIGURE 15–23 Gravity thrust fault. The block broke away and slid down the slope.

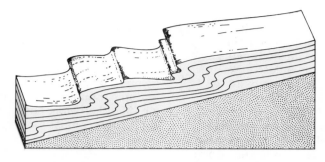

FIGURE 15–24 Gravity folding. Plastic sediments tend to flow downslope, producing folds.

floated, at least partly, by water in the rocks. This process seems possible, and the high water pressures encountered in some wells are also suggestive. If the upper plate were unweighted in this way, the friction force would be reduced and movement could occur with a very small force. (See Fig. 15–25.)

Vertical movements on the continents are easy to prove. Marine sedimentary rocks high on mountain ranges certainly prove such movements. Igneous and metamorphic rocks that formed deep in the earth and are now exposed in mountain ranges also show uplift. Other evidences of vertical movements are the uplifted, wave-cut beaches on many seacoasts, as discussed in Chapter 11. Areas below sea level, such as Death Valley, imply downward movement. The only agent that could have eroded Death Valley below sea level is the wind, and the wind is much too feeble an eroder to have done this.

Although most of our evidence of stress in the crust comes from study of ancient rocks, some movements can be seen directly. Recent earthquake faults show rapid movements, both horizontal and vertical. Slow creep measured by surveying also is currently occurring on some faults, as described in Chapter 13.

Dating folding and faulting is discussed in Chapter 16, but the principles can be seen in question 25 at the end of this chapter.

Other types of deformation are the igneous and metamorphic processes considered earlier. These processes are commonly part of the mountain-building processes that create folds and faults. Some of the relationships among these processes will be seen in the next section.

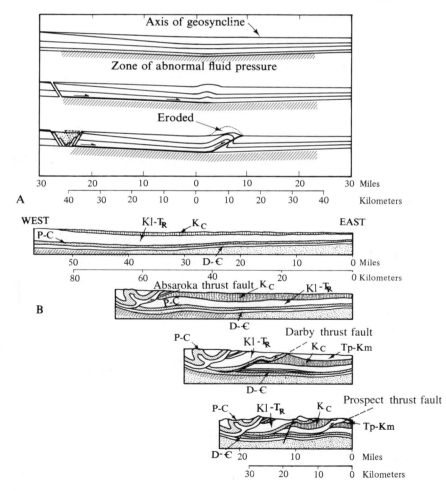

FIGURE 15–25 Possible role of fluid pressure in overthrust faulting. A. Idealized diagram showing zone of abnormal fluid pressure and incipient fault. B. Suggested example of a thrust belt in western Wyoming that formed as a result of abnormal fluid pressure. The letters designate various sedimentary rock units. From W. W. Rubey and M. K. Hubbert, Geological Society of America, *Bulletin*, Vol. 70, 1959.

LARGER FEATURES OF CONTINENTS

The most obvious features of the continents are relatively flat continental platforms and linear belts of mountains. (See endpapers.) The earliest idea on the origin of the continents was that they formed during the initial solidification of the earth. Study of the older parts of the continents disclosed that the oldest rocks are metamorphosed sedimentary rocks that are themselves the products of erosion of a still earlier terrane. Radioactive dates have revealed that, in a general way, the oldest rocks are in long belts. Because each of these belts has structures similar to parts of some of our present mountain ranges, they may be the eroded stumps ("roots") of older mountain ranges. (See Fig. 15–26.) Thus, mountains may be the key to understanding continental structure.

This discovery suggests that the present processes going on at the surface have been going on since the formation of the crust, and that the continents may have formed by accretion. The oldest rocks so far discovered are about 4000 million years old, which suggests a total age of the earth of 5000 million years. Thus, the process of continent formation may still be going on. We can look for evidence of this in the mountains where much of the present volcanic and earthquake activity occurs.

MOUNTAINS

A mountain is a topographic feature that rises above the level of the surrounding area, more or less abruptly. Various types of mountains are shown in Fig. 15–27. The definition and the illustration do not emphasize the most important geologic aspect of mountains, which is the deformation of the rocks that form the mountains. Most mountains occur in ranges of the complex type, and although single isolated mountains of the other types shown in Fig. 15–27 do occur, they are relatively rare.

Thus, typical mountains have complex internal structures formed by folding, faulting, volcanic activity, igneous intrusions,

and metamorphism. In a sense, complex mountains are residual, too. The origin of the internal mountain structure is different from the origin of the topographic expression of mountains. Topographic mountains are formed by uplift and erosion, and so are residual. *Thus most mountain*

ranges have complex internal structure formed at an earlier time, and the topographic expression of mountains is the result of later uplift and erosion—a two-step history. When the work of erosion is completed, only deformed rocks remain as evidence of an earlier mountain range.

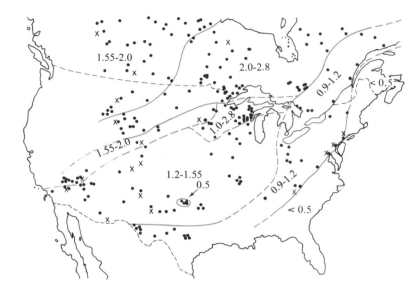

FIGURE 15–26　Distribution of ages in crystalline rocks from the central part of North America. Circles: ages that are within the limits specified for a given zone on the map; crosses: ages that are outside the limits. Age limits are given in thousands of millions (1,000,000,000) of years. After G. R. Tilton and S. R. Hart, *Science* (April 26, 1963), Vol. 140, p. 364.

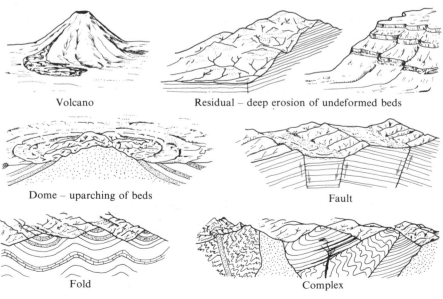

Volcano

Residual – deep erosion of undeformed beds

Dome – uparching of beds

Fault

Fold

Complex

FIGURE 15–27　Types of mountains.

Dating mountain building is difficult because three dates are involved:

The age of formation of the rocks,
The age of the internal structure, and
The date of uplift when erosion formed the topographic mountains.

Commonly the internal structure is formed in several stages, all of which should be dated. Determining each of these three dates in actual cases involves many difficulties.

GEOSYNCLINES—EARLY OBSERVATIONS

In 1859, James Hall, an American geologist who made important studies of fossils, made an important observation. He discovered that rocks in the northern part of the Appalachian Mountains were much thicker and more clastic than were rocks of the same age located further in the interior of the continent. In other words, the present mountains are on the site of a much deeper, earlier basin than is the area to the west. (See Fig. 15–28.)

This observation was extended by James Dwight Dana, who recognized that it was true of the whole Appalachian Mountains. In 1873 he applied the name *geosyncline to the great, elongate basin of deposition that was later folded, faulted, and uplifted to form a mountain range.* Similar observations have been made all over the world, namely, that many of the present mountain ranges were once elongate basins, or geosynclines.

These observations have been combined to form the geosyncline theory. The geosyncline begins as a broad, elongate downwarp, generally at the edge of a continent, that is flooded by seawater. Because the geosyncline acts as a trap for sediments, the sedimentation stage begins with this downwarp. In general, the part of the

geosyncline near the ocean is more active than the rest, is depressed deeper, and is the site of volcanic activity. These geosynclines are large, the order of many hundred miles long and up to several hundred miles across.

In spite of the great thicknesses of rocks in geosynclines, there is abundant evidence that the water was at no time very deep. The cross-bedding, ripple-marks, mud cracks, and types of fossil life displayed in these rocks are all features that form today in relatively shallow water. Thus, the reason for the great thicknesses of sedimentary rocks must be that the geosynclines continued to sink as they were filled with sediment. In some places the features in the volcanic, or at least more active, part of the geosyncline suggest that the water may have been

deep there at times. In addition to the evidence of deep-water fossils, such features as graded bedding and the channeling due to mud and muddy water sliding down slopes suggest some deep water. Such slides are known to occur today on the slope between the continental shelf and the deep ocean.

Differences in environment are shown by differences in sedimentation. As shown in Fig. 15–29, the geosyncline gradually tapers off to form the shelf that borders the continent. In most cases the continent has little relief; this is shown by the sediments (shale, limestone, and fairly pure quartz sandstone) that form on the shelf. Such sediments reveal rather complete weathering. This same suite of sedimentary rocks also develops nearby in the geosyncline, but here the rocks are much thicker

FIGURE 15–28 Sedimentary rocks of the same age are thicker and more clastic in mountain ranges than in the interior of continents.

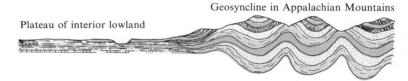

1. Initial sedimentation stage (above).

2. Orogeny accompanied by plutonic activity in the volcanic or more active part of the geosyncline (not shown).
Sedimentary rocks are folded, thrust faulted, and uplifted.

3. A new depositional basin is produced by downwarp of folded rocks in front of the orogenic area. Clastic fan in this foredeep is deposited in part during deformation (above).

4. Main orogenic phase accompanied by more plutonic activity.
5. Uplift and erosion.

FIGURE 15–29 Stages in development of a geosyncline.

because of the greater downsinking. The volcanic part of the geosyncline, on the other hand, is the scene of more rapid sedimentation as a result of deeper and more rapid downwarping. In addition to volcanic rocks, the sediments here are rapidly deposited, poorly sorted, quickly buried clastic rocks such as impure shales, dark sandstones with clay matrix (graywacke), and coarser clastic rocks.

The source of these "dumped-in" clastics is a problem. Some of the material may come from the continent, but the volume generally requires a source on the other side of the geosyncline. The problem is further complicated, as we shall see, by the later igneous and metamorphic activity, which obscures the early situation in this part of the geosyncline.

The *sedimentation phase* of a geosyncline ends generally after 9150 to 15,250 meters (30,000 to 50,000 feet) of sedimentary rocks have been deposited. The next phase is the *orogenic (mountain-making) phase* during which (1) the sediments are deformed to make mountain structure and (2) the volcanic part of the geosyncline may be changed by igneous and metamorphic activity into a rigid mass of granitic composition, thus becoming continentlike.

The orogenic phase of the geosyncline is probably the most interesting, most important, most studied, and until recently least understood phenomenon in geology. In this phase the geosyncline is folded, apparently by compressive forces. In the volcanic part of the geosyncline, the folding may be accompanied by metamorphism and emplacement of granitic rocks. In this way part of the geosyncline may be transformed into a rigid mass. It should be noted that granitic rocks and metamorphism are not always confined to only the volcanic part of the geosyncline, and, at places, parts of the volcanic geosyncline may not be affected at all. At a few

places, batholiths, generally small ones, occur in nongeosynclinal areas. At this stage a new depositional basin generally develops over the border between the old geosyncline and the shelf or continent, and a new sedimentation cycle may ensue. This may be followed by other episodes of folding, faulting, and thrust faulting in this or other parts of the geosyncline. Metamorphism and emplacement of batholiths may or may not accompany these orogenic events. These events may be spread over a few hundred million years. Geosynclinal folding and faulting occur, probably episodically, over long lengths of time, but emplacement of batholiths appears to be much more rapid. There is an unfortunate tendency to date mountain building from the easily obtained radioactive dates of batholith crystallization, even though many batholiths are clearly earlier or later than the deformation.

After the deformation the geosynclinal areas generally are uplifted to form mountain ranges. The geosynclinal deformation produces mountain structures, but topographic mountains do not exist until this uplift occurs. The uplift is probably caused by the development of a low-density granitic root by igneous and metamorphic processes. Then the lower-density rocks tend to float up under the influence of isostasy.

Once a mountain range exists, isostasy may control its future to a large extent. As soon as it rises, a mountain range is subjected to erosion. As erosion removes weight from the mountains, the buoyancy of the root probably causes the range to rise, as shown in Fig. 15-30. This process should go on until the root has risen to the level of the rest of the continent and the topographic mountains have been reduced by erosion to the level of the rest of the continent. In this manner a new

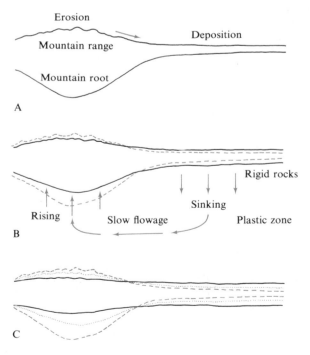

FIGURE 15-30 Removal of a mountain root by erosion and isostasy. Erosion removes material from the mountains, thereby lightening them. This can result in isostatic uplift that reduces the depth of the root. The place where the eroded material is deposited may also be depressed by the additional weight. Several stages in the process are shown in the diagram.

section of the continental platform may be formed. The rocks of this section would display mountain structures, and this process may produce an age distribution similar to that shown in Fig. 15–26.

The continents are made of bands of deformed rocks with the oldest bands generally toward the interior. These bands of deformed rocks appear to be the eroded stumps of mountain ranges. The younger mountain ranges formed from geosynclines, and it seems likely that the older ones also did. Therefore, geosynclines seem to be part of the process of continent formation. Geosynclines have many problems associated with them, such as the source of the volcanic and sedimentary rocks on the oceanic side. More important is the lack of a reason for their formation and later deformation. Plate tectonics answers many of these problems.

PLATE TECTONICS AND THE EVOLUTION OF CONTINENTS

One of the reasons that plate tectonics has been widely accepted by geologists is that most of the earlier observations are explained by this theory. Mountain building and origin of continental crust are problems in classical geosynclinal theory that are better understood in terms of plate tectonics. We will begin with a further consideration of geosynclines, this time using plate tectonics. Then we will explore the deformation, metamorphism, and intrusion that occur at convergent plate boundaries.

GEOSYNCLINES AND PLATE TECTONICS

Uniformitarianism is the cornerstone of modern geology. If geosynclines are important in the geologic past, we should be able to recognize and study present-day geosynclines. We should be able to see where sediments are accumulating, and where they are being deformed.

Geosynclines form at the margins of continents. Two types of continent-ocean border are common, and two contrasting types of geosynclines have been recognized. Some of the present geosynclines are at the sedimentation phase, and others are at various stages in the deformation or orogenic phase.

The simplest model of a geosyncline is the continental shelf-slope type of continent-ocean border (Fig. 15–31). This type of geosyncline can only form where the continent-ocean border occurs within a plate. Other geosynclines form at continent-ocean borders that are also margins of plates. At typical continental shelf-slope areas, sandstone, limestone, and some shale are deposited on the continental shelf; these rocks, with more shale, form the thick sedimentary accumulations of the continental slope. All of these rocks are deposited in shallow water. The sediments on the ocean side of the slope are mainly deposited in deep water by turbidity currents moving down the continental slope.

Geosynclines at plate boundaries are more complex because of the plate motion and volcanic activity that can result in a number of forms (Fig. 15–32). The configuration closest to the classical geosyncline is the offshore volcanic

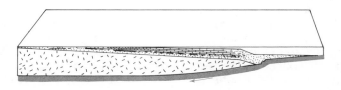

FIGURE 15–31 A geosyncline at a continental shelf and slope.

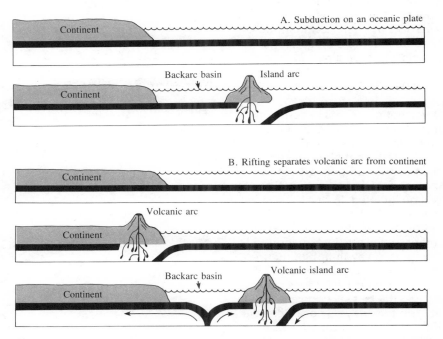

FIGURE 15–32 Origin of backarc basins. A. Subduction zone forms on oceanic plate. B. Subduction zone forms at continent-ocean boundary, and later rifting forms backarc basin.

313

island arc. This configuration can develop in at least two ways, and they and the other forms that subduction zones can take will be considered in the next sections. *The sea between the island arc and the continent is called a backarc basin.* On the continent side of this basin, typical continental shelf and slope sediments, such as sandstone, shale, and some limestone, will accumulate. Near the active volcanic island arc, volcanic rocks and sediments composed of volcanic fragments will accumulate rapidly. On the ocean side of the volcanic islands, similar sediments will be trapped in the trench. The ocean-floor sediments on the descending plate may be scraped off the plate and so also accumulate near the trench.

At geosynclines at active plate margins, the sedimentation phase is very difficult to separate from the deformation or orogenic phase. The separation into these phases is only conceptual because sedimentation and deformation are ongoing processes. In the following sections, our attention will be concentrated on deformation, now that we have established that the rocks at many continent-ocean borders resemble those of the classical geosyncline.

SUBDUCTION

Deformation occurs at converging plate boundaries. If one of the plates is composed of oceanic lithosphere, that plate will descend and a subduction zone with its associated magmatic activity will be initiated. In this section we will consider a variety of active subduction zones and then attempt to use this information to interpret the evolution of mountain ranges and continents.

The volcanic island arcs considered in the last section can form in at least two ways. The simplest is if the subduction zone forms initially well offshore (Fig. 15–32A). Why subduction would occur at such a location is not understood.

The Aleutian Islands appear to have formed in this way because the islands are younger than the floor of their backarc. This dating was done using the magnetic stripes in the Bering Sea.

The other way that some volcanic island arcs formed requires two steps. The subduction zone initially forms at the continent-ocean border where it appears most likely to form. Then a rift develops that separates the volcanic arc from the continent (Fig. 15–32B). Honshu Island and the Sea of Japan probably formed in this way because parts of Honshu are much older than the volcanic arc. Note that unlike the previous case, the crust under the Sea of Japan is younger than the volcanic arc.

Along the west coast of South America, a subduction zone has formed at a plate boundary separating oceanic and continental lithosphere. The subduction zone slopes down under the continent, and the volcanic arc has formed on the continent (Fig. 15–33). The volcanoes are mainly andesite; this rock type was named for the

Andes Mountains. The magmatic rocks, both volcanic and intrusive, that are associated with most subduction zones are generally of this composition. Here the sediments initially resemble those of a classical geosyncline, with continental shelf sediments at the continent and deeper-water sediments in the trench. Material from the volcanic arc and sediment scraped from the descending slab of lithospheric plate soon overwhelm the initial sediments (Fig. 15–33B). Many earthquakes, including some very deep and some of high magnitude, are associated with this South American subduction zone.

Not all subduction zones are as active as the one just discussed. Those with backarc basins caused by currently active spreading centers in the backarc basin behave differently. The Marianas volcanic island arc is an example (Fig. 15–34). There the subduction is much steeper, the sediments are not scraped off the descending slab, and no high-magnitude earthquakes occur. The apparent rea-

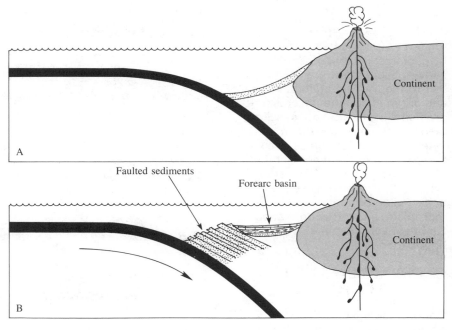

FIGURE 15–33 Volcanic arcs at continent-ocean boundary. A. Volcanic arc forms on continent edge. B. As subduction proceeds, sediments are scraped off the oceanic plate, forming a ridge and forearc basin.

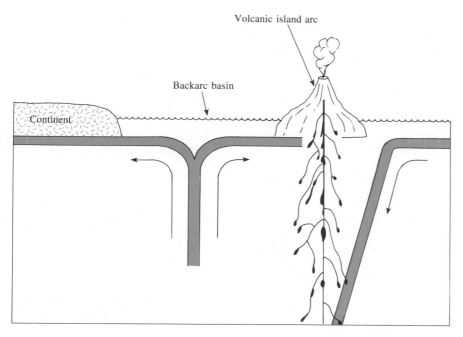

FIGURE 15–34   Subduction is much steeper where the backarc basin is caused by an active spreading center, such as at the Marianas Islands.

son for these differences is that the compression is less at the Marianas Islands than at the west coast of South America. The spreading center in the Marianas backarc is apparently moving with the western plate away from the subduction zone and so reduces the compression. Further studies may modify these conclusions, but for whatever reasons, all subduction zones do not behave in the same way.

At some subduction zones, the sediments scraped off the oceanic slab as it descends may accumulate and form a submarine ridge or an island (Fig. 15–33B). The sedimentary rocks are broken into thin slices in thrust contact with each other, or they may be a heterogeneous mixture of blocks and slabs enclosed in a matrix of rock that was soft sediment at the time of formation. The term *mélange* has been applied to such mixed rocks. At some subduction zones (Indonesia is an example) islands formed in this way create a basin between the trench and the volcanic arc. Such basins are termed *forearc basins.* Where exposed on

continents, the rocks in the ridges formed by this tectonic accumulation are difficult to study because of their broken nature. The Franciscan rocks of the California coast, which have puzzled geolo-

gists for decades, probably formed in such an environment.

Igneous and metamorphic activity also occurs near subduction zones. The descending slab is partially melted and is the source of the magma. The rising magma and probably the active hot fluids derived from moist sediments on the melting plate cause metamorphism of the rocks above the subduction zone. In this way rocks of continental composition are formed. If the subduction zone exists for a long enough time, the volume of the rising magmatic material becomes large and sedimentary rocks further inland are deformed by folding and faulting, especially thrust faulting. The rising magma may form an arch, causing thrust faulting by gravity gliding even farther inland (Fig. 15–35). The deformation phase ends when, for some reason, subduction ceases.

COLLISION

The deformation at a convergent plate boundary can be carried

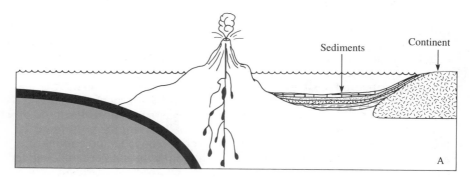

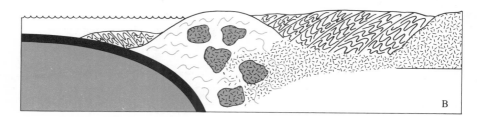

FIGURE 15–35   Subduction may cause deformation of geosynclinal rocks far inland from the volcanic arc.

315

even further. If, as in the last description, subduction occurs at the edge of a continent, the events described above will occur. If the plate being subducted is narrow and carries a continent on it, that continent will collide with the subduction zone (Fig. 15–36A).

Continent-continent collisions cause more intense compressional deformation than does subduction. The collision of Africa and Europe produced the thrust structures in the Alps and in North Africa. The collision of India with Asia produced the Himalaya Mountains.

Other types of plate collisions are possible, and they produce subduction zones. Crustal deformation occurs where two rigid plates, moving in opposite directions, meet. If the collision is between two plates of oceanic crust, one will move downward, forming a subduction zone (Fig. 15–36B). If one of the plates is continental crust, it will not move down because of its buoyancy; the oceanic plate will move downward, forming a subduction zone (Fig. 15–36C). If two continental plates meet, a high mountain range, such as the Himalaya Mountains, will form because of the compressive thickening of the crust. The heat and friction of such a collision could also cause metamorphism and generation of magma. In this way new continental crust could be created.

Where two continents collide they are welded together, forming a new, larger continent. In such cases, the youngest part of the new, larger continent will be the area of collision. We saw before that generally the oldest parts of continents are in their interiors. Collisions, then, can account for

many places where younger rocks are in continental interiors, surrounded by older rocks. Rifting, or the breaking-up of a continent by the formation of a new spreading center, can also change the arrangement of deformed belts.

So far we have seen that the magmatic and metamorphic activities that accompany subduction and continental collision are the processes by which continents are formed. Continent-continent collisions and rifting of continents have changed the continents throughout geologic time. Collisions between North America and both Africa and Europe have caused orogeny that produced parts of the present Appalachian Mountains. After these collisions, rifting reopened the Atlantic Ocean; but at some places, the separation did not occur at the old shoreline. As a result, some areas in New England and Newfoundland appear to have once been parts of other continents. This conclusion is based on rock types and fossils. At other places on both coasts of North America and elsewhere, parts of the present continents also appear to be foreign, but their sources are problems.

Mountains and other orogenic features are associated with some of these foreign terranes, suggesting that collisions between the terranes and the larger continent may have occurred. The source of the small continents or terranes is a problem because no continent-continent collisions are believed to have occurred along most of the west coast of North America. A possibility is that small continent-like areas in the oceans may have collided with North America. In the Pacific and the other oceans,

there are areas with underlying layers that appear more like continental than oceanic rocks. The origin of these areas is not known; they are recognized by their shallower depth and the seismic velocities of the rocks.

EXAMPLES OF CONTINENTAL STRUCTURE

We will now look at the deformed rocks in a few areas of North America and attempt to relate their origin to the principles just discussed. The structural provinces considered are shown in Fig. 15–37, which should be compared with the landforms map (see endpapers).

The *Appalachian Mountains* have a long and varied history. The eastern side is dominated by igneous and metamorphic rocks, and the west is mainly sedimentary rocks. In the southern Appalachian Mountains, metamorphic rocks at low elevations form the Piedmont, and similar rocks at higher elevations form the Blue Ridge. In this area the sedimentary rocks are mainly thrust faulted in the Ridge and Valley area. (See Fig. 15–38.) Farther north, in Pennsylvania, the sedimentary rocks are mainly folded. (See Fig. 15–39.) West of the Ridge and Valley area is the Appalachian Plateau where the sedimentary rocks are only mildly deformed. New England has mainly igneous and metamorphic rocks, and the folded and thrust belt is narrow.

The *Atlantic Coastal Plain* is a thin cover of undeformed sedimentary rocks that presumably

A. Continent-continent plate collision

B. Ocean-ocean plate collision

C. Continent-ocean plate collision

FIGURE 15–36   Types of plate collisions. A. Continent-continent. B. Ocean-ocean. C. Continent-ocean.

FIGURE 15–37 Geologic provinces of the United States, based on geologic structure and surface topography. The boundaries of such provinces are necessarily arbitrary. Cross-sections showing geologic structures are located on the map.

FIGURE 15–38 Cross-section through the southern Appalachian Mountains. The Piedmont and Blue Ridge areas contain mainly metamorphic rocks, and the Valley and Ridge Belt is largely thrust-faulted sedimentary rocks. From P. B. King, *Guides to Southeastern Geology,* Geological Society of America, 1955.

overlies eroded metamorphic and batholithic rocks similar to those of the Piedmont.

The northern part of the Appalachian Mountains was deformed by collision, probably with Africa and Europe at about 430 million years ago and again about 340 million years ago. (See Fig. 15–40.) In the southern Appalachians in this same time span, two collisions also occurred, and there the metamorphic rocks of the Piedmont were thrust over sedimentary

rocks on the continental shelf (Fig. 15–41). The origin of this thrust plate was apparently a small continent in the ocean between North America and Africa. The present Appalachian Mountain structures were created around 300 million years ago when Africa, Europe, and North America collided to form one supercontinent. The structure of the southern Appalachian Mountains is known from seismic studies, using widely spaced instruments to obtain in-

formation about deeply buried rocks.

The structures in the rocks from the Pacific Coast through the Rocky Mountains are varied, but their deformation can be related to plate tectonics. We will describe the structures and then attempt a brief synthesis.

The *Pacific Coast Province* is a very complex, geologically young area. It is geologically active with much seismic activity and with the only active volcanoes in contermi-

317

Appalachian Plateau | Ridge and Valley | Piedmont | Coastal Plain

A

FIGURE 15–39   A. Cross-section showing the structures in the Appalachian Mountains in Pennsylvania. Here the sedimentary rocks are folded rather than thrust as in Fig. 15–38. After Douglas Johnson. B. The Appalachian Mountains in Pennsylvania from 918 kilometers (570 miles) high. The plunging folds of the ridge and valley area are in the upper left. The Susquehanna River flows from upper left to lower right. ERTS photo.

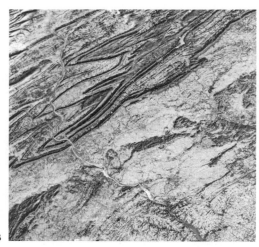

B

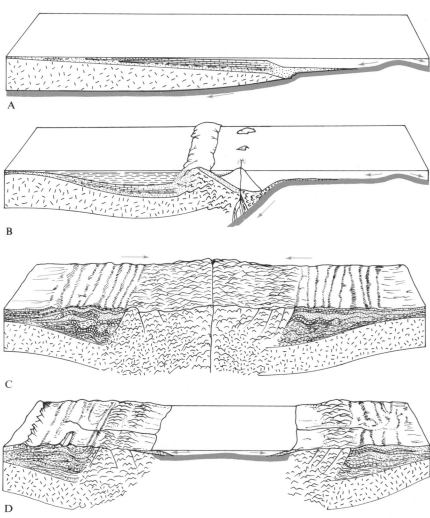

A

B

C

D

FIGURE 15–40   Deformation of the Appalachian geosyncline to form the Appalachian Mountains.

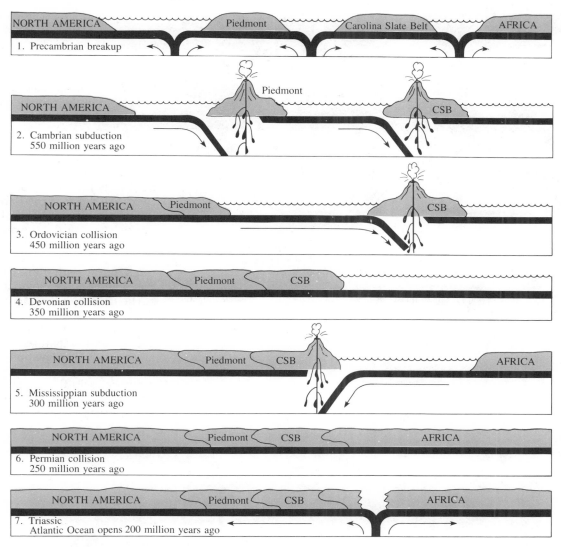

FIGURE 15-41 The development of the southern Appalachian Mountains. After F. A. Cook, et al.

nous United States. The Coast Ranges, consisting of complexly folded and faulted rocks, form the western part of this province. The San Andreas fault (Fig. 15–42) passes through the Coast Ranges. It is probably a transform fault that separates the North American plate from the Pacific plate. The central part of the province is a lowland except in northern and southern California. The eastern part of the province is, in California, the Sierra Nevada mountains, which are related to the Basin and Range Province. North of the Sierra Nevada is the Cascade Range. This range is composed of

volcanic and continental sedimentary rocks capped by very young, high volcanoes. The volcanic rocks of the Cascade Mountains have more in common with the Northwest Volcanic Province to the east than with the Pacific Coast Province, but other structural and sedimentary features are related to the rest of the province.

The *Basin and Range Province* is an area of many relatively small block-faulted mountain ranges. (See Figs. 15–43 and 15–44.) The rocks that make up these ranges are greatly deformed by folding, faulting, and igneous and metamorphic processes. Thrust faults

with many miles of displacement can be traced from range to range. The farthest west of these ranges is the Sierra Nevada in California that is the batholithic part of the geosyncline in this area.

The northern *Rocky Mountains* are folded and thrust faulted like the western part of the Appalachians (see Figs. 15–45 and 15–46), and the many thrust faults indicate considerable shortening. The western part of the northern Rocky Mountains is the Idaho batholith and the batholiths of northeastern Washington. The rest of the Rocky Mountains is different. In Wyoming the mountains

319

A

B

FIGURE 15–42 San Andreas fault. A. Young sedimentary rocks deformed at the San Andreas fault near Palmdale, California. B. The San Andreas fault cuts across the Carrizo Plains in southern California. Photos from U.S. Geological Survey.

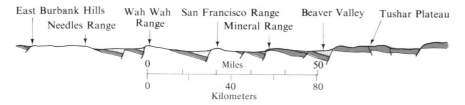

FIGURE 15–43 Diagrammatic cross-section of part of the Basin and Range Province. From J. H. Mackin, *American Journal of Science,* Vol. 258, 1960.

are mainly uplifts with some faulting on the margins, and there are large sedimentary basins between the mountain ranges. These structures are something like those of the Basin and Range Province but differ in that the ranges and the basins are much larger in Wyoming. The age of faulting, and the rock types and the internal structures exposed in the mountains, are also very different. In the central Rocky Mountains ancient rocks like those of the Canadian Shield are exposed. In Colorado, unlike Wyoming, there are more ranges than basins. (See Fig. 15–47.) In New Mexico these structures merge with those of the Basin and Range Province.

The *Colorado Plateau* may be the cause of the narrowing of the Rocky Mountains. This is an unusual structural feature in that it is an area at high elevation with a fairly thick accumulation of sedimentary rocks that is relatively undeformed. Why this province is undeformed is unknown because

FIGURE 15–44 Oblique aerial view of a range in the westernmost part of the Basin and Range Province in southern California. Erosion has removed much of the range front, and recent movement on the fault has revealed its location far in front of the range. Note the difference in the dissection on both sides of the fault scarp. Compare with Fig. 9–4. Photo from U.S. Geological Survey.

the rocks surrounding it are deformed. The sedimentary rocks are flat-lying and are broken by a few nearly vertical faults and monoclines. (See Fig. 15–48.)

The structural development of western North America begins with accumulation of sediment in a continental-shelf type geosyncline. This geosyncline was well to the east of the present west coast.

A subduction zone developed near the present west coast. The sediments scraped from the descending slab formed a ridge, and the forearc basin between this ridge and the continent accumulated sediments eroded from the magmatic arc. Continuing subduction further deformed the ridge and the forearc basin. The ridge became part of the Coast Ranges,

and forearc basin sediments form the present Great Valley of California. Continued development of the magmatic-volcanic arc created a rise and caused deformation by folding and faulting, especially thrust faulting of the rocks of the old geosyncline. The basin and range faulting occurred near the end of the deformation. Deformation ceased when the subduction zone became inactive. As one might imagine, this history is generalized, and many episodes of deformation are omitted in this brief summary.

The *Northwest Volcanic Province* consists of volcanic rocks of several ages. (See Fig. 15–49.) The Columbia River Basalt covers southeastern Washington and adjacent Oregon with several thousand feet of basalt flows. Much of Oregon and the Snake River Valley in Idaho are covered by younger volcanic rocks. In this area are exposed many recent volcanic features not yet old enough to be much affected by weathering. The volcanic rocks extend into the northwest corner of the Basin and Range Province and are here considered part of that province because they have been deformed by basin and range faulting. The Blue Mountains in Oregon are composed of much older rocks that give our only indication of the structures of the geosynclinal rocks in this region.

The *Continental Interior* is the stable part of the continent between the geosynclinal mountains. For the most part this province is low lying and is covered by a thin veneer of sedimentary rocks. To the northeast the sedimentary rocks are thin and the ancient metamorphic rocks of the Canadian Shield are exposed. To the west the elevation increases on the Great Plains. The Continental Interior is, for the most part, only slightly deformed and contains a number of gentle basins and domes. (See Fig. 15–50.) Faulting is minor, folds are very gentle, and igneous rocks are few

FIGURE 15–45 Cross-section through the northern Rocky Mountains near the International Boundary. The sedimentary rocks here are thrust faulted, and the total movement may be many miles. Two interpretations of the structure are shown. A assumes that the thrusts join, and the deformation is mainly in the sedimentary rocks. Recent seismic data support this interpretation. B assumes that the basement rocks are involved in the deformation.

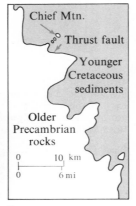

FIGURE 15–46 A. Chief Mountain, Glacier National Park, Montana. Chief Mountain is an erosional remnant of a thrust plate of older rocks thrust over much younger rocks. B. Map showing thrust fault near Chief Mountain. The minimum displacement on the thrust is the distance between the farthest east and the farthest west that the thrust fault is exposed, a distance of about 19 kilometers (12 miles) (not shown here). The movement was slightly north of east as shown by gouges on the fault plane. Part A from National Park Service. Part B from C. P. Ross and others, *Geologic Map of Montana,* U.S. Geological Survey.

321

FIGURE 15–47 Looking north at the Rocky Mountain front south of Denver, Colorado. The ranges on the left are composed of metamorphic and igneous rocks. In the foreground and middleground the eroded remnants of sedimentary rocks are prominent. Thus the basic structure shown is one flank of a large anticline. In the distance on the right is Table Mountain, formed by a resistant lava flow. Photo by T. S. Lovering, U.S. Geological Survey.

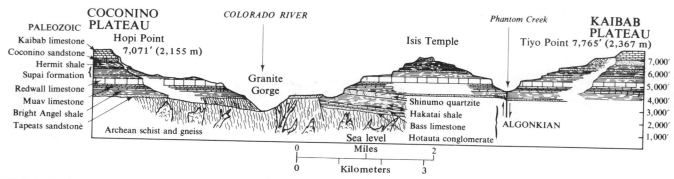

FIGURE 15–48 Cross-section of Grand Canyon from Hopi Point, on the south rim, to Tiyo Point, on the north rim. From U.S. Geological Survey.

and far between. As the mountains in either direction are approached, the sedimentary rocks become thicker, folds and faults can be easily seen in the outcrops, and intrusive and volcanic rocks become more common.

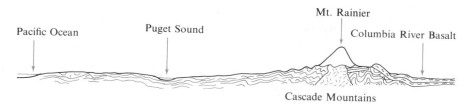

FIGURE 15–49 Cross-section through central Washington. The Cascade Mountains are composed of volcanic and continental sedimentary rocks that grade into marine rocks to the west. Small batholiths invade the Cascade Mountain rocks, and the large volcanoes such as Mount Rainier were built on top of the existing range. The Columbia River Basalt covers the area to the east.

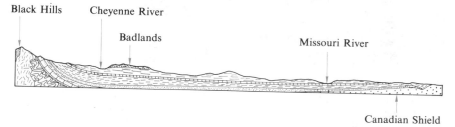

FIGURE 15–50 Cross-section across the Great Plains from the Black Hills of South Dakota to the Canadian Shield in eastern South Dakota. After David White and others, U.S. Geological Survey Bulletin 691, 1919.

SUMMARY

Isostasy is the name given to the theory that the crust of the earth floats in a viscous layer.

The difference in elevation between ocean floors and continents results from each floating at a different height because of differences in composition and thickness.

The continents are made of granite, and the oceans are floored by a thinner and denser layer of basalt.

Mountains are higher than their surroundings because they have deep roots.

The *low-velocity zone,* about 100 kilometers (60 miles) down, is believed to be the level at which floating occurs.

Folded rocks generally record compressive forces.

Synclines are basinlike folds and are best characterized by having younger rocks in their cores. See Fig. 15-8.

Anticlines are hill-like folds and are best characterized by having older rocks in their cores. See Fig. 15-8.

Reverse faults are caused by compression. See Fig. 15-12.

Normal faults are caused by tension or uplift. See Fig. 15-12.

Thrust faults are nearly flat reverse faults.

Lateral faults are nearly vertical and have horizontal movements.

Some folding and thrust faulting may be the result of downslope movement.

The oldest rocks so far found are 3980 million years old.

Radioactive dates reveal that, in a general way, rocks of similar age occur in bands, and the oldest rocks are near the centers of continents.

Mountains are of several types, and, in general, the age of the deformation of the rocks is earlier than the time of the uplift that creates the mountains.

Geosynclines begin as elongate basins of deposition, and they are subsequently deformed by folding, faulting, intrusion of batholiths, and metamorphism. Later uplift creates mountains.

The simplest model of a geosyncline is a continental shelf and slope.

Subduction zones at active plate boundaries may also be geosynclines. The sea between a volcanic island arc and the continent is called a *backarc basin.*

At subduction zones, the geosynclinal rocks are deformed, and metamorphism and intrusion occur.

Collisions between plates can cause deformation. If two oceanic plates collide, one of them will descend, forming a volcanic arc. If an oceanic and a continental plate collide, the oceanic plate will descend, forming a volcanic arc. Subduction is not caused by two continental plates colliding.

Continents may have formed by the intrusion of granitic batholiths and metamorphism in the subduction process.

The *Appalachian Mountains* are composed of igneous and metamorphic rocks to the east and folded and thrust faulted sedimentary rocks to the west.

The *Atlantic Coastal Plain* is a thin cover of sedimentary rocks that overlies igneous and metamorphic rocks of the Appalachian Mountains.

The *Pacific Coast Province* is a complex area with the Coast Ranges on the west, large valleys further east, and the Sierra Nevada and Cascade Mountains forming the eastern portions.

The *Basin and Range Province* is an area of relatively small, block-faulted mountain ranges.

The northern *Rocky Mountains* are similar to the Appalachians, with folded and thrust sedimentary rocks, and igneous and metamorphic rocks.

In Wyoming, the Rocky Mountains are ranges with large sedimentary basins separating the ranges.

The *Colorado Plateau* is an area of slightly deformed sedimentary rocks surrounded by deformed rocks.

The *Northwest Volcanic Province* is composed of young volcanic rocks. The Columbia River Basalt covers southeastern Washington and adjacent Oregon with several thousand feet of basalt flows.

In the *Continental Interior,* a thin veneer of nearly undeformed sedimentary rocks overlies ancient igneous and metamorphic rocks that are exposed on the Canadian Shield to the north.

QUESTIONS

1 The continents are composed of _____.
2 The oceans are underlain by _____.
3 What is the evidence for mountain roots?
4 How is the crust of the earth supported?
5 Why do the continents stand higher than the ocean floors?
6 Is it the strength of rocks that enables mountains to stand above the general level of the continents? Prove it.
7 Describe the similarities between island arcs and geosynclines.
8 What are the similarities and differences between geosynclines and continental slopes?

9   What are the ages of the basement rocks forming North America and how are they distributed?
10 Why have geologists concentrated much of their attention on mountain ranges?
11 What steps are involved in the formation of mountains?
12 How are mountains dated?
13 Outline briefly the history of a geosyncline.
14 Discuss the origin of continents.
15 What type of stress produces folding?
16 What type of stress produces normal faults?
17 What type of stress produces reverse faults?
18 What is an anticline? Discuss several ways one can be distinguished from a syncline.
19 What is a thrust fault?
20 What evidence would you look for in the field to tell whether a crack in the rocks is a fault or a joint?
21 What determines whether a rock folds or faults if it is subjected to a compression? List all factors.
22 How could you tell in the field whether a fault is still active?
23 How does one determine the amount of movement on a fault?
24 Prove to your friend, who has taken no science courses, that crustal movements have occurred.
25 The following diagram is a cross-section through an area. Starting with the earliest event, list the sequence of events that have occurred in this area. You may wish to review your answers after reading Chapter 16.

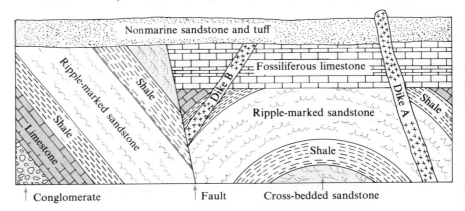

SUPPLEMENTARY READINGS

Bailey, E. B., *Tectonic Essays Mainly Alpine*. Oxford: Clarendon Press, 1935, 200 pp. Reissued by University Microfilms, Ann Arbor, Michigan, 1967.

Ben-Avraham, Zvi, "The Movement of Continents," *American Scientist* (May-June 1981), Vol. 69, No. 3, pp. 291–299.

Cook, F. A., L. D. Brown, and J. E. Oliver, "The Southern Appalachians and the Growth of Continents," *Scientific American* (October 1980), Vol. 243, No. 4, pp. 156–188.

Dietz, R. S., "Geosynclines, Mountains and Continent-Building," *Scientific American* (March 1972), Vol. 226, No. 3, pp. 30–38.

Dietz, R. S., and J. C. Holden, "The Breakup of Pangaea," *Scientific American* (October 1970), Vol. 223, No. 4, pp. 30–41.

Engel, A. E. J., "Geologic Evolution of North America," *Science* (April 12, 1963), Vol. 140, No. 3563, pp. 143–152.

James, D. E., "The Evolution of the Andes," *Scientific American* (August 1973), Vol. 229, No. 2, pp. 60–69.

Jordan, T. H., "The Deep Structure of the Continents," *Scientific American* (January 1979), Vol. 240, No. 1, pp. 92–107.

Kay, Marshall, "The Origin of Continents," *Scientific American* (September 1955), Vol. 193, No. 3, pp. 62–66. Reprint 816, W. H. Freeman and Co., San Francisco.

King, P. B., *The Evolution of North America,* revised edition. Princeton, N.J.: Princeton University Press, 1977, 197 pp.

Molnar, Peter, and Paul Tapponnier, "The Collision between India and Eurasia," *Scientific American* (April 1977), Vol. 236, No. 4, pp. 30–41.

Moorbath, Stephen, "The Oldest Rocks and the Growth of Continents," *Scientific American* (March 1977), Vol. 236, No. 3, pp. 92–104.

Oliver, Jack, "Exploring the Basement of the North American Continent," *American Scientist* (November-December 1980), Vol. 68, No. 6, pp. 676–683.

Windley, B. F., *The Evolving Continents.* New York: John Wiley, 1977, 385 pp.

The result, therefore, of our present enquiry is, that we find no vestige of a beginning,—no prospect of an end.

James Hutton, 1788

Historical geology, the history of the earth, is perhaps the most important goal of geology. In this chapter the methods used to determine geologic history will be described. Some of these methods have already been discussed, such as interpretation of the formation of a rock based on study of the rock itself and its relationship to other rocks. Simple examples are the interpretation of a sandstone composed of well-rounded quartz grains as probably the product of deep weathering and long transport, and recognition that an intrusive rock is younger than the rock it intrudes. Geologic history is determined from interpretation of the way rocks were formed and the dating of these events. This chapter is concerned mainly with methods of dating so that rocks and events can be placed on a time scale.

RELATIVE AND ABSOLUTE DATES

Rocks can be dated both relatively and absolutely. *Relative dates* are of two types. The simpler of these is used to determine the sequence of events in a limited area. This method uses superposition and cross-cutting relationships. Beds are dated by *superposition* from the knowledge that in undisturbed sedimentary rocks, the oldest beds are at the bottom of the sequence and the youngest at the top (Fig. 16–1A). *Cross-cutting features,* such as faults and discordant intrusions like dikes and batholiths, are younger than the rocks they cut (Fig. 16–1B). This type of relative date is determined from the relationships among the rocks, and these relationships are determined by geologic mapping. Geologic mapping is the main method of geologic study, and consists of plotting the locations of rock units on an accurate map. *The second*

way of relative dating is by referring them to a geologic time scale based on the study of the fossils in the rocks. *Absolute dating* is a more recent development and *is based on radioactive decay of elements in rocks;* this method has been used to calibrate the relative geologic time scale developed from the study of fossils.

GEOLOGIC TIME

The recognition of the immensity of geologic time and the development of methods for subdividing geologic time are among the great intellectual accomplishments of humans. It is difficult to discuss the development of ideas in generalities, but a brief outline is presented here. Discoveries made near the end of the eighteenth century led to rapid progress, although there had been earlier studies. The culmination of this early work was the theory now known as *uniformitarianism,* published by the Englishman James Hutton in 1785 and 1795. Uniformitarianism simply means that the present is the key to the past; that is, the history of old rocks can be interpreted by noting how similar rocks are being formed today. Hutton's book was poorly written, and the theory did not gain wide acceptance until it was popularized by his colleague Playfair in 1802. Uniformitarianism is a simple idea, but it is the keystone of modern geology. A simple example will show the power of this theory. Many varieties of shell-bearing animals can be seen in the mud along the seashore at low tide. A similar mudstone with similar shells encountered in a canyon wall is interpreted, using uniformitarianism, as a former sea bottom that has been lithified and uplifted. If this seems too elementary, we need only remember that just a

few years before Hutton's time, only a few people recognized that fossil shells were evidence of once-living organisms.

Recently uniformitarianism, as just presented, has been criticized as being oversimplified. The objection is that geological processes have modified the earth; therefore, at least in some cases, these processes do not operate now in the same ways that they did in the past. Thus, the present may not be the key to the past. For example, it will be shown that after the origin of life, organisms changed the composition of the atmosphere, making further creation of life impossible. Thus, though uniformitarianism in its simplified version can explain many geologic phenomena, the doctrine has limitations that must be taken into account.

Geologic history is based largely on the study of sedimentary rocks and their contained fossils. Such studies have led to the development of a geologic time scale showing the relative ages of rocks. In recent years it has been possible, using measurements of the radioactive decay of certain elements in rocks, to determine the absolute ages of some rocks. This ability to date igneous and metamorphic rocks has expanded the range of historical geology.

After several false starts, historical geology developed in the last part of the eighteenth century and the first half of the nineteenth. In 1782 the great French chemist Antoine Lavoisier demonstrated that

Unconformity between conglomeratic sandstone and overlying flat limestone. Antigonish-Mabou Basin, Nova Scotia. Photo from Geological Survey of Canada.

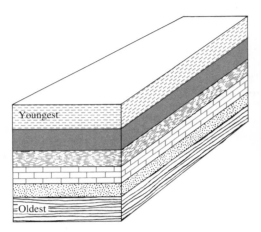
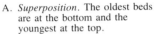
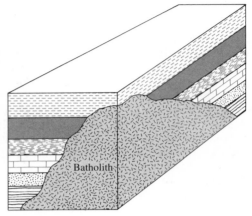

A. *Superposition.* The oldest beds are at the bottom and the youngest at the top.

B. *Cross-cutting.* The batholith is younger than the beds that it intrudes.

FIGURE 16–1 Relative dating.

near Paris the quarries dug for pottery and porcelain clay all exposed the same sequence of sedimentary rocks. Georges Cuvier and Alexandre Brongniart published maps in 1810 and 1822 showing the distribution of the various rock types around Paris, and in 1815 William Smith published a geologic map of England. Cuvier and Brongniart studied the fossils in each of the sedimentary layers that they mapped and discovered, as did Smith, that each layer contained a different group of fossils. They discovered that a sedimentary bed could be identifed by its fossils. That work set the scene for the next 20 years, for during this period the geologic time scale was developed.

The geologic time scale, or column, was the result of individual work by a number of people in western Europe. These men studied the rocks and their fossils at well-exposed places, generally near their homes, and described these rocks and their fossils in books and papers. They called these sequences of rocks *systems* and named each of the systems. As the studies proceeded in this haphazard manner, the general sequence of older to younger was recognized and the gaps were filled in. It was also shown that rocks in similar stratigraphic posi-

tion, although far removed from the type area, contained the same or very similar fossils. Thus it took work by many people spread over a wide area to demonstrate that fossils can be used to date rocks, not only in local areas, but all over Europe. In this way a *geologic column,* or *time scale,* was developed as a standard of reference; fossils from other areas could be correlated or dated by comparison with fossils on which the time scale was based. The geologic time scale in current use is shown in Fig. 16–2. Almost all of the systems or periods were originally defined in the first half of the nineteenth century.

The units used in historical geology are of several types. The *systems just described are the rocks deposited during a time interval*

called a period. Thus the Cambrian System was deposited during the Cambrian Period. The systems and periods are arbitrarily defined, and their recognition away from the type area where they were defined is based on interpretation of fossils. *A geologist mapping an area locates on the map the occurrence of easily recognized, objective units called* ***formations.*** Formations are rock units named for a location, a type area, where they are well exposed, such as the Austin Chalk and the Tensleep Sandstone. All parts of a formation are not necessarily the same age. The sandstone bed in part B of Fig. 16–9, for example, might be defined as a formation. Thus there are three kinds of units: time units such as periods, time-rock units such as

TABLE 16–1 Stratigraphic units.

A. Time and Time-Rock Units

Time Units	Time-Rock Units
Eon	—
Era	Erathem
Period	System
Epoch	Series
Age	Stage
(Phase)	Chronozone

B. Rock Units

Group
Formation
Member

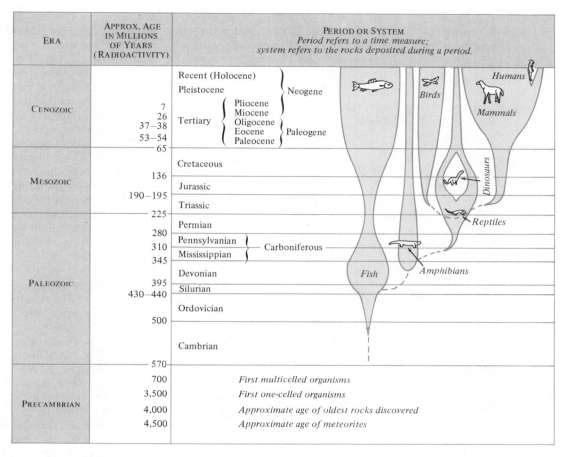

ERA	APPROX. AGE IN MILLIONS OF YEARS (RADIOACTIVITY)	PERIOD OR SYSTEM *Period refers to a time measure;* *system refers to the rocks deposited during a period.*
CENOZOIC	7 26 37–38 53–54	Recent (Holocene) Pleistocene } Neogene Tertiary { Pliocene / Miocene / Oligocene / Eocene / Paleocene } Paleogene
	65	
MESOZOIC	136	Cretaceous
	190–195	Jurassic
	225	Triassic
PALEOZOIC	280	Permian
	310	Pennsylvanian } Carboniferous
	345	Mississippian }
	395	Devonian
	430–440	Silurian
	500	Ordovician
	570	Cambrian
PRECAMBRIAN	700	*First multicelled organisms*
	3,500	*First one-celled organisms*
	4,000	*Approximate age of oldest rocks discovered*
	4,500	*Approximate age of meteorites*

FIGURE 16–2 The geologic time scale. Shown to the right is a very simplified diagram showing the development of life. Not included on the diagram are many types of invertebrate fossils such as clams, brachiopods, corals, sponges, snails, and so forth, which first appeared in the Cambrian or Ordovician and have continued to the present.

systems, and rock units such as formations. Table 16–1 summarizes the terms used.

The changes in marine invertebrates during geologic history are obvious to even a casual observer; for this reason such fossils can be used to date the enclosing rocks. (See Fig. 16–3). The changes between periods are not as great but are just as obvious to a trained observer. The most marked changes between adjacent periods are between Cambrian and Ordovician, Permian and Triassic, and Cretaceous and Cenozoic. Marked changes also occur at the end of the Ordovician, Devonian, and Triassic.

The geologic column or time scale provided only a relative time scale to which rocks could be referred. It also showed the general development of life and so set the stage for Darwin, who published his theory of evolution in 1859.

Although the idea of evolution was not new to Darwin, his book started a controversy that is still continuing in nonscientific circles. Until his time no controversy existed between geology and religion because even the geologists interpreted geologic history in biblical terms. Biblical scholars thought that the earth was about 6000 years old, based on their interpretation of the Old Testament. However, in the hundred years before Darwin, some geologists began to recognize that a much longer time was necessary for the deposition of the sedimentary rocks and other geologic events than students of the Bible would allow. Before evolution was accepted, the changes in fossil life throughout the geologic column were explained by a catastrophic dying out of all life and a new creation of the animals found as fossils in the succeeding bed. This idea had its roots in the biblical Noachian Flood and the widely held theory of Werner, a late eighteenth century German, who believed that all surface rocks, including igneous rocks, were precipitated in a universal sea. Darwin had a pronounced influence on the development of geologic thought, but his theory also needed the long duration of geologic time envisioned by the uniformitarianists.

Thus the central point became the age of the earth. The geologists and the evolutionists needed a very old earth. Joly in 1899 calculated the age at 90 million years by dividing the total amount of salt in the ocean by the amount of salt annually brought into the ocean by the rivers. Estimates of the age of the earth based on the time necessary to deposit the rocks of the geologic time scale were made by many geologists in the latter part of the nineteenth century; and although they varied between 3 and 1584 million years, about 100 million years was an average figure. The lack of precision and the subjective nature of the assumptions made these estimates suspect. Lord Kelvin, one of the most influential physicists of the day, entered the fray and attempted to calculate the age of the earth from thermodynamics. He assumed that the earth began as a melted body and has been cooling ever since. He calculated that 20 to 40 million years had passed from the time when the earth had cooled enough for life to exist to the present. About the turn of the century, when Kelvin was an old man but still refining his calculations, radioactivity was discovered. As noted earlier, radioactivity is a source of heat within the earth, and so Kelvin's heat-flow calculations were based on an incorrect assumption.

Radioactivity was discovered by Becquerel in 1896. In 1902 Rutherford and Soddy showed that by radioactive decay an element is changed in part to another element. This sounded much like the quest of the alchemists, who tried to change lead into gold, and was not immediately accepted by many scientists. However, in 1905 Boltwood noted the association of lead and uranium in uranium minerals and that the ratio of lead to uranium was higher in the older minerals. Boltwood used this information to calculate the age of these minerals, and one of them had a computed age of over 2000 million years. This greatly extended the possible age of the earth and was quickly accepted by a few geologists. Present methods of radioactive dating show the earth to be about 5000 million years old. These methods will be described later in this chapter. Thus, in a single stroke, radioactivity provided the energy source not taken into account by Kelvin and a means to find the age of the earth.

A

B

FIGURE 16–3   Brief comparison of these two fossiliferous slabs of rock show the great changes in life that have occurred during geologic time. A. Pennsylvanian fossils from Kansas. Bryozoa and brachiopods are most abundant. Photo from Ward's Natural Science Establishment, Inc., Rochester, N. Y. B. Miocene fossils from Virginia. Snails and clams are most abundant. Photo by W. T. Lee, U.S. Geological Survey.

UNCONFORMITIES

Once the time scale was set up, geologists traced the systems into previously unmapped areas. One problem that soon appeared was that the systems were originally defined, in many cases, as the rocks between two important breaks in the geologic record. In other areas there were no breaks, but continuous deposition. Thus, between the original systems were intervals not included in either system; and arguments about the assignment of the neglected intervals are, in some cases, still in progress. The breaks are of several types, such as change from marine to continental deposition, influx of volcanic rocks, and unconformities. (See Fig. 16–4.) The latter is the most important and requires discussion.

All breaks in the geologic record that indicate a time interval for which there is no local record are called unconformities (see Figs. 16–5 and 16–6); they may take several different forms.

If sedimentary rocks overlie metamorphic or granitic rocks, the sediments were deposited after the time of metamorphism or intrusion. Granitic and metamorphic rocks form deep in the earth, so the erosion that uncovered them required time. In this case, there are no sedimentary rocks representing the time of intrusion or metamorphism and the time required for erosion. This type of unconformity is called a nonconformity. *Nonconformities occur where sedimentary rocks are deposited on igneous or metamorphic rocks that formed at some depth in the crust.* (See Fig. 16–5C.)

Another type of unconformity called a *disconformity* may consist of a change in fossils, representing a short or a long *period of time, that occurs between two parallel beds in the sedimentary section.* Such an unconformity may be due to erosion of previously deposited beds or to nondeposition, and may not be at all conspicuous until fossils are studied.

The third type of unconformity is more obvious and consists of

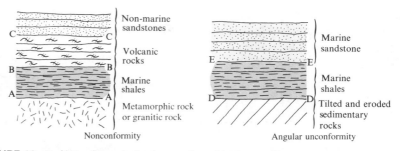

FIGURE 16–4 Natural breaks in the stratigraphic record. A-A and D-D are unconformities because the time necessary for erosion is not represented by sedimentary rocks. The other breaks may or may not be unconformities, depending on whether there is a time interval between the different rock types or continuous deposition. D-D is an angular unconformity.

folded and eroded sedimentary rocks that are overlain by more sedimentary rocks. Such an unconformity is called an *angular unconformity because the bedding of the two sequences of sedimentary rocks is not parallel.* In this case, the time of folding and erosion is not represented by sedimentary rocks.

CORRELATION AND DATING OF ROCKS

So far we have considered how the geologic systems were defined, and it is obvious that they are recognized in previously unstudied areas by comparing the contained fossils with the fossils from originally defined type areas. In this manner the systems defined in Europe were recognized in North America and the other continents. This type of correlation by fossils is also most important in establishing the time equivalence of nearby beds.

Another type of *correlation* is to establish that two or more outcrops are parts of a once continuous rock body. The simplest case of such rock correlation is actually to trace the beds from one area to the other by walking out the beds. In other cases, distinctive materials in a bed, such as fragments of an unusual rock type or a distinctive or uncommon mineral, or assemblage of minerals, may be used to identify a bed. Another method uses the sequence of beds. (See Fig. 16–7.)

By correlating from one area to another, the total thickness of sedimentary rocks in a region can be determined as shown in Fig. 16–8.

Note that the methods of correlation, with the exception of the use of fossils, establish only the continuity of a sedimentary bed. Fossils establish the fact that both beds are the same age. The difference between these two types of correlation can be seen by considering an expanding sea, a not uncommon occurrence in geologic history. As the sea advances, the beach sands, for example, form a continuous bed of progressively younger age. This is shown diagrammatically in Fig. 16–9.

This diagram also illustrates another difficulty in the use of fossils. Note that limestone, shale, and sandstone are all being deposited at the same time in different parts of the basin. Each of these different environments will attract and be the home of different types of organisms. Thus, the fossils found in each of these environments will be different, even though they are of the same age. This means that environment must be taken into account in establishing the value of a fossil or group of fossils in age determination. For this reason, free-swimming animals make the best fossils because their remains are found in all environments. *Fossils that establish clearly the age of the enclosing rocks are very useful and have been called index fossils.* Ideally, an index fossil should be widespread in all environments, be abundant, and have a short time span. Very few fossils meet all of these requirements, and generally, groups of fossils are used to establish the age of a bed. (See Figs. 16–10 and 16–11.) *Time-lines,* or surfaces, can be established in sedimentary rocks by physical means alone in some rare instances. For example, a thin ash bed from a volcano may be spread over a very large area by

A

B

FIGURE 16–5 Unconformities. A. An angular unconformity in the Grand Canyon. The rocks beneath the unconformity are Precambrian in age and those just above it are Cambrian. Photo by L. F. Noble, U.S. Geological Survey. B. An erosional unconformity in Silurian rocks in Niagara Gorge, N. Y. Photo by G. K. Gilbert, U.S. Geological Survey.

C

FIGURE 16-5 Unconformites (cont.). C. Cambrian sedimentary rocks overlying Precambrian metamorphic and igneous rocks. Note that the bedded sediments were deposited on a hilly terrain. Photo was taken during construction of Taum Sauk project, about 161 kilometers (100 miles) south of St. Louis, Missouri. Photo from Union Electric.

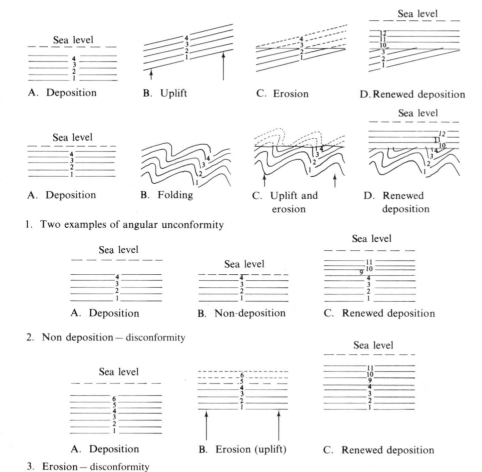

FIGURE 16-6 Development of unconformities. In each case a time interval is not recorded in the bedded rocks. It is commonly difficult to distinguish between cases 2 and 3. In each case, for simplicity, sea level is assumed to be the base level that controls whether erosion or deposition takes place.

wind. The deposition of such an ash bed may occur over a number of days or even weeks, but in terms of geologic time it is instantaneous. These rare occurrences help in local correlation problems and show the validity of the methods of dating and correlation described here.

The only rocks that have reasonably abundant fossils are marine sedimentary rocks, and not even all of these have enough fossils to establish their age clearly. On the previous pages, we have discussed dating of fossiliferous marine rocks. Similar techniques can be used to date continental sediments, but here the number of fossils is generally much smaller. The fact is easy to understand if one compares the number of easily fossilized animals exposed at low tide at the seashore with the much smaller number of easily fossilized animals living in a similar-sized area of forest or grassland, and if one considers how easy it is for a clam shell to be buried and so preserved, especially when compared to the chance that a plant leaf or a rabbit skeleton will be buried and so preserved as a fossil. A land organism, even if buried, may not be preserved, as oxidizing conditions that destroy organic material exist even at some depth. Also, a dead

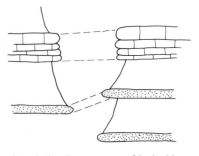

Correlation by sequence of beds. Note the changes in thickness in the two areas.

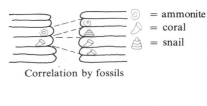

= ammonite
= coral
= snail

Correlation by fossils

xxxx = distinctive volcanic ash

Correlation by lithologic similarity. Also shows use of a keybed (an easily recognized bed)

FIGURE 16–7 Types of correlation.

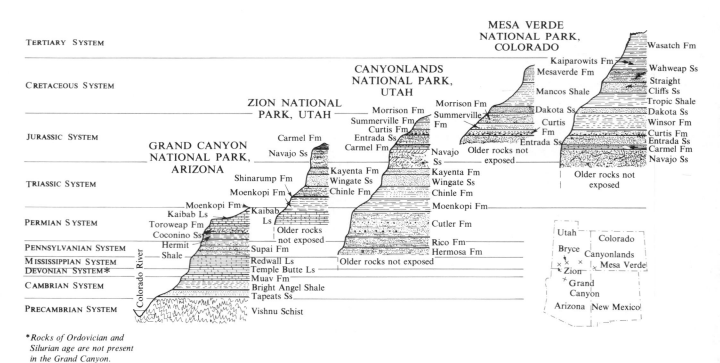

*Rocks of Ordovician and Silurian age are not present in the Grand Canyon.

FIGURE 16–8 Correlation of the strata at several places on the Colorado plateau reveals the total extent of the sedimentary rocks. After U.S. Geological Survey.

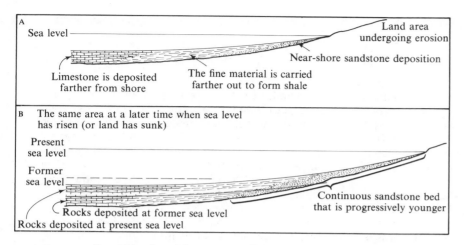

FIGURE 16–9 Deposition in a rising or expanding sea. A. The deposition at the start. B. Deposition at a later time after sea level has risen.

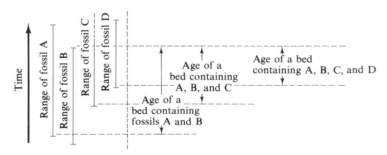

FIGURE 16–10 The use of overlapping ranges of fossils to date rocks more precisely than can be accomplished by the use of a single fossil.

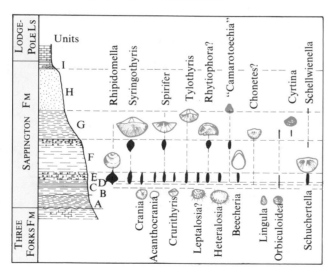

FIGURE 16–11 An example of the distribution of brachiopods in the Sappington Formation in western Montana. The relative number of each fossil found in the beds is indicated by the thickness of the black lines. The occurrence of these animals was at least in part controlled by the sedimentary environment. From R. C. Gutschick and Joaquin Rodriquez, American Association of Petroleum Geologists, *Bulletin,* Vol. 51, 1967.

rabbit is apt to be eaten and dismembered by scavengers. Another problem in the use of continental fossils is that they are controlled even more by climate than are marine organisms. The progression of types of plant and animal life seen during a mountain climb will illustrate this point.

Dating igneous and metamorphic rocks presents even more problems because such rocks almost never contain fossils. They must be dated by their relationships to fossil-bearing rocks. Volcanic rocks are dated by the fossiliferous rocks above and below and, in some cases, by interbedded sedimentary rocks. Intrusives must be younger than the rocks that they intrude; they cannot be dated closer unless the intrusive has been uncovered by erosion and, then, unconformably covered by younger rocks. (See Fig. 16–12.)

Metamorphic rocks are even more difficult to date because both the age of the original rock and the date of the metamorphism should be obtained. The age of the parent rock may be found by tracing the metamorphic rock to a place where fossils are preserved. Finding the date of metamorphism may be more difficult because, in general, an unconformable cover, if present, gives only the upper limit.

RADIOACTIVE DATING

Some of these problems in dating igneous and metamorphic rocks can be overcome by modern radioactive methods. Such methods produce an *absolute age* in years and so have been used, in addition, to date the geologic time scale in years. (See Fig. 16–13.) In spite of the seeming precision of radioactive dating, it is not yet possible to date a rock as accurately this way as with fossils. This is because of inherent inaccuracies in the measurements of the

DATING GHOST TOWNS AND OLD MINING CAMPS

The dating of old abandoned settlements is an example of how geologic dating works. Instead of fossils, the refuse from the settlement, especially the bottles and tin cans, is used. Establishing dates in this way is common practice for archaeologists, who refer to dumps as *middens*.

Searching for old bottles and jars, as well as other artifacts, has become a widespread hobby as the popularity of metal detectors attests. Bottles and tin cans are like index fossils, except that each type was manufactured over a long time span. The use of an assemblage of several artifacts can, like a group of fossils, establish a smaller time zone.

The figure shows the approximate ranges of each of the styles of bottles and cans, as well as other things. Like fossils, these artifacts must have been preserved to be useful, and tin cans are less apt to be preserved in humid climates. Not included in the figure is the ubiquitous Coca-Cola bottle, with its distinctive shape, that will surely be an index fossil of our times. The aluminum beer can, with its changing type of opener, could also be added to date more recent dumps. Most of this information came from "Dating of Mining Camps with Tin Cans and Bottles," by Charles B. Hunt, *Geotimes,* May-June, 1959.

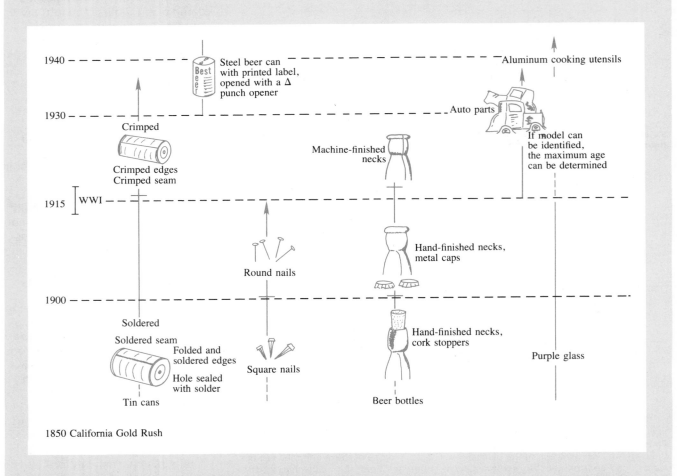

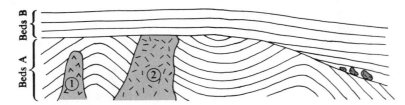

FIGURE 16–12 Dating of intrusive rocks. Intrusive body ① is younger than beds A; no upper limit can be determined. Intrusive ② is younger than beds A and older than beds B. Closer dating may be possible if the conglomerate composed of pebbles of ② in the basin at the right contains fossils. The contact between beds A and B is an angular unconformity.

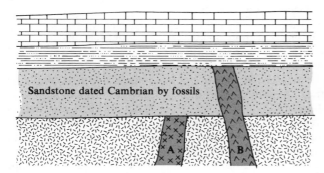

FIGURE 16–13 Determination of absolute age of one of the geologic periods. The sandstone contains Cambrian fossils. Dikes A and B have been dated by radioactive methods. The Cambrian Period is younger than dike A and older than dike B. Can the absolute age of the Cambrian be determined from a single occurrence such as this?

amounts of the elements produced by radioactive decay. These inaccuracies produce an uncertainty in the radioactive date that is more than the span of zones based on fossils, especially in those parts of the geologic column that contain abundant short-lived fossils. As methods of analysis are improved, this limitation may be overcome and the present high cost of dating may also be reduced.

Radioactive dates are obtained by studying the daughter elements produced by a radioactive element in a mineral. The principles of radioactivity were covered in Chapter 2. Sometimes more than one element is studied in an effort to provide a check on accuracy. A radioactive date tells the time of formation of that mineral and so tells the time of crystallization of an igneous rock, or the time of metamorphic recrystalliza-

tion in the case of a metamorphic rock. In some favorable cases, in which microscopic studies show that different minerals formed at different times in a rock that has undergone more than one period of metamorphism, it is possible to date the periods of metamorphism by dating elements in minerals formed during each period of metamorphism. In many radioactive decays a gas is one of the products, and heating during metamorphism may drive off the gas, thus resetting the radioactive clock to give the time of metamorphism.

A radioactive element changes to another element by spontaneously emitting energy. The rate of this decay is unaffected by temperature or pressure; hence, if we know the rate of formation of the daughter products, all we need to find is the ratio of original element to daughter product to be able to

calculate the time of crystallization of the mineral containing the original radioactive element. The main assumptions, then, in using radiodates are that the decay rate is known and constant, and that no daughter or parent elements are lost. The latter assumption is the least sound because of weathering and metamorphism.

Several radioactive elements are used to date rocks, but only the most important are reviewed here. The methods in current use are tabulated in Appendix F (see also Chapter 2). A few minerals contain uranium, and these generally also contain thorium, which is about the same atomic size. These minerals are rare and with a few exceptions are almost completely confined to pegmatite veins in batholiths. Because such veins are probably the last to crystallize, these occurrences give only the upper limit of the age of the batholith. Each of the three radioactive isotopes that these minerals contain—the two of uranium as well as thorium—produces a series of radioactive daughter products; each member of each series decays to the next daughter until, finally, a stable isotope of lead is produced. (See Fig. E–1, Appendix E.) These many steps constitute a possible source of inaccuracy, as it is possible for any one of the daughters to be removed by leaching or some other process. The elements and the final products are:

Uranium 238 → Lead 206
Uranium 235 → Lead 207
Thorium 232 → Lead 208

Thus, in this case, it is possible to measure three ratios which should all agree in age.

A problem with age determinations involving lead is that many rocks contain some lead of nonradioactive origin. The amount of nonradiogenic lead is quite small but may require an appreciable correction in calculating the age of very old rocks. This ordinary lead contains the isotope lead 204, so its presence can be easily recognized. The problem is that along with the minor amounts of lead 204 are the three isotopes, lead 206, lead 207, and lead 208, which also occur radiogenically, and the ratio of all four isotopes varies. Thus to correct for the pre-existing ordinary lead, one must analyze nonradioactive rocks in the area to find the ratio between lead 204 and each of the other isotopes. In this way the nonradiogenic lead 206, lead 207, and lead 208 can be excluded from that due to radioactive decay. Meteorites contain lead 204, and the correction for the original lead is a problem. As discussed later, meteorites are used to date the formation of the earth, so this problem results in some uncertainty in this important date.

Rubidium is a radioactive element that is generally present in small amounts in any mineral that contains potassium, because both elements are about the same size. Rubidium decays in a single step to strontium (rubidium 87 → strontium 87). Because potassium minerals that contain rubidium are so common (feldspar and mica) and only a single radioactive decay is involved, this process may become the most useful method of radioactive dating. However, at present the half-life of this decay cannot be measured accurately because the low energy of the beta particles emitted makes them difficult to detect. Some studies have suggested that the rubidium and strontium isotopes in some rocks may be inherited from their parent materials, and the rubidium-strontium ages determined for such rocks would tend to be too old.

Potassium also has a radioactive isotope that is useful in dating, especially since it is such a common element. The small amount of radioactive potassium decays to form two daughters:

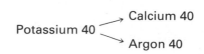

Calcium 40 is ordinary calcium and so is present in most minerals. Therefore, the method used is to measure the amounts of potassium 40 and argon 40. The main problem is finding minerals that retain the argon gas. It appears that micas, in spite of their cleavage, retain most of the argon but feldspar loses about one-fourth of the argon. Potassium 40 is the most commonly used method of radioactive dating.

Another type of radioactive dating is to measure the damage done to a crystal lattice. The crystal structure of minerals containing alpha-emitting elements can be damaged or destroyed by the alpha particles, especially if the mineral has weak bonding. In other cases alpha radiation produces discoloration, and this can cause gray (smoky) quartz, violet fluorite, and blue or yellow halite. Beta particles can also cause discoloration in some minerals. Another common example of alpha damage is the halos or rings that commonly surround tiny crystals of radioactive minerals. These small crystals generally occur as inclusions in later-formed minerals, and dark halos are common in biotite. Uranium and thorium minerals emit alpha particles with various amounts of energy, and because the alpha particles do most of their damage near the end of their movement, traveling a distance determined by the energy of each particle, they form a series of concentric halos. (See Fig. 16–14.) The amount of damage done to a mineral (the thickness and darkness of the halo) can be used to calculate the age of the mineral, but so far such dating methods have met with only limited suc-

FIGURE 16–14 Radiation halo. Such halos are commonly seen in biotite crystals where they surround tiny crystals of radioactive minerals such as uraninite. Typical size is less than one-tenth of a millimeter.

cess. Uranium 238 also decays by spontaneous fission at a very much slower rate than it does by alpha decay. The high-energy fission particles released in this way do much damage to the enclosing crystal. Under high magnification with an electron microscope, the paths of such particles can be seen in some minerals after etching with acid. The path is a small tube, not unlike a bullet hole. (See Fig. 16–15.) By counting the tubes, the number of fission disintegrations that have occurred can be determined. The number of remaining uranium atoms, the source of the particles, can be determined by instrumental analysis, and thus the age of the mineral can be calculated. This method is useful for dating material that is a

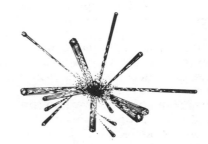

FIGURE 16–15 Fossil fission tracks. The tubes are caused by high-energy radioactive particles escaping from the mineral at the center. The tubes are much like bullet holes. The longer tubes are about one one-hundredth of a millimeter in length.

few tens of years old to the oldest objects in the universe.

Radiation can increase the energy of an atom by moving its electrons into different orbital shells and by trapping within the atom other electrons from beta particles or from other atoms. Remember that each electron belongs in a certain shell. This energy can be released by freeing the electrons so that they can escape or move back to the original shell. One way to do this is to excite the atom by heating it. The energy is released in the form of light. This phenomenon can be seen easily by heating a mineral such as fluorite in a frying pan in a dark room. The complete explanation of this phenomenon would require much deeper knowledge of atomic structure. Because the amount of energy stored depends on the length of time a mineral has been exposed to radiation, the age of some minerals can be determined by measuring the amount of stored energy. So far this has not proved to be an accurate method of dating minerals.

NEUTRONS—COSMIC RAYS

To complete our understanding of radioactive dating, it is necessary to consider another particle—the neutron. The main source of natural neutrons is cosmic rays, and only a few of the possible types of neutron reactions will be considered here.

Primary cosmic rays are nuclei (atoms that have lost their electrons) moving at nearly the speed of light. They are mainly hydrogen and helium, but most other elements are also present in small quantities. That they come from outer space is shown by their increase with altitude. The number does not change appreciably from night to day, so they do not come from the sun. The atoms are be-

lieved to come from exploding stars (supernovae), and are accelerated to very high energy levels by the associated magnetic fields.

The primary cosmic rays collide with the atoms of gas that form the earth's atmosphere, producing secondary cosmic rays. These secondary cosmic rays are particles knocked out of the atoms; they consist of neutrons, protons, and other particles. Neutrons outside a nucleus are not stable, but decay, with a half-life of 12.8 minutes, to a proton and a beta particle. Each primary cosmic ray can produce several secondary cosmic rays, and each of the secondaries can produce more cosmic rays, and so on. Each successive generation has less energy; however, primary cosmic rays and certain of the early-formed particles have great penetrating power. Neutrons have great penetrating power because they have no charge, and thus are not deflected by protons or electrons.

Of the cosmic rays, only the neutron is important geologically. A neutron colliding with a nitrogen atom knocks a proton out, converting the nitrogen to carbon 14:

$$^{14}_{7}N + {}^{1}n \rightarrow {}^{14}_{6}C + {}^{1}_{1}H$$

Carbon 14 is radioactive and decays by beta emission to nitrogen 14:

$$^{14}_{6}C \rightarrow {}^{14}_{7}N + \beta^{-}$$

These reactions are utilized in the carbon-14 method of dating that is effective in dating carbonaceous material, such as wood or coal, that is less than 40,000 to 50,000 years old. This method depends upon the fact that all living matter contains a fixed ratio of ordinary carbon 12 and radioactive carbon 14. Cosmic rays react with nitrogen in the atmosphere to form the carbon 14, which ultimately gets into all living matter via the food cycle. The carbon 14 decays back

to nitrogen. Therefore, the ratio of carbon 12 to carbon 14 can be used to calculate how long an organism has been dead, because a dead organism takes in no more carbon 14. Extensive use of this method is made in archaeology and in glacial studies.

Cosmic rays also make possible the dating of meteorites. When cosmic rays hit iron atoms, helium 3 is produced. Thus the amount of helium 3 reveals how long the iron meteorite has undergone cosmic-ray bombardment. This effect was first noticed when attempts were made to date meteorites by measuring the amount of helium produced by alpha decay. More helium than expected was found, and analysis of the helium revealed both helium 4 from alpha activity and helium 3 from cosmic-ray activity. It was also discovered that the helium 3 concentration was greatest near the surface of a meteorite and became less dense in the interior, as would be expected from cosmic-ray activity. Similar effects were noted with other isotopes formed by cosmic-ray bombardment. Thus ordinary radioactive dating can tell the time of crystallization of a meteorite, and cosmic-ray dating tells when the body broke up to form a meteorite.

Ages of rocks can also be determined by chemical means. At present these methods yield mainly relative ages, but new advances may result in absolute ages. An obvious example is the relative dating of glacial moraines on the basis of weathering. Rates of weathering are dependent on so many factors that absolute ages are not possible in this way and great care must be exercised in application. Another example is the crystallization of volcanic glass. Obsidian, or any other glass, is not stable but tends to crystallize; however, the rate of crystallization depends on the composition and amount of water in the glass, as well as on other factors. A promising chemical method of dating is the breakdown of amino acids in fossils. The breakdown is dependent on many factors, such as temperature and composition of formation water, but gives fairly accurate results in young rocks.

AGE OF THE EARTH

The age of the earth itself is a geological problem to which radioactive dating has been applied. So far, the oldest rocks found are about 3800 million years old and are found at Isua, West Greenland. This, then, is the youngest that the earth can possibly be, and is probably about the time the crust formed; the earth itself is probably older. Meteorites were discussed earlier and are thought to be the material from which the solar system was formed. Radioactive dates from meteorites give ages of around 4550 million years. This suggests that the earth may have formed 4500 to 5000 million years ago.

The first abundant fossils occur in rocks of the Cambrian system although a few fossils have been found in much older rocks. The base of the Cambrian has been dated radioactively as about 600 million years. Why abundant, relatively advanced animals appeared suddenly just prior to this time is a problem, but these fossils may record only the development of easily preserved hard parts such as shells. (See Fig. 16–2.) Because fossils are so much easier to use than radioactive methods in dating and correlating rocks, we know many more details of geologic history from Cambrian time onward than from Precambrian time. We know only the broadest outlines of the history of the Precambrian, even though it includes almost nine-tenths of the earth's past.

INTERPRETING GEOLOGIC HISTORY

Deciphering the history of the earth or any part of it is a difficult reiterative process. It has many steps, and each affects every other step. Multiple working hypotheses must be used to test each step. The process begins with study of the rocks themselves and their relationships to other rocks. This leads to some ideas about their mode of origin. The rocks must be dated and correlated, using the methods described in this chapter. If the rocks have been deformed by folding and faulting, their original relationships and positions must be determined. Each layer or rock unit can now be reconstructed to show the geography at its time of formation. Remember that the rock record is incomplete because of erosion and because many rock layers are exposed only at a few places. The final step is to reconstruct the events that caused the changes between each of the steps.

This brief outline shows how the incomplete rock record is interpreted. Note the various levels of abstraction that are involved. It is no wonder that the history of the earth is not completely known. Interpreting the history of the earth or a small area is difficult, and revision is constantly necessary as more data are obtained. Thus geology is an ongoing process that offers great challenge; many fundamental problems remain to be solved.

INTERPRETING ANCIENT ENVIRONMENTS

The environment of the geologic past can be determined at least partially from study of rocks. Such features as coral reefs, evaporites, desert deposits, and glacial deposits are a few obvious examples.

Some of the features that indicate environment were described in earlier chapters. Fig. 16–9A shows in a generalized way how the rock type changes from sandstone near shore to shale and limestone farther from shore. Fig. 4–43 is an example of a reconstruction of an ancient sedimentary basin from the study of scattered outcrops. Many of the features used in such studies are also described in Chapter 4. The grains that compose a sedimentary rock, such as a sandstone, also reveal much environmental information. The composition of the grains in a sandstone is influenced by both the nature of the source rocks and the amount of weathering. The topography and the climate greatly influence the weathering. The size and shape of the grains are influenced by the topography, and the distance and method of transportation. Obviously, only some of the factors involved in interpretation of ancient environments have been mentioned here, and a little thought will reveal many more aspects.

Isotopes can be used to determine ancient temperatures as well as age. Oxygen isotopes are used in temperature determination. Oxygen has three isotopes: oxygen 16, oxygen 17, and oxygen 18. When water evaporates, the lightest isotope—common oxygen 16—evaporates at a slightly higher rate than the heavier isotopes. This difference in evaporation rates is proportional to temperature. The ratio of oxygen 16 to oxygen 18 can be measured and used to determine the temperature. In practice, this ratio is measured in fossil shells made of calcium carbonate ($CaCO_3$). There are many pitfalls in this method, but so far the results have agreed fairly well with other geological evidence.

Other means of estimating temperatures are coral reefs, evaporites, desert deposits, and glacial deposits. Fossils, both plant and animal, have also been used very successfully to reconstruct climates and ecology.

SUMMARY

Rocks can be dated both relatively and absolutely.

Relative dates can be determined by the relationships of the rock units to each other using *superposition* and *cross-cutting* relationships, *and* by referring the rocks to a geologic time scale based on the fossils in the rocks.

Absolute dating is based on radioactive decay of elements in the rocks and has been used to calibrate the relative geologic time scale based on fossils.

Uniformitarianism is the basis of geological interpretation and means that ancient rocks are interpreted as having formed in the same ways that similar rocks are presently forming, or "the present is the key to the past."

Systems are the rocks deposited during a time interval called a *period.*

Formations are rock units that can be mapped easily. The other stratigraphic units are summarized in Table 16–1.

The *geologic time scale* is shown in Fig. 16–2.

The most marked changes in life occurred at the end of the Cambrian, Permian and Cretaceous Systems.

The development of life through geologic time was important evidence for Darwin's evolution theory and the recognition of the immensity of geologic time.

Unconformities are breaks in the geologic record that indicate a time interval for which there is no local depositional record. If the older rocks were deformed so that their bedding planes are not parallel to the younger rocks, an *angular unconformity* results. If the bedding planes are parallel above and below the unconformity, a *disconformity* exists. Where sedimentary beds overlie granitic or metamorphic rocks, a *nonconformity* exists.

Correlation and *dating* of rocks can be accomplished in many ways. Comparison of fossils can establish the time equivalence of beds. Rock correlation can establish that two outcrops are parts of a single rock body. This can be accomplished if the bed has distinctive properties that can be recognized, or from a similar sequence of beds, or by actually tracing the beds on the ground.

A lithic body, such as a sandstone unit, is not necessarily the same age throughout.

Index fossils should be found in all environments, be abundant, and have a short time span. Very few fossils meet these requirements, so most dating uses groups of fossils. Fossils are most abundant in marine sedimentary rocks.

Igneous rocks are dated by their relationships to sedimentary rocks or by radioactive methods. An intrusive rock is younger than the rocks that it intrudes.

In dating metamorphic rocks, the age of the parent rock and the age of metamorphism should be determined. They are dated by tracing them to places where they are not metamorphosed, by their relationships to sedimentary rocks, and by radioactive methods.

A radioactive element spontaneously changes into another element at a rate that is unaffected by temperature and pressure.

The more important methods of radioactive dating are:

Uranium 238 → Lead 206
Uranium 235 → Lead 207
Thorium 232 → Lead 208
Rubidium 87 → Strontium 87
Potassium 40 → Calcium 40 / Argon 40
Carbon 14

Cosmic rays are nuclei (atoms that have lost their electrons), mainly hydrogen and helium, moving at nearly the speed of light. They come from outer space. They collide with atoms in the atmosphere and knock out neutrons, protons, and other particles.

Neutrons from cosmic rays colliding with nitrogen in the atmosphere produce carbon 14.

Meteorites are about 4550 million years old, suggesting that the earth formed about 4500 to 5000 million years ago.

The oldest rocks so far found on the earth are about 3800 million years old, suggesting that the crust formed about this time.

The first abundant fossils are found in Cambrian rocks which are about 600 million years old.

Geologic history is interpreted by determining how the rocks were formed and when, how they have been deformed and when, and how and when they have been eroded. Multiple hypotheses must be used to test each step in the reconstruction of events.

Ancient environments are interpreted from the rocks themselves, the fossils, and the ratio of oxygen 16 to oxygen 18.

QUESTIONS

1 How does one interpret the depositional environment of a layer of ancient rocks?
2 Describe in general terms how the geologic column was established and discuss whether it consists of natural subdivisions in North America.
3 What are unconformities? Describe the various types.
4 How was the value of fossils in determining geologic age established?
5 What is meant by correlation?
6 List and briefly describe the methods of correlating rocks.
7 Why is it generally easier to date marine rocks than continental sediments?
8 How are metamorphic rocks dated? What dates are required?
9 Which method of radioactive dating is best and why?
10 What are index fossils?
11 What effect would a small amount of modern root material mixed with a sample to be dated by the carbon 14 method have?
12 How can an angular unconformity be distinguished from a low angle thrust fault?
13 Fossils aid in determining the age of a rock unit. How else do they aid in geologic interpretation?
14 Prove to your friend, who is an art major and has not taken any science, that the earth is very old.
15 Distinguish between relative and absolute dating in geology.

SUPPLEMENTARY READINGS

Cohee, G. F., M. F. Glaesner, and H. D. Hedberg, eds., *Contributions to the Geologic Time Scale.* Studies in Geology No. 6. Tulsa, Okla.: American Association of Petroleum Geologists, 1978, 388 pp. (paperback).

Eicher, D. L., *Geologic Time,* 2nd ed. Englewood Cliffs, N.J.: Prentice-Hall, Inc., 1976, 152 pp. (paperback).

Eiseley, L. C., "Charles Lyell," *Scientific American* (August 1959), Vol. 201, No. 2, pp. 98–106. Reprint 846, W. H. Freeman and Co., San Francisco.

Emiliani, Cesare, "Ancient Temperatures," *Scientific American* (February 1958), Vol. 198, No. 2, pp. 54–63. Reprint 815, W. H. Freeman and Co., San Francisco.

Engel, A. E. J., "Geologic Evolution of North America," *Science* (April 12, 1963), Vol. 140, No. 3563, pp. 143–152.

Faul, Henry, *Ages of Rocks, Planets, and Stars.* New York: McGraw-Hill Book Co., 1966, 109 pp. (paperback).

Hay, E. A., "Uniformitarianism Reconsidered," *Journal of Geological Education* (February 1967), Vol. 15, No. 1, pp. 11–12.

Macdougall, J. D., "Fission-track Dating," *Scientific American* (December 1976), Vol. 235, No. 6, pp. 114–122.

McAlester, A. L., *The History of Life,* 2nd ed. Englewood Cliffs, N.J.: Prentice-Hall, Inc., 1977, 160 pp. (paperback).

Mintz, L. W., *Historical Geology,* 3rd ed. Columbus, Ohio: Charles E. Merrill Publishing Co., Inc., 1981, 611 pp.

Nichols, R. L., "The Comprehension of Geologic Time," *Journal of Geological Education* (March 1974), Vol. 22, No. 2, pp. 65–68.

Ralph, E. K., and H. N. Michael, "Twenty-five Years of Radiocarbon Dating," *American Scientist* (September 1974), Vol. 62, No. 5, pp. 553–560.

17

EVOLUTION AND FOSSILS

In the following chapters the development of life is traced from its origin to the present. In this chapter the way that people recognized the meaning of the diversity of both fossils and present life is described. Indeed, it was not always recognized that fossils are the remains of once-living organisms. Even after this was generally conceded, separate creations were invoked to account for the diversity and for the numerous gaps in the record.

EVOLUTION

Evolution is a biologic process, but it is also important to geology. The evidence for evolution comes both from fossils—the province of the geologist—and from living organisms—the province of the biologist. Biological evolution is the continuing change in populations of organisms that is now occurring and has occurred in the geologic past. In a sense, evolution is the biologic counterpart of the continuing changes that have occurred and continue to occur on the earth. Because of its importance in human intellectual development, evolution will be discussed historically.

At one time people considered themselves at the apex of a static earth, located at the center of the universe. Copernicus indicated that the earth is dynamic and not at the center of the universe. Darwin, whose name is almost synonymous with evolution, showed that in the broad view humans occupy only a small, probably temporary, place on the earth. Understandably, such revolutionary changes in viewpoint have not gone unchallenged.

LINNAEUS—SPECIES AND GENUS

An important step occurred in 1737 when Carolus Linnaeus published his book *Systema Naturae*. In this book he attempted to classify all living organisms in a logical system. He used two names,

genus and species, for each kind of organism, and this system is still in use. The tenth edition of this book, published in 1758, is the generally accepted start of this system of classification. The value of this system is that it indicates that there are degrees of similarities and differences among organisms rather than random variations.

Although there is no completely satisfactory definition of species, it can be considered as a group of similar individuals that can interbreed to produce fertile offspring. A genus is a group of closely related species. Thus the cats belong to the genus *Felis;* the common house cat is *Felis catus,* and the ocelot is *Felis pardalis* (which is written *F. pardalis* if the genus has been mentioned nearby). Note that the first letter of the generic name (genus) is capitalized, but the specific name (species) is in lower case, and that both genus and species are in italics. The other terms in the system, *family, order, phylum,* and so on, are shown in Table 17–1. To biologists working with living organisms, a species is a group that can interbreed, and the whole system attempts to show the genetic relationships among organisms. Geologists working with fossils have the same goal, but they must work with dead remains of organisms. A typical fossil is a clam or a snail shell, so geologists must define a species and deduce the relationships to other species on the basis of shell morphology alone. As will be described in later chapters, in rare cases they may have more evidence on which to base

their conclusions. Because they work with form and structure alone, geologists define a species on the basis of a single fossil or a group of fossils called the *type specimen(s).* All other fossils referred to this species must be compared to the type(s). This is done by using descriptions, pictures, plaster casts, other specimens from the same location as the type, and direct comparison with type(s). Linnaeus laid this foundation for Darwin about one hundred years before Darwin left on the voyage of the *Beagle,* and in this intervening time the classification was refined and expanded.

LAMARCK AND CUVIER

Geology developed rapidly near the beginning of the nineteenth century, when the geologic time scale was set up. Evolution was an idea mentioned by a few workers. In 1809 Jean-Baptiste Pierre Antoine de Monet, Chevalier de Lamarck, described his theory of evolution. He believed that all organisms evolved from a single ancestor, and that the environment created the need for change

Table 17–1.   Classification of the more important fossil-forming organisms.

A. Classification scheme and an example.

Kingdom	Animalia
Phylum	Chordata
Class	Mammalia
Order	Primate
Family	Hominidae
Genus	*Homo*
Species	*sapiens*
Individual	John Doe

Recovering fossil bones at Dinosaur National Monument, Colorado. Photo from National Park Service.

or evolution. His followers interpreted this to mean that the offspring inherited the acquired characteristics of the parents and, in turn, passed them on to their children. For example, Lamarckists believed that a blacksmith who developed large biceps would pass these muscles on to his children. We now know that traits are not inherited in this way.

One of Lamarck's critics was Georges Léopold Chrétien Frédéric Dagobert Cuvier, who was mentioned earlier in connection with his studies in the Paris basin. Cuvier studied the vertebrate fossil bones found in the sedimentary beds of the Paris basin. He was an antievolutionist and attacked Lamarck on two grounds. Lamarck believed that organisms became more complex through evolution, but Cuvier noted that the fish that he found in ancient rocks were already very complex animals. He also could see no evidence of gradual evolution in the fossils that he studied. The bones in each successive bed were different from those in the underlying bed, and there was no trace of any of the intermediate steps necessary in evolution. For these reasons, he believed in a separate creation for each bed, and that the change in rock type from bed to bed was caused by the revolution that killed the organisms and so set the stage for the next creation. Cuvier did some excellent work in his studies of fossil vertebrates. He founded modern comparative anatomy, and it is ironic that comparative anatomy is one of the more important lines of evidence for evolution. If the skeletons of vertebrates are compared, similar bones are found in each, although they may serve a somewhat different function in each. In some cases, the correspondence is very close (Fig. 17–1). It is highly unlikely that most vertebrates would have the same bones unless all evolved from a single common

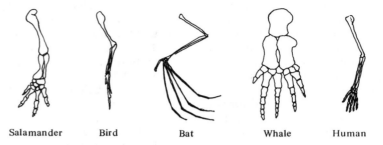

Salamander   Bird   Bat   Whale   Human

FIGURE 17–1 Similarity of bones in forelimbs of vertebrates. Similar bones serve different purposes in each of the examples. It seems very unlikely that such different limbs should have similar bones without evolution.

ancestor. Surely the most efficient design of a whale fin, a bird wing, and a human hand would not all use the same bones for such different functions.

DARWIN

The stage is now set for Charles Robert Darwin, who was born on the same day as Abraham Lincoln. In 1831, when he was 22, he sailed on the *Beagle* as naturalist on a five-year voyage around the world. During the early days of the voyage, while fighting seasickness, he read Charles Lyell's *Principles of Geology,* which had been published the preceding year. In his book, Lyell had gathered much evidence for uniformitarianism and had shown that the earth is much older than was then currently thought. This book had a great influence on Darwin, and it started him thinking of slow biologic changes over an immense length of time.

During the voyage, Darwin made many observations, both biologic and geologic, but it is his study of finches in the Galápagos Islands that has become best known. The Galápagos Islands are 800 to 970 kilometers (500 to 600 miles) off the coast of Ecuador, too far for these birds to fly. Darwin found 13 species of finches that are unknown elsewhere in the world, although they are closely related to South American finches. The 13 species vary mainly in the

size and shape of their beaks, and this reflects differences in their food. (See Fig. 17–2.) Darwin recognized the importance of his discovery at the time, but he did not develop his theory until later. Darwin believed that in some accidental way, such as being blown by a storm, a few mainland finches reached the islands, and from these birds the various species evolved, each species being adapted to a different food supply. The finches were able to take over the various food sources because no other birds except ocean birds are found in the Galápagos Islands; thus there was no competition.

In 1838, long after the voyage, Darwin read Thomas Malthus' essay on population that had been published forty years before. Malthus' main point was that populations grow faster than their food supplies, and so food supply will ultimately control the world's population, an idea that is causing much concern today. Malthus used the phrase "struggle for existence" and noted that most organisms produce more young than can easily survive. Darwin reasoned that those young with any slight advantage over the others would be the most likely to survive and produce offspring. He called this "Natural Selection" by survival of "the fittest."

As soon as Darwin realized that species are not fixed, he knew that he must answer all possible ques-

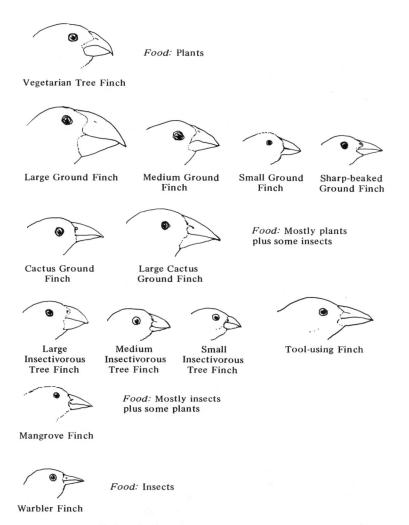

Food: Plants

Vegetarian Tree Finch

Large Ground Finch

Medium Ground Finch

Small Ground Finch

Sharp-beaked Ground Finch

Cactus Ground Finch

Large Cactus Ground Finch

Food: Mostly plants plus some insects

Large Insectivorous Tree Finch

Medium Insectivorous Tree Finch

Small Insectivorous Tree Finch

Tool-using Finch

Mangrove Finch

Food: Mostly insects plus some plants

Warbler Finch

Food: Insects

FIGURE 17–2 Darwin's finches of the Galápagos Islands. These finches apparently evolved from a common ancestor. Notice the differences in the shapes of their beaks and in the size of the birds.

tions and produce an airtight case because he was sure to be attacked. He began to write, documenting the evidence for evolution. He produced a 35-page abstract in 1842, and in 1844 he allowed a few friends to read a 230-page essay. By 1858, after finishing his studies of the materials collected on the voyage of the *Beagle,* he had eleven chapters written of what would be *The Origin of Species* and was proceeding at a leisurely rate. At that point, he received an essay from a much younger man, Alfred Russel Wallace, that contained the main points of Darwin's work. Darwin was ready to abandon his work in favor of Wallace, but his friends urged a joint presentation. Darwin finally published *The Origin of Species* late the next year. The attack that Darwin had expected came; it was bitter and still, in a few places, continues today. Until Darwin's time, most geologists were trying to fit their geological history into the then-current interpretation of biblical history. Darwin finally broke this tradition. He was attacked because he did not believe that all species came from the Ark, and because many interpreted his work as meaning that humans evolved from apes. Darwin's ideas on humans were finally published in 1871 in the book *The Descent of Man, and Selection in Relation to Sex.*

MENDEL—THE LAWS OF INHERITANCE

One reason for the attacks on Darwin was that though he had documented the evidence for evolution, he had no process or mechanism to cause evolution to occur. He leaned toward Lamarckism, but he also recognized its weakness. Had Darwin known it, the mechanism had been discovered in 1865 and published in 1866 by Gregor Johann Mendel, an obscure Austrian monk. Mendel experimented over an eight-year period with garden peas. He used several characteristics, such as whether the peas were round or wrinkled. He discovered that in cross-pollinating round and wrinkled peas, the first generation are all round peas. These hybrid round peas are not all the same, however, because the following generation is in the ratio of three round peas to one wrinkled pea. (See Fig. 17–3.) In succeeding generations, the wrinkled pea produces only wrinkled peas. In the same way, one-third of the round peas produce only round peas in succeeding generations. The remaining two-thirds of the round peas produce in the next generation round and wrinkled peas in the ratio of three round to each wrinkled one. Mendel called characteristics such as roundness "dominant" and wrinkledness "recessive." He continued his experiments, using two characteristics such as color and roundness, and discovered again that the offspring occur in fixed ratios. He showed that traits are inherited as discrete units.

347

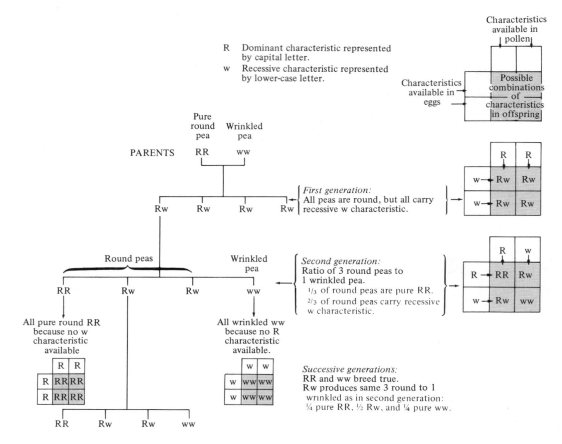

R Dominant characteristic represented by capital letter.

w Recessive characteristic represented by lower-case letter.

Characteristics available in pollen

Characteristics available in eggs

Possible combinations of characteristics in offspring

Pure round pea   Wrinkled pea

PARENTS   RR   ww

	R	R
w	Rw	Rw
w	Rw	Rw

First generation: All peas are round, but all carry recessive w characteristic.

Rw   Rw   Rw   Rw

Round peas   Wrinkled pea

RR   Rw   Rw   ww

Second generation: Ratio of 3 round peas to 1 wrinkled pea.
¹/₃ of round peas are pure RR.
²/₃ of round peas carry recessive w characteristic.

	R	w
R	RR	Rw
w	Rw	ww

All pure round RR because no w characteristic available

	R	R
R	RR	RR
R	RR	RR

All wrinkled ww because no R characteristic available.

	w	w
w	ww	ww
w	ww	ww

Successive generations: RR and ww breed true. Rw produces same 3 round to 1 wrinkled as in second generation: ¼ pure RR, ½ Rw, and ¼ pure ww.

RR   Rw   Rw   ww

FIGURE 17–3   Mendel's inheritance law.

Mendel's genius was that he could apply mathematics to breeding. He was ahead of his time because no one else understood the importance of his discoveries. Darwin himself experimented with cross-breeding and noticed the same three-to-one ratio, but the meaning and importance escaped him. Mendel's work lay buried until 1900 when three separate workers rediscovered both his findings and then his published work. The Darwin-Wallace simultaneous discovery and the triple rediscovery of Mendel's work are two examples of parallel discoveries that occur from time to time in science.

Mendel discovered that characteristics are passed to offspring by discrete units that are now called genes. At first, the mechanism was believed to be mutations that caused big steps. Darwin had referred to these as "sports." Fur-

ther study and reflection show that most big changes are detrimental and the mutant generally dies before producing any offspring. Evolution is caused by mutation; but the changes are typically very small steps, and natural selection of these small changes is the mechanism of evolution. Genes change spontaneously, but, of course, at a very low rate. Changes are also caused by radiation. Natural radioactivity and cosmic rays can cause changes in genes, as can radiation such as X rays. Many feel that such changes are the real danger from atomic bombs, and this is why atomic power plants are so carefully designed and monitored.

Evolution can occur because in any species, especially if there are a large number of individuals, there is a large pool of genes. These genes are varied enough that favorable characteristics are

available to meet almost any change in environment. Even if the mutation is only slightly favorable, it will, over a number of generations, replace the original characteristic and become the normal characteristic of the population. Thus it is the population that undergoes evolution.

PUNCTUATED EQUILIBRIUM VS. GRADUALISM

The geologic record is important in all theories of evolution, but that record is not complete. The remains of most organisms are not preserved by fossilization; and even the few that are preserved may be destroyed later by erosion, metamorphism, or rock deformation. In spite of these problems, at a few places the fossil record is reasonably complete. At those places we find one species being replaced by another species, and that species replaced by still another, and so forth. The problem is the almost complete lack of the intermediate populations that must have existed if one species evolved into another species. The geologic record shows long intervals of nonchanging species, or, in other terms, successful species do not change.

Most theories of evolution assume that changes in species occur slowly, but that external changes, such as climatic change, may at times cause more rapid evolutionary changes. *Slow evolution by small steps is called gradualism.* Gradualism is not supported by the geological record because of the lack of intermediate stages. The question is whether the lack of intermediate stages is real or whether it is caused by a lack of data. Gradualists believe that lack of data is the problem, but others believe that the intermediate stages are not found for other reasons.

The other possibility is that *at times evolution is rapid, but most of the time very little change occurs.* This is called *punctuated equilibrium.* In this case evolution occurs in small, stressed populations at the fringes of the habitat. The changes occur rapidly in small populations and so are not apt to be preserved. After such a short burst of evolutionary change, perhaps caused by a climatic or other environmental change, the new species is better adapted and so challenges and rapidly replaces the old species.

PATTERNS OF EVOLUTION

DIVERGENCE (RADIATION)

As the history of the earth is outlined, it will be clear that some organisms have changed and others have not. Sponges, for instance, have not changed much during geologic history. They are very well suited to their form of life. Other organisms have changed or *adapted* to take advantage of new or changing conditions. These changes are generally a *spreading out* or *divergence (radiation) to different habitats.* Radiation occurs again and again in the geologic record. Darwin's finches (Fig. 17–2) are an example of divergence.

CONVERGENCE

A striking feature of evolution is the way that *different organisms adapting to the same habitat develop similar characteristics.* This is called *convergence,* and an example is the similarities between the present placental mammals and the marsupials of Australia. There is a marsupial in Australia to match almost every placental mammal habitat. For instance, there is a marsupial "mouse," a marsupial "mole," and the Tasmanian "wolf." Another example is the similar body shapes of the shark, a fish; the dolphin, a mammal; and the ichthyosaur, an extinct reptile (Fig. 17–4).

EXTINCTION

Some of the most pronounced changes in life occurred at times when the environments changed, such as when the seas retreated from the continents. This could be more apparent than real, however, because of the unrecorded time represented by the resulting unconformity. When the environment changes, a species can do one of three things: it can migrate, it can adapt, or it can die out. The reasons for extinction are never simple and, generally, must be speculations because of the lack of evidence.

Disease epidemics have been suggested as the cause of the mass extinctions of the geologic past. The pathogenic organisms that cause disease generally only attack a single species, so this cannot explain the extinction of many species, such as the dying out of all the dinosaurs at the end of the Cretaceous.

Climatic change has probably been the most popular explanation for mass extinctions. At many of the times that the mass extinctions of the geologic past oc-

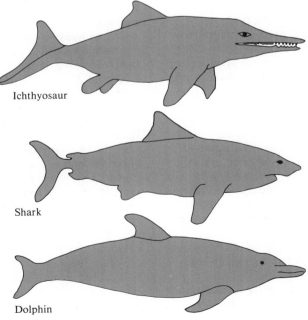

FIGURE 17–4 Convergence is shown by the similar body shapes of the shark, a fish; the dolphin, a mammal; and the ichthyosaur, an extinct reptile.

349

curred, mountain building also occurred, and seas withdrew. These changes would, of course, drastically change the climate. Movements of continents on tectonic plates can also change climates. Many species would be unable to adapt or migrate and would die out, and new species, better adapted to the environment, would evolve. Close examination reveals flaws in this theory. At the times of great animal extinctions, the animal population changes drastically but not the plants, and plants seem to be at least as sensitive to climate as animals, and probably more so. The last glacial events record a climate change, but the associated extinctions, which will be described later, occurred long after the main glacial advances.

Extinctions of some organisms and reversals of the earth's magnetic field correspond very closely. It was thought that radiation from space caused mutations that resulted in extinction. At the times when the earth's magnetic field reverses, the temporary lack of a magnetic field would allow high-energy particles from the sun to reach the earth. If this effect did exist at times in the past, however, it would speed up evolution by causing more mutations. This effect should be most pronounced on terrestrial life. As noted before, plants are least affected at the times of mass extinction; and at most of these times marine life, somewhat protected from radiation by the sea, is most affected. The extinctions may occur because some organisms are adversely affected by a low-intensity magnetic field such as would exist at the time of reversal. Recent experiments have indicated that low-intensity magnetic fields have deleterious effects on a wide variety of organisms.

Another possibility is that many species must adapt or die if, in some way, an element on which they depend for food is destroyed or consumed by a more efficient competitor. This reduces the problem to finding a reason for the death of only one species instead of whole populations. Many of the theories just offered could explain the death of a single species. Many of the mass extinctions are associated with withdrawal of the sea, suggesting that such an event might shrink the environment of a key species, causing the demise of many, perhaps somewhat marginal, species. New species would develop when the seas again advanced, expanding the environment.

A variant of this theory is that most of the life of the ocean would be affected if the plankton died because the plankton, directly or indirectly, are the food for most forms of life in the ocean. The record suggests such a demise of at least some of the plankton. Chalk deposits of late Cretaceous age, composed almost entirely of the shells of plankton, are common in many parts of the world. These and other late Cretaceous deposits suggest that at many places this was a time of little erosion on the continents. It has been suggested that this lack of erosion reduced the amounts of inorganic nutrients in the ocean and so caused the dying of the plankton.

Changes in the amount of oxygen may have caused some extinctions. It was recently shown that the present-day representatives of organisms that underwent extinction need more oxygen than those that did not undergo extinction. The correlation is very high. Near the end of the Cretaceous Period, when a great many organisms died out, a new group of photosynthesizers expanded greatly. They are the coccoliths, and their remains form great chalk deposits, such as the White Cliffs of Dover, England.

The extinctions, such as of the dinosaurs, near the close of the Cretaceous have always been a geological problem. Apparently an asteroid collided with the earth about that time and may have been partly responsible for the extinctions. The evidence is described near the end of Chapter 22.

In every case of extinction of a large group, some other group expands and so takes over the habitat. In most cases it is not possible to say why some groups died and others flourished.

FOSSILIZATION

Fossils are any evidence of past life, and this evidence may be preserved in many different ways. When most animals die, the body is rapidly destroyed; scavengers consume the body or bacteria cause it to decay. Only a very few escape this destruction, possibly to become fossils. Generally, if the body has hard parts, those parts are most likely to be preserved; and if the body is rapidly buried, its preservation is more likely. Burial may also bring on physical and chemical changes that may cause destruction. Later uplift is necessary to expose the fossil, and weathering and erosion may destroy it then. Thus fossilization is a relatively rare event, and this is why the record of fossil life is so incomplete. For these reasons, the most common fossils are the shells of shallow-water marine animals such as clams.

The incompleteness of the fossil record makes reconstructions of past life very difficult. It has been estimated that only about 1 out of 44 species of the fossilizable marine invertebrates of the past is known. It is also estimated that of the 10,000 or so species found on a tropical river bank, only 10 to 15 are likely to be preserved.

Fossils form in the following ways:

Preservation of part of the animal, such as a clam shell.
Petrifaction by replacement, recrystallization, or permineralization of the original shell or other part of the animal. This is a very common type of fos-

A

B

C

D

FIGURE 17-5 Types of fossilization. A. Mold and cast. Left is cast, right is mold of *Maclurites,* an Ordovician snail from New York [7.6 centimeters (3 inches)]. B. Internal cast of *Turritella,* a Cretaceous snail [length, 5.1 centimeters (2 inches)]. C. A leaf preserved as a carbonaceous film. *Liquidambar* from Miocene of Oregon. D. Insect preserved in amber, Oligocene, East Prussia; insect preserved as a carbonaceous film. Parts E, F, G, H, I, and J on the following page.

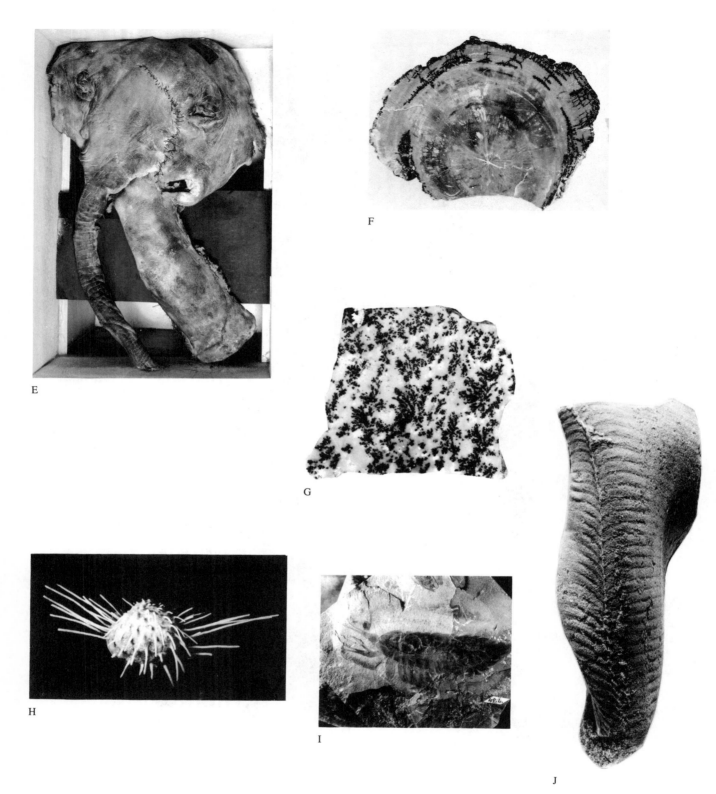

E

F

G

H

I

J

FIGURE 17–5 (cont.) E. Baby wooly mammoth found in frozen ground in Alaska. F. Petrified wood. Silica has filled the voids in the wood, thus preserving it. G. False fossil. The structures in this specimen of moss agate are inorganic. H. Spiny brachiopod recovered by dissolving enclosing limestone with acid. I. Delicate features are preserved in this worm-like animal by carbonization. J. Crawl marks. Parts A and B from Ward's Natural Science Establishment, Inc., Rochester, N.Y. Parts C, D, F, G, H, and I courtesy of the Smithsonian Institution. Part E courtesy of the American Museum of Natural History. Part J courtesy of P. E. Cloud, Jr.

silization, and most fossil shells are either recrystallized or replaced. Ground water is the agent that causes the replacement in most cases. Silica and calcite are the main replacing minerals, but many other minerals are also found as replacements. In some cases, very delicate internal structures may be preserved. Some of the best fossils with internal structures and surface ornamentations preserved are replacements by silica in limestone. In such cases, the fossils can be very easily recovered by dissolving the limestone in acid. (See Fig. 17–5H.) In permineralization the permineralizing mineral

fills the voids in the fossil, as silica does in petrified wood.

Molds of fossils form if the fossil is dissolved, leaving an empty space. (See Fig. 17–5A.)

Casts form if molds are filled with minerals or sediments. This is probably the most common way that marine fossils are preserved. (See Figs. 17–5A and B.)

Carbonization occurs if a fossil such as a leaf is squeezed, generally on a bedding plane, and the fluids are removed, leaving a thin carbon film. Worms and other animals without hard parts can be preserved in this way. Very tiny, delicate structures can be preserved by carbonization. (See

Figs. 17–5C and I.)

Tracks and *burrows* are also evidence of life. (See Fig. 17–5J.)

Coprolites, or fossil excrement, can tell much about the eating habits of fossil life.

Frozen mammoths have been uncovered with every part of the animal intact. Such occurrences are, of course, very rare. (See Fig. 17–5E.)

Mummified animals are also quite rare. Mummification could occur through the trapping and sinking of an animal in a tar seep; the tar would inhibit decay of the body. Peat bogs have yielded specimens with soft parts well preserved. Mummification can also occur in very dry climates.

SUMMARY

Biological *evolution* is the continuing change in populations of organisms that is now occurring and has occurred in the geologic past.

A *species* is a group of similar individuals that can interbreed to produce fertile offspring.

A *genus* is a group of closely related species.

Species of fossil organisms must be based on form and structure of a specimen.

A *type specimen* is used by geologists to define a species.

Linnaeus was first to classify organisms, using species and genus to show the degree of relationship.

Lamarck was an early evolutionist. Lamarckism has come to mean that offspring inherit the acquired traits of parents.

Cuvier did not believe in evolution, but his comparative anatomy is important evidence for evolution. That is, it is highly unlikely that most vertebrates would have equivalent bones unless all evolved from a common ancestor.

Darwin marshalled the evidence for evolution. He believed in "Natural Selection" by survival of "the fittest," but he could find no mechanism of inheritance.

Mendel, experimenting with cross-pollinated peas, discovered how traits are inherited.

Genes are the discrete units by which characteristics are passed to offspring. They change spontaneously at a very slow rate and more rapidly when exposed to radioactivity.

Evolution occurs because in any large population there are genes that favor almost any change in environment; and the mutation, even if it is only slightly favorable, will, over a number of generations, replace the original characteristic and become the normal characteristic of the population.

The lack of intermediate stages in evolution has led to the concept of *punctuated equilibrium,* the concept that at times evolution is rapid, but most of the time very little change

occurs.

Adaptation is a change by an organism to take advantage of new or changing conditions.

Radiation is a spreading out into different habitats.

Convergence is the development of similar characteristics by different organisms in the same habitat.

When the environment changes, a species can migrate, adapt, or die out.

It is not yet possible to explain the large extinctions of the geologic past.

Fossils are evidence of past life.

Fossilization is a rare event, so the record of fossil life is incomplete.

Most fossils are hard parts that have been preserved by *petrifaction,* generally by recrystallization or replacement by silica or calcite.

Fossilization may also occur by preservation, molds, casts, carbonization, freezing, and mummification.

Tracks, burrows, and coprolites (fossil excrement) are also fossils.

QUESTIONS

1 List the evidence for biologic evolution.
2 Who first used genus and species in naming organisms? When?
3 What is a species?
4 What is a genus?
5 How does a geologist's use of species differ from a biologist's use?
6 What is a type specimen?
7 What were Lamarck's ideas about evolution?
8 What was Cuvier's evidence for separate creations? Is it good evidence?
9 What led Darwin to his ideas on evolution?
10 What was the chief weakness in Darwin's *Origin of Species?*
11 What is Mendel's law of inheritance?
12 How do genes control evolution?
13 What is biologic radiation?
14 List and describe the types of fossilization.
15 Discuss the relationship between geologic changes and evolution of new species.
16 Describe several possible explanations for the extinction of species.
17 What is meant by punctuated equilibrium?

SUPPLEMENTARY READINGS

Bramlette, M. N., "Massive Extinctions in Biota at the End of Mesozoic Time," *Science* (June 25, 1965), Vol. 148, No. 3678, pp. 1696–1699.

Clarke, B., "Causes of Biological Diversity," *Scientific American* (August 1975), Vol. 233, No. 2, pp. 50–61.

Dorf, Earling, "The Petrified Forests of Yellowstone Park," *Scientific American* (April 1964), Vol. 210, No. 4, pp. 106–114.

Ehrlich, Paul, and Anne Ehrlich, *Extinction. The Causes and Consequences of the Disappearance of Species.* New York: Random House, 1981, 305 pp.

Fridriksson, Sturla, *Surtsey: Evolution of Life on a Volcanic Island.* New York: John Wiley and Sons, Inc., 1975, 198 pp. Describes the development of life on a newly formed island.

Kimura, Motoo, "The Neutral Theory of Molecular Evolution," *Scientific American* (November 1979), Vol. 241, No. 5, pp. 98–126. This paper presents an alternative to Darwinism.

Kolata, G. B., "Paleobiology: Random Events over Geologic Time," *Science* (August 22, 1975), Vol. 189, No. 4203, pp. 625–626, 660. Describes a different view of evolution.

Newell, N. D., "Crises in the History of Life," *Scientific American* (February 1963), Vol. 208, No. 2, pp. 76–92. Reprint 867, W. H. Freeman and Co., San Francisco.

Newell, N. D., "The Evolution of Reefs," *Scientific American* (June 1972), Vol. 226, No. 6, pp. 54–65.

Russell, D. A., "The Mass Extinctions of the Late Mesozoic," *Scientific American* (January 1982), Vol. 246, No. 1, pp. 58–65.

Scientific American (September 1978), Vol. 239, No. 3. Entire issue devoted to evolution.

Simpson, G. C., *Life of the Past.* New Haven, Conn.: Yale University Press, 1961, 198 pp. (paperback).

Stanley, S. M., *Macroevolution. Pattern and Process.* San Francisco: Freeman, 1979, 332 pp.

Stone, Irving, *The Origin. A Biographical Novel of Charles Darwin.* Garden City, N.Y.: Doubleday, 1980, 744 pp.

Walter, M. R., "Interpreting Stromatolites," *American Scientist* (September–October 1977), Vol. 65, No. 5, pp. 563–571.

18

EARLY HISTORY
OF THE EARTH—
ASTRONOMICAL
ASPECTS OF GEOLOGY

The kindest words regarding these entertainments are those of H. H. Read, quoted by L. R. Wager (1958) in his address at the 1958 meeting of the British Association for the Advancement of Science: ". . . just as things too absurd to be said can yet with perfect propriety be sung, so views too tenuous, unsubstantial and generalized for ordinary scientific papers can yet appear with some measure of dignity in presidential addresses." This is then an appropriate occasion to review some speculations, by no means original, on the large-scale evolution of the Earth. Views on the early history of the Earth and on happenings in the deep interior can hardly fail to be tenuous and generalized, so I think they may qualify under Read's dictum. Nevertheless, we cannot hope to account for the development of the surface film which naturally engages most of our attention without considering the enormously larger interior from which this film was derived and which continues to determine its evolution. . .

Introduction to Francis Birch's presidential address to The Geological Society of America, "Speculations on the Earth's Thermal History," Bulletin, Geological Society of America, Vol. 76, 1965, pp. 133–154.

The Earth[1] is part of the solar system, so we must look to astronomy for clues to the Earth's origin. Our main interest is not astronomy, so this chapter will be descriptive and much information will be presented without proof. Details will be discussed only where they have important application to geology.

Our sun is not the only star thought to have planets. Although the other stars are too far away to see the other planets, even with our largest telescopes, their existence is inferred from the wavering motion of three stars and the as yet unconfirmed wavering motions of five more stars.

SOLAR SYSTEM

The solar system is composed of the sun and the planets that orbit around the sun, the satellites of the planets, and comets, meteors, and asteroids. All of the orbital motions are explained by the law of gravity.

The sun is an ordinary, average-size star, composed mainly of hydrogen and helium, but containing most of the elements found on Earth. The sun's energy, which makes life possible on Earth and ultimately causes most geologic processes, comes from nuclear reactions. The sun has 99.86 percent of the mass of the solar system, and the Earth has about 1 percent of the remaining 0.14 percent.

The nine planets all revolve in the same direction around the sun. The orbits are all elliptical, very close to circular, and lie close to a single plane, except Pluto whose orbit is inclined. (See Fig. 18–1.) Seen from the Earth, the sun and the planets follow the same path across the sky. The planetary data are summarized in Table 18–1. The inner planets, Mercury, Venus, Earth, and Mars, form a group in which the individual members are more or less similar in size. The next four planets, Jupiter, Saturn, Uranus, and Neptune, form a similar grouping of much larger planets. Pluto is much smaller than the other outer planets and has a tilted, eccentric orbit, all of which suggest that it may not be a true planet. It may once have been a satellite of Neptune and escaped into its own orbit. Between Mars and Jupiter are several thousand small bodies, the asteroids, that are probably the origin of meteorites.

Venus and Mars, our closest neighbors, are of great interest to us and have been studied by telescope and with space probes. Because they are our closest neighbors, they are the planets on which life might occur. For life, such as we know it, to exist, liquid

---

[1] In this section the astronomical convention of capitalizing the names of all planets will be followed.

Scientist-astronaut Harrison Schmitt standing next to a huge split lunar boulder. Photo from NASA.

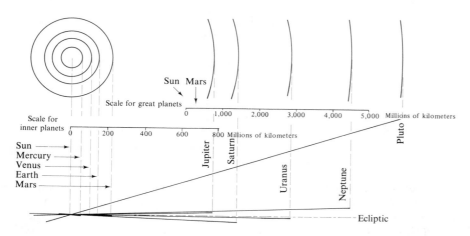

water is probably necessary, and only Mars has temperatures in this range.

We will consider the moon and the inner planets in order of their sizes beginning with the smallest, our moon.

MOON

Just as the planets orbit the sun, most of the planets have moons, or satellites, that revolve around them. Most of the moons revolve in the same direction as the planets, but some move in the opposite direction. Most of the moons are much smaller than their planets. Ours is much closer in size to the Earth so that the Earth-moon system should probably be con-

FIGURE 18–1 The orbits of the planets. Because Pluto is almost 40 times farther away from the sun than is the Earth, it is not practical to draw this diagram to scale. The inner planets, the circles, are drawn to scale, and the distances among the great planets are drawn to a different scale. Eccentricity of the orbits is neglected. The edge view shows that, with the exception of Pluto, the orbits lie nearly in the same plane (ecliptic).

TABLE 18–1 Summary of solar system data. These statistics are approximations and vary somewhat in the different sources from which they were gathered.

Body	Distance from Sun (average)		Radius		Mass (Earth = 1)	Specific Gravity	Eccentricity of Orbit	Inclination of Spin Axis
	Millions of Kilometers	Millions of Miles	Kilometers	Miles				
SUN	—	—	695,088	432,000	332,000	1.4	—	—
Mercury	58	36	2,414	1,500	0.05	5.4	0.206	10° (?)
Venus	108	67	6,114	3,800	0.81	5.2	0.007	6°
Earth	150	93	6,376	3,963	1.00	5.5	0.017	23.5°
Mars	227	141	3,379	2,100	0.11	4.0	0.093	25.2°
Jupiter	777	483	69,992	43,500	318	1.3	0.048	3.1°
Saturn	1,426	886	59,533	37,000	95	0.7	0.056	26.7°
Uranus	2,869	1,783	25,342	15,750	14.6	1.2	0.047	98°
Neptune	4,494	2,793	26,549	16,500	17.3	1.7	0.009	29°
Pluto	5,899	3,666	2,900 (?)	1,800 (?)	0.1 (?)	?	0.249	?

[1]Data from various sources varies up to ten percent.
[2]Venus rotates clockwise, opposite to other planets.
(Uranus does also because of its 98° tilt of its spin axis.)

sidered a double planet. The moons that orbit in the opposite direction and have very inclined orbits may have been captured by the planets instead of having been formed with the planets.

The moon orbits the Earth and rotates on its axis at the same rate and so only one side of the moon is visible from the Earth. Our only knowledge of the far side of the moon comes from space probes that have photographed the far side. They have revealed a cratered surface with almost no smooth areas. (See Fig. 18–2.)

The visible face of the moon has large, dark, smooth areas and lighter-colored, more rugged areas. The early astronomers called the dark areas seas [maria (pl.), mare (sing.)] and the lighter

areas, lands, in a fanciful analogy to the Earth. Both have craters but the lighter uplands have more craters than the lower maria. The craters range from hundreds of miles across down to a few inches in diameter. Most of these craters probably were formed by meteorite impacts. The moon, unlike the Earth, is not protected by an atmosphere, so more meteorites hit the moon than the Earth. The relative ages of various parts of the moon's surface can be determined by the density of the craters, the older surfaces having more craters. The present estimated rate of meteorite impact is not enough to have formed all of the moon's craters.

Some of the craters have rays. (See Fig. 18–3.) The rays are be-

lieved to be material splashed from the crater by impact. They are visible from Earth at full moon and appear lighter colored in such illumination. The rays cover older craters, and the rays have more small, sharp, young craters than their surroundings. At places these young craters form in rows radial to the main crater, suggesting that their origin is from impact of material thrown out of the main crater.

Evidence of volcanic activity on the moon is abundant. Some of the craters may be volcanic in origin, and volcanic activity probably modifies many of the craters. Figures 18–3 and 18–4 show domes; and these domes could be volcanoes, laccoliths, or hills of other origins, partially buried by vol-

TABLE 18–1 (continued)

Period of Revolution around Sun	Period of Rotation	Temperature (°C)			Atmospheric Pressure at Surface (Earth = 1)	Composition of Atmosphere
		Theoretical[3]	Actual			
			Top of Atmosphere	Surface		
	25 days at equator 34 days at poles	—	—	5,500°	—	—
88 days	58.6 days	Similar to actual	—	425° sun; 17° dark	0.003	None
225 days	242.9 days[2]	−35°	−38°	536°	100	Mainly carbon dioxide, minor water
365.25 days	23 hours 56 min.	−20°	—	15°	1	78% nitrogen 21%oxygen
687 days	24.5 hours	−65°	−50°	−42°[4]	0.006	Mainly carbon dioxide, possibly minor water
11.9 years	9.9 hours	−168°	−123°	2,000° est.	200,000	60% hydrogen, 36% helium, 3% neon, 1% methane and ammonia
29.5 years	10.2 hours	−195°	−167°	2,000° est.	200,000	Same as Jupiter
84 years	23 hours[2]	−185°	−185° approx.	?	8	Similar to Jupiter, with more methane, no ammonia
165 years	18.2 hours	−185°	−185° approx.	?	?	Same as Uranus
248 years	6.4 days(?)	−212°	Large variation because of eccentricity		?	?

[3]Theoretical temperatures are calculated from the energy input from the sun. Actual temperatures are generally higher because of absorption in the atmospheres.
[4]Average varies from 20° to −95° depending on latitude. Actual measurements range from −123° to 17°.

canic flows or other surface layers. Probably all of these features exist on the moon, but Figs. 18–3 and 18–4 show a number of domes with pits or craters at the top, strongly suggesting volcanoes.

Other probable volcanic features are the wrinkle ridges shown in Fig. 18–4. Ridges like these are several hundred feet high and extend for hundreds of miles. They are almost entirely confined to the maria. Their origin is unknown, and it has been suggested that they may be caused by downsinking of the maria, or they may be the location of dikes that fed the surface flows.

FIGURE 18–2   Hidden side of the moon photographed by *Lunar Orbiter III* from an altitude of about 1450 kilometers (900 miles), looking south from a point about 400 kilometers (250 miles) south of the lunar equator. The prominent crater is about 242 kilometers (150 miles) in diameter, and its rim shows evidence of inward slumping like many of the moon's craters. Photo from NASA.

Some very slow erosion does occur because the older craters are more rounded than younger ones. This rounding is believed to be caused mainly by the impact of micrometeorites.

The moon has a rugged surface with many mountains; however, these mountains have apparently been formed by impacts and volcanic activity. Thus, they are very different from the compressive folded and faulted mountains of the Earth.

Several types of rocks have been returned from the moon. Those from the maria resemble Earth basalt, but have much more titanium and much less of the more volatile elements such as sodium and potassium. Gas holes are common. The main minerals are plagioclase feldspar and titanium-rich pyroxene, so that the mineralogy and texture are much like Earth rocks. (See Fig. 18–5.) The plagioclase feldspar, unlike in most Earth rocks, is almost pure calcium plagioclase. Many of these rocks are breccias and were formed by meteorite impact. The fragments in these breccias are basalt, gabbro, rock (breccia), glass, anorthosite (feldspar rock), and peridotite. Much of this material is dust size. (See Fig. 18–6.)

FIGURE 18–3 Oblique view of the crater Kepler taken by *Orbiter III* from a height of 52.6 kilometers (32.7 miles). Kepler has rays extending radially in all directions. The rays are believed to be the material splashed out by the impact and are visible from Earth only near full moon. The light color streaks in the lower right may be ray material, especially as the associated smaller craters form lines radial from Kepler. Many domes are visible; and one in the foreground has a crater at its summit, suggesting that the domes may be volcanic. The line in the foreground may be a fault. Photo from NASA.

361

FIGURE 18–4   An array of domes near the older impact crater Marius. The domes are from 3.2 to 16 kilometers (2 to 10 miles) in diameter and from 305 to 458 meters (1000 to 1500 feet) high. Many have pits on the summit and so resemble volcanoes. The wrinkle ridges may be volcanic or subsidence features. Photo from NASA.

FIGURE 18–5   Thin section of basalt from the moon. The texture closely resembles terrestrial basalt. White is plagioclase, gray is pyroxene, and black is ilmenite, a titanium mineral. Photo from NASA.

Melting and fracturing from impact shocks are visible in most samples, and spherical glass particles and iron-nickel spheres of impact origin are common. (See Figs. 18–7, 18–8.) The mare basaltic rocks all have radioactive dates between about 3900 and 3100 million years, and are the youngest rocks so far found on the moon.

The rocks from the rugged uplands of the moon are different. In composition many are anorthosite, a rock composed of between 70 and 90 percent plagioclase, and in texture they are breccias. This suggests that the uplands may be underlain by anorthositic (plagioclase-rich) gabbro that has been brecciated by meteorite impacts.

Many of these breccias were partially melted. Some of the highland rocks are lavas. The highland rocks all have radioactive ages of about 4000 million years.

Carbon compounds have been found in the moon rocks, but no trace of life has been found.

Study of the orbits of spacecraft around the moon has shown that some parts of the moon are much denser than others. The areas of high mass concentrations (called *mascons*) are the large circular maria.

The seismographs left on the moon reveal that most of the moon's "earthquakes" are caused by tidal attraction and occur when the moon is closest to the Earth. In comparison to the Earth, the moon is an almost dead planet. Most of the moonquakes have magnitudes less than two on the Richter Scale and most appear to originate from two active zones about 966 kilometers (600 miles) long. The foci of the moonquakes are between about 805 and 1126 kilometers (500 and 700 miles) deep, which is much deeper than earthquakes. Moonquakes build up and decay much slower than earthquakes, probably because of the thick surface layer of breccia

on the moon. Meteorite falls have also been recorded by the seismographs.

Seismic studies reveal that the moon is layered. The surface to about 19 kilometers (12 miles) is

apparently rubble. The base of the crust is at about 64 kilometers (40 miles), and there the velocity increases. The mantle extends to about 966 kilometers (600 miles). At that depth the shear waves are somewhat reduced in amplitude, suggesting that the core may be partly melted.

The history of the moon begins with its formation, probably about 4550 million years ago. This is the same date as the origin of the Earth and the solar system. The moon probably was formed by accumulation of material in the same manner as the rest of the solar system—a topic that will be covered later. Much of the outer part of the moon probably melted at this time. The infall of material and the resulting compaction would cause heating. This melting would produce the plagioclase-rich crust of the moon which crystallized about 4000 million years ago. After the crust formed, the interior of the moon would begin to heat. This heating would be caused by radioactive elements. It would first melt the layer below the crust and could be the origin of the mare basalt that formed between 3900 and 3100 million years ago. Melting would then progress downward as the upper layers cooled. This process could produce the partially melted core predicted by the seismic data. After 3100 million years, the main process acting on the surface has been meteorite impacts.

This history agrees well with all of the data except the remanent magnetism of some moon rocks. At present the moon has almost no magnetic field, but some of the moon rocks have remanent magnetism that could only have formed in a much stronger magnetic field. If the moon once had a magnetic field, that field might have originated in a liquid iron core, just as the earth's magnetic field does. This would imply that in the past the moon had a hot liquid core that has since cooled; this is the opposite of the history just outlined.

FIGURE 18–6 The surface of the moon. In the foreground is the orange "soil" found by the crew of *Apollo 17*. The tripodlike object is used as a photographic reference and to establish sun angle. Note the footprints in the fine surface material. Photo from NASA.

FIGURE 18–7 Glass spheres from the moon's surface. The largest is 0.5 mm in diameter. Colors are clear, yellow, red, and black. Their origin is probably from melting during meteorite impact. Photo from NASA.

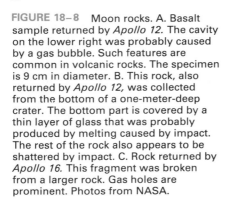

B

FIGURE 18-8   Moon rocks. A. Basalt sample returned by *Apollo 12.* The cavity on the lower right was probably caused by a gas bubble. Such features are common in volcanic rocks. The specimen is 9 cm in diameter. B. This rock, also returned by *Apollo 12,* was collected from the bottom of a one-meter-deep crater. The bottom part is covered by a thin layer of glass that was probably produced by melting caused by impact. The rest of the rock also appears to be shattered by impact. C. Rock returned by *Apollo 16.* This fragment was broken from a larger rock. Gas holes are prominent. Photos from NASA.

C

MERCURY

Mercury is too small and too close to the sun for detailed telescope study even though it has no atmosphere to obscure observation. A space probe has revealed cratered highland areas and presumably younger, uncratered basin areas. In this, Mercury is similar to Mars and the moon. The basins are much like those on the moon but have many more cracks and wrinkle ridges. (See Fig. 18–9.) The rocks filling these basins appear to be volcanic, like the similar rocks on the moon; however, unlike Mars and the moon, none of the photos of Mercury show any volcanoes. As on Mars, the distribution of cratered and basin areas is not uniform.

The big basins on Mercury are apparently impact craters caused by huge meteorites. At points on the opposite side of the planet from such basins are what is called *weird terrain.* (See Fig. 18–10.) It is believed that the shock effects of great impacts are concentrated at such places. The moon has similar big basins, and somewhat similar weird terrain is found there also.

Mercury has many scarps over 483 kilometers (300 miles) long and 3 kilometers (1.75 miles) high. (See Fig. 18–11.) These scarps cut both smooth and cratered areas and may have been formed by faulting. They may be compressional features, and, if so, they are the only compressional features seen on any planet except Earth.

Mercury's magnetic field is about one percent that of Earth's, but even this weak field is much greater than that of Venus or Mars. This suggests that Mercury may have or may once have had a liquid iron core, or that the present field is caused by remanent magnetism in the rocks. Mercury's specific gravity is almost as high as Earth's, further suggesting that it may have an iron core. All of this suggests that Mercury melted, at least partially, early in its history.

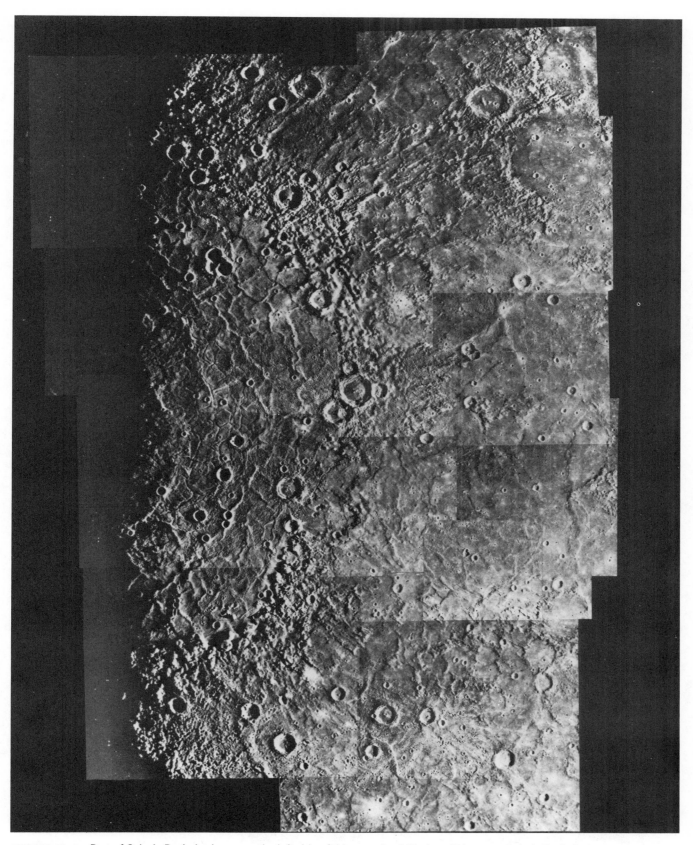

FIGURE 18-9 Part of Caloris Basin is shown on the left side of this mosaic of *Mariner 10* images. Caloris Basin is the largest feature so far discovered on Mercury and is 1300 kilometers (800 miles) in diameter. The floor of the basin has many cracks and ridges. Photo courtesy Jet Propulsion Laboratory.

FIGURE 18–10   This unusual terrain, seen only on Mercury, is termed *weird terrain.* The hills and ridges cut across both smooth and cratered areas. This image is on the opposite side of Mercury from Caloris Basin shown in Fig. 18–9. Photo courtesy Jet Propulsion Laboratory.

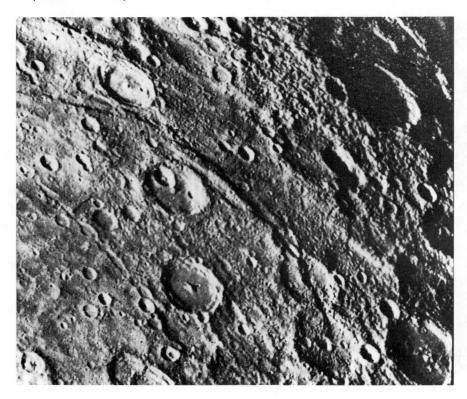

FIGURE 18–11   The scarps running from upper left to lower right in this image of Mercury are about 300 kilometers (185 miles) long. They may be compressional features. Photo courtesy Jet Propulsion Laboratory.

MARS

Mars has a reddish color and so was named for a god of war. The surface is covered with fine material and boulders that appear to be weathered basalt. It is farther from the sun than is Earth, and so has lower temperatures. Its rotational axis is tilted with respect to its orbital plane, and so it has seasons like the Earth. Telescopic observation reveals that polar ice caps form and disappear each year, and the color of the planet changes from bluish-green in spring to brown in the fall, which once suggested seasonal growth to some observers. (See Fig. 18–12.) *Mariner 9* photographs reveal dust clouds, so the observed color changes may be caused by dust storms and clouds. Mars' atmosphere is composed mainly of carbon dioxide and a little water vapor. *Mariner* spacecraft revealed that this inhospitable atmosphere is less than one percent as dense as the Earth's. The average temperature at the equator at noon is probably 4°C (40°F), and the temperature may fall to −130°C (−200°F) at night. Thus the conditions are not very favorable for life on Mars, but some may exist. The pictures returned by *Mariner* space probes show evidence of both internal and surface activity.

Dust storms have been observed on Mars. As the storm subsides, first the higher elevations come into view and finally, the lowlands. In spite of the thin atmosphere, winds up to 483 kilometers (300 miles) per hour are possible and 275-kilometer-per-hour (170-mile-per-hour) winds are not uncommon. The dust may result from meteorite impacts or other processes. The wind may move the dust to lower elevations, and in this way, create the featureless plains of Mars. The pits and basins of Fig. 18–13 could also be caused by wind erosion. The cliff, up to a mile high, that surrounds Olympus Mons (see Fig. 18–14) may also have been produced by wind erosion.

The volcanic area has Olympus Mons (see Fig. 18–14), the largest volcano ever seen. It is about 8 kilometers high and 483 kilometers (300 miles) across at the base; it has a broad complex crater, and its flanks are covered with what appears to be flow material. Nearby are three big volcanoes in a row, one of which is over 24 kilometers (15 miles) high.

The equatorial region near the volcanic area is a plateau cut by what appear to be faults. (See Fig. 18–15.) The valleys shown in Fig. 18–15A are part of a parallel system that extends more than 4830 kilometers (3000 miles). They are interpreted as tensional features, perhaps associated with uplift. Fig. 18–15B shows another valley that also may be of fault origin. Note the lines of craters that parallel the main valley.

Mars also has features that suggest erosion by running water, although very little water is present in the atmosphere. Figure 18–16 shows an example of a long, sinuous gully that resembles a river course. The water might have come from melting part of the polar caps; that will be discussed later. It is also possible that water in the form of ice is present in the "soil." The pits and basins shown in Fig. 18–13 are near the south pole, and so could have formed by the melting of subsurface ice. The chaotic terrain of short ridges and small valleys shown in Fig. 18–17 is similar to landslide and collapsed areas on Earth caused by melting subsurface ice. Clouds on some of the big volcanoes suggest the possibility that they may be emitting gases, possibly water vapor, that might at times be abundant enough to cause rain and so erosion.

Craters are common in many areas, and, although some are of volcanic origin, many are caused by meteorite impact. This is not unexpected because Mars is near the asteroid belt and has a thin atmosphere. The effects of meteorite impacts are clearly shown on Mars' moon Phobos. (See Fig. 18–

FIGURE 18–12   The northern hemisphere of Mars. The shrinking north polar ice cap is at the top, and the huge volcanoes are near the bottom. Photo from NASA.

FIGURE 18–13   A dune field within a 150-kilometer- (93-mile-) wide crater on Mars. Photo from NASA.

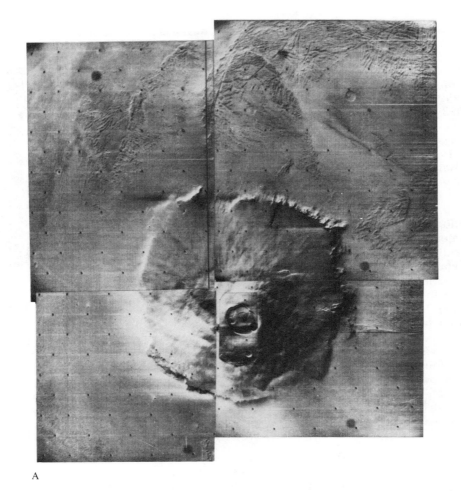

A

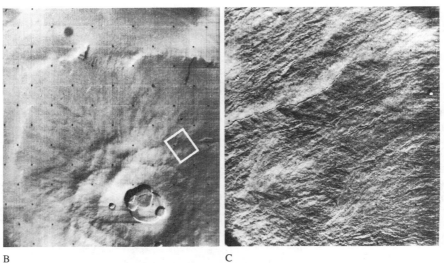

B   C

FIGURE 18–14 Olympus Mons, a gigantic volcanic mountain on Mars. A. The mountain is 500 kilometers (310 miles) in diameter across the base. The complex crater is 64 kilometers (40 miles) in diameter. B. Closer view of the summit. C. Telephoto view of the flank showing features that resemble flow material on Earth. Part C shows an area 43 by 56 kilometers (27 by 34½ miles). Photos from NASA.

368

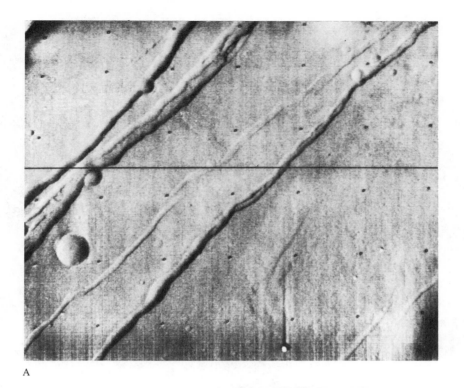

A

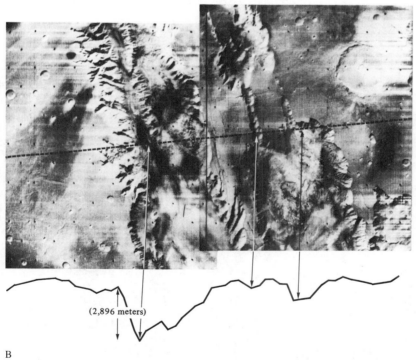

(2,896 meters)

B

FIGURE 18–15   Valleys of probable fault origin on Mars. A. These structures are part of a system of parallel fissures extending more than 4830 kilometers (3000 miles). The widest valley is about 1.6 kilometers (1 mile) across and has a smaller valley inside it. The photo covers an area 34 by 42 kilometers (21 by 26 miles). B. A valley on Mars much deeper than the Grand Canyon. This valley is 120 kilometers (75 miles) across. Photos from NASA.

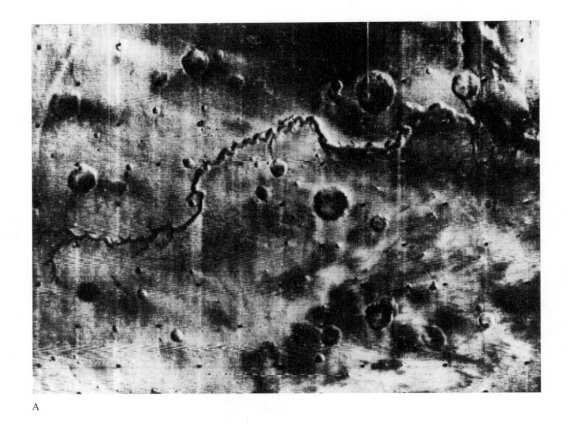

A

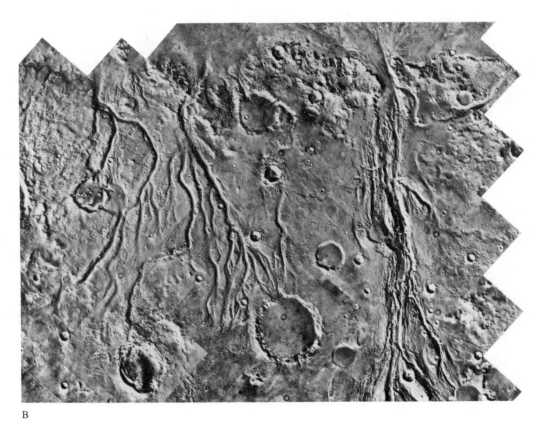

B

FIGURE 18-16 Erosional features on Mars. A. River-like gully suggests erosion by running water. B. The pattern of these gullies suggests erosion by floodwater.

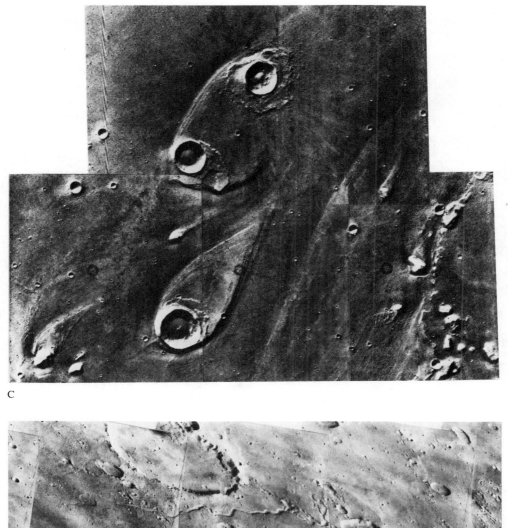

C

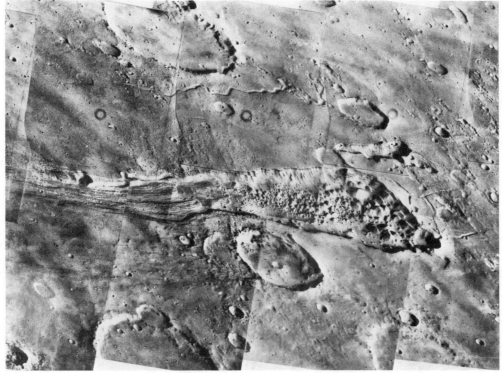

D

FIGURE 18–16 (cont.)   C. These teardrop-shaped islands are also suggestive of erosion by floodwater. D. The chaotic terrain at the head of this gully suggests that melting of subsurface ice may have been the source of the fluid that eroded the gully. Photos from NASA.

FIGURE 18–17   Chaotic terrain on Mars is shown on the right. The chaotic terrain is a jumble of ridges, each of which is 2.4 to 3.2 kilometers (1.5 to 2 miles) wide and 1.6 to 8 kilometers (1 to 5 miles) long. *Mariner 6* photo from NASA.

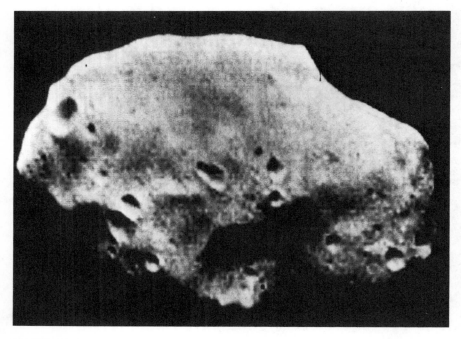

FIGURE 18–18   *Mariner 9* photograph of Mars' moon Phobos. Phobos is about 21 by 26 kilometers (13 by 16 miles). The age and strength of Mars' innermost moon is shown by the impact crater. Photo from NASA.

18.) Mars' equator almost separates cratered from uncratered areas. Why such a hemispherical division occurs is not known.

The polar ice caps appear to be dual ice caps. Much of the ice is frozen carbon dioxide, and this is the part that disappears and reforms seasonally. The rest is composed of water ice. The laminated structure near the ice caps suggests that windblown dust accumulates in the ice at times. These deposits seem to be at least 100 meters (300 feet) thick.

VENUS

Venus is named for the goddess of love because it is visible only near sunrise and sunset. It has a thick atmosphere, composed almost entirely of carbon dioxide, and the carbon dioxide causes a greenhouse effect, which accounts for Venus' high temperature. Also present in minor amounts are hydrogen, water vapor, carbon monoxide, and hydrofluoric and hydrochloric acids. The surface temperature is near 465°C (870°F), making life a very remote possibility. The atmosphere has a solid cloud cover 72 to 97 kilometers (45 to 60 miles) thick so the surface is not visible. In spite of its slow rotation, the surface temperature does not change as much as expected from day to night, suggesting a circulation in the atmosphere that transfers the heat. Venus is the only planet that rotates on its axis in a clockwise direction (see discussion of Uranus below). No magnetic field has been detected, suggesting that it, too, lacks a liquid core, although its rate of rotation may be too slow to develop a magnetic field. The photographs sent back to Earth by soft-landers that operated for about an hour before the high surface temperatures incapacitated them showed angular rocks and no sand or dust. Radar observations reveal mountain ranges, plateaus, and craters, but little can be said about their type or origin, although tectonic and volcanic features appear to be present.

GREAT PLANETS

The great planets, Jupiter, Saturn, Uranus, and Neptune, are so called because they are much larger than the other planets. Although larger in diameter and mass, they are much less dense than the other planets. They all have thick atmospheres, which are, of course, all that we can see. Their rapid rotation causes pronounced bulges of the gaseous exteriors. The upper surfaces of their atmospheres are very cold because of their distances from the sun. They all have methane in their atmospheres, with ammonia, also, in Jupiter and Saturn; but most of their atmospheres are hydrogen and helium. Their low overall densities suggest that they, like the sun, are composed mainly of hydrogen and helium. Saturn, Jupiter, and Uranus are known to have rings. Saturn's rings are best known and probably are dust similar in composition to a moon. (See Fig. 18–19.) This material is too close to the planet to consolidate and form a moon. The rings may be left over from the formation of the planet, or they might be satellites that approached the planet too closely and were broken up when the gravitational attraction overcame their strength. Uranus is unique in having its rotational axis tilted 98 degrees to its orbital plane; alternately, it could be considered tilted 82 degrees and revolving in the opposite direction to the majority of the planets. Its moon's orbit is in the same direction as its rotation.

LARGER MOONS OF JUPITER AND SATURN

The larger moons of Jupiter and Saturn range in size from slightly smaller than our moon to some-

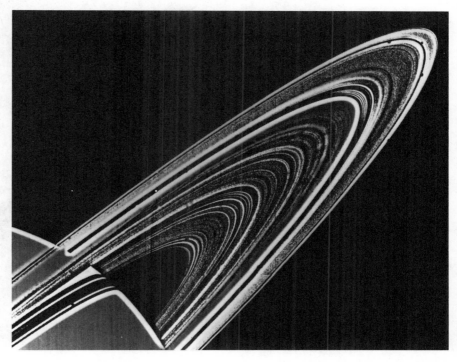

FIGURE 18–19 Saturn's ring system. The shadow of the rings on the planet is seen in the lower left. *Voyager* photograph provided by National Space Science Data Center.

what larger than Mercury. Jupiter and Saturn have at least 15 satellites each, but only 4 of Jupiter's and one of Saturn's are similar in size to the inner planets. Only one other moon, Neptune's Triton, is of similar size, but very little is known about it.

The most interesting of these bodies is Jupiter's Io, which is slightly larger, 3640 kilometers (2260 miles) in diameter, than our moon and has about the same density. Spacecraft images have revealed eight active volcanoes (Fig. 18–20). The volcanic eruptions are caused by tidal flexing by the other nearby large satellites, especially Europa. The gravitational attraction of the other satellites deforms Io and so heats it by friction. The material erupted by the volcanoes is sulfur and sulfur dioxide; these materials are responsible for Io's colors—red, orange, black, and white. Io and the other large satellites are obviously much different from the inner planets.

The other large satellites of Jupiter are composed of rock and water. Europa [3050 kilometers (1890 miles) diameter] has a density of 3.03 grams per cubic centimeter, which is somewhat less than our moon's. It is believed to be composed of rock with a layer of ice about 100 kilometers (60 miles) thick at the surface. The surface shows a tangle of lines, perhaps cracks in the ice (Fig. 18–21). Ganymede [5270 kilometers (3270 miles) diameter] has a density of 1.93 grams per cubic centimeter and is believed to be about half rock and half water. Callisto [5000 kilometers (3000 miles) diameter] is less dense and contains more water than rock. Ganymede's surface has dark cratered areas and lighter grooved, perhaps faulted, areas (Fig. 18–22). Callisto's surface is covered with craters (Fig. 18–23).

Saturn has only one large moon, Titan, 5000 kilometers (3000 miles) in diameter. Titan has a thick atmosphere, so little is

FIGURE 18–20   Io is the only object, except Earth, on which volcanic eruptions have been observed. *Voyager* photograph provided by the National Space Science Data Center.

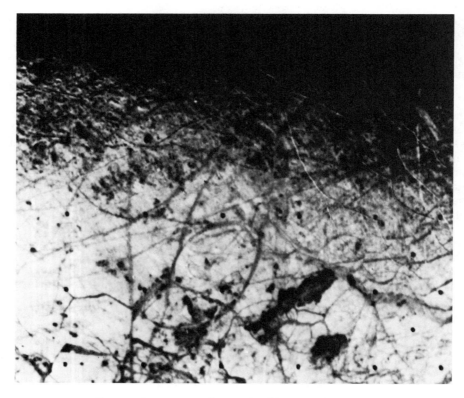

FIGURE 18–21   Europa shows a complex tangle of lines. *Voyager* photograph provided by National Space Science Data Center.

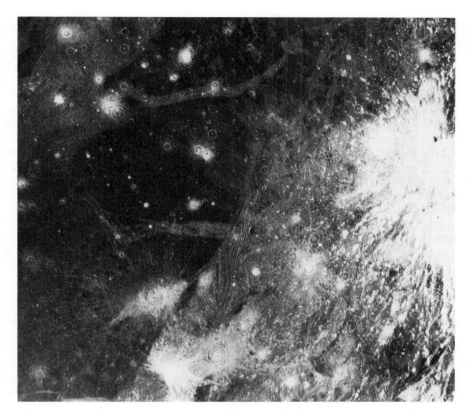

FIGURE 18–22   Ganymede is cratered. It has light-colored grooved terrain and dark, heavily cratered areas. *Voyager* photograph provided by National Space Science Data Center.

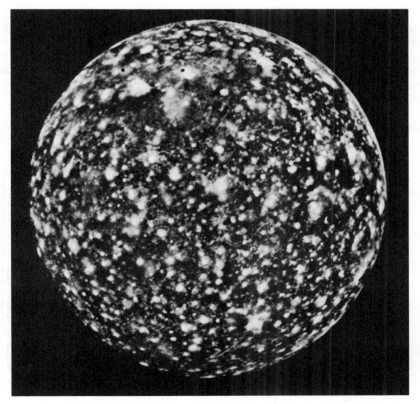

FIGURE 18–23   Callisto has many bright craters. *Voyager* photograph provided by National Space Science Data Center.

375

known about its surface. Although it is only slightly larger than Mercury, Titan is able to retain an atmosphere because it is so cold. The atmosphere is, like the Earth's, composed largely of nitrogen. The other constitutents of Titan's atmosphere are hydrocarbons, especially methane. The surface temperature is estimated to be about −184°C (−300°F), and liquid nitrogen may be present on the surface. The action of the sun's energy on this atmosphere has produced brown gases that are probably similar to photochemical smog on Earth, which forms as the result of the sun's actions on hydrocarbons in our atmosphere.

Images of some of Saturn's smaller inner satellites have also been returned (Fig. 18–24). They are all much smaller, ranging in size from Mimas, 390 kilometers (240 miles) in diameter to Rhea, 1530 kilometers (950 miles) in diameter. They are probably all composed largely of water ice, and all except Enceladus are cratered. Mimas has a crater with a diameter about one-quarter of the diameter of the satellite. Tethys is fractured with a valley 70 kilometers (40 miles) wide and 800 kilometers (500 miles) long. Smaller sinuous, branching valleys are seen on Dione.

COMPARATIVE GEOLOGY OF THE INNER PLANETS AND THE MOON

We have studied the inner planets using spacecraft, and their structures and processes appear somewhat like Earth's. Samples have been returned from the moon, and close-up images of it and the other inner planets have been obtained. Images have also been returned of some of the moons of Jupiter and Saturn, and, although similar in size, they are different from the inner planets.

The most obvious difference between the Earth, and the moon and inner planets is the absence of or greatly reduced amount of erosion on the latter because, with the exception of Venus, they have very tenuous or no atmospheres. This aids our interpretation of their internal processes because surface forms are not destroyed by erosion. *All of the inner planets and their moons have been bombarded by meteorites, and the resulting craters are conspicuous on all except Earth where erosion has destroyed most of the craters.*

Sea-floor spreading and plate tectonics are the main internal processes on Earth, and somewhat similar processes appear to operate on some of the other planets. The energy for these internal processes must come from within the planet. *The amount of internal energy apparently is proportional to the size of the planet.* The other planets are smaller than Earth, and they show some aspects of plate tectonics.

The similarities among the moon and the inner planets so far studied by space probe are striking. The similarity of surface features strongly suggests similar histories for the moon and these planets—Mars and Mercury. Rocks have been returned only from the moon, so the dating of events on Mars and Mercury depends on correlation with the moon on degree of cratering. Their history begins with their formation about 4550 million years ago. From then until about 4000 million years ago, cratering occurred at a rapid rate, forming the cratered highlands. Apparently the inner planets experienced a high influx of meteorites during this interval. Erosion has largely obliterated evidence of this event on Earth. From 4000 million to about 3100 million years ago, the number of impacts diminished rapidly before the less cratered volcanic lowlands formed.

The moon's cratered uplands differ somewhat in composition from the volcanic maria. This is

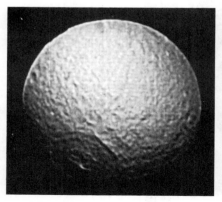
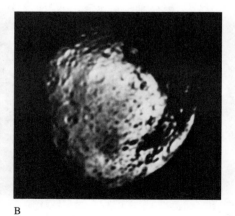
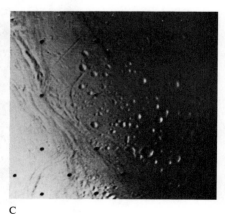

A   B   C

FIGURE 18–24 Satellites of Saturn. A. Tethys. B. Iapetus. C. Surface detail on Enceladus. *Voyager* photographs provided by National Space Science Data Center.

seen from Earth as a difference in reflectivity. Similar differences in reflectivity are apparent on Mars and Mercury. This suggests that they, like Earth, underwent some melting early in their histories. The resulting layering is revealed by seismic studies on the Earth and the moon. The energy released by crystallization of the Earth's fluid core may be the energy that causes deformation of the crust. The early crystallization of the cores of the smaller planets may account for their lack of compressional features and for the great age of their surface features. The larger of the small planets, Mars and Mercury, have some surface deformation. Mars has great fractures, and Mercury has long scarps, suggesting again that size and internal energy are related.

The smaller planets also differ from Earth in that their surface features are not evenly distributed. Mars' equator is almost the dividing line between a cratered hemisphere and a smooth one. Mercury also has cratered and smooth halves. The side of the moon opposite Earth is more cratered than the side facing us. Why such a distribution of upland and lowland exists is not clear, although at present there is more continent in the Earth's northern hemisphere than in the southern. At times in the geologic past, the Earth's continents are believed to have been joined into a single large continent. Thus had the Earth run out of internal energy earlier, as the smaller planets did, a land hemisphere and an ocean hemisphere might have resulted.

METEORITES

Meteorites are objects from space that hit the Earth. Their composition and some other features were discussed earlier in relation to the layered structure of the Earth. The iron-nickel meteorites are believed to be similar in composition to the core of the Earth, and the stony meteorites similar to the mantle. One interesting aspect of meteorite recognition is that over 90 percent of those seen to fall are stony meteorites, but only a third of the finds not seen to fall are stony. Only a few aspects of meteorites will be discussed here; much more information can be found in the references at the end of this chapter.

Meteorite specimens range from dust to 264,000 kilograms (60 tons). Those that have produced the large impact craters on the Earth were much larger, weighing perhaps hundreds or thousands of tons. (See Figs. 18–25 and 18–26.) The shock wave caused by such impacts can cause metamorphism of the rocks as well as shattering. A few very high pressure minerals form in this way and are found only at suspected impact craters. Meteorite craters on the Earth are less than 16 kilometers (10 miles) across; however, on the moon some craters are more than 242 kilometers (150 miles) in diameter. Perhaps such large craters are not recognizable on the Earth. Many of the smaller meteorites do not survive the heat generated by friction in passing through the Earth's atmosphere, producing meteors, or shooting or falling stars. It is estimated that between 1000 and 1,000,000 tons of meteorites fall on the Earth each year.

Meteorites are believed to have formed at the same time as the rest of the solar system (discussed in the next section). Thus, they are used to date the formation of the solar system and as models of the Earth's structure. Meteorites are fragments of much larger bodies. This is shown by the coarse crystalline nature of many, suggesting slow cooling that is possible only in a large body. Many, however, have complex crystallization histories. As shown in the next chapter, the Earth developed its layered structure by melting after initial formation. Some meteorites are also differentiated bodies, and this melting can only occur in bodies large enough to generate the necessary heat, again suggesting that meteorites are fragments of large bodies. Meteorites are generally believed to be fragments from the asteroid belt. The asteroids are a group of about 30,000 small bodies ranging from about 805 kilometers (500 miles) to about 1.6 kilometers (1 mile) in diameter. Thousands of millions more down to dust size are also thought to exist. The asteroids are believed to have formed by collisions between two or more small planets.

FIGURE 18–25   Meteor Crater (Barringer Crater), Arizona. A similar crater would not be as well preserved in a more humid climate. Photo from Yerkes Observatory.

The orbits of many meteors have been recorded by radar and are consistent with an asteroid origin.

However, some of the meteor swarms that the Earth passes through each year are very different in that they are predictable and are not associated with large meteorite falls. They are in the or-bits of now dead comets. Small particles of meteorite material, called micrometeorites, probably originate from such meteor swarms and so have their origin in comets. Micrometeorites of this or other origin are believed to cause the erosion of the moon. Comets are described next.

Cosmic ray bombardment in space can cause the formation of new elements, and the amounts of such elements record the time that the object has been in space. Large nuclei cosmic rays can also produce tracks like those shown in Fig. 16–15, and the number of such tracks is a measure of the time of exposure. The various types of meteorites have different exposure ages, again suggesting that meteorites have complex histories.

COMETS

Comets are among the most awe-inspiring spectacles in nature. (See Fig. 18–27.) In the past they were considered harbingers of evil times. We now know their makeup

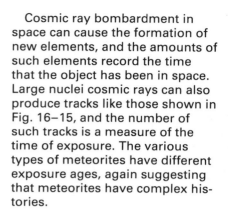

Moenkopi formation
Kaibab limestone
Coconino and Toroweap formations

1. Meteorite approaches ground at 15 km/sec.

fused rock

2. Meteorite enters ground, compressing and fusing rocks ahead and flattening by compression and by lateral flow. Shock into meteorite reaches back side of meteorite.

shock front   fused rock

3. Rarefaction wave is reflected back through meteorite, and meteorite is decompressed, but still moves at about 5 km/sec into ground. Most of energy has been transferred to compressed fused rock ahead of meteorite.

shock front

4. Compressed slug of fused rock and trailing meteorite are deflected laterally along the path of penetration. Meteorite becomes liner of transient cavity.

shock front

5. Shock propagates away from cavity, cavity expands, and fused and strongly shocked rocks and meteoritic material are shot out in the moving mass behind the shock front.

reflected rarefaction   shock front

6. Shell of broken rock with mixed fragments and dispersed fused rock and meteoritic material is formed around cavity. Shock is reflected as rarefaction wave from surface of ground and momentum is trapped in material above cavity.

reflected   shock front   rarefaction

7. Shock and reflected rarefaction reach limit at which beds will be overturned. Material behind rarefaction is thrown out along ballistic trajectories.

8. Fragments thrown out of crater maintain approximate relative positions except for material thrown to great height. Shell of broken rock with mixed meteoritic material and fused rock is sheared out along walls of crater; upper part of broken rock is ejected.

9. Fragments thrown out along low trajectories land and become stacked in an order inverted from the order in which they were ejected. Mixed broken rock along walls of crater slumps back toward center of crater. Fragments thrown to great height shower down to form layer of mixed debris.

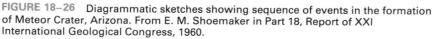

0   1,000   2,000 Feet

0   600   Meters

FIGURE 18–26 Diagrammatic sketches showing sequence of events in the formation of Meteor Crater, Arizona. From E. M. Shoemaker in Part 18, Report of XXI International Geological Congress, 1960.

FIGURE 18–27 Photo of Halley's comet. Photo from Mount Wilson and Palomar Observatories.

in general terms and can predict their reappearance, but we know almost nothing about how they form. Comets have very small masses. One passed through Jupiter's moons without noticeable effect on the moons, and the Earth has passed through the tail of a comet in historic time. However, in 1908 a great explosion, heard 805 kilometers (500 miles) away, occurred in Siberia that was believed to have been caused by a meteorite. When the area was visited in 1927, although the forest was knocked down for over 16 kilometers, no large crater was found. It has been postulated that the explosion might have been caused by a gaseous comet rather than a solid meteorite. *Comets are believed to be masses of frozen gas with some very small particles similar in composition to meteorites.* This is sometimes called the "dirty snowball" theory. In the case of the Siberian explosion the ice of the comet is believed to have turned to gas very rapidly by frictional heating in the atmosphere and the resulting shock wave caused the explosion and damage.

When distant from the sun, a comet is very small, perhaps 0.8 kilometer (0.5 mile) in diameter. At this distance it is seen only by the sunlight that it reflects. When closer to the sun, the sun's radiation melts some of the ice and ionizes some of the resulting gases, making the comet more visible. Thus, comets disintegrate by passing near the sun and die after a number of passes near the sun. During the process of disintegration, the spectra of water, carbon dioxide, methane, and ammonia can be seen. At this stage an average comet may have a head 128,800 kilometers (80,000 miles) in diameter, made mostly of very thin gas. The solar wind, to be discussed in the next section, pushes this gas outward from the head to form a tail up to 80 million kilometers (50 million miles) long. This is the spectacular stage of a comet

and occurs only when the comet is near the sun. One interesting point is that the solar wind pushes the tail away from the sun without regard to the direction in which the comet is moving. Most comets have very elliptical orbits, which explains why they reappear at intervals. As with the planets, the sun is at one focus of the ellipse, but when distant from the sun, comets are too small to see. Some comets have orbits that bring them back only after thousands of years; some may be parabolic and so never return. (See Fig. 18–28.)

The origin of comets is not known; they may form at the fringe of the solar system or may come from outside the solar system. Their composition is not too unlike the material from which the solar system is believed to have formed (discussed in the next section). Thus they may be fragments "left over" from the origin of the solar system.

ORIGIN OF THE SOLAR SYSTEM

The origin of the solar system is a problem that we have attempted to solve for many years. No completely satisfactory theory has yet been offered, and objections exist to all theories, including the one presented here. The best theory offered to date states that the planets and the other orbiting objects formed at the same time the sun formed. Other theories assumed that the planets formed from material pulled from the existing sun, generally by the gravitation of another star that happened to pass nearby. The many physical objections to this latter theory suggest that it is impossible. Only a brief outline of the newer theory can be given here.

Stars are believed to form by the condensation of a cosmic dust or gas cloud called a *nebula* (cloud), composed mainly of hydrogen and helium and perhaps frozen like snow. This condensation is caused largely by gravitational attraction. Condensation causes the temperature to rise; and when the temperature rises high enough, nuclear reactions begin and the new star begins to radiate energy.

The planets are believed to have formed in a similar way during the origin of the sun. During condensation, such a gas cloud would almost certainly rotate because of the initial velocities of the particles brought together. As condensation continued, the rotation became more uniform as the random velocities cancelled out. Ultimately, the cloud flattened and became disk-shaped with most of the mass concentrated near the center. Further contraction and flattening made the disk unstable, and it broke into a number of smaller clouds. These smaller clouds were all in the same plane.

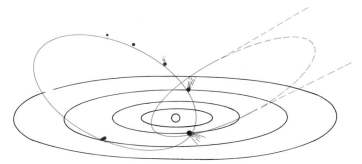

FIGURE 18–28 The solar system showing some possible comet orbits. The comet is shown on one orbit. Note that the tail always points away from the sun and that the comet gets larger where it is heated by the sun. The other orbit shows possible elliptical and parabolic orbits.

Condensation occurred in each of these clouds, concentrating the heavier elements at the center. Because the nebula had about the same composition as the sun, the ancestral planets were surrounded by envelopes of hydrogen and helium. At about this stage, condensation had heated the ancestral sun to the point that nuclear reactions began. The sun's radiation ionized the remaining part of its nebula; and these ionized particles interacted with the sun's magnetic field, reducing the sun's angular momentum (spin).

Thus, the sun was slowed to its present slow rate of rotation, and the particles in the dust cloud gained in energy. Ionized particles radiated from the sun swept away the remaining gaseous dust surrounding the sun and the planets in the same way that the sun's radiation makes the tail of a comet point away from the sun, regardless of the direction of motion of the comet. In this process the planets nearest the sun lost the most material. They are much smaller and denser than the more distant planets. As each planet condensed, it rotated faster, just as spinning skaters spin faster when they bring their arms in to their bodies. This theory explains the size and density of the planets, the distribution of momentum in the solar system, and why most of the planets revolve in the same direction.

The asteroid belt between Mars and Jupiter may have originated from a number of small centers of accumulation rather than from a single center, as did the other planets. Collisions among these bodies have increased their total number, and it is the resulting fragments from larger bodies that are the meteorites.

The moons formed close to the planets in the same way. When the planets lost much of their original masses because of the sun's radiation, the gravitational attraction between the planets and their moons was reduced. As a result, the moons moved further away from the planets. Some of these satellites may have escaped from their original planets to be captured by other planets. Such a transposition could account for the moons that revolve in the opposite direction. Pluto may once have been a moon of Neptune. Some of these erratic moons may even be captured asteroids.

This theory accounts for many other features of the solar system. The composition of the crust of the earth was our starting point in Chapter 2, and this can now be discussed further. The estimated composition of the universe is shown in Table 18–2; this is probably representative of the composition of the nebula from which the solar system condensed. Note that hydrogen predominates, that helium forms most of the rest, and that all the other elements constitute less than one percent of the total. Most astronomers believe that originally the universe consisted entirely of hydrogen, the simplest atom, and that all other elements have been made by combining hydrogen atoms by a process called fusion. Such fusion can occur at the high temperatures, millions of degrees, inside stars. The development of the elements is believed to begin with the condensation of a gas cloud, or nebula, composed of hydrogen. The resulting star produces energy by atomic fusion of hydrogen to helium. As the star ages, a stage is reached when the atomic "fuel" in the core has been nearly used up. At this time the temperature increases greatly and energy is produced by the formation of heavier elements from helium. At

TABLE 18–2 Comparison of the estimated composition of the crust, total solid earth, and universe. The figures for the total earth and the universe are much more uncertain than those for the crust. The compositions of the whole solid earth and the universe are discussed in Chapters 13 and 18 respectively.

Element	Atomic Number	Approx. Atomic Weight	Abundance (weight percent)		
			Crustal Rocks[1]	Total Solid Earth[2]	Universe[3]
Hydrogen	1	1	0.14	—	61.0
Helium	2	4	0.0000003	—	36.8
Carbon	6	12	0.02	—	0.25
Nitrogen	7	14	0.002	—	0.08
Oxygen	8	16	46.60	30.0	1.0
Silicon	14	28	27.72	15.0	0.06
Aluminum	13	27	8.13	1.1	0.006
Iron	26	56	5.00	35.0	0.02
Calcium	20	40	3.63	1.1	0.006
Sodium	11	23	2.83	0.57	0.003
Potassium	19	39	2.59	0.07	0.0003
Magnesium	12	24	2.09	13.0	0.05
Titanium	22	48	0.44	0.05	0.0002
Phosphorus	15	31	0.105	0.10	0.0006
Manganese	25	55	0.095	0.22	0.0008
Nickel	28	59	0.0075	2.4	0.004
Sulfur	16	32	0.026	1.9	0.02
Chromium	24	52	0.010	0.26	0.001
Cobalt	27	59	0.0025	0.13	0.0002

[1]From Table 2–1.
[2]From Brian Mason, *Principles of Geochemistry*, John Wiley & Sons, 3rd ed., 1966, p. 53. (These data are estimates, and because only two significant figures are used the total is 101 per cent.)
[3]Data after A. G. W. Cameron.

this stage, energy may be produced in the core faster than it can be radiated from the surface and an explosion results. The explosion may disintegrate the star, making it a *supernova* (a very bright, short-lived "star"); or the star may lose its material in less spectacular ways. The complete history of stars is complex and not entirely known; however, it is known that the history of a star depends greatly on its mass, and that larger stars move through this cycle much more rapidly than do average-size stars, such as our sun. In any case, a star formed initially of hydrogen produces heavier elements and scatters these elements into the universe. The solar system formed from a dust cloud that contained heavy elements as well as hydrogen.

The correspondence between the composition of the universe and the composition of the earth's crust is shown in Table 18–2. The light elements, hydrogen and helium, are most abundant but were lost during the formation of the earth, as just described. The next most abundant element in the universe is oxygen, which is most abundant in the crust.

EARTH-MOON SYSTEM

Interaction between the earth and the moon is believed to be slowing the earth's initially rapid rotation. This process is a continuing one, and the length of day is increasing. The year, the period of the earth's trip around the sun, is not affected because it is determined by the distance between the sun and the earth. Thus the number of days in the year is decreasing. The present average rate of slowing of the earth's rotation is estimated at between 1 second per day per 120,000 years to 2 seconds per day per 100,000 years. This is enough to produce apparent errors if the times of eclipses computed back in terms of our present day are compared with the ancient records. An ingenious

check on this lengthening of the day in the geologic past has been devised. Some corals grow by adding layers much like tree rings. Thinner layers, apparently daily, occur within the annual layers. These daily layers are commonly not well preserved, but the few that can be counted suggest that in Middle Devonian time, about 350,000,000 years ago, the year had 400–410 days and in Pennsylvanian time, about 280,000,000 years ago, 390 days. (See Fig. 18–29.) The figures are in good agreement with the estimated slowing of the earth's rotation. Other types of fossils also show daily, monthly, tidal, and annual cycles, and it may soon be possible to determine the early history of the earth-moon system from such studies.

The slowing of the earth's rotation is caused by the tidal drag of the moon. The total rotational energy of the earth-moon system is conserved, so the energy lost by the earth is gained by the moon. The moon moves farther from the earth, and the length of the month, the moon's period around the earth, increases. Calculations show that ultimately the earth's rotation and the moon's orbital period will be the same, so that one part of the earth will always face the moon just as only one part of the moon now faces the earth. Similar calculations can be made to find the moon's past position and orbit. The results of such calculations depend to a large extent on the assumptions made. This is unfortunate, because otherwise the problem of

FIGURE 18–29 Growth lines in coral. The fine bands are daily growth layers. By counting the number of such bands within the yearly growth bands, it is possible to determine the number of days in the year at the time the coral lived. Photo courtesy of S. K. Runcorn.

381

the moon's origin and early history probably could be solved.

The origin of the moon cannot be determined, and there are serious objections to all theories so far proposed. The theory that the earth and the moon formed at about the same time at different centers of accumulations in a dust cloud was just described. The main objection to this is that both earth and moon should have the same composition if they formed in this way, but their average specific gravities are different. The average specific gravity of the earth is 5.5, and that of the moon is 3.3. Capture hypotheses were also mentioned in that earlier section, especially in regard to moons of other planets. Some of the recent calculations of the moon's early orbit strongly suggest a capture origin. However, if the earth captured the moon, the orbital energy (kinetic energy) of the moon would have heated the earth, probably to near its melting point, an event not suggested by the earth's thermal history. Many other ideas have been suggested, but all have physical objections.

SUMMARY

The *solar system* is composed of the sun and the planets, comets, meteors, asteroids, and the satellites of the planets.

The *inner planets,* Mercury, Venus, Earth, and Mars, are smaller and denser than the more distant great planets.

The *moon* orbits the Earth and rotates on its axis at the same rate, so only one side is visible from Earth. The dark, smooth areas are called *maria* (*mare,* singular) and the lighter areas are uplands.

Most of the moon rocks are breccias formed by the meteorite impacts that also cause the many craters. The mare rocks are basaltic in composition, and those of the uplands are feldspar-rich rocks.

Mascons are areas of concentration of mass under the large circular maria.

Most *moonquakes* are low magnitude and are caused by tidal effects. They originate from a single zone and reveal a layered moon with about 19 kilometers (12 miles) of surface rubble with velocity increases at about 19 and 64 kilometers (12 and 40 miles) deep. The core, starting at about 966 kilometers (600 miles), may be partially melted because seismic waves are somewhat attenuated there.

The moon probably formed about 4550 million years ago and probably melted soon after this. The oldest rocks so far found are from the highlands and are about 4000 million years. The maria basalts were extruded between 3900 and 3100 million years. Since then the moon has been quiet, with only meteorite impacts and perhaps some limited local volcanic activity.

Mars has dust storms, seasonal ice caps, huge volcanoes, evidence of faulting, apparent erosion by running water, and impact craters, as well as featureless and chaotic terrains. It is the other planet most likely to have life.

Venus has a thick atmosphere and is very hot.

None of the inner planets have mountain ranges like the Earth's.

The *great planets,* Jupiter, Saturn, Uranus, and Neptune, are composed mainly of hydrogen and helium.

Meteorites are objects from space that hit the Earth. Iron-nickel meteorites are believed to be similar in composition to the earth's core. Over 90 percent of the meteorites seen to fall are stony. Meteorites range from dust to 264,000 kilograms (60 tons). Large meteorites cause craters. Their origin is probably the asteroids.

Comets are masses of frozen gas with some small particles. They orbit the sun and are visible when they are close enough to the sun that their gases are ionized by the sun's energy.

Growth rings on corals and other marine animals show that the earth's rotation has been slowing. This is caused by the tidal drag of the moon.

QUESTIONS

1 List the components of the solar system.
2 How do the inner planets differ from the great planets?
3 How can the different numbers of moons that each planet has be explained?
4 Which planets might contain life? Why?
5 What is the origin of the moon's surface features?
6 How do comets differ from meteorites?
7 How long would it take for meteorite falls to produce the earth? Is this a reasonable theory for the origin of the earth?
8 The uranium-lead and thorium-lead methods are used to date meteorites in spite of the problems in correcting for original lead. Why are not other radioactive methods used?

SUPPLEMENTARY READINGS

Planets—General

Head, J. W., and S. C. Solomon, "Tectonic Evolution of the Terrestrial Planets," *Science* (July 3, 1981), Vol. 213, No. 4503, pp. 62–76.

Head, J. W., S. E. Yuter, and S. C. Solomon, "Topography of Venus and Earth: A Test for the Presence of Plate Tectonics," *American Scientist* (November-December 1981), Vol. 69, No. 6, pp. 614–623.

Head, J. W., and others, "Geologic Evolution of the Terrestrial Planets," *American Scientist* (January-February 1977), Vol. 65, No. 1, pp. 21–29.

King, E. A., *Space Geology: An Introduction.* New York: John Wiley & Sons, Inc., 1976, 400 pp.

Lewis, J. S., "The Chemistry of the Solar System," *Scientific American* (March 1974), Vol. 230, No. 3, pp. 51–65.

Murray, Bruce, M. C. Malin, and Ronald Greeley, *Earthlike Planets. Surfaces of Mercury, Venus, Earth, Moon, Mars.* San Francisco: Freeman, 1981, 388 pp.

Sagan, Carl, "The Solar System," *Scientific American* (September 1975), Vol 233, No. 3, pp. 22–31. This entire issue is on the solar system.

Mars

Arvidson, R. E., A. B. Binder, and K. L. Jones, "The Surface of Mars," *Scientific American* (March 1978), Vol. 238, No. 3, pp. 76–89.

Carr, M. H., "The Volcanoes of Mars," *Scientific American* (January 1976), Vol. 234, No. 1, pp. 32–43.

Horowitz, N. H., "The Search for Life on Mars," *Scientific American* (November 1977), Vol. 237, No. 5, pp. 52–61.

Murray, B. C., "Mars from *Mariner 9*," *Scientific American* (January 1973), Vol. 228, No. 1, pp. 48–69.

Venus

James, J. N., "The Voyage of *Mariner II*," *Scientific American* (July 1963), Vol. 209, No. 1, pp. 70–84.

Moon

Goldreich, Peter, "Tides and the Earth-Moon System," *Scientific American* (April 1972), Vol. 226, No. 4, pp. 43–52.

Hammond, A. L., "Lunar Science: Analyzing the Apollo Legacy," *Science* (March 30, 1973), Vol. 179, No. 4080, pp. 1313–1315.

Hartmann, W. K., "Cratering in the Solar System," *Scientific American* (January 1977), Vol. 236, No. 1, pp. 84–99.

Mason, Brian, "The Lunar Rocks," *Scientific American* (October 1971), Vol. 225, No. 4, pp. 48–58.

Runcorn, S. K., "Corals as Paleontological Clocks," *Scientific American* (October 1966), Vol. 215, No. 4, pp. 26–33.

Schmitt, H. H., "Apollo 17 Report on the Valley of Taurus-Littrow," *Science* (November 16, 1973), Vol. 182, No. 4113, pp. 681–690. A report on the geology of the moon, written by a geologist who was there.

Meteorites

Lawless, J. G., and others, "Organic Matter in Meteorites," *Scientific American* (June 1972), Vol. 226, No. 6, pp. 38–46.

Urey, H. C., "Meteorites and the Moon," *Science* (March 12, 1965), Vol. 147, No. 3663, pp. 1262–1265.

Comets

Whipple, F. L., "The Nature of Comets," *Scientific American* (February 1974), Vol. 230, No. 2, pp. 49–57.

Great Planets

Hubbard, W. B., "Interiors of the Giant Planets," *Science* (October 9, 1981), Vol. 214, No. 4517, pp. 145–149.

Ingersoll, A. P., "Jupiter and Saturn," *Scientific American* (December 1981), Vol. 245, No. 6, pp. 90–108.

Owen, Tobias, "Titan," *Scientific American* (February 1982), Vol. 246, No. 2, pp. 98–109.

Science (January 29, 1982), Vol. 215, No. 4532. The majority of this issue (pp. 499–594) is devoted to reports on the *Voyager 2* encounter with Saturn and its satellites.

Science (April 10, 1981), Vol. 212, No. 4491. A number of articles in this issue cover the *Voyager* mission to Saturn.

Soderblom, L. A., "The Galilean Moons of Jupiter," *Scientific American* (January 1980), Vol. 242, No. 1, pp. 88–100.

Soderblom, L. A., and T. V. Johnson, "The Moons of Saturn," *Scientific American* (January 1982), Vol. 246, No. 1, pp. 100–116.

Waldrop, M. M., "The Puzzle That is Saturn," *Science* (September 18, 1981), Vol. 213, No. 4514, pp. 1347–1351.

19

THE PRECAMBRIAN—
THE OLDEST ROCKS
AND THE
FIRST LIFE

The conclusion cannot be avoided that the iron-formations of the pre-Cambrian, not only of the Lake Superior region but of the world, were the unique result of some special combination of conditions that has not since been repeated.

C.K. Leith, R.J. Lund, and Andrew Leith
Pre-Cambrian Rocks of the Lake Superior Region
U.S. Geological Survey Professional Paper 184, p.23, 1935.

One of the most important horizons in geologic history is the base of the Cambrian System where for the first time easily recognized fossils became abundant. This event occurred about 600 million years ago; and all of the time prior to that event, almost 90 percent of geologic time, is called the Precambrian Era. The discussion of the origin of the earth in Chapter 18 was largely based on theoretical speculation and covered much of Precambrian time. This chapter is concerned with the classical meaning of Precambrian—that is, the history interpreted directly from study of old crustal rocks. All interpretation in geology is at least somewhat speculative, but Precambrian history is more speculative than that of younger parts of the geologic column. The reasons for this are many. The rocks are mainly metamorphic, making their origin difficult to decipher. The have been intruded by batholiths. They are generally deformed and may have been involved in more than one period of mountain making. (See Fig. 19–1.) At many places, only fragments of once-extensive rock units are preserved. Perhaps most important, dating is difficult, generally requiring extensive laboratory work. Thus Precambrian history is even more sketchy, and more subject to major revision as new data become available, than the geologic history of later times.

The study of Precambrian areas began later than that of younger terrains. The first attempts at deciphering the history of large areas of metamorphic rocks were made in Canada. (See Figs. 19–2 and 19–4.) The work began in 1842 with the establishment of the Geological Survey of Canada by William Edmond Logan (1798–1875). The great age of these rocks had been long recognized, but Logan was the first to subdivide the rocks into units of different ages. He established that the metamorphic rocks were clearly unconformably below the oldest fossiliferous rocks. He eventually recognized three groupings within the metamorphic rocks, each with a different history. Later work has shown that the history of these rocks is much more complex, but Logan's pioneering work was done under very difficult conditions; even today the detailed history of these rocks is in doubt.

Canada's Precambrian rocks extend from the Great Lakes to the Arctic Ocean to Labrador. In the bush in the south, canoes were used for transportation; in the north, at times dogsleds were used. The difficulties of the early workers can easily be imagined. They had to make their own maps and were expected to describe the flora and fauna as well as map the geology. In recent years the airplane and helicopter have made the work easier. Unlocking the history of the Precambrian was not the only reward of this exploration, because Canada's Precambrian rocks contain great mineral wealth.

CRYSTALLIZATION OF THE INTERIOR OF THE EARTH

The oldest crustal rocks so far dated by radioactive methods are about 3800 million years old and are in Greenland. This is probably close to the time that the crust began to form. The earth is probably appreciably older than this. Meteorites yield dates of 4550 million years; this is the time that they crystallized. As noted earlier, meteorites are dated using the uranium-lead and thorium-lead methods, and the correction for original lead is a problem. Meteorites are believed to have the same origin as the earth and to have the same composition as the interior of the earth. Thus the 4550-million-year date is perhaps the time that the interior of the earth crystallized. As will be seen shortly, this event is believed to have occurred sometime after the formation of the initially homogeneous earth. How much after is the problem because the size of the planet probably is important in determining the rate at which recrystallization occurred. At any rate, 4550 million years is the time at which the bodies that became meteorites crystallized, and the solar system, including the earth, formed sometime earlier.

The earth, as originally formed, was probably nearly homogeneous and relatively cool. The contraction caused some heating, and radioactivity also must have caused heating. The radioactive heating must have been much greater than at present because much more radioactive material was present on the earth. (See Table 2–5.) This follows because after each half-life, half of the original radioactive element is gone. Meteoritelike material hitting the earth probably also caused local heating. The heating of the earth caused melting, and, under the influence of gravity, differentiation of the earth into core, mantle, and crust began. The heavier materials

Sugarloaf, which rises above the harbor of Rio de Janeiro, is composed of resistant Precambrian rock. Photo courtesy Pan American World Airways.

FIGURE 19-1 Deformed Precambrian metamorphic rocks in Northwest Territories, Canada. Such rocks are much harder to map than most sedimentary rocks. Photo by J. A. Donaldson, Geological Survey of Canada, Ottawa.

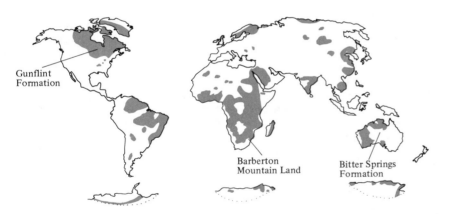

FIGURE 19-2 Areas where Precambrian metamorphic rocks are exposed.

earth is the largest of the inner planets, and for this reason may be the only one with a liquid core.

The actual formation of the earth is believed to have occurred 5000 to 6000 million years ago, perhaps closer to 6000 million years. This estimate is reached from at least two lines of reasoning, but neither method gives more precise figures. The age of the sun is estimated to be about 6000 million years. This figure is based on knowledge of stellar structure and rate of energy production. This does put an upper limit on the age of the earth, however, because the whole solar system is believed to have originated at the same time. The second line of reasoning is based on estimates of the time necessary for melting and recrystallization to occur. About 500 to 1000 million years is a reasonable time for these processes, so this figure should be added to the 4550-million-year date of meteorites.

ORIGIN OF OCEANS AND ATMOSPHERE

So far we have accounted for the main events between the origin of the earth, 5000 to 6000 million years ago, and 4550 million years ago when the interior crystallized. Conventional geology begins with the oldest crustal rocks—about 3800 million years old. We know very little of the events in this interval between 4550 and 4000 million years ago. The crust had at least begun to form. Other events occurred during this time, however; the earth's initial rapid rotation was being slowed, as described earlier, and the atmosphere and the oceans began to evolve.

The present atmosphere and oceans have developed since the earth solidified, but the development is known only in general terms. Although the earth and the sun formed from the same dust

such as iron and nickel melted and moved toward the center, and the lighter silicates moved outward. The silicate mantle solidified from the bottom outward. The core, unable to lose its heat, remains liquid. The inner core may be crystallized. The radioactive elements that caused most of the heating—uranium, thorium, and potassium—all moved outward. These are all large atoms that do not fit into the iron-magnesium silicates, so they stayed with the melt as crystallization progressed from the core outward. The concentration of the radioactive elements near the surface enables their heat to be radiated at the surface. If they were still deep in the earth, their heat would be able to cause remelting.

The resulting layered structure of the earth has great geological significance. The fluid iron core makes possible the earth's magnetic field. The heat of the core may be responsible for convection currents in the mantle. As earlier noted, the other inner planets do not have magnetic fields and mountain ranges. The moon also does not have folded mountains. Perhaps it is significant that the clouds, the earth's overall composition now differs from that of the sun. Therefore, it is clear that sometime after the initial formation of the earth by accumulation from a dust cloud, the earth lost much of its original thick gaseous envelope of mainly hydrogen and helium. We must account for the loss of these gases by processes not now operating because the earth is at present losing very little of its atmosphere. Much of the original atmosphere was lost when it was ionized by the sun's radiation and then swept away by moving ionized particles from the sun, as earlier described. The remaining gas nearer the earth was probably lost as a result of the rapid rotation of the earth, mentioned earlier, and as a result of the heating that melted the earth and formed the core and mantle.

It is difficult to reconstruct the next stage in the development of the early atmosphere; it may have consisted of such gases as ammonia, methane, hydrogen, and steam. This is the composition of the atmospheres of the great planets such as Jupiter (where the water is ice and helium is also present). In the case of the great planets, the source is believed to be the initial dust cloud. The earth, being nearer the sun, had lost by this stage the gases of the original dust cloud; however, frozen chunks of these gases left in the earth would have melted and been released during the later remelting of the earth to form this atmosphere. Evidence for the existence of this atmosphere is very weak. If such an atmosphere ever existed on earth, it probably was lost quickly because of the prevailing high temperature. Ammonia would be quickly decomposed by the sun's ultraviolet energy. Methane would cause fixation of carbon in the early sedimentary rocks, but there is no evidence that this occurred. The present atmosphere developed mainly from volcanic emanations; and the

cooler, more slowly rotating earth retained these gases. Methane, ammonia, hydrogen, and water would form an atmosphere in which life could have begun. However, as outlined here, the temperature was probably too high for life at the time the atmosphere had this composition. Life could also form in a volcanic-derived atmosphere. The origin of life will be discussed in more detail later. Returning to our main theme, volcanoes today emit mainly steam, carbon dioxide, and nitrogen, together with some sulfur gases; there is no reason to believe that earlier volcanoes were different. Thus the atmosphere became steam, carbon dioxide, and nitrogen. With the cooling of the earth some of the steam condensed. The oxygen now in the atmosphere came from photosynthesis by plants after life began, and life probably could not form until the temperature was below 93°C (200°F). At the present time the atmosphere is 78 percent nitrogen and 21 percent oxygen, with all other gases forming about 1 percent. The disposition of the volcanic gases probably follows this pattern: The steam has become water and is mainly in the oceans; the nitrogen is in the atmosphere; and the carbon dioxide has been partly broken down by photosynthesis to carbon in life and to oxygen in the atmosphere, and is partly in carbonate rocks.

This development of the atmosphere seems reasonable, and it is possible to work out a crude timetable. By about 4000 million years, the crust had formed. By 3700 million years, water was present because sedimentary rocks of this age have been found in southern Africa. Prior to that time, anaerobic life may have existed. Oxygen, however, was not present in the atmosphere because up to about 1800 million years, clastic fragments of pyrite and uraninite are found in sedimentary rocks. If oxygen had been present, these min-

erals would have been oxidized. Another point suggesting that the early atmosphere had no oxygen is that organisms had to develop oxygen-mediating enzymes, because without them, oxygen would destroy the organisms. Life developed, probably in the ocean, and oxygen was being released by photosynthesis by about 3200 million years. This oxygen stayed in the ocean, and not much of it reached the atmosphere.

As long as there was no oxygen, the iron in the ocean was in solution. This iron came from weathering of rocks and submarine volcanic activity. As oxygen was released into the ocean by photosynthesis by primitive organisms in the ocean, it combined with the iron (and silica) to form banded iron formation. *Banded iron formation* is a chemically precipitated rock with alternate thin layers of chert and iron oxide. It was formed only in the time from about 3200 million to 2000 million years ago, with most forming between 2500 million and 2000 million years ago. After about 2000 million years ago, apparently all of the iron in the oceans was precipitated, and oxygen could escape from the oceans and begin to accumulate in the atmosphere.

Free oxygen appeared in the atmosphere at about 2000 million years because from that time on, red beds—that is, clastic rocks in which the iron is oxidized—are common. Thick, extensive red beds appear in the range 1800 to 2000 million years and so slightly overlap the last of the banded iron formation. Apparently at this time, oxygen-mediating enzymes formed, allowing life to expand rapidly. The life, probably algae, produced much oxygen; and as the oxygen accumulated in the atmosphere, iron was oxidized in weathering, preventing further development of banded iron formation.

Banded iron formation is found on every continent and is a valuable mineral resource. This unique ore could only form during the time that iron was in solution in the oceans. Only in the absence of oxygen can the iron released by weathering be transported in solution and exist in the oceans in solution. As soon as oxygen was present in the oceans, the iron there was oxidized and precipitated as banded iron formation; later, when oxygen appeared in the atmosphere, iron was oxidized in the weathering process, cutting off the supply of iron to the oceans.

Up to this time, ultraviolet from the sun was lethal to life at the surface, so the organisms must have lived in water deep enough to shield them.

Some of the oxygen in the atmosphere was changed into ozone by ultraviolet radiation from the sun. Ordinary atmospheric oxygen is in the diatomic form, O_2, in which two oxygen atoms share electrons to form stable outer electron shells. In the ozone molecule, O_3, three oxygen atoms share electrons. Ozone is not stable and ultimately returns to ordinary diatomic oxygen. The formation of ozone absorbs much ultraviolet radiation that would otherwise reach the surface of the earth. Higher forms of life could not originate until the ozone layer had formed.

PRECAMBRIAN ROCKS AND HISTORY

Precambrian time includes almost 90 percent of geologic time. This section is about the rocks and the physical history that can be determined from them. This should be the longest and most interesting chapter in geologic history. However, the difficulties of dating these rocks, the repeated deformation many have had, and their fragmentary nature due to long erosion or younger covering rocks

make their study one of the most difficult and demanding aspects of geology. The story begins after the formation of the crust and extends to the base of the Cambrian System where extensive fossils begin and dating rocks becomes easier. All time correlation in Precambrian rocks must be done with radioactive dating. Locally, lithologic similarities or similar methods can be used to establish rock correlations. For these reasons, the decipherable history consists of scattered episodes.

From earlier chapters, it is obvious that geologic history, viewed very broadly, is a sequence of sea-floor spreading and the associated geosynclines. Such events have shaped the earth's surface. It seems likely that such events occurred in Precambrian time as well. However, the beginning had to be different. Somehow the first continent had to form. The record of these events is not clear, and so much speculation is involved in the following story.

Because Precambrian time is so long, some subdivisions are desirable; but subdivision was not easy before radioactive dating became available. The traditional subdivision is into an older Archean and a younger Proterozoic Eras. When these terms were first introduced, workers thought that the metamorphic rocks were older than the sedimentary rocks, so Archean and Proterozoic were used in that sense. Radioactive dating showed that many metamorphic rocks are younger than some sedimentary rocks. This led to much confusion and the use of other terms in local areas. The twofold subdivision is useful and is now again being used, but the division now is made on the basis of radioactive dating. Rocks older than 2500 million years are assigned to the Archean, and the younger Precambrian rocks are assigned to the Proterozoic.

The Archean rocks, those older than 2500 million years, are ar-

ranged differently from younger rocks. The oldest rocks yet found are 3800 million years old at Isua, West Greenland. Areas of Archean rocks are made of high-metamorphic-grade rocks, mainly gneiss of granitic composition, in which belts of greenstone are found. The greenstone belts are composed of deformed and metamorphosed volcanic and sedimentary rocks. They are called *greenstones* because basalt, the dominant rock, develops green-colored minerals when metamorphosed. The greenstones are lower-grade metamorphic rocks than are the high-grade gneisses. Batholithic intrusive rocks and metamorphism permeate the Archean terrain, so relationships among the rocks are not clear. The greenstone belts appear to be relics of larger units, however. The greenstone belts range from small to up to 100,000 square kilometers (38,500 square miles) in area. They range in age from 3500 to 1800 million years, with many at 2700 million years.

The origin of these Archean greenstone and high-grade gneiss terrains is not clear. The occurrence of high-grade and lower-grade metamorphic rocks in close contact suggests vertical movements of the crust. One possibility is that the greenstones originated at volcanic island arcs. At the time that these terrains formed, the earth must have been more active because it had much more radioactive energy than at present (Table 2–5). This suggests the possibility of active convection currents in a warmer mantle.

At places convection in the mantle would cause eruption of mantle rocks through the thin crust. The weight of the resulting volcanic pile would cause the area to sink isostatically into the hot mantle, resulting in metamorphism and perhaps melting of the deeply buried rocks. The top of such a pile might extend above sea level, and the resulting erosion and deposition would produce sedimentary rocks at places above the volcanic rocks. Whatever processes acted, the Archean must have been a time of many active areas of accumulation of volcanic and sedimentary rocks.

Near the end of the Archean, the crust stabilized. Many large granitic batholiths were emplaced around 3000 million years, and the orogeny that ended the Archean was generally between 2800 and 2500 million years. These events did not occur at the same time everywhere.

The effect of this stabilization of the crust was to change the style of both sediment accumulation and later deformation. Beginning about 3000 million years ago, a series of basins formed on granitic crust in Swaziland in southern Africa. The oldest rocks were basaltic volcanics with younger sedimentary rocks. They were followed by conglomerates, sandstones, and volcanic rocks about 2800 million years old. These rocks are covered unconformably by the Witwatersrand Supergroup 2700 to 2500 million years old. The Witwatersrand is composed of shale, quartz sandstone, banded iron formation, volcanic rocks, and conglomerate. It is famous as the earth's most productive gold source, and it also contains uranium mines. The sedimentation in this area continued until about 2300 million years ago; thus it continued into the Proterozoic. Apparently, by 3000 million years ago, the continents had become big enough for large basins to develop, and from then on they did on all continents.

The Proterozoic rocks are generally similar to those just described—large basins with thick accumulations of sedimentary and volcanic rocks. These large basins could develop because the continents were much larger than they had been during most of the Archean. Deformation was apparently localized at the margins of the continents, suggesting that plate tectonics at least somewhat like that of later times was occur-

ring. About 2000 million years ago, banded iron formation ceased being deposited, and red beds began to appear. This occurred in response to the production of oxygen by photosynthetic algae; and it may be that the algae flourished because for the first time, widespread shallow seas such as continental shelves appeared.

The conditions during the Proterozoic did differ from later times; this is illustrated by certain intrusive rocks. At about 1900 million years ago, very large intrusive bodies of gabbroic composition were emplaced at many places around the earth. These bodies crystallized slowly, with the crystals falling to the bottom and forming layers. These layered complexes are the sites of many platinum and chromium mines. Examples are the Bushveld complex in South Africa, the Great Dyke in Zimbabwe, and the older Stillwater Complex in Montana. Around 2200 million years ago, huge basaltic dike swarms were emplaced in the Lake Superior region, Scotland, India, Australia, and Africa. In the period between 1600 and 1200 million years ago, about 75 percent of the earth's anorthosite bodies were emplaced. Anorthosite is a coarse-grained rock composed almost completely of plagioclase. The Adirondack Mountains in New York and many areas in eastern Canada are examples.

By the later part of the Proterozoic, conditions were probably similar to those of later times. Geosynclines formed at a number of places.

Some idea of the main episodes of the Precambrian history of

North America can be obtained from a tabulation of radioactive dates. (See Fig.19–3.) Such tabulation must be treated with caution because the rocks that have been dated are far from a random sample. However, the Precambrian rocks of North America have been fairly well studied so that the tabulation gives reasonable insight. Figures 15–26 and 19–4 show the areal distribution of the different ages. Note in Fig. 19–4 that at places, younger rocks cut across the trends of older rocks. It is also apparent that Precambrian rocks underlie the central interior ot North America where they are covered by younger rocks. This is shown by the Precambrian rocks exposed in the uplifted Rocky Mountains. Precambrian rocks also appear to underlie parts of the Appalachian Mountains.

The oldest rocks in North America appear to be about 3600 million years and are in the Minnesota River Valley. Periods of extensive emplacement of granitic rocks and mountain building occurred around 2700, 1800, 1400, and 1000 million years, suggesting four major sequences of Precambrian rocks in North America. The 2700, 1800, and 1000 peaks may be significant because they mark times of granite emplacement on other continents.

Extensive exposures of Precambrian rocks are found in central, southern, and eastern Canada. (See Fig.19–4.) Glaciation has exposed many of these rocks, and their mineral wealth has encouraged study.

The growth of what is now North America is shown in Fig.19–5. The extent of Precambrian rocks in both the Rocky Mountains and

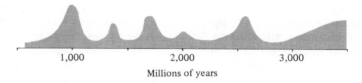

1,000   2,000   3,000
Millions of years

FIGURE 19–3 Ages of Precambrian granites of North America. After A. E. J. Engel, 1963.

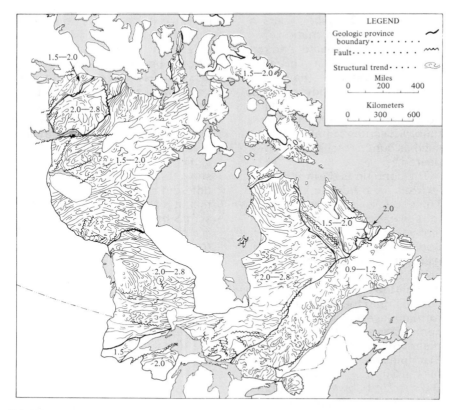

FIGURE 19-4 The Precambrian metamorphic rocks of Canada. The structural trends (foliation) of these much-deformed rocks are shown. The numbers are the ages in thousands of millions of years. The extensions of the age provinces southward into the United States are shown in Fig. 15–26 where these rocks are exposed mainly in the cores of the uplifted Rocky Mountains. Thus it appears that similar metamorphic rocks underlie much of the interior of the United States. The blank areas are places where sedimentary rocks are at the surface. Data from C. H. Stockwell, 1965.

in the Appalachian Mountains can be determined in that figure. The size of the continent is only the minimum size at each time.

At many places orogeny occurred near the end of the Precambrian, but the Precambrian sedimentary rocks of western North America were not deformed then. Unmetamorphosed Precambrian sedimentary rocks occur in both the Rocky Mountains and in the Appalachian Mountains. In the west these rocks are very thick. In the northern Rocky Mountains, over 21,350 meters (70,000 feet) of sedimentary rocks were deposited in Idaho, Montana, and Alberta. These rocks have been dated at between 850 and 1450 million years. (See Fig.19–6.) Unconformably above these is another sedimentary sequence over 1830 meters (6000 feet) thick. The second sequence is covered without notable unconformity by Cambrian rocks at some places and unconformably at other places by rocks as young as Silurian. Other thick sequences of Precambrian sediments occur in northern Utah in the Grand Canyon and south of there in Arizona. At other places, Precambrian sediments grade upward to Cambrian rocks without apparent unconformity and are described in the next chapter with the Cambrian rocks with which they are more closely allied.

On all continents except Antarctica glaciation may have occurred in very late Precambrian time (750 to 850 million years). The evidence for this is rocks that resemble glacial deposits in that they contain conglomerates and overlie striated rocks. Distinguishing between glacial deposits and other conglomerates is difficult. Similar rocks 950 and 2000 to 2400 million years old suggest much earlier Precambrian glaciations.

ORIGIN OF LIFE

All discussions on the origin of life are necessarily speculations. Life must have begun very early in the history of the earth, probably almost as soon as it became cool enough. The role of early life in the development of the atmosphere was described earlier.

One of the first people seriously to consider the problem of the origin of life was A. I. Oparin, a Russian biochemist. He speculated in the 1920s and 1930s that life could originate only in an oxygen-free environment containing the elements needed to form amino acids and larger organic molecules. He suggested an atmosphere containing ammonia, methane, hydrogen, and water vapor. At that time, this was the composition thought to have existed on the early earth. In 1953, Stanley L. Miller, an American graduate student, put these materials in a closed container in which an electric spark could be caused. The spark acted like lightning in the atmosphere. He obtained a number of amino acids, suggesting that life could have originated in this way. This experiment was repeated in 1957 by Philip Abelson, who used mainly carbon monoxide, carbon dioxide, and nitrogen. This mixture is much closer to the probable atmosphere of the early earth. He obtained most of the amino acids necessary for life.

The steps involved in the formation of life are believed to start with the formation of amino acids. Next, these amino acids are brought together to form proteins, nucleic acids, and other compounds. The most difficult step is the transformation into living, self-replicating structures.

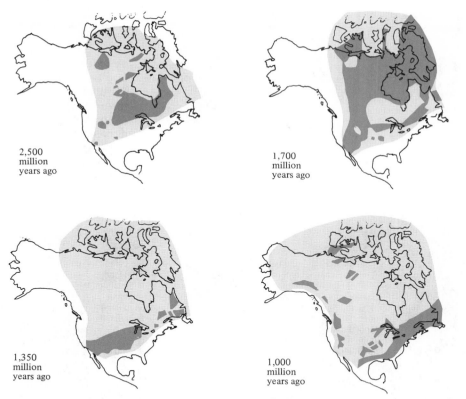

2,500
million
years ago

1,700
million
years ago

1,350
million
years ago

1,000
million
years ago

Dark color: known to be underlain by rocks of that age
Light color: probable minimum size of continents

FIGURE 19–5 Growth of North America in Precambrian time. The maps show the minimum size based on radioactive dating of the rocks. After W. R. Muehlberger, R. E. Denison, and E. G. Lidiak, 1967.

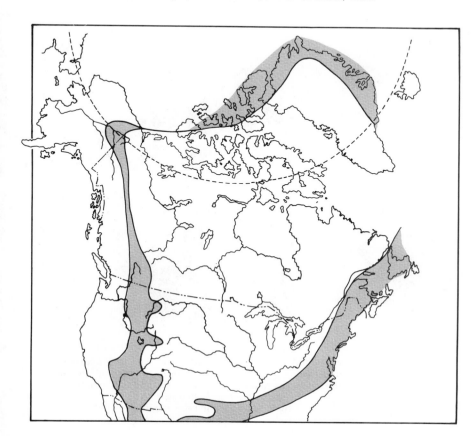

FIGURE 19–6 Late Precambrian sedimentary rocks were deposited in the areas shown. In parts of the western area, they accumulated to thicknesses up to 21,350 meters (70,000 feet). This map was constructed from many scattered outcrops. These rocks may have been deposited elsewhere, but subsequent erosion has removed them.

The amino acids and other organic molecules probably collected in the oceans, forming what has been called a "thin soup." The early life probably got its food directly from the ocean, using material dissolved in the ocean water. The next advance was probably the development of chlorophyll by some of the organisms. This en-

abled them to use the sun's energy and carbon dioxide in their life cycles and produced oxygen as a byproduct. From that point on, oxygen became available in the oceans and atmosphere.

Life could only have formed in an oxygen-free environment. If oxygen had been present, it would have destroyed the amino acids

by oxidation. After life formed and oxygen was produced by photosynthesis, no new life could be formed. Thus, all of the present life on the earth must have evolved from that first population of living organisms.

Clearly, if life formed in the manner described, it could form anywhere where these conditions

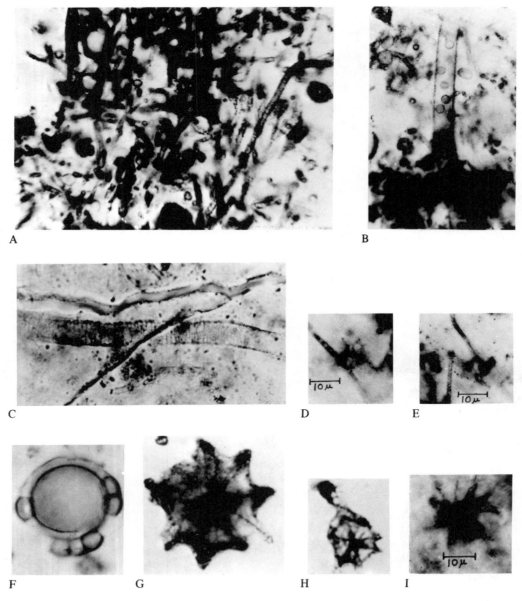

FIGURE 19-7 Microfossils from the Precambrian Gunflint Chert. A. Tangled filaments, mainly *Gunflintia minuta;* and sporelike bodies, *Huroniospora.* Diameter of field 0.1 mm. B. *Entosphaeroides amplus* with sporelike bodies within the filament. Diameter of field 40 μm. C. *Animikiea septata* probable algal filament with transverse septae. Diameter of field 50 μm. D. and E. Filaments with well-defined septae similar to iron-precipitating bacteria and blue-green algae. F. *Eosphaera tyleri.* Diameter of field 32 μm. G. and H. *Kakabekia umbellata.* Diameter of field of G is 20 μm and H is 12 μm. I. Radiating structure similar to manganese- and iron-oxidizing colonial bacteria. [One micrometer (μm or μ) is one-millionth of a meter.] Photos A, B, C, F, G, and H courtesy of E. S. Barghoorn; D, E, and I courtesy of P. E. Cloud, Jr.

exist. That is, it could have formed on another planet in the solar system or elsewhere in the universe. The conditions needed are the elements carbon, nitrogen, and hydrogen, plus water in the liquid state. Such conditions must occur at many places in the vast universe, and so there must be other life in the universe. In the space program, one of the aims was the search for life on Mars, the only planet with anything like the right temperature range. The materials returned from the moon have been carefully searched for evidence of life. Amino acids and other hydrocarbons were found in small amounts on moon rocks. They are believed to be of inorganic origin. On September 28, 1969, a carbonaceous meteorite fell at Murchison, Victoria, Australia. This meteorite that was dated as 4500 million years old contains amino acids. These discoveries suggest that the conditions necessary to form life are not uncommon and that amino acids may have formed in pregeologic time in the solar system.

THE OLDEST FOSSILS

Until the late 1950s, very few Precambrian fossils were found. Many of the reported fossils found before that time are probably of inorganic origin. In recent years, microfossils have been found at many localities around the world, and larger fossils have been found in upper Precambrian rocks in a dozen or so places. The record is complete enough to be able to continue the story that began in the last section.

The first living organisms probably could not produce their own food but lived on material dissolved in the ocean. This situation could only exist for a short time; otherwise, the life would use all of the available nutrients. The development of photosynthesis probably made the first self-sustaining organism.

The oldest possible evidence of life is in highly metamorphosed sedimentary rocks at Isua, West Greenland. These rocks, dated at 3800 million years, contain carbon isotopes that suggest biologic origin.

The oldest probable fossils are found in the 3500 million-year-old Warrawoona Group of Western Australia. The cherts in these rocks contain well-preserved microfossils. The slightly younger (3400 million years old) Swaziland Supergroup in eastern South Africa also contains microfossils. The microfossils in these old rocks are similar to modern blue-green algae and bacteria, and are accepted as evidence of life by most who have studied them. They could be nonbiologic because somewhat similar structures were produced in experiments, like those discussed in the last section, that synthesized amino acids and other organic molecules. The next youngest rocks with microfossils are found in Australia, Canada, and Rhodesia. These rocks are 2700 to 2800 million years old, and these microfossils are generally accepted as genuine.

Abundant Precambrian fossils from the Gunflint Chert on the north shore of Lake Superior in Canada have been studied. (See Fig. 19–7.) Again, blue-green algae and bacteria are found; these organisms were clearly photosynthetic. The rocks are about 2000 million years old.

The organisms found at these localities are primitive. Blue-green algae and bacteria are the only living things whose reproduction does not involve genetic material from two parents, so their form does not change. Thus these organisms in rocks over 3000 million years old are much like living species. Because redistribution of parental genes is not involved, any mutation that occurs, in general, dies out in a few generations. This is not at all like the evolution described in Chapter 17.

The first organisms that have genes are found in the 1300 million-year-old Pahrump Group in California, and in the 900–1000-year-old rocks at Bitter Springs in the Northern Territory of Australia. The organisms are green algae, and all green algae differ from blue-green algae in having genes. Some of the fossils recovered from this chert show cell division. The development of these early forms of life is summarized in Fig. 19–8. Younger Precambrian fossils are shown in Fig. 19–9.

These discoveries suggest that true evolution could not begin un-

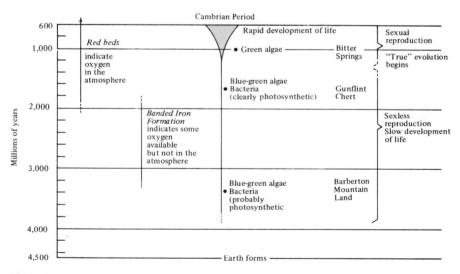

FIGURE 19–8 Development of Precambrian life and oxygen in the atmosphere.

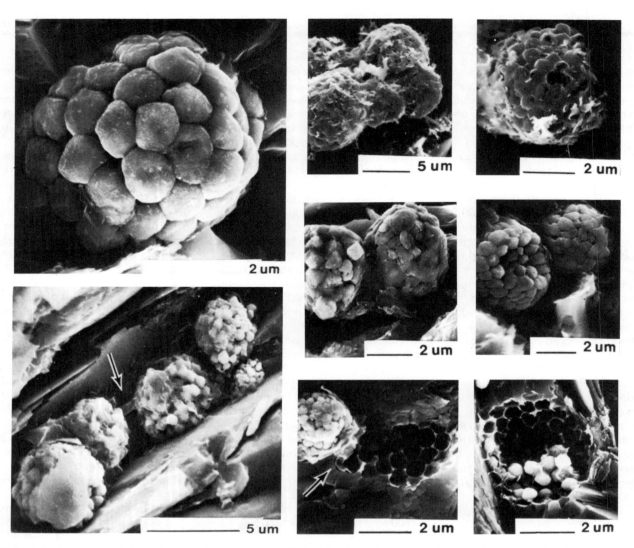

FIGURE 19–9 Late Precambrian microflora from southern British Columbia. These microphotographs were made with a scanning electron microscope and reveal more surface detail than is possible with a light microscope. These fossils were found in hard black shale. One micrometer (μm) is one-millionth of a meter. Photos courtesy of B. J. Javor and E. W. Mountjoy, and the Geological Society of America.

FIGURE 19–10 Map showing locations where the unusual late Precambrian fauna shown in Fig. 19–11 have been found. Other Precambrian faunas have been found elsewhere. After M. Glaessner.

til about 1400 million years ago. Up to that time, the organisms were too primitive and evolutionarily too conservative to change very much. After that time, evolution could proceed, and apparently did, because about 600 million years ago, life became abundant. Thus, in less than 800 million years, all of the many types of life found in Cambrian rocks may have evolved from green algae. Representatives of every phylum capable of fossilization are found in Cambrian or early Ordovician rocks.

Very little evidence of this evolution has yet been found. Perhaps the reason is that the organisms had few hard parts capable of fossilization. At the three localities just described, the algae were preserved in chert beds and were only found after searching with both optical and electron microscopes. Thus intensive search may unearth more evidence. Another possibility is that although the number of organisms was increasing rapidly during this time, only

by the start of the Cambrian Period was life abundant enough to form abundant fossils. This is suggested by the discovery at ten places of a Precambrian fauna in rocks between 600 and 700 million years old. (See Figs. 19–10 and 19–11.) This suggests that by late Precambrian, life was becoming abundant but had few preservable hard parts.

This late Precambrian fauna (Fig. 19–11) is important also because the organisms are not microscopic. They are believed to be the first organisms with genes and

so may signal more rapid evolution. Bigger organisms may have become possible about 700 million years ago because by then, oxygen in the atmosphere had increased to about 10 percent of its present amount. This may have aided respiration, but, more important, the ozone layer may have formed and protected the organisms from lethal ultraviolet in the sunlight.

The most common Precambrian fossils are algal mounds or columns (*stromatolites*) similar to those made by living blue-green

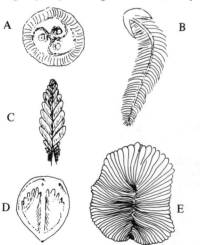

FIGURE 19–11 This unusual Precambrian fauna is found at the places shown in Fig. 19–10. These organisms are found as impressions in sandstone. A. Circular form of unknown affinities, possibly with coiled arms. About 1.9 centimeters (0.75 inch). B. Probable worm. About 2.5 centimeters (1 inch) long. C. Leaflike form of unknown affinities. About 17.8 centimeters (7 inches). D. Oval form of unknown affinities. About 1.9 centimeters (0.75 inch). E. Wormlike form. About 4.5 centimeters (1.75 inches).

A

B

FIGURE 19–12 Algal structures in rocks about 1000 million years old in Glacier National Park, Montana. A. Top view, and B. side view of *Collenia undosa*. Photos by R. Rezak. U. S. Geological Survey.

algae. These algal structures can be used in a general way to make broad correlations of Precambrian rocks. (See Fig. 19–12.) Figure 19–13 shows some unusual markings that may be worm tubes, or they may be of inorganic origin. Figure 19–14 shows another type of marking found in Precambrian rocks. The example shown is thought by some to be a jellyfish. Crawl and scratch marks are found at many places in both young and old rocks. Figure 19–15 shows both Precambrian and Cambrian crawl marks, and the similarities suggest that trilobites were present in some late Precambrian rocks.

FIGURE 19–13 Possible marks or tubes of a bottom-dwelling or burrowing organism. Some worms make similar markings. These markings are in sedimentary rocks 2000 to 2500 million years old. Scale in centimeters. Photo by Geological Survey of Canada, through the courtesy of H. J. Hoffman, who first described these occurrences.

FIGURE 19–14 *Brooksella canyonensis,* a supposed jellyfish from Precambrian rocks in the Grand Canyon. Diameter of field is 8.9 centimeters (3.5 inches). Photo from Smithsonian Institution.

FIGURE 19–15 Crawl and scratch marks, probably of trilobites. Marks such as these are commonly found both in Cambrian and younger rocks and in rocks well below (older than) the first trilobites. Parts A and C are from rocks of the Nama System, South West Africa, from well below the first trilobites. Parts B, E, F, and G are from the Deep Spring Formation, White Mountains, California, well below the first trilobites. Part D is from the Middle Cambrian Tapeats Sandstone, Grand Canyon, Arizona. Photos courtesy of P. E. Cloud, Jr.

SUMMARY

The earth's crust formed about 3800 million years ago.

Meteorites crystallized about 4550 million years ago.

After formation, the earth melted about 4550 million years ago to form its layered structure. Compaction and radioactivity were the probable energy sources.

The earth formed about 5000 to 6000 million years ago.

The oceans and the atmosphere formed in the interval 4500 to 4000 million years ago.

The early atmosphere may have been ammonia, methane, hydrogen, and steam, like the present great planets.

The present atmosphere and the oceans came from volcanic gases.

Oxygen came from carbon dioxide, the process being photosynthesis by plants after life had formed.

The oldest sedimentary rocks found so far are about 3700 million years old, so water was present by then.

Oxygen was not present in the atmosphere until about 1800 million years ago because fragments of pyrite and uraninite are found in sedimentary rocks.

Oxygen formed between 3200 and 2000 million years ago produced banded iron formation.

Red beds became common between 1800 and 2000 million years ago, suggesting that oxygen was in the atmosphere.

As soon as oxygen was present, the ozone layer formed, shielding lethal ultraviolet and so allowing higher forms of life to evolve.

Time correlation of Precambrian rocks must be done radioactively.

The Precambrian is subdivided into the Archean Era, more than 2500 million years, and the younger Proterozoic Era.

The oldest rocks yet found are 3800 million years old at Isua, West Greenland.

Most Archean areas are made of high-grade granitic gneiss, with lower-grade greenstone belts. The greenstones are made of metamorphosed volcanics with some sediments. The greenstone belts may be ancient volcanic arcs.

At the end of the Archean many batholiths were emplaced, making the continents larger and more stable, and sedimentary rocks accumulated in basins. Deformation of Proterozoic rocks seems to have been concentrated at continental margins, as it was in later times.

During the Proterozoic, huge gabbroic intrusive bodies were emplaced at many places.

Much of the interior of North America is underlain by Precambrian igneous and metamorphic rocks.

Precambrian sedimentary rocks, some very thick, are found in the Rocky Mountains and the Appalachian Mountains.

Glacial deposits occur in rocks 2000 to 2400 million years old, about 950 million years old, and between 850 and 750 million years ago.

The formation of life probably began with formation of amino acids.

Amino acids can be produced by an electrical discharge in a mixture of water, methane, ammonia, and hydrogen, or of carbon monoxide, carbon dioxide, and nitrogen.

To form life, amino acids must join to form proteins and nucleic acids, and these must be transformed into living, self-replicating structures.

Life could only form in an oxygen-free environment.

Life probably originated in the oceans.

Life could form on any planet, given the same conditions.

Fossil bacteria and blue-green algae are found in rocks as old as 3400 million years.

Blue-green algae and bacteria do not evolve rapidly.

The first organisms with genes are green algae 1300 million years old in California and about 1000 million years old at Bitter Springs, Australia.

True evolution began about 1400 million years ago, and all phyla had evolved by the start of the Cambrian about 600 million years ago. Very little record of the evolution has been found, perhaps because the early organisms had no hard parts capable of fossilization.

A few fossils of soft-bodied late Precambrian animals have been found, but algal columns, crawl marks, and burrows are the only nonmicroscopic Precambrian fossils.

QUESTIONS

1 Explain why the initial melting of the earth occurred sometime after the formation of the earth.
2 How did the earth's layered structure form?
3 When did the first life form?
4 Outline the steps by which the atmosphere is believed to have formed.
5 What is the significance of banded iron formation?
6 Can the beginning of Cambrian time be recognized at all places where sedimentary rocks of this age occur?
7 Where are the oldest rocks in North America found? What is their age?
8 What is the age of the oldest rocks so far found? Where are those rocks found?

9 Where are Precambrian metamorphic rocks exposed on North America?

10 Where are Precambrian sedimentary rocks exposed on North America? How thick are they?

11 How is life believed to have originated on the earth? Can this process account for the development of new types of life?

12 List the types and ages of the Precambrian fossils so far found.

13 What are some of the explanations for the sudden appearance of abundant fossils at the base of the Cambrian?

14 Why is Precambrian history so difficult to interpret?

15 How is the Precambrian subdivided?

16 Describe the Precambrian greenstone belts. How did they form?

17 How do Proterozoic rocks differ in general from those of the Archean?

SUPPLEMENTARY READINGS

Barghoorn, E. S., "The Oldest Fossils," *Scientific American* (May 1971), Vol. 224, No. 5, pp. 30–42.

Calvin, Melvin, "Chemical Evolution," *American Scientist* (March-April 1975), Vol. 63, No. 2, pp. 169–177.

Cloud, Preston, "Evolution of Ecosystems," *American Scientist* (January-February 1974), Vol. 62, No. 1, pp. 54–66.

Cloud, Preston, James Wright, and Lynn Gover, III, "Traces of Animal Life from 620-million-year-old Rocks in North Carolina," *American Scientist* (July-August 1976), Vol. 64, No. 4, pp. 396–406.

Engel, A. E. J., "Geologic Evolution of North America," *Science* (April 12, 1963), Vol. 140, No. 3563, pp.143–152.

Engel, A. E. J., "The Barberton Mountain Land: Clues to the Differentiation of the Earth," Economic Geology Research Unit, University of the Witwatersrand, Information Circular No. 27, 1966. Revised, 1969. This paper reprinted in Preston Cloud, ed., *Adventures in Earth History,* pp. 431–445. San Francisco: W. H. Freeman and Co., 1970 (paperback).

Frieden, E., "Chemical Elements of Life," *Scientific American* (July 1972), Vol. 227, No. 1, pp. 52–64.

Glaessner, M. F., "Pre-Cambrian Animals," *Scientific American* (March 1961), Vol. 204, No. 3, pp. 72–78. Reprint 837, W. H. Freeman and Co., San Francisco.

Goodwin, A. M., "Precambrian Perspectives," *Science* (July 3, 1981), Vol. 213, No. 4503, pp. 55–61.

Groves, D. I., J. S. R. Dunlop, and Roger Buick, "An Early Habitat of Life," *Scientific American* (October 1981), Vol. 245, No. 4, pp. 64–73.

Hargraves, R. B., "Precambrian Geologic History," *Science* (July 30, 1976), Vol. 193, No. 4251, pp. 363–371.

Harland, W. B., and M. J. S. Rudwick, "The Infra-Cambrian Ice Age," *Scientific American* (August 1964), Vol. 211, No. 2, pp. 28–36.

Kerr, R. A., "Origin of Life: New Ingredients Suggested," *Science* (October 3, 1980), Vol. 210, No. 4465, pp. 42–43.

King, P. B., *Precambrian Geology of the United States: An Explanatory Text to Accompany the Geologic Map of the United States.* U. S. Geological Survey Professional Paper 902, Washington, D. C.: U. S. Government Printing Office, 1977, 85 pp.

Murray, G. E., M. J. Kaczor, and R. E. McArthur, "Indigenous Precambrian Petroleum Revisited," *Bulletin,* American Association of Petroleum Geologists, (October 1980), Vol. 64, No. 10, pp. 1681–1700.

For a first survey, I had got the upper grauwacke, so called, into my hands, for I had seen it in several situations far from each other, all along the South Welsh frontier, and in Shropshire and Herefordshire, rising out gradually and conformably from beneath the lowest member of the Old Red Sandstone. Moreover, I had ascertained that its different beds were characterized by peculiar fossils, . . . a new step in British geology. In summing up what I saw and realised in about four months of travelling, I may say that it was the most fruitful year of my life, for in it I laid the foundation of my Silurian system. I was then thirty-nine years old, and few could excel me in bodily and mental activity.

Roderick Impey Murchison
Quoted in Life of Sir Roderick I. Murchison
Sir Archibald Geikie, Vol.i; pp.183, 192; 1875

ESTABLISHMENT OF THE LOWER PALEOZOIC SYSTEMS

In western Europe the late Paleozoic and younger sedimentary rocks are distinctive and relatively easy to trace. Following the early discoveries of Cuvier, Brongniart, and Smith mentioned earlier, these rocks were studied and this part of the geological time scale was established. Beneath them were deformed sedimentary rocks that were difficult to study, and below those rocks were ancient metamorphic rocks. Those deformed sedimentary rocks were a challenge to two Englishmen.

Roderick Impey Murchison, a gentleman of independent means, and Adam Sedgwick, a Cambridge professor, who had worked and traveled together for some years in geologic studies, began a study of those contorted rocks in Wales

in 1831. They worked independently, with Murchison beginning at the top just below the well-known, easily recognized Old Red Sandstone, while Sedgwick began his work somewhere in the middle in an area of complexly contorted rocks. Murchison found fossiliferous beds which he was able to trace easily. He carefully described the beds and their fossils in a book, *The Silurian System,* published in 1838. In the meantime Sedgwick began to unravel the folded and faulted rocks that he was studying. He did a marvelous job in deciphering the structures, but the rocks were so badly deformed that he was unable to trace beds any distance. In such circumstances it is understandable that he did not study the fossils in any detail. He named his rocks the Cambrian System.

It soon became apparent that Sedgwick's upper Cambrian included some of Murchison's lower Silurian. This led to a falling out

between the two friends and a long-standing controversy. Arguments went on for years as to which system to refer the disputed strata. In 1879 Charles Lapworth proposed the Ordovician System for the disputed beds, and after many years that solution has been widely accepted.

In defining a system or any other stratigraphic unit, a type section is a necessity. Other workers must have an objective standard to which they can refer other rocks. In defining a system, both fossils and rocks of the type section must be described. This was done by Murchison, but not by Sedgwick. Had Sedgwick described a type section, including its fossils, this unfortunate controversy would not have happened. Later workers did describe a type section for the Cambrian and so set the scene for Lapworth's peacemaking.

Mapping is the main tool of the geologist. The distribution and re-

Folded Ordovician shale, Port au Port, Newfoundland. Photo courtesy Geological Survey of Canada (120457).

lationships of the various rock types exposed in the area under study are revealed through mapping. The rocks mapped are dated by comparison (correlation) with the type sections of the geologic column. The rocks under study are interpreted, using uniformitarianism. The resulting interpretations are integrated with similar interpretations of all other rocks of the same age to show the paleogeography (old geography) of that time. The interpretations of all parts of geologic time obtained in this manner are then synthesized to produce the earth's history. This is a very simplified sketch of how geologic history is interpreted. Note especially the many levels of interpretation. The evidence—the rocks—is commonly fragmentary, and establishing its contemporaneity may be very difficult.

PATTERNS IN EARTH HISTORY

The development of geology shows how the idea of patterns or cycles came about. All sciences had chaotic beginnings, and geology was more chaotic and slower to develop than were the other sciences because wide areas had to be studied before patterns began to emerge.

The first step was the establishment of the geologic time scale. The first systems to be defined were natural groupings; that is, they were rocks bounded by unconformities or other distinct horizons. It soon developed that the systems were, in general, natural subdivisions only at the places where they had been defined, and their boundaries transgress time as do all unconformities. At other places no natural boundary existed, and disputed zones were found where rocks had been deposited during the time of the unconformity at the type section. The type sections were not all in the same area, and this also led to

correlation problems and other disputed zones. As a result of these problems, the type sections and extent of each system were defined arbitrarily. This established a workable time scale on which geology developed, but it is not in any sense a natural time scale. Because the time scale developed in Europe, it is even more unnatural in North America.

The recognition of geosynclines as important elements in geologic history was the next step. Geosynclines seem to provide a cyclic element in the history of continents, although they occurred at more or less different times on each continent. For the first time, though, a pattern emerged. The development of geosynclines is, as we now know, related to plate tectonics. The mountain-building episodes attracted the most attention in geosynclinal areas. Because of the batholithic intrusions, the metamorphism, and the later erosion, mountain building is difficult to date closely. However, it appeared that mountain building had occurred at roughly the same time at widely separated areas. When a closer look was taken, it became obvious that in many cases major mountain-building events at one place were being correlated with minor unconformities at other places. It also soon became apparent that at some place an unconformity cound be found at almost any time in geologic history. Viewed on a worldwide scale, though, some parts of geologic time were very active and other times were fairly quiet. The quiet times were Cambrian and most of Triassic, and the more active times were Ordovician, Devonian, Pennsylvanian, and late Cretaceous. Thus a broad, fuzzy, worldwide pattern seemed to be emerging.

Another development has been the recognition of broad patterns of deposition and unconformity in the stable continental interior of North America. This pattern will be the framework of the brief outline of the geologic history of North America presented here. It

has been possible to recognize these cycles in the continental interior because there the rocks are relatively undeformed, are fossiliferous, and have been extensively studied in the quest for oil. Such a pattern requires a relative change in sea level. Deposition occurs when the continent is below sea level, and nondeposition and erosion occur when the continent is above sea level. Either the continents move up and down, or sea level changes, or both occur. If the continent rises and falls, the resulting cycle of erosion and deposition would be local to that continent and changes in sea level would occur worldwide. The flat-topped seamounts and atolls show changes in sea level in parts of the oceans, suggesting that the shape of the ocean basins may change, causing a change in volume and thus a change in sea level. Such changes are probably associated with plate tectonics. During a time of rapid plate motion, the mid-ocean ridges or spreading centers would be active and hot; therefore, because of thermal expansion, the ridges would have a large volume. This would cause sea level to rise worldwide, thus causing flooding of the continents. Conversely, at times of slow plate movement, the spreading centers would be much smaller and so the ocean basins would have a larger volume.

Many of the North America cycles correspond closely with the worldwide cycles that have so far been recognized. The times of worldwide unconformities that separate cycles are indicated in the right hand column of Fig. 20–1. Dating of these unconformities is an ongoing endeavor.

The pattern of geologic history of North America, then, is the advance and retreat of seas causing cycles of deposition followed by unconformity in the interior, and the geosynclines on the margins. Had the geologic time scale been defined in the interior of the North American continent, these cycles probably would have been the geologic periods. The present continental shelf, slope, and rise may be geosynclines, and Hudson Bay may be an example of a continental sea.

The study of fossils has also revealed some worldwide changes. Extinctions of large numbers of families occurred toward the end of the Cambrian, Devonian, Permian, Triassic, and Cretaceous Periods. Soon after these extinctions, new forms appeared. The greatest extinctions occurred near the end of the Cambrian, Permian, and

Cretaceous Periods. The latter two mark the close of two of the eras, and it is these changes in life that were used to designate the eras. The Paleozoic (old life) Era ends with the Permian, and the Mesozoic (middle life) Era ends with the Cretaceous Period. The extinctions, although they are abrupt in the broad sense, took place over a considerable span of time. The cause of these extinctions is not known, but they may have been caused in part by changes in the extent of the seas. Some of these

extinctions broadly correspond with the cycles recognized in North America.

WORLD GEOGRAPHY IN THE EARLY PALEOZOIC

It is not yet possible to reconstruct the positions and movements of the continents in detail in the early Paleozoic. When those data are available, it will be possible to describe the history of the whole earth in a unified way. From Permian onward, the reconstructions are somewhat easier although disagreements do exist among those who have studied the problems closely. Much of what is said in these sections will require revision as more data are gathered.

Very little is known of the positions of the continents throughout most of the Precambrian. Paleomagnetic data suggest that the continents came together in the late Precambrian, about 800 million years ago. (See Fig. 20–2.)

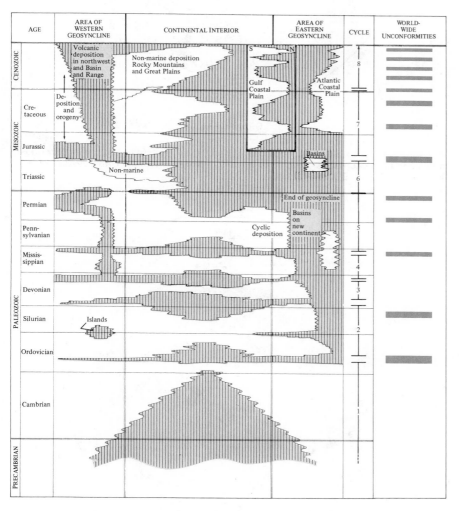

FIGURE 20–1 The geologic cycles of deposition in Central North America. This diagram shows in a general way where and when deposition occurred. Geosynclinal deposition in Texas and Oklahoma is omitted. Erosion and deformation cause many difficulties in constructing such a diagram. This is especially true in the lower Paleozoic of the far west. To gain a real understanding of geologic history, the student should add the rock types and their sources to the depositional sequences in the diagram. In general, the times of nondeposition in the geosynclines correspond to orogenic periods. The vertical lines indicate nondeposition.

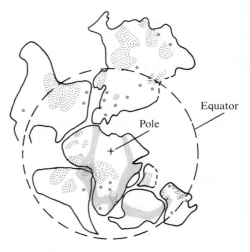

FIGURE 20–2 Possible late Precambrian positions of the continents. The stippled areas have ages greater than 1700 million years (after Hurley and Rand, 1969). Late Precambrian mountain building is shown in color. The dots are places where glacial deposits of late Precambrian age have been reported. Those near the equator are probably conglomerates of nonglacial origin.

The evidence that parts of this supercontinent were together in the Precambrian is the matching of ages and structures such as shown in Fig.14–22. The locations of the equator and the south pole are from paleomagnetic data. Also shown are the locations where late Precambrian glacial deposits have been reported. Many of these reported glacial tills are merely conglomerates and not of glacial origin. Those near the pole may be glacial, but those near the equator most likely are not. The fold belt of Precambrian age shown is an excellent match between South America and Africa.

In the Cambrian there were five large continents: North America, Europe, Siberia, Asia, and Gondwanaland (Fig. 20–3). The southern continent called Gondwanaland consisted of South America, Africa, Antarctica, Australia, and India. Gondwanaland remained a single continent throughout the Paleozoic. The North American continent, which we will follow most closely, probably contained parts of the present Europe, and the Europe of the Cambrian Period probably included parts of the present eastern North America. Much of southeastern United States and probably Mexico were part of Gondwanaland, as was part of present-day southern Europe and Asia.

In the Cambrian Period, the North American continent had continental shelf-slope geosynclines on all sides. Europe was close by, probably separated by a mid-ocean ridge. In the Ordovician, a subduction zone with a volcanic arc formed along the east coast of North America (Fig. 20–4). In the Silurian Period, Europe collided with North America; and on the west coast, a subduction zone and volcanic arc developed (Fig. 20–5). The effects of these events on North America are discussed in the next section. In the Silurian Period, Siberia and Asia also collided (Fig. 20–5).

From late Precambrian through Ordovician the south pole re-

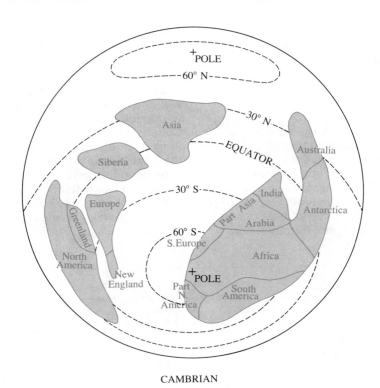

CAMBRIAN

FIGURE 20–3   The continents of the Cambrian Period. This map and the similar ones in this and later chapters are modified from L.W. Mintz, *Historical Geology*, 3rd ed. Columbus: Charles E. Merrill, 1981.

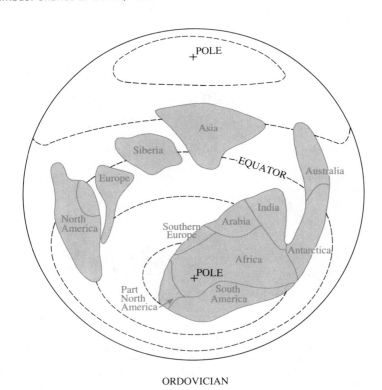

ORDOVICIAN

FIGURE 20–4   The continents in the Ordovician Period. A subduction zone has formed between Europe and North America.

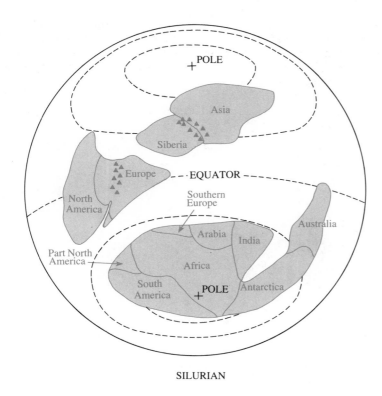

SILURIAN

FIGURE 20–5 The continents in the Silurian Period. Gondwanaland has moved so that a pole is in southern Africa; Europe and North America, and Siberia and Asia collided.

mained in North Africa although it did move somewhat. Glacial deposits of Ordovician age have been found in the Sahara, confirming the location of the paleomagnetic pole. In the Silurian Period Gondwanaland moved so that the south pole was in southern Africa. The equator passed through North America in a north-south direction in the Cambrian, and southwest-northeast direction in the Silurian.

LATE PRECAMBRIAN TO EARLY ORDOVICIAN— CYCLE I

The base of the Cambrian rocks with their abundant fossils is one of the most important points in geologic history. The base of the Cambrian System is generally defined as the first occurrence of certain fossils (trilobites); however, some place it at the first occurrences of easily recognized fossils. Thus recognition of this important time line depends on chance preservation of fossils. At many places the first Cambrian fossils occur in the midst of thick sedimentary sections so that dating the start of the cycle is impossible.

A general reconstruction of early Paleozoic geography is shown in Fig. 20–6. On this map and the other paleogeographic maps, the present outline of the continent and some of the present rivers are shown for reference. This does not imply that they existed at the times depicted.

At the beginning of the Cambrian Period, the main framework of North American geology was established. Geosynclines occupied both sides of the continent as well as the far north. As the cycle progressed, the seas spread over more and more of the continental interior; by Early[1] Ordovician time, most of the interior was awash. The seas then withdrew, first from the interior, and finally from the geosynclines, ending the cycle. Cycle I is the simplest of the cycles.

INTERIOR

Cycle I begins with the interior dry but surrounded by geosynclines. The seas gradually expanded over much of the southern, western, and far northern parts of the interior. (See Fig. 20–6.) Much of northeastern Canada and most of Greenland remained dry land, as did a broad arch from roughly Lake Superior to the Gulf of California. This pattern of land and sea was to be repeated but with a variety of minor variations in the later cycles. The land areas were undergoing erosion and provided the clastic sediments deposited in the seas. This is indicated by the fact that the sediments become coarser as these land areas are approached. The present continental interior is underlain by Precambrian metamorphic rocks, such as those presently exposed north of the Great Lakes. Erosion reduced these rocks to relatively smooth surfaces that underlie the Cambrian rocks. At some places, such as the Ozark Mountains of Missouri, Upper Cambrian rocks lap up against the flanks of the ancestral Ozark Mountains, showing that highlands did exist in the interior. The movement of the seas onto the interior was not a uniform advance, but probably had a number of short retreats and readvances. This is suggested by both rock types and changes in fossils.

At most places in the interior, the oldest rocks are sandstones. These rocks are generally less than three hundred meters thick,

[1]The terms *Early, Middle,* and *Late* are used when referring to time. *Lower, Middle,* and *Upper* are used when referring to rocks. All of these terms are capitalized when used in the formal sense.

although they are much thicker at places. Except at the margins of the interior platform, the oldest rocks are Upper Cambrian in age. Lower and Middle Cambrian rocks are, for the most part, confined to the margins of the interior. The Lower Ordovician rocks conformably overlying these rocks are largely limestones with some shale. This change from clastic rocks to carbonates suggests that the land areas had been reduced to low relief in early Ordovician time near the end of Cycle I.

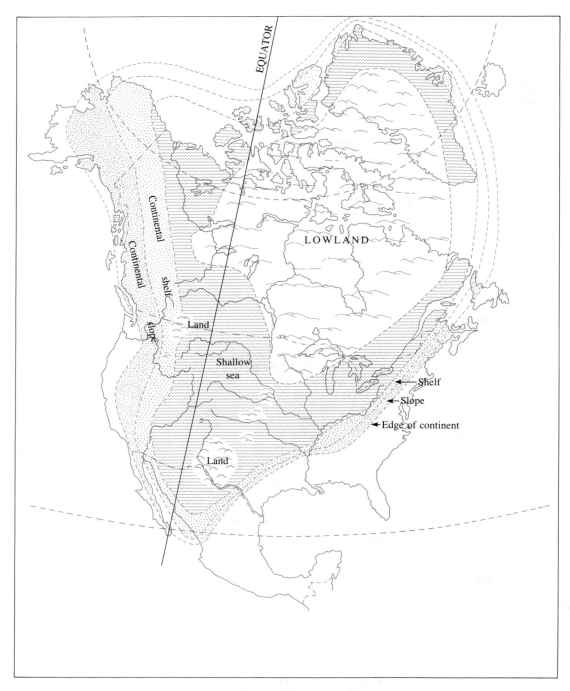

FIGURE 20-6   Generalized map of North America in the Cambrian Period. Continental shelf-slope geosynclines surround the continent. The extent of the shallow sea in the interior changed during the period. This general paleogeography of marginal geosynclines and shallow interior seas persisted through most of the Paleozoic. In eastern Canada and New England, the edge of the continent shown on this map is the dividing line between two different trilobite faunas of the same age.

EAST

In the late Precambrian, a geosyncline formed in the east. The oldest rocks are in the southern Appalachian Mountains. They consist of a sequence of 9150 meters (30,000 feet) of clastic rocks. Overlying them is a 1830-meter (6000-foot) quartz-sandstone and shale unit, in which the first trilobites are found near the top. The two units are more or less conformable, although some workers consider the contact unconformable. At any rate, it seems clear that sedimentation began in Precambrian time. The source of these rocks was the interior of the continent. Overlying them are about 2135 meters (7000 feet) of Middle and Upper Cambrian shales and limestone. The amount of carbonate increased upward as the seas covered more and more of the interior, moving the source of the clastics farther away. Above these are about a thousand meters of Ordovician carbonates. The transition from Cambrian to Ordovician is in this conformable sequence of carbonates.

In the northern part of the geosyncline, volcanic rocks of Cambrian age are found with clastic rocks, suggesting that volcanic island arcs may have been present. Cambrian shale and sandstone up to 2440 meters (8000 feet) thick are found in Vermont. These are overlain by thick Lower Ordovician clastic rocks. Later mountain building has produced complex structures in New England and to the north, making interpretation difficult.

The trilobites in eastern New England and adjacent Canada are different from those in the rest of the eastern geosyncline. The boundary separating the two trilobite faunas is the edge of the Cambrian continent shown in Fig. 20–6. The eastern trilobites are similar to those found in Europe. In the past, this distribution of fossils was interpreted as indicating that a barrier existed in the geosyncline and separated the faunas. We now believe that the rocks containing the European trilobites originated in Europe and were welded to North America when Europe collided with North America in the middle and late Paleozoic.

WEST

The early Paleozoic also began in the west with the establishment of a geosyncline, probably a trench at the edge of the Pacific Ocean. The seas moved into the area from both north and south. A barrier existed between these two seas in Idaho until Middle Cambrian time. Although a much younger batholith obscures the evidence for the barrier, its presence is indicated by a difference in fauna between the two seas and by coarser clastic rocks near the old shoreline.

A very thick section of Precambrian and lower Paleozoic rocks is found in the southwestern ranges of the Great Basin near the California-Nevada border. Here sedimentation began in Precambrian time. The oldest sedimentary rocks are a 2440-meter (8000-foot) sequence of clastics with some carbonates.

Unconformably above those rocks are the Cycle I rocks. They begin with 460 meters (1500 feet) of dolomite, followed by up to 3050 meters (10,000 feet) of clastic sediments, largely quartz sandstones. Conformably above these rocks is another unit of similar clastic rocks about 1525 meters (5000 feet) thick. Near the top of this unit, the first Cambrian fossils are found. The Lower Cambrian rocks here are about 3050 meters (10,000 feet) thick and are mainly sandstone, grading into shale near the top. The Middle and Upper Cambrian rocks are close to 2440 meters (8000 feet) thick and are mainly carbonates. They are overlain by similar thick Lower Ordovician carbonates.

The Cambrian clastic rocks had their source in the continental interior, as was the case with the other areas described. As the cycle progressed, the seas moved eastward onto the interior. As the seas transgressed onto the continent, the basal Cambrian deposits were a distinctive beach sand. This unit can be traced from southern Nevada through Utah and Montana to the Midwest. This sandstone is progressively younger to the northeast, ranging from Precambrian to middle Cambrian in age. (See Fig. 20–7.)

Farther north in the western geosyncline in Canada, similar rock accumulated in great thicknesses. The history is similar, except that at most places, the Lower Cambrian rocks appear to be unconformably on the late Precambrian sediments. The Precambrian sedimentary rocks are about ten thousand meters thick here.

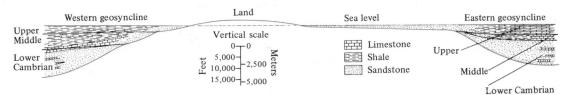

FIGURE 20–7 Generalized east-west restored section of Cambrian rocks. The section is drawn from the California-Nevada border to Virginia. Note that the basal Cambrian sandstone was deposited in a sea that gradually encroached on the continent, so the sandstone becomes younger as the interior of the continent is approached. In northern United States the late Cambrian sea was continuous across the continent. A section such as this is an interpretation from many fragmentary occurrences of deformed rocks.

FAR NORTH

Geosynclinal rocks are found in the Arctic islands of Canada. They apparently connect with similar rocks in Greenland and in northern Alaska. This is a very difficult area in which to study geology. Very few Cambrian rocks have been found. Ordovician rocks are widespread. To the north in the geosyncline, they are up to 3050 meters (10,000 feet) thick and consist of volcanic and clastic rocks. Farther south, they are thick shales; and still farther to the south, on the edge of the Canadian Shield, they grade into carbonates up to 915 meters (3000 feet) thick. In the geosyncline in East Greenland, about 12,200 meters (40,000 feet) of clastic rocks of Precambrian age are found. Overlying these are up to 2440 meters (8000 feet) of Cambrian and Ordovician carbonates. The Ordovician rocks are unconformably overlain by Devonian clastics.

MIDDLE ORDOVICIAN TO EARLY DEVONIAN— CYCLE II

In Middle Ordovician time the seas returned. Cycle II differs from Cycle I in that mountain building occurred in the east, and the rest of the continent was less stable.

INTERIOR

The seas returned in Middle Ordovician time, first to the geosynclines, and then across the interior. The pattern of land and sea was much like that of the Cambrian. The new cycle began with the deposition of a remarkably widespread, thin, quartz sandstone over most of the interior. This sandstone is up to 92 meters (300 feet) thick at places and covers almost two million square kilometers. Its source was apparently the Canadian Shield. Overlying

this sandstone are carbonates west of the Mississippi Valley and shale to the east. The carbonates are mainly dolomite that extends from western United States to Arctic Canada and is only about a hundred meters thick. The shale in the eastern interior grades into carbonate in the Silurian. The source of shale was a highland to the east in the area of the present Appalachian Mountains. Mountain building had begun in the geosyncline.

During the Ordovician, the interior changed slowly from a flat lowland on which uniformly thick sediments were deposited to a number of very broad domes and basins. (See Fig. 20–8.) The domes and basins were pronounced by Devonian time, when they dominated the deposition pattern. The seas withdrew from the interior in late Silurian time and from the geosynclines in early Devonian time, ending the cycle. In Late Silurian time, a remnant of the sea was isolated in one of the basins in Michigan, Pennsylvania, and New York. Here a thick section of evaporites, especially rock salt (halite) and gypsum, was deposited.

EAST

The eastern geosyncline was an area of active mountain building throughout this cycle. This was the beginning of a new pattern here that culminated in the formation of the Appalachian Mountains. In the Ordovician, a volcanic arc and subduction zone developed along northeastern North America, causing deformation and mountain building there. In the Silurian, Europe collided with North America, forming mountain ranges on both continents. During the Ordovician and Silurian, mountain building took place at different times and places throughout the geosyncline. None of the events described occurred simultaneously through the whole geosyncline. The mountain build-

ing and associated igneous and metamorphic events have destroyed much of the evidence, so reconstruction of the history is difficult.

Cycle II begins with clastic rocks of Middle Ordovician age. The source of these rocks was clearly the eastern part of the geosyncline because they thicken and become coarser to the east. Large fragments of Cambrian and Lower Ordovician limestones are found in thick clastic sections in the northern Appalachian Mountains, suggesting that uplifted and probably deformed rocks of Cycle I were eroded to form Cycle II deposits. This is shown diagrammatically in Fig. 20–9B. Volcanic rocks in eastern New England suggest volcanic island arcs there. A similar middle Ordovician clastic fan had its source in eastern Tennessee.

Mountain building continued throughout the Ordovician, filling the geosyncline with clastics and spreading shale far to the west. From West Virginia to New York, Late Ordovician clastics unconformably overlie Middle Ordovician rocks. These clastics were shed so rapidly that the geosyncline was filled and the sediments accumulated above sea level. Some of these rocks are red beds, so called because their iron has been oxidized to hematite, which is believed to happen when the rocks are subjected to seasonal wetting and drying.

This orogeny reached a peak near the end of the Ordovician. Folding, faulting, and uplift were widespread. The Taconic Mountains in Massachusetts and Vermont were formed at this time. This range is a block of Cambrian and Ordovician shales 242 by 48 kilometers (150 by 30 miles) that were thrust at least 48 kilometers (30 miles) to the west. Middle and Upper Ordovician rocks were folded at this time.

The Silurian record begins with a widespread quartz sandstone that is unconformable on Upper Ordovician rocks in the Appalachian Mountains. (See Figs. 20–9C and 20–10.) Farther to the west, Silurian carbonates overlie Upper Ordovician carbonates conformably. Uplifts in the highlands to the east occurred throughout the Silu-

rian Period, but the resulting clastic sediments did not spread west of the geosyncline. Thus the mountains were lower than during the Ordovician. In parts of New England and in Maritime Canada,

mountain building continued, resulting in thick sections of volcanic and clastic rocks. These rocks were folded and metamorphosed, and batholiths were emplaced. Elsewhere, the Silurian rocks were

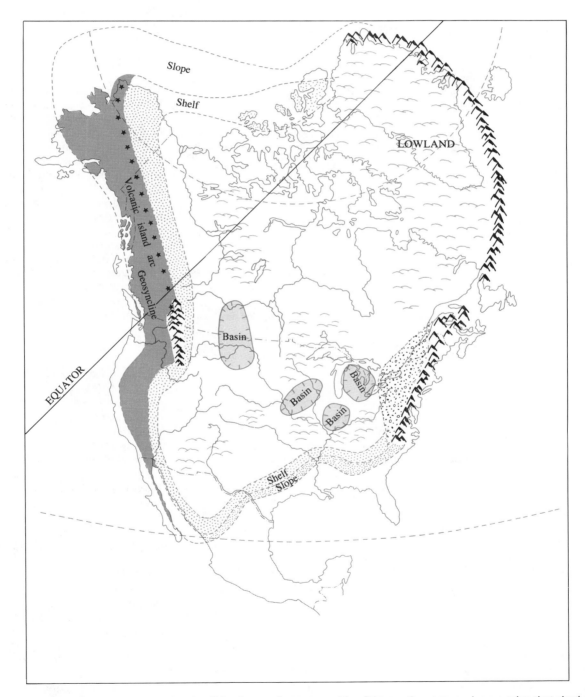

FIGURE 20–8 Ordovician paleogeography. A subduction on the eastern side of the continent caused mountains that shed clastics. Basins and domes began to form in the interior and were pronounced by Devonian time.

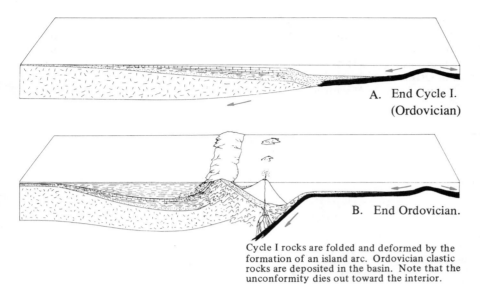

A. End Cycle I.
(Ordovician)

B. End Ordovician.

Cycle I rocks are folded and deformed by the formation of an island arc. Ordovician clastic rocks are deposited in the basin. Note that the unconformity dies out toward the interior.

EUROPE

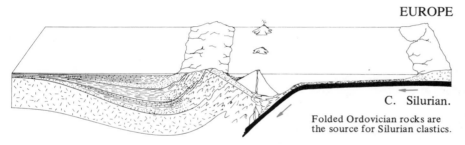

C. Silurian.

Folded Ordovician rocks are the source for Silurian clastics.

FIGURE 20-9   Development of the eastern geosyncline. A. At the end of Cycle I, the geosyncline consisted of a continental shelf, slope, and rise. B. Cycle II began after a volcanic island arc formed and the Cycle I rocks were deformed. C. Renewed activity deformed the Ordovician rocks.

not affected by this mountain building.

The Lower Devonian rocks are a continuation of the Silurian patterns. Thick clastic deposits are found to the east, and thin limestones to the west. In the northeast, volcanic rocks occur with clastics, many of which are nonmarine. The seas withdrew from the geosyncline in early Devonian time, ending the cycle.

WEST

Fewer rocks of Ordovician and Silurian age are found in the west, so the history is not so detailed. A subduction zone apparently formed along the west coast of North America, causing more activity in the western geosyncline.

In this geosyncline in central Nevada, 6100 meters (20,000 feet) of Ordovician coarse clastics with volcanic rocks are found. In Idaho, 3050 meters (10,000 feet) of Ordovician clastics are found. In Alaska, thick Ordovician clastics with volcanic rocks occur. Late Ordovician batholiths and an unconformity between middle and upper Ordovician rocks suggest mountain building there. These western clastics grade eastward into the carbonates of the interior. In late Ordovician and early Silurian time, a chain of low islands in central Nevada may have separated these rock types.

Very few rocks of Silurian age are found in the west. About 1220 meters (4000 feet) of shale are found in central Nevada. Much

thinner carbonate rocks are found toward the interior. In Alaska, thick clastic, carbonate, and volcanic sections are found.

This pattern continues to Early Devonian—that is, clastic rocks to the west, grading into thin carbonates in the interior. The cycle ends with the withdrawal of the sea in Early Devonian time.

FAR NORTH

The Arctic Islands contain thick Ordovician rocks as described earlier. These rocks grade from volcanic and clastic rocks in the geosyncline to the north through shale to thin limestone to the south near the shield. The Silurian and Lower Devonian rocks are similar. In at least one place, deformation took place in the Silurian, and Late Silurian and Early Devonian rocks overlie younger Silurian rocks. At other places, Early Devonian rocks are folded.

In northern Greenland, dark limestone and coarse clastics of Silurian age are found.

LIFE OF THE EARLY PALEOZOIC

The first abundant, easily recognized fossils occur abruptly at the base of the Cambrian System. Because well-developed, advanced forms of life appear suddenly at this horizon, the implication is clear that life must have existed for a long time previously. In the last chapter, some possible reasons for this were discussed. All of the phyla (the major subdivisions of life) (see Appendix H) appear in the Cambrian. (See Fig. 20-11.) This suggests that a great radiation and diversification had occurred in the 500 million or so years since green algae had appeared. Accidents of preservation probably also bias our knowledge of life. This is suggested by the discovery of a number of excellently preserved, flattened carbonaceous films, even with fine hairs intact, of soft-bodied animals of Middle Cambrian age at a single locality on the crest of the Canadian Rockies near Field, British Columbia. This occurrence is the only fossil record in the whole world of some of these animals, and in most cases it is the only place where soft parts of the bodies are preserved. It took an unusual set of conditions to preserve these animals. (See Fig. 20–12.)

Life began in the sea, and the first abundant fossils are marine. More than half of all Cambrian fossils are trilobites, and they and brachiopods together are about 90 percent of Cambrian fossils. Trilobites were arthropods. (See Fig. 20–13.) Trilobites underwent a great adaptive radiation during the Cambrian, and they are good index fossils for that period. They

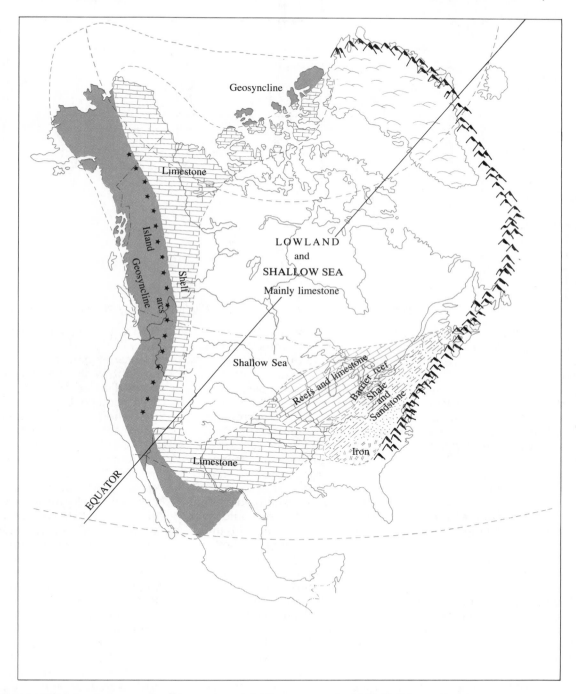

FIGURE 20–10 Silurian paleogeography. There were subduction zones on both sides of the continent.

411

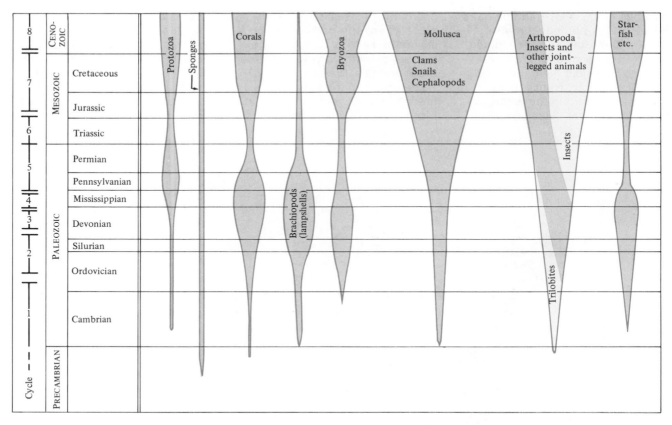

FIGURE 20-11 Geologic occurrence of invertebrate fossils. The width of the areas indicates relative abundance.

were bottom dwellers and swimmers, and probably ate algae and soft-bodied creatures that are not preserved. They were probably scavengers, too. Trilobites became less abundant after Early Ordovician time as other invertebrates developed, and the trilobites became extinct near the end of the Paleozoic.

A reconstruction of a Cambrian sea bottom is shown in Fig. 20-14. Similar reconstructions are also shown for most of the other periods. The crowding together of so many organisms shown in all of these probably did not ever occur in nature but is necessary to show the varieties of life in each period. The reconstruction in Fig. 20-14 is based on the collections made at the site shown in Fig. 20-12. The fossils are the evidence; the reconstruction is an interpretation.

Near the end of Cycle I, in Early Ordovician time, the fauna

changed somewhat. Brachiopods became more abundant (see Fig. 20-15), as did the other shell-bearing groups such as corals, bryozoa, crinoids, snails, and straight-shelled nautiloids. This change may have occurred in response to the development of predators. Few predators are found in Cambrian faunas, but they are found in Ordovician and later rocks. Straight-shelled nautiloids probably resembled present-day squids, and some could swim rapidly and catch prey with their tentacles. It may be that shells and skeletons were developed mainly for support in the Cambrian, but evolved into protection in the Ordovician. The development of spiny shells in late Paleozoic adds credence to this idea.

The life of Cycle II is shown in Fig. 20-16, a typical Middle Ordovician sea floor. The many differences between this scene and the Cambrian one are obvious. In

the Ordovician, corals and bryozoans expanded rapidly, and, to a lesser extent, snails, clams, straight-shelled nautiloids, and crinoids did also. (See Fig. 20-17.) This trend continued into the Silurian when corals became very abundant, forming reefs as in Figs. 20-18, 20-19, 20-20.

Graptolites are important fossils in Cycle II, especially in the Ordovician. The floating graptolites evolved rapidly from their first appearance in Late Cambrian to their extinction in the Early Devonian. They are good index fossils because they changed rapidly and are found in many environments because they floated. They tend to be found more commonly in black shale, probably because they sunk after death onto muddy bottoms where not much other life occurred. For this reason, it is common to speak of graptolite-bearing rocks versus shell-bearing rocks of Ordovician age. They were colo-

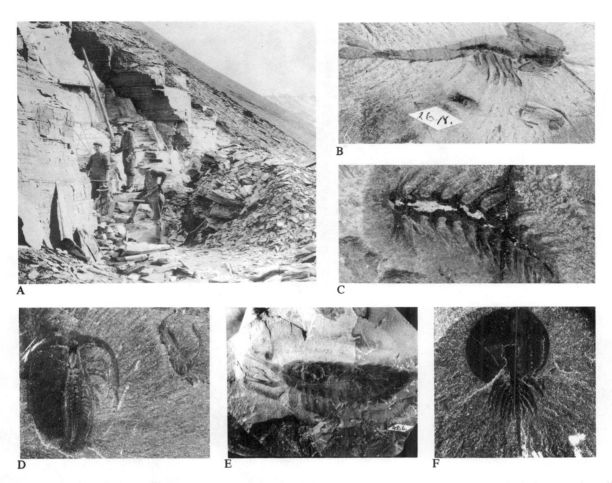

FIGURE 20–12 Burgess Shale on Mount Wapta, British Columbia. This Middle Cambrian formation has yielded extremely well-preserved fossils of 130 species. A. Collecting fossils. The first fossils were found by chance in 1909. B. *Waptia fieldensis,* a shrimplike arthropod. C. *Canadia setigera,* a worm. D. *Marrella splendens,* a trilobitelike arthropod. E. *Leanchoila superlata.* F. *Burgessia bella.* Photos from Smithsonian Institution.

FIGURE 20–13 Trilobite, *Elrathia kingi,* Middle Cambrian, about 2.5 centimeters (1 inch) long. Photo from Ward's Natural Science Establishment, Inc., Rochester, N.Y.

FIGURE 20-14 Reconstruction of Middle Cambrian (Cycle I) sea floor in western North America. The fauna shown include the soft-bodied animals in Fig. 20-12. A. Spongelike animal. B. *Marrella,* an arthropod. C. A worm. D. A jellyfish. Several different trilobites are also shown. From Field Museum of Natural History.

A   B   C

FIGURE 20-15 Brachiopods. Brachiopods are abundant in many Paleozoic rocks. They have two shells, and most were attached to the bottom by a stalk that came out through a hole near the hinge line. They are traditionally figured with the hinge line up. A. *Rafinesquina alternata,* Ordovician, from Ohio, about 55 millimeters (2 inches). B. *Lepidocyclus capax,* also Ordovician, from Ohio, about 25 millimeters (1 inch). C. *Eocoelia hemispherica,* Silurian, from Quebec, about 10 millimeters (0.4 inch). Photos A and B from Ward's Natural Science Establishment, Inc., Rochester, N.Y.; C. from Geological Survey of Canada (113250-C).

FIGURE 20-16 Middle Ordovician (Cycle II) sea floor in north-central United States. A. Straight-shelled nautiloids (somewhat like squids). B. Trilobites. C. Snails. D. Colonial corals. E. Solitary corals (with tentacles). F. Brachiopods. G. Bryozoa. H. Seaweed. From Field Museum of Natural History.

414

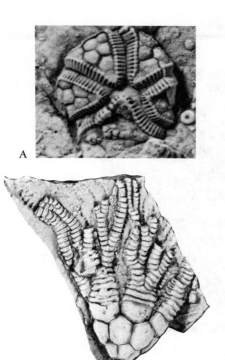

A

B

FIGURE 20–17 Ordovician echinoderms. A. *Edrioaster,* a primitive echinoderm that attached to the bottom, about twice natural size. B. *Cupulocrinus jewetti,* a crinoid, natural size. Both from Ontario. Photos courtesy Geological Survey of Canada (112910-B, 112339).

FIGURE 20–18 Bryozoa. These colonial animals build intricate structures on the sea floor in which a separate animal occupies each hole. Some build lacy and others stony or twiglike structures. Fossils are generally fragmentary because the structure is broken after death of the colony. Silurian bryozoa, Rochester Shale, Lockport, New York. A few brachiopod shells can be seen in this slab with several types of bryozoa. Photo from Smithsonian Institution.

FIGURE 20–19 Middle Silurian (Cycle II) sea floor in north-central United States. A. Cystoid. B. *Favosites,* honeycomb coral. C. *Halysites,* chain coral. D. *Syringopora,* tube coral. E. Several types of nautiloids. F. Two types of brachiopods. G. Trilobite. Several other types are also in the view. H. Two solitary corals. From Field Museum of Natural History.

415

nial animals, with many small individuals living in the stalks that hung down from the float. The fossils look like pencil marks on the enclosing shales (see Fig. 20–21), hence their name. In late Cambrian and early Ordovician, they were multibranched; in middle and upper Ordovician, they had two branches; and in Silurian, they had one branch.

Vertebrates first appear in the Cambrian. They, too, began in water. The first evidence of their existence is scales and bone fragments found in an upper Cambrian sandstone. Few other fish fossils are found until Late Silurian, although in Cycle III they are dominant. Thus there is little evidence for the stages in the development of the fish. The early fish may have had predators. Their predators may have been straight-shelled nautiloids or, probably more likely, the eurypterids. In any case, the first fish were armored, suggesting that they had predators.

Eurypterids became abundant in Silurian rocks. (See Figs. 20–22 and 20–23.) They are not found in normal marine rocks but in what probably were brackish waters or perhaps restricted saline waters. Some were as much as 2.75 meters (9 feet) long, and they must have ruled their environment.

Scorpions are also found in Silurian rocks. (See Fig. 20–24.) They may have been the first animals to breathe air. The oldest definitely terrestrial scorpion known is Early Devonian. Scorpions, eurypterids, and trilobites were all arthropods.

Plants on the land must have preceded any animal life. The earliest land plants so far discovered are lower Silurian in age. Plant spores (microfossils) have been recovered from rocks as old as Middle Ordovician so plants were present by then. Plants started from algae, and the first larger plants were marine. The widespread lowlands of Late Silurian time probably were a favorable

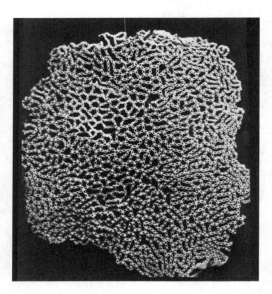

FIGURE 20–20 Chain coral, *Halysites,* Silurian, Louisville, Kentucky. A separate animal lived in each hole in this colonial coral. Photo from Smithsonian Institution.

FIGURE 20–21 Graptolites. The fossils of these curious animals look like pencil marks on rocks. They were colonial animals, and a separate animal lived in each of the saw teeth. The stalks hung down from a float, and because they floated, the fossils are found in all environments. They evolved very rapidly: In Late Cambrian and Early Ordovician they had many branches, in Middle and Upper Ordovician they had two branches, and in Silurian they had one branch. They became extinct in Early Devonian. Shown is *Didymograptus,* Middle and Upper Ordovician. Photo from Ward's Natural Science Establishment, Inc., Rochester, N.Y.

environment for the development of plants. Plants were abundant from Devonian on. Beginning in Late Devonian, coal appears in every geologic period. The only earlier coal is in the Precambrian in Michigan; and although it may be of organic origin, it probably was not formed by land plants.

FIGURE 20-22   Late Silurian (Cycle II) sea floor in New York. Several types of eurypterids (see Fig. 20-23) are shown with snails. Eurypterids probably lived in somewhat unusual environments such as restricted, saline, shallow seas. A. *Carcinosoma,* a scorpion. From Field Museum of Natural History.

FIGURE 20-23   Silurian eurypterid, *Eurypterus remipes.* This specimen came from the same formation as the scorpion shown in Fig. 20-24, but from a different part of New York. Photo from Smithsonian Institution.

FIGURE 20–24 *Carcinosoma scorpionis,* Silurian scorpion. This specimen came from the same formation as the eurypterid shown in Fig. 20–23, but from a different part of New York. Photo from Smithsonian Institution.

SUMMARY

In late Precambrian the continents came together, causing mountain building in Gondwanaland. From Precambrian through Ordovician, the south pole was in northern Africa. During the Silurian Gondwanaland moved so that the south pole was in southern Africa.

The geologic systems are natural subdivisions only at the places where they were defined.

Geosynclines are important in the history of continents.

Mountain building does not seem to be synchronous over the earth, although the Cambrian and most of Triassic were quiet times, and Ordovician, Devonian, Pennsylvanian, and late Cretaceous were more active.

The interior of North America has experienced times of deposition followed by withdrawal of the seas. These cycles are the framework of this book.

Large extinctions occurred near the end of the Cambrian, Devonian, Permian, Triassic, and Cretaceous Periods.

In the Ordovician mountains resulted from a subduction zone, and in the Silurian they were caused by a collision with Europe.

At the beginning of Cycle I, continental shelf-slope geosynclines occupied both sides and the north of North America, and at times the interior was a lowland covered by shallow seas.

Cycle I—Late Precambrian to Early Ordovician

Interior: Sandstone grading upward into carbonates. Source was land area in northern and central Canada.

East: Clastics grading upward into carbonates. To the north, mainly clastics with some volcanic rocks.

West: Mainly clastics that transgressed northeastward on the interior.

Far North: Mainly clastics. Very few Cambrian rocks.

Cycle II—Middle Ordovician to Early Devonian

Interior: Starts with widespread thin sandstone. Carbonates to west, shale in east. Domes and basins form.

East: Mountain building is source of clastics in geosynclines and interior.

West: Mainly clastics. A volcanic island chain may have formed in Nevada.

Far North: Clastics in the geosyncline grade to limestone in the interior.

Cambrian life was not primitive, suggesting a long evolution.

Our only knowledge of early soft–bodied animals comes from a single location in the Canadian Rocky Mountains.

Trilobites and brachiopods are the main Cambrian fossils.

In the Ordovician, brachiopods, corals, bryozoa, crinoids, snails, and nautiloids became more abundant.

Graptolites are good index fossils in Ordovician and early Silurian.

The first vertebrates were fish in the late Cambrian.

Silurian scorpions may have been the first animals to breathe air.

The earliest plants are Middle Ordovician in age.

QUESTIONS

1 List as many of the assumptions made in interpreting geologic history as you can.

2 Describe the various patterns recognized or assumed in geologic history.

3 Is the Cambrian-to-Ordovician boundary a natural one in North America?

4 What events separate Cycle I from Cycle II?
5 What occurred in the eastern part of North America during the Silurian Period?
6 In what continent was the "south" pole in early Paleozoic time?
7 Where was the equator in North America in early Paleozoic time?
8 What is the importance of the Burgess Shale fauna?
9 What are typical Cambrian fossils?
10 In what ways do Cycle II fossils differ from those in Cycle I?
11 Why are graptolites important?
12 What were the first vertebrates, and when did they appear?
13 When did the first land plants appear?

SUPPLEMENTARY READINGS

General

Bambach, R. K., C. R. Scotese, and A. M. Ziegler, "Before Pangea: The Geographies of the Paleozoic World," *American Scientist* (January-February 1980), Vol. 68, No. 1, pp. 26–38.

Cloud, Preston, *Adventures in Earth History.* San Francisco: W. H. Freeman and Co., 1970, 992 pp. A collection of papers by many authors.

Engel, A. E. J., "Geologic Evolution of North America," *Science* (April 12, 1963), Vol. 140, No. 3563, pp. 143–152.

King, P. B., and H. M. Beikman, *The Paleozoic and Mesozoic Rocks: A Discussion to Accompany the Geologic Map of the United States.* U.S. Geological Survey Professional Paper 903, Washington, D.C.: U.S. Government Printing Office, 1976, 76 pp.

Morris, S. C., and H. B. Whittington, "The Animals of the Burgess Shale," *Scientific American* (July 1979), Vol. 241, No. 1, pp. 122–133.

Palmer, A. R., "Search for the Cambrian World," *American Scientist* (March-April 1974), Vol. 62, No. 2, pp. 216–224.

Pojeta, John, Jr., and Bruce Runnegar, "*Fordilla troyensis* and the Early History of Pelecypod Mollusks," *American Scientist* (November-December 1974), Vol. 62, No. 6, pp. 706–711.

Patterns in Geologic History

Johnson, J. G., "Timing and Coordination of Orogenic, Epeirogenic, and Eustatic Events," *Bulletin,* Geological Society of America (December 1971), Vol. 82, No. 12, pp. 3263–3298.

Kerr, R. A., "Changing Global Sea Levels as a Geologic Index," *Science* (July 25, 1980), Vol. 209, No. 4455, pp. 483–486.

Loutit, T. S., and J. P. Kennett, "New Zealand and Australian Cenozoic Sedimentary Cycles and Global Sea-Level Changes," *Bulletin*, American Association of Petroleum Geologists, (September 1981), Vol. 65, No. 9, pp. 1586–1601.

Schleh, E. E., "Review of Sub-Tamaroa Uncomformity in Cordilleran Region," *Bulletin,* American Association of Petroleum Geologists (February 1966), Vol. 50, No. 2, pp. 269–282.

Sloss, L. L., "Sequences in the Cratonic Interior of North America," *Bulletin,* Geological Society of America (February 1963), Vol. 74, No. 2, pp. 93–114.

Sloss, L. L., "Areas and Volumes of Cratonic Sediments, Western North America and Eastern Europe," *Geology* (May 1976), Vol. 4, No. 5, pp. 272–276.

Wheeler, H. E., "Post-Sauk and Pre-Absaroka Paleozoic Stratigraphic Patterns in North America," *Bulletin,* American Association of Petroleum Geologists (August 1963), Vol. 47, No. 8, pp. 1497–1526.

21 LATE PALEOZOIC – DEVONIAN, MISSISSIPPIAN, PENN-SYLVANIAN, PERMIAN

The late Paleozoic strata in much of England are bounded above and below by distinctive red sandstones. The late Paleozoic Coal Measures, or Carboniferous System, were, of course, recognized very early because of their economic importance. The coal beds are in the terrestrial rocks in the upper part of the Carboniferous, and the lower beds are marine rocks. A similar twofold division in North America led to the subdivision of the Carboniferous into the Mississippian and Pennsylvanian Systems.

The Old Red Sandstone is a distinctive unit in much of England, Scotland, and Wales that lies below the Carboniferous. The Old Red Sandstone is a nonmarine deposit and contains fossil plants, fish, and impressions of footprints. In 1841 Hugh Miller, a stone mason, wrote a book, *The Old Red Sandstone,* which became very popular. The interpretation of natural history for the working class was a popular movement of the times, and Miller went on to write other popular books about geology.

In the years 1836–1839, Sedgwick and Murchison, who had studied the Cambrian and Silurian Systems, were still friends, and they were working together on the deformed rocks in Devonshire. They established that the rocks were under the Carboniferous. From 1837 on, William Lonsdale, a retired army man who had been studying fossil corals, stated that the Devonshire rocks were intermediate in age between the Carboniferous and the Silurian. Eventually Sedgwick and Murchison were convinced, and in 1839 they named these rocks the Devonian System. These marine rocks in Devonshire were the equivalent of the Old Red Sandstone. Thus they recognized that the character of rocks of the same age may differ if their environment of deposition is different.

Murchison then went to Europe to seek extensions of the Devonian there. The search led him to Russia in 1840 and 1841. There he found his system well developed and overlying it undescribed strata. He named these rocks the Permian System for the Province of Perm on the west flank of the Ural Mountains. Permian rocks are not well developed in England, and so were not recognized there first.

WORLD GEOGRAPHY IN THE LATE PALEOZOIC

In the late Paleozoic, the continents came together, forming one supercontinent. The collisions caused mountain building, and the mountains together with the large landmass profoundly affected the climate and therefore the life of the earth. The evidence that the southern continent, Gondwanaland, was a single continent in the late Paleozoic was presented in Chapter 14, and that evidence was very important in the establishment of the reality of continental drift.

In the Devonian, the collision between Europe and North America continued (Fig. 21–1); and in the Pennsylvanian or Permian Periods, Africa collided with southeastern North America (Figs. 21–2 and 21–3). These collisions formed the Appalachian Mountains as well as other mountains on all of the continents involved. In western North America, subduction and the resulting deformation continued through the late Paleozoic.

In the northern supercontinent, the mountains formed by the Silurian orogeny shed similar red clastics in Europe, eastern North America, and northern North America.

The Ural geosyncline between Europe and Asia was deformed by the collision of those two continents, forming the Ural Mountains.

A major change that occurred in late Paleozoic is the movement of the south pole from North Africa

to near its present position in Antarctica. This means that the southern continent moved about 90 degrees north during the Paleozoic, assuming that the geographic pole remained in the same place. The Permian glacial deposits described and figured in Chapter 14 resulted from this drift. North America was rotated in this move, so that the equator ran closer to east-west instead of north-south as it had earlier. This change may account for the aridity indicated by the late Paleozoic and Mesozoic rocks in western North America.

MIDDLE DEVONIAN TO LATE DEVONIAN—CYCLE III

INTERIOR

The seas returned to the interior in Middle Devonian time, and at many places rocks of that age overlie Middle Silurian beds. In the eastern part of the interior, shale was the main rock type deposited. Its main source was highlands in the eastern geosyncline. Near the domes of the interior and the highlands in the geosyncline, some sand was deposited. The western interior, extending to the

El Capitan in the Guadalupe Mountains, Texas, is composed of Permian reef limestone. Photo courtesy U. S. Geological Survey.

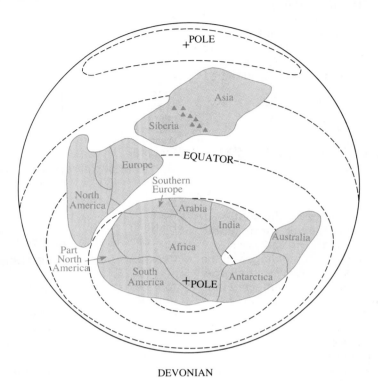

DEVONIAN

FIGURE 21-1   The continents in the Devonian Period. North America and Europe are in contact.

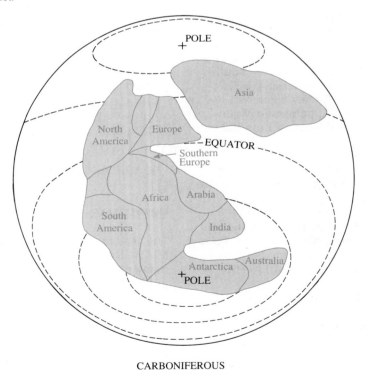

CARBONIFEROUS

FIGURE 21-2   The continents in the Pennsylvanian Period. Africa and North America collide. Gondwanaland moved so that the southern pole was in Antarctica.

Arctic, was the site of limestone and some shale deposition. Reefs developed in this limy sea; in Alberta, these reefs are reservoirs for important oil fields. The domes and basins of the interior became more pronounced. In Montana and Alberta and at other places in Canada, evaporites formed in some of these basins that were cut off from the seas. The seas withdrew in the late Devonian, ending this short cycle.

EAST

In Middle and Late Devonian time, orogeny continued in the eastern geosyncline. The collision with Europe that started in the Silurian continued, with mountains forming farther south (Fig. 21–4). The intrusive and metamorphic processes that accompanied this collision transformed the old geosyncline into new continent. The southeastern border of North America was probably a subduction zone because Africa (Gondwanaland) was coming closer.

The area of the old eastern geosyncline became an upland. This upland was rapidly eroded, and the resulting clastic sediments were deposited in the area just west of the old geosyncline. In this depositional area first one place and then another would downsink to trap the debris. Thus, in Middle and Late Devonian, over 3050 meters (10,000 feet) of deltaic rocks were deposited in Pennsylvania and adjacent New York (Catskill Mountains) and West Virginia. These deposits were similar to the Ordovician and Silurian deltas but were much thicker and more widespread, and this type of deposition continued into the Pennsylvanian. In Mississippian time the clastic fan spread farther south along the Appalachian Mountains, suggesting that orogeny spread southward. This pattern, too, as we shall see, continued throughout the remainder of the Paleozoic with the addition of smaller basins

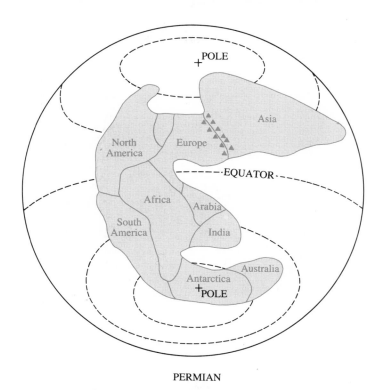

PERMIAN

FIGURE 21–3   The continents in the Permian Period. Asia and Europe collide, forming a single continental mass.

and adjacent highlands in the area of the old geosyncline. The seas again withdrew during part of Late Devonian time, ending the cycle.

WEST

In the far west, few Devonian rocks are found, but the history can be reasonably deciphered. The subduction zone along the west coast continued to be active. A volcanic island arc is believed to have existed in the vicinity of the California-Nevada border. The evidence for this is that thick sections of clastic and volcanic rocks are found in western Nevada. Farther east, these rocks interfinger with carbonates that, in turn, extend onto the western interior. Similar sequences are found in the Canadian Rockies in Alberta, where the carbonate rocks are very thick and contain both reefs and evaporites. In latest Devonian time orogeny began in Nevada. This orogeny

continued to the Pennsylvanian and will be described later.

FAR NORTH

The Devonian Period began quietly in the far north. The carbonate deposition of the interior extended to the Arctic. In the geosyncline, a thousand or more meters of carbonates with some shale and siltstone, especially near the top, make up the Middle Devonian rocks. These rocks thin to the south on the interior platform. The Upper Devonian rocks are sharply different. Uplift occurred to the north, forming a highland that shed clastics into the geosyncline. These nonmarine clastic rocks are up to 3050 meters (10,000 feet) thick and contain some coal beds. They closely resemble the rocks of this same cycle in New York and Pennsylvania.

In East Greenland thick deposits of nonmarine clastic rocks were

also deposited in this cycle. The rocks are up to 6100 meters (20,000 feet) thick and were deposited in basins adjacent to uplands, and so record activity here also. The rocks are red and gray in color and resemble the Old Red Sandstone, a formation of the same age in England. The source of all of these clastic rocks was probably the highland created by the collision of Europe and North America.

LATEST DEVONIAN TO LATE MISSISSIPPIAN— CYCLE IV

INTERIOR

This cycle is confined mainly to the Mississippian Period and begins with the return of the sea to the interior in latest Devonian time. The interval between Cycles III and IV was short, and it may be that at places the seas did not completely withdraw. The pattern of this cycle in the interior is much like the last cycle, but the rocks are quite different. (See Fig. 21–5.) The Mississippian System was named for the exposures in the area drained by the upper Mississippi River. The Lower Mississippian rocks in that area are black, organic-rich shales. The shale is not the same age everywhere and is rarely more than about ten meters thick but is generally present over a wide area. It is difficult to reconstruct the conditions under which this shale was deposited. The organic content suggests stagnant water, but the fossils and scour channels suggest shallow water. The shale contains enough uranium that it may in the future be an economic resource. The source of the shale was highlands that had formed in the geosyncline to the east and south.

Middle Mississippian limestones overlie the black shale. Some of

423

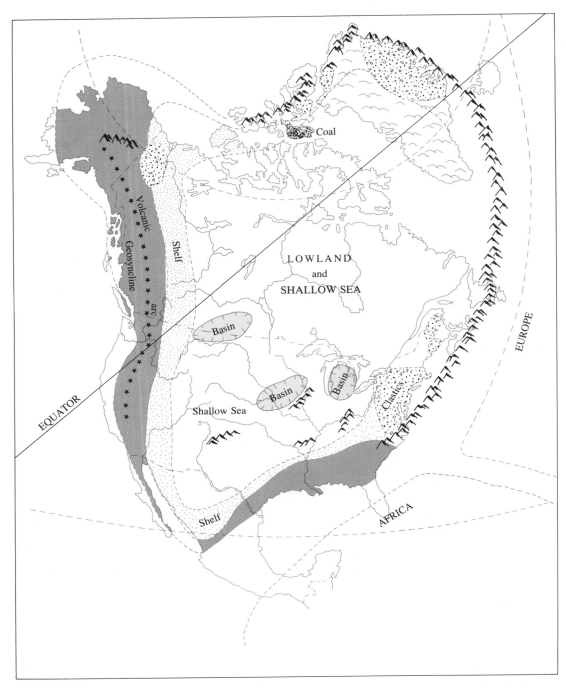

Labels within the map:

Coal

Volcanic

Geosyncline

arc

Shelf

LOWLAND
and
SHALLOW SEA

Basin

Basin

Basin

EUROPE

EQUATOR

Shallow Sea

Clastics

Shelf

AFRICA

FIGURE 21–4 North America in the Devonian Period. Collision with Europe caused mountains that shed clastics in the east and in Greenland, as well as in Europe.

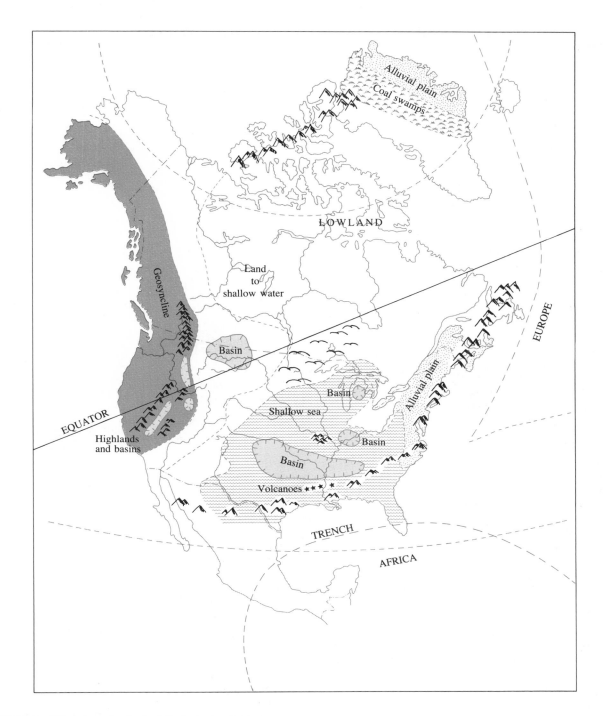

FIGURE 21-5 North America in the Mississippian Period.

these limestones are composed largely of fossil fragments and some of spherical calcite grains (oölites). They are quarried for building stone. In Late Mississippian time, limestone and sandstone layers alternate. The source of the sands was highlands to the east. These are the first sandstones in the interior since Ordovician time.

The western interior was the site of limestone deposition through this cycle. These carbonates extend well into Canada. There the amount of shale increases to the north and in the younger rocks.

The domes and basins of the interior were active and, to a large extent, controlled the thicknesses of all of the units described. The basin in Montana-Dakota was the site of evaporite deposition in Middle Mississippian time.

The cycle ended with the withdrawal of the seas in Late Mississippian time. Erosion followed over most of the continent. The erosion was more pronounced in the west than in the east. The east was probably near sea level, setting the stage for the Pennsylvanian coal swamps.

EAST

The area of the old geosyncline was again uplifted, producing another clastic fan like that of Devonian time. The center was farther south in Pennsylvania, and the total extent was less than in the Devonian. At the same time, in the area of the old geosyncline in New England and northeastern Canada, a number of basins developed. These basins persisted from Mississippian to Permian time and developed great thicknesses of clastic rocks eroded from the surrounding highlands, together with some carbonates, evaporites, and red beds.

In Texas, Oklahoma, and Alabama, thick deposits of clastic

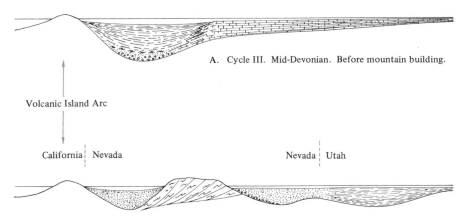

A. Cycle III. Mid-Devonian. Before mountain building.

Volcanic Island Arc

California ⁝ Nevada   Nevada ⁝ Utah

B. Cycle IV. Early Mississippian. After mountain building.

rocks developed in Middle and Late Mississippian time, suggesting orogeny in that area. Subduction apparently continued there, and Africa came nearer. The collision occurred in late Mississippian or Pennsylvanian time.

WEST

In Mississippian time, the pattern of the far west changed. Mountain building occurred in the geosyncline in central Nevada, and the resulting upland was to persist in some form throughout most of the later periods. The first evidence of the orogeny is found in late Devonian clastic rocks in the eastern half of Nevada. Prior to that time, clastic and volcanic rocks accumulated farther west, and the source of these rocks was apparently a volcanic island arc near the California-Nevada border. The orogeny began in Late Devonian time and extended into Pennsylvanian. The new mountain structures formed in the area where the older clastic rocks of western Nevada graded into the carbonate rocks of eastern Nevada. (See Fig. 21–6.) The orogeny thrust the western clastic rocks over the eastern carbonates. The rocks are estimated to have moved almost 160 kilometers (100 miles). The resulting mountain range extended all the way across Nevada and

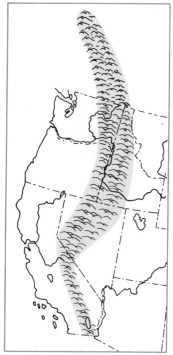

C. Extent of Devonian to Pennsylvanian mountain building.

FIGURE 21–6 Devonian and Mississippian events in Nevada. Thrust faulting in Late Devonian and Early Mississippian time created two areas where clastic sediments were deposited. After R. J. Roberts.

into Idaho where younger igneous rocks obscure its extent. The mountains were never very high and underwent erosion as they were uplifted, shedding clastics both east and west. They were probably a series of islands connected by shallow water. To the east were deposited about a thousand meters of conglomerate and sandstone that grade into the thinner limestones of the interior. To the west, geosynclinal sediments accumulated. This pattern continued into the Pennsylvanian.

FAR NORTH

In Alaska, uplift caused deposition of thick clastic rocks that grade into the thin limestones of the interior. In Arctic Canada, mountain building apparently took place because Middle Pennsylvanian rocks overlie Upper Devonian sediments on a major angular unconformity. The lack of Mississippian rocks makes it impossible to decipher the details. In East Greenland, the deposition of nonmarine clastics that had begun in Devonian time continued.

LATEST MISSISSIPPIAN TO PERMIAN—CYCLE V

The seas returned to parts of the interior in very latest Mississippian time, and they remained there through the Permian Period. During this interval, orogeny occurred at many places in North America and the rest of the world. This was a time of great changes when all of the continents came together and formed one supercontinent.

INTERIOR

The oldest rocks of this cycle are late Mississippian and early Pennsylvanian shale and sandstone that are found east of the Mississippi River valley. These rocks were apparently deposited in the deltas of rivers flowing from highlands to the east and south. By

Middle Pennsylvanian time, the seas again covered the interior, and great coal swamps had formed from the Mississippi valley eastward to the Appalachian Mountains. Thicker deposits formed in the basins of the interior. The climate was apparently warm, and the plants that grew in the swamp were buried before they decayed, becoming the very important coal deposits of eastern North America. (See Fig. 21–7.) Both the coal deposits and the marine Pennsylvanian rocks of the interior are remarkable for the great number of repeated sequences of rocks called *cyclothems,* which are believed to be the result of periodic changes in sea level. (See Fig. 21–8.) The cause of these changes in sea level is not known, but they could be due to climatic changes such as glaciation that occurred elsewhere or to vertical movements of the continents or the ocean bottoms. A typical cyclothem begins with coarse, nonmarine sediments that become finer grained as the sea approaches. Coal is deposited during a near-shore swampy time and is followed by marine sediments that grade upward from shale to limestone as the water depth increases. The sea then retreats; then, generally after a short period of erosion, the sea readvances, starting a new cycle.

Permian rocks in the east are few and consist of clastic rocks in the area of the Pennsylvania-Ohio-West Virginia border. The upper Pennsylvanian rocks grade upward into the Permian. These nonmarine Permian rocks, less than 305 meters (1000 feet) thick, are the youngest deposits associated with the eastern geosyncline. This will be discussed further in the next section.

There are no Triassic rocks in the eastern interior.

The Cycle V rocks of the western interior are very different from those of the east. In this cycle, the

paleogeography of North America changed drastically. One of these changes was the development of a number of highlands and deep basins in the general area of the central Rocky Mountains and extending to Oklahoma and Texas. (See Fig. 21–9.) The Pennsylvanian rocks of the western interior are mainly quartz sandstone with some carbonates. In the deep basins adjacent to the highlands, thick clastic sections were deposited. The older broad basin in Dakota-Montana was the site of carbonate and evaporite deposition with some shale. Because of the changes in paleogeography, it will be more convenient to include the later history of the western interior under the heading "West."

EAST

In Cycle V, deposition in the eastern geosyncline ends. In the east the geosyncline was a highland at the start of the cycle, and rivers flowing from the highland brought sediments to the interior as was just described. Basins formed in Massachusetts-Rhode Island, and in eastern Canada. Thick clastic sections containing coal accumulated in these basins.

Farther to the south, thick clastic deposits indicate renewed orogeny as Africa neared North America. In the Texas-Oklahoma area, the geosyncline received thick clastic deposits, suggesting continual orogeny to the south. In this area the geosyncline itself was deformed into a series of linear highlands and basins. These basins received sediments eroded from the adjacent highlands. Another orogeny affected this area, probably in latest Pennsylvanian time, when nonmarine clastic rocks were deposited and finally thrust plates moved northward over these rocks. In western Texas the thrusting was somewhat earlier, and the end of the Pennsylvanian was a time of erosion. (See Fig. 21–10.)

In the late Paleozoic the Appalachian geosyncline was the site of orogeny that formed the Appalachian Mountains. The orogeny was probably caused by the collision of Africa (Gondwanaland) and North America (Fig. 21–11). This collision caused mountain building in southeastern North America and also farther north where parts of Europe were deformed. Our present Appalachian Mountains contains parts of both North Africa and Europe, and perhaps other much smaller continental plates.

Figure 21–12 shows the extent of intrusive rocks emplaced then and during previous Paleozoic orogenic activity. The only rocks of Permian age are Early Permian clastics, similar to the underlying Pennsylvanian rocks that were described in the last section. The ex-

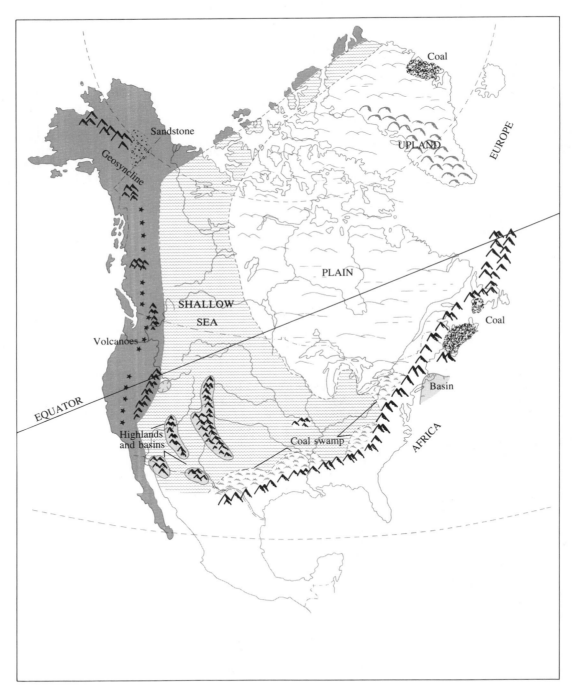

FIGURE 21–7 North America in the Pennsylvanian Period.

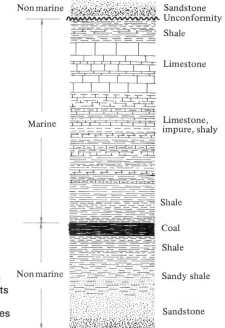

FIGURE 21-8 A typical cyclothem begins with nonmarine sandstone that becomes finer grained as the sea advances. Coal is deposited in a near-shore swamp. As the water depth increases, the marine rocks grade from shale to limestone. The sea then retreats to start the cycle again. Glaciation elsewhere may have caused the changes in sea level.

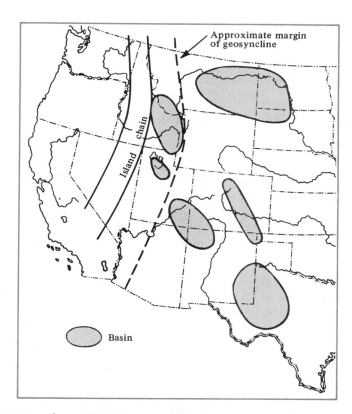

FIGURE 21-9 Some late Paleozoic events in the west. In Late Devonian and Mississippian time orogenic events produced a chain of islands in the western geosyncline that was the source of clastic sediments. In the geosyncline the Early and Middle Pennsylvanian rocks are clastic west of the islands and limestone to the east. In late Pennsylvanian time renewed activity in the island chain resulted in clastic sediments east of the islands.

Starting in Mississippian, and especially in Pennsylvanian and Permian, a number of deep basins, some with adjacent highlands, developed both in the area of the geosyncline and in the interior sea.

429

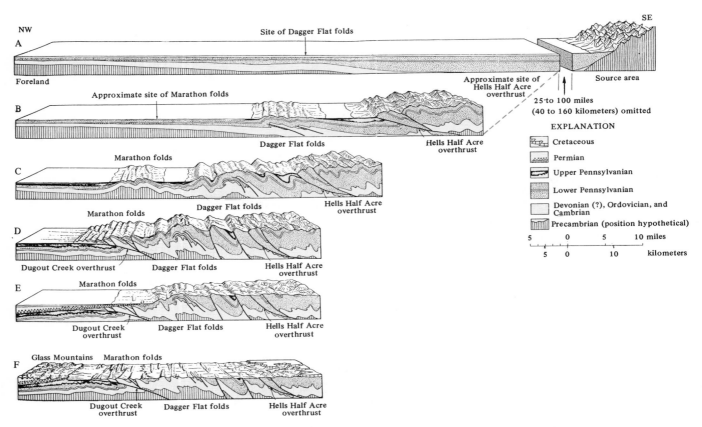

FIGURE 21–10 Hypothetical block diagrams showing the development of the Marathon Mountains in west Texas. Undeformed Permian rocks overlie deformed Upper Pennsylvanian rocks and so date the final orogenic phase. Note the inferred shortening of the rocks. From P. B. King, U.S. Geological Survey Professional Paper 187, 1938.

act age of this orogeny that turned the remaining parts of the old geosyncline into continental structure is not known. The Early Permian rocks were folded, presumably at this time, and Late Triassic to Early Jurassic sediments farther east were not affected and appear to lie on an erosion surface cut on the older folded rocks. Thus it appears that the orogeny probably took place in the Permian and may have extended into Triassic time, and then, after a period of erosion, the Triassic rocks were deposited.

The Late Triassic-Early Jurassic rocks just mentioned are red clastics interbedded with basalt. In Late Triassic time these rocks accumulated to thicknesses up to 6100 meters (20,000 feet) in a number of fault basins within and just east of the present Appalachian Mountains from Connecticut to North Carolina. (See Fig. 21–13.) It is possible that the fault basins may have been formed by the tension that separated the continents. (See Fig. 21–11C.) These are the only Triassic rocks in the east, so they are described here with their related rocks. The next depositional event in the east was the Cretaceous to Recent deposition on the coastal plain.

WEST

The west, too, was active in the late Paleozoic. The western geosyncline was dominated by an active subduction zone and its associated volcanic arc. In the western geosyncline, to the east of the island chain that had appeared earlier, the Early and Middle Pennsylvanian rocks are limestones unlike the Mississippian clastics. West of the islands, conglomerate and sandstone were deposited. In Pennsylvanian time the island chain was the site of more orogeny and uplift. (See Fig. 21–14.) The Late Pennsylvanian deposits, in part eroded from this uplift, are sandstones and limestones and overlie earlier Pennsylvanian rocks with an angular unconformity. These accumulated in a very thick [12,200-meter (40,000-foot)] basin in eastern Utah and adjacent Colorado. (See Fig. 21–9.)

The area east of the geosyncline has, up to this point, been included with the interior; but after early Permian time, the seas never again covered all the interior. Only the western interior was covered by seas, and because these seas were extensions of those covering the western geosyncline, it is better to include the closely related western interior with the western geosyncline.

The Pennsylvanian rocks of the western interior are quartz sandstones and carbonates. In southwestern Colorado a highland de-

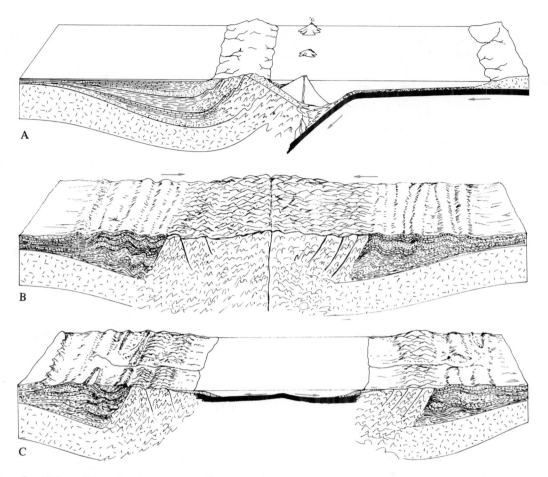

FIGURE 21-11 Formation of the Appalachian Mountains. This figure completes the history shown in Figs. 15–40 and 20–9. A. Silurian Period. B. Collision of continents in late Palezoic time produces the Appalachian Mountains. C. Reopening of the Atlantic Ocean in the Mesozoic.

FIGURE 21-12 Paleozoic granitic rocks of the United States. Most of these batholiths were emplaced during mountain building in middle and late Paleozoic time. From P. B. King and H. M. Beikman, U.S. Geological Survey Professional Paper 903, 1976.

veloped, and thick accumulations of clastic rocks were shed into the adjacent basins. The basin on the Colorado-Utah border (see Fig. 21–9) received Early Pennsylvanian clastics, followed by thick evaporites and Middle and Late Pennsylvanian clastics and carbonates; the total thickness is 3050 meters (10,000 feet).

The Permian was an active time in the western geosyncline. In the far west thick deposits of clastic and volcanic rocks are found. Orogeny occurred in western Nevada; and in Oregon, the Triassic overlies the Permian on an angular unconformity. In central Nevada the orogenic island chain was uplifted and deformed and was again the site of thrust faulting. (See Fig. 21–14.) East of this

431

area, the main deposits were quartz sandstone and carbonate. In the last half of Permian time, in parts of Idaho, Montana, Wyoming, and Utah, a restricted basin formed in which peculiar phosphate-rich rocks were deposited. They are an important economic source of phosphate. (See Fig. 21–15.) Farther to the east, all of these marine rocks interfinger with clastic red beds whose source was the mountains in Colorado that had formed during the Pennsylvanian.

Parts of the Permian rocks of the western interior just mentioned have been studied very closely. At the start of the Permian Period, a shallow sea covered the area from Nebraska to west Texas. The main deposits were carbonates and shale along with clastics from the highlands of Colorado and Texas. This sea retreated to the south, so that by late Permian time, only west Texas was the site of marine deposition, and the highlands were largely eroded away by the end of the period. As the sea retreated, marine deposition was replaced by nonmarine clastic red beds. These red beds interfinger with the marine deposits. They suggest that arid conditions prevailed there in Permian time, and similar Mesozoic rocks suggest that this climate persisted into later periods. Because of the retreating sea, almost all of the Permian System is represented by marine rocks in west Texas. It is the most complete Permian section in North America and would be studied for this reason alone; however, many environments are represented in these rocks, making their study even more valuable. This area had a number of basins separated from each other by shallower seas. Reefs developed, evaporites formed in restricted basins, and all of these rocks interfingered with nonmarine red beds. In very Late Permian time, nonmarine red beds were deposited, ending the deposition.

Near the end of the Permian, the seas retreated elsewhere as well. This retreat was not complete in the western geosyncline, and the Triassic rocks reflect a continuation of the Permian pattern.

FAR NORTH

The Devonian and Mississippian orogeny formed geosynclinal

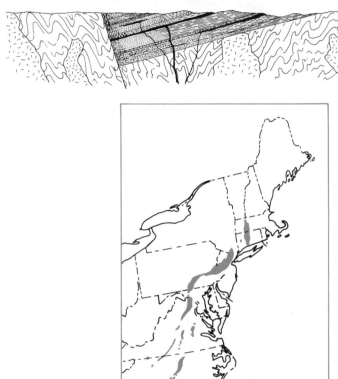

FIGURE 21–13 Late Triassic to Early Jurassic sediments formed in faulted basins in the metamorphic rocks east of the present Appalachian Mountains. The rocks are coarse clastic red beds that grade into finer clastics away from the faults, and basaltic volcanic rocks. Dinosaur footprints are found in these sedimentary rocks at some places. The Basin and Range fault structures described in Chapter 23 are somewhat similar. A. Typical cross-section. B. Map, from U.S. Geological Survey.

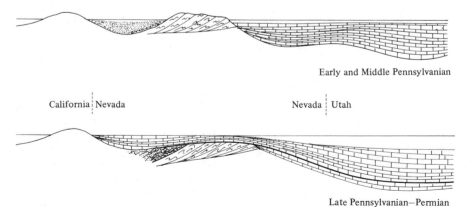

Early and Middle Pennsylvanian

California ¦ Nevada   Nevada ¦ Utah

Late Pennsylvanian—Permian

FIGURE 21–14 Pennsylvanian and Permian events in Nevada. These cross-sections continue the history shown in Fig. 21–6.

mountains to the north, composed of metamorphosed volcanic and clastic rocks. To the south of these is a basin in which most of the later rocks were deposited. (See Fig. 21–16.) To the south of the basin is a fold-mountain belt.

South of the fold belt is the northern interior on which basins and arches formed just as in the interior farther south in the United States.

In the basin just described, Pennsylvanian rocks were depos-

ited. They are mainly limestone and chert and are found at widely scattered points. Clastic rocks are also found, and they may be more or less continuous with similar rocks in the western geosyncline. Few Permian rocks have been

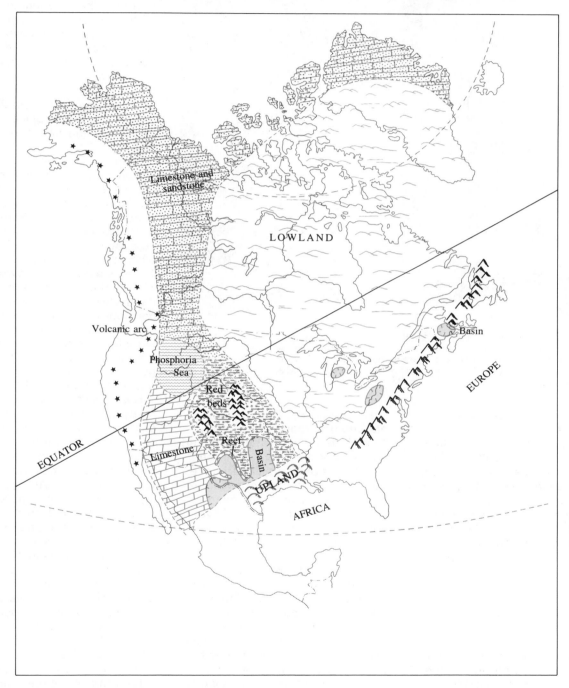

FIGURE 21–15   North America in the Permian Period.

FIGURE 21–16   Oblique aerial view of Melville Island in the Canadian Arctic. The rocks exposed in the center of the prominent anticline are conformable Ordovician, Silurian, and Devonian sedimentary rocks that were folded in the late Devonian. The rocks are overlain unconformably by Pennsylvanian sedimentary rocks, and all of these rocks were folded near the end of the Paleozoic. Photo from Geological Survey of Canada.

found. Clastics, carbonates, and 1525 meters (5000 feet) of volcanics have been reported, but their relationships are not known. In Triassic time, shale, siltstone, and sandstone were deposited. Their source was to the south.

In Greenland, the Triassic begins with marine rocks and ends with continental deposits, all of which accumulated in faulted basins.

LIFE OF THE LATE PALEOZOIC

The marine invertebrates continued to evolve rapidly so that the life of the late Paleozoic differed from that of the early Paleozoic.

Brachiopods (see Fig. 21–17), corals, and bryozoans (see Figs. 21–18 and 21–19) were abundant, and snails and clams were common. Graptolites died out, and trilobites declined, eventually becoming extinct at the end of the Permian. Figure 21–20 shows a late Devonian sea bottom. Note the spines on the trilobite. Fish ruled the Devonian sea, as we will see, and spines on trilobites and brachiopods developed, probably to protect them from the fish. The first spined brachiopods appeared in the Early Silurian. Also shown in this Devonian scene is the coiled cephalopod, which developed by the coiling of a straight-shelled nautiloid. The coiled cephalopods developed very rapidly and are

the most important fossils for dating the late Paleozoic and Mesozoic rocks. Both the external ornamentation and the shape of the margins of the partitions between the chambers changed rapidly. The cephalopods will be described further in the next chapter.

The Mississippian scene in Fig. 21–21 shows another aspect of late Paleozoic life. Crinoids and blastoids (see Fig. 21–22) expanded greatly in the Mississippian after some other orders of echinoderms died out in the mid-Paleozoic. At most places, however, the fauna was more balanced.

Pennsylvanian and Permian sea bottoms are shown in Figs. 21–23 and 21–24. The development of the cephalopods and spiny brachiopods shows clearly. Some excellently preserved examples of the Permian spiny brachiopods, on which the reconstruction in Fig. 21–24 is based, are shown in Fig. 21–25. Another aspect of the Pennsylvanian and Permian is the development of fusulinids. (See Fig. 21–26.) They are a type of large foraminifera and were very abundant at that time. They are good index fossils in Pennsylvanian and Permian rocks.

Near the end of the Paleozoic, many groups became extinct. No completely satisfactory explanation is available for this great dying. Trilobites and blastoids became extinct. Almost no corals or foraminifera survived, and several types each of bryozoans, brachiopods, and crinoids died out. The survivors, however, populated the Mesozoic seas. The only reasonable explanation is that many groups died when their shallow-sea habitat disappeared. An interesting point is that some organisms that died out at most places lived on somewhere. The Permian rocks of Timor contain echinoderms that became extinct elsewhere in the Mississippian.

Near the end of the Paleozoic, the continents came together,

FIGURE 21–17   Paleozoic brachiopods. A. *Athyris spiriferoides,* Devonian, New York, about 2.5 centimeters (1 inch). B. *Mucrospirifer mucronatus,* Devonian, about 3.1 centimeters (1.2 inches). Photos from Ward's Natural Science Establishment, Inc., Rochester, N.Y.

FIGURE 21–18   Honeycomb coral, *Favosites,* Devonian, Arkona, Ontario. Photo from Smithsonian Institution.

FIGURE 21–19   Permian bryozoa from Glass Mountains, Texas. Photo from Smithsonian Institution.

435

forming a single supercontinent. The evidence for this supercontinent and its subsequent breakup was described earlier. Most of the

invertebrates that died out about this time lived in shallow seas, and the joining together of the continents would, of course, de-

stroy much of the area of shallow seas. Thus this major subdivision of geologic eras on the basis of fossils is a natural one. The great change in life was obvious to the early workers who defined the eras. In southwestern United States, deposition continued across this important boundary; however, the seas were very restricted, and most of the rocks are nonmarine.

The Devonian is called the age of fish because fish apparently ruled the seas. As mentioned in the last chapter, the first fish are found in Late Cambrian rocks, and they first had any abundance in Late Silurian. The first fish were not at all like those of today. They were less than a foot long, jawless and finless, and were armored with bone. (See Fig. 21–27.) Without jaws, they apparently fed in the bottom mud. Without fins, they were poor swimmers. They probably could rise from the bottom by wiggling their tails like a tadpole, but without fins they could not change direction. All of these reasons, plus the armor, suggest that they were essentially bottom dwellers. Almost all were extinct by the end of the Devonian, although some of Pennsylvanian age are known and the living lampreys and hagfish are close relatives.

The jawless fish were replaced by the jawed fish (placoderms) in the Devonian. They, too, were armored, but they had paired fins and so were good swimmers. Some were quite large, and this, together with their jaws and swimming ability, made them fierce predators. (See Fig. 21–28.) The eurypterids that had been the predators declined in the Devonian as the jawed fish developed. The jawed fish spread rapidly because they were so well adapted to predation. Sharks, spiny fish, bony fish, as well as other types of fish that will be discussed later, all appeared in the Silurian or Devonian. Sharks and bony fish have continued to the present. The placoderms died in the Mississippian,

FIGURE 21–20   Late Devonian (Cycle III) sea bottom in western New York. A. Bryozoa. B. Crinoid. C. Colonial corals. Several types are shown, and corals are the main element in the fauna. D. Solitary coral. Several types are shown. E. Coiled cephalopod. F. Straight nautiloid. G. Spiny trilobite. Spines probably developed as a defense against fish that had developed by this time. Two smaller trilobites are also in the view. H. Brachiopods. Other types are attached to the coral. J. Sponge to the left of the letter. Also shown are snails. From Field Museum of Natural History.

FIGURE 21–21   Mississippian (Cycle IV) sea bottom in Central United States. Crinoids and blastoids form a local "sea lily" garden. A starfish is also shown. At other places a more balanced fauna is present. From Field Museum of Natural History.

436

FIGURE 21–22 Blastoid *Pentremites,*
Mississippian of Illinois, 1.25 centimeters
(0.5 inch). Members of the phylum
Echinodermata are recognized by their
fivefold symmetry. Blastoids are
attached to the sea floor by a stalk. Photo
from Ward's Natural Science
Establishment, Inc., Rochester, N.Y.

FIGURE 21–23 Pennsylvanian (Cycle V)
sea bottom in north-central Texas.
A. Sponges, both solitary and colonial.
B. Crinoid. C. Solitary coral. D. Snail.
Several other types are also shown.
E. Brachiopods and clams. F. Cephalopod.
Several other types are also shown.
From Smithsonian Institution.

FIGURE 21–24 Permian (Cycle V) sea
bottom, Glass Mountains, Texas.
A. Several varieties of sponges. B. Two
types of cephalopods. C. Spiny
brachiopods. Many other brachiopods
and clams can be seen. Corals with
tentacles can also be seen. From Field
Museum of Natural History.

437

FIGURE 21–25   Spiny brachiopods from Permian limestones in Glass Mountains, Texas. These fossils were replaced by silica and were recovered, with the spines and other delicate features intact, by dissolving the enclosing limestone with acid. Photo from Smithsonian Institution.

and the spiny fish in the Permian. The relationships are shown in Fig. 21–29.

The plant life on the land also developed rapidly in the late Paleozoic. From a few Ordovician occurrences came the abundant flora of the Devonian. (See Fig. 21–30.) The Early Devonian plants were small and primitive and

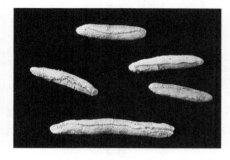

FIGURE 21–26 Fusulinid. Permian *Parafusulina*, Marathon, Texas. About 1 centimeter (0.4 inch) long. Fusulinids are very large foraminifera that are excellent index fossils in the Pennsylvanian and Permian. They died out near the end of the Permian. Photo from Smithsonian Institution.

seem to have grown in marshes or swamps. The plants were successful and evolved rapidly. By Middle Devonian time, forests existed. (See Fig. 21–31.) A Pennsylvanian coal swamp is shown in Fig 21–32, and Fig. 21–33 shows typical Pennsylvanian plant fossils. The lush Pennsylvanian coal swamps disappeared in the Permian. This was the result of the climatic change to arid or semiarid conditions that is recorded by the red-bed and evaporite deposits and the wind-deposited sandstones. Continental drift may have caused the climatic change. The southern supercontinent had the distinctive *Glossopteris* flora in late Paleozoic; and in the northern supercontinent, three floras are found in the late Paleozoic.

The first insects appeared in the Devonian, and they became large and relatively abundant during the Pennsylvanian. (See Fig. 21–32.) As noted earlier, arthropods may have been the first animals to breathe air. They have an imper-

vious surface and so were probably among the first to move in fresh water and finally out of water altogether.

As soon as plants grew on the lands, terrestrial plant eaters could develop. This, too, occurred in the Devonian. The first land vertebrates developed from the lobe-finned fish. This probably occurred in a freshwater environment because the change from ocean to fresh water solved some of the problems of living out of the water. The body fluids of most animals contain about the same amount and type of salts as seawater. Some way to enclose these fluids to prevent dilution was necessary for the animals to move into fresh water. Once that was accomplished, only a way to get oxygen from the air was necessary. It seems likely that the ability to travel out of the water was a great advantage if the pool a fish was in dried out. It could then crawl to the next pool. In the course of moving from pool to pool, it might find something tempting to eat and so begin to live more and more out of the water. In this way, perhaps, the first amphibians developed.

The oldest amphibian fossils so far found are in the upper Devonian rocks of Greenland and eastern Canada. (See Fig. 21–34.) They developed from the lobe-finned fish. (See Figs. 21–35 and 21–29.) The fins of these fish developed into clumsy legs. Lungs developed from swim bladders that are found in Devonian fish. Swim bladders enable a fish to adjust its buoyancy and so are a help in swimming. The lobe-finned fish also developed another adaptation that enabled them to live to the next rainy season if their pool dried up. The lungfish (see Fig. 21–36) can burrow in the mud and breathe air while in a hibernation-like state.

The development of amphibians was a great step toward inhabitation of the continents, but they were still tied to the water. They

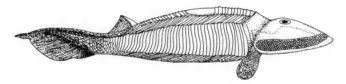

FIGURE 21–27 Early fish. The first fish were the Ostracoderms, or jawless fish. The lamprey is the only living close relative. *Hemicyclaspis* of Devonian age is shown here [0.3 meter (about 1 foot) long].

FIGURE 21–28 Devonian jawed, armored fish. *Dunkleosteus intermedius* from northern Ohio. This fish is about 1.2 meters (4 feet) high. Other members of this group that died out in the Mississippian were up to 9.1 meters (30 feet) long with jaws over 1.8 meters (6 feet) high. Photo from Smithsonian Institution.

439

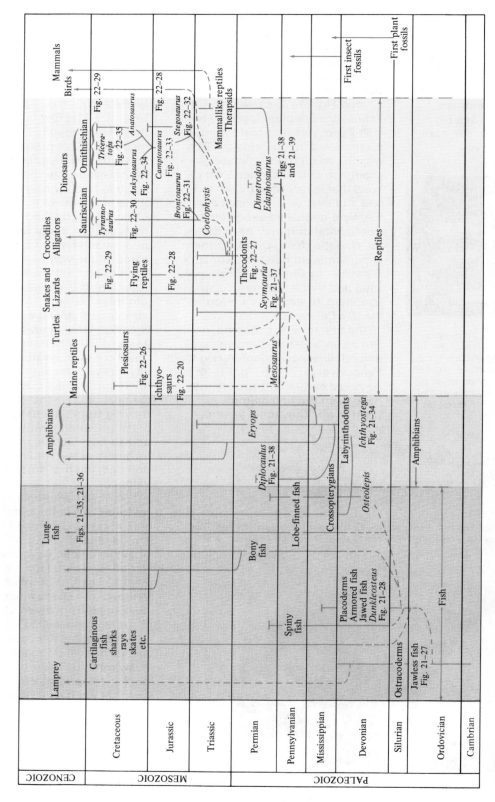

FIGURE 21–29 Evolution of the vertebrates. Although this chart appears formidable, the main theme can be seen along the bottom where progressive developments form a diagonal from primitive fish in the lower left to mammals in the upper right. Only the more important vertebrates are indicated. Figures showing these animals are shown.

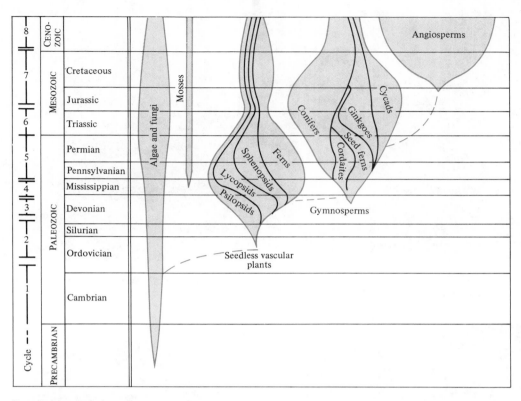

FIGURE 21–30 Development of plants.

FIGURE 21–31 Middle Devonian forest in western New York. The Early Devonian plants were generally small, primitive marsh- or swamp-dwelling plants. A. *Psilophyton* was one of the larger that carried over into the first true forests of Middle Devonian age. B. An early tree that grew up to 15.25 meters (50 feet) high. C. Horsetail rushes. D. *Archaeosigillaria*. Painting by C. R. Knight, Field Museum of Natural History.

had to return to the water to lay their eggs. In spite of this, they expanded very rapidly. The reptiles were the next step, and they first appeared in the Pennsylvanian. The reptiles are characterized by the amniote egg that can hatch on land. It is difficult to distinguish between reptiles and amphibians on the basis of skeletons alone because they are so similar. *Seymouria* (see Fig. 21–37), found in Lower Permian rocks in Texas, is known to be a reptile only because the fossil eggs are found with it. *Seymouria*'s line has now been traced to the Pennsylvanian.

Many branches of reptiles developed in the Late Pennsylvanian and Permian. (See Fig. 21–29.) The Mesozoic is the age of rep-

441

FIGURE 21–32   Reconstruction of a Pennsylvanian coal swamp. A. *Lepidodendron.* B. *Sigillaria.* Note the large cockroach. C. *Calamites.* D. Large dragonfly. From Field Museum of Natural History.

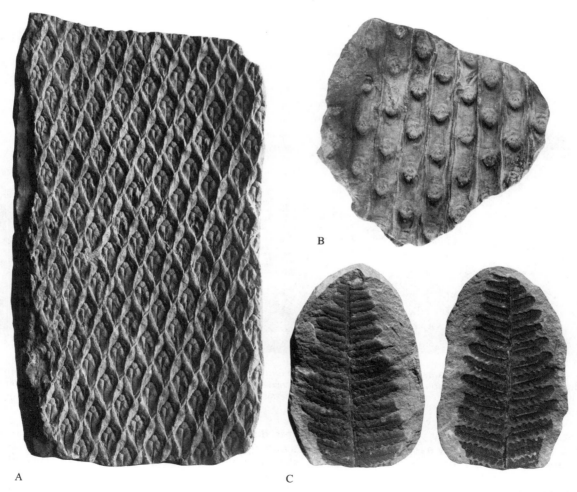

A

B

C

FIGURE 21–33   Pennsylvanian plant fossils. A. *Lepidodendron.* B. *Sigillaria.* Parts A and B are both parts of the trunk showing leaf scars. Photos from Field Museum of Natural History. C. *Asterotheca miltoni,* a fern leaf. Photo from Ward's Natural Science Establishment, Inc., Rochester, N.Y.

tiles, and they will be discussed in the next chapter. The Permian "sail lizards" did not continue into the Mesozoic. These unusual animals (see Figs. 21–38 and 21–39) had large fins. The purpose of the fins is not known. They may have been some sort of protection, or they may have been used to regulate their temperature. Reptiles are so-called cold-blooded animals, and temperature regulation is a problem for them.

The land dwellers continued on from the Permian into the Mesozoic and expanded rapidly. Thus whatever caused the great dying among the invertebrates had little effect on the vertebrates.

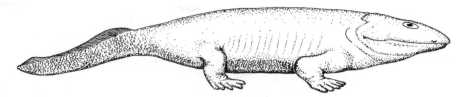

FIGURE 21–34   *Ichthyostega,* one of the earliest amphibians, from upper Devonian of Greenland. About 60 centimeters (2 feet) long.

FIGURE 21–35   *Latimeria.* A "living fossil" found near Madagascar. Lobe-finned fish similar to this developed into the early amphibians. About 1.5 meters (5 feet) long. Photo from Field Museum of Natural History.

FIGURE 21–36   Australian lungfish. The ability to breathe air shown by lungfish was probably a step in the development of amphibians from fish. Up to 1 meter (40 inches) long. Photo from Field Museum of Natural History.

FIGURE 21–37   *Seymouria* closely resembles the amphibians and would be classified as an amphibian if eggs were not found in the same rocks. From this stem reptile evolved all other reptiles and higher vertebrates. About 0.75 meter (2.5 feet) long. Photo from Field Museum of Natural History.

443

FIGURE 21–38 Permian "sail lizards." The one showing teeth is *Dimetrodon,* a carnivore. The smaller headed one is *Edaphosaurus,* a vegetarian. Both were about 2.75 meters (9 feet) long. The fin may have been used to control body temperature. In the right foreground is *Diplocaulus,* a 0.6-meter- (2-foot-) long, bottom-dwelling amphibian. Painting by C. R. Knight, Field Museum of Natural History.

FIGURE 21–39 Skeleton of *Dimetrodon* shown in Fig 21–38. Photo from Field Museum of Natural History.

SUMMARY

In the Devonian, Europe and North America collided. In the Pennsylvanian or Permian Period, North America collided with Europe and Africa, forming the Appalachian Mountains.

In the late Paleozoic, all of the continents came together, forming one supercontinent. This affected the climate and therefore the life of the earth.

Cycle III—Middle to Late Devonian

Interior: Clastics near the domes and the eastern geosyncline. Car-

bonates in the west and north.

East: Mountain building caused by collision with Europe was the source of thick clastic deltas.

West: A volcanic island arc was the source of volcanics and clastics that spread east to interfinger with carbonates on the interior.

Far North: Carbonates, then thick nonmarine clastics, formed a highland to the north.

Cycle IV—Latest Devonian to Late Mississippian

Interior: Carbonates to the west. Shale followed by limestones and sandstones to the east.

East: Mountain building was the source of thick clastic fans in Pennsylvania, Alabama, Texas, and Oklahoma.

West: Mountain building in Nevada shed clastics toward the east.

Far North: Clastic rocks record uplift and mountain building.

Cycle V—Latest Mississippian to Permian

Interior: Coal swamps developed to the east. Basins and highlands formed to the west.

East: Mountain building occurred that ended the sedimentation phase of the geosyncline in the Permian.

West: Orogeny occurred in Nevada in the Pennsylvanian and Permian, and in the far west in the Permian. A Permian sea extended from Texas to Nebraska.

Far North: Pennsylvanian limestone and chert were deposited in a basin between mountain ranges.

Brachiopods, corals, and bryozoans were abundant, and snails and clams were common.

Many animals developed spines, apparently to protect them from fish.

Fusulinids are good index fossils of the Pennsylvanian and Permian.

Many invertebrates, such as trilobites, blastoids, and many bryozoans, brachiopods, and crinoids, became extinct near the end of the Paleozoic.

In the Devonian, fish became abundant.

Forests developed by the Middle Devonian, and coal swamps were common in the Pennsylvanian.

The first insects appeared in the Devonian, and they became large and abundant in the Pennsylvanian.

The first land vertebrates were amphibians that developed from the lobe-finned fish in the Devonian.

The oldest reptiles are found in Pennsylvanian rocks.

QUESTIONS

1 Where is the thickest section of Devonian rocks in North America? What is the evidence as to their source area?

2 Sketch a hypothetical cross-section through a dome of the type found in the interior during Paleozoic time. Pay particular attention to the changes in thickness of the sedimentary rocks, and contrast these thicknesses with those in a dome formed by a much younger domal uplift.

3 What occurred in the far west during Cycle III?

4 What do the Cycle III rocks in the far north resemble?

5 Describe the Cycle IV rocks in the east.

6 What occurred in Nevada during Cycle IV?

7 How is the orogeny that produced the Appalachian Mountains dated?

8 What events were occurring in the west during this time?.

9 What was the climate in western North America during Permian time? What is the evidence?

10 Where is the most complete section of Permian marine rocks in North America?

11 Describe the movement of the southern continent in the late Paleozoic.

12 Which of the present continents were in contact at the end of the Paleozoic?

13 Why do we know so little about pre-Devonian plants?

14 When did the amphibians first appear? From what did they evolve?

15 When did the reptiles first appear?

16 What are some differences between reptiles and amphibians?

17 When did the first insects appear?

18 In what ways did late Paleozoic life differ from early Paleozoic life?

19 Name some organisms that became extinct near the end of the Paleozoic.

20 Were the jawless fish good swimmers?

SUPPLEMENTARY READINGS

Colbert, E. H., "The Ancestors of Mammals," *Scientific American* (March 1949), Vol. 180, No. 3, pp. 40–43. Reprint 806, W. H. Freeman and Co., San Francisco.

Colbert, E. H., *Evolution of the Vertebrates.* New York: John Wiley & Sons, Inc., 1969, 542 pp.

Craig, L. C., et al., *Paleotectonic Investigations of the Mississippian System in the United States.* U.S. Geological Survey Professional Paper 1010. Washington, D.C.: U.S. Government Printing Office, 1979, pp. 1–559.

Romer, A. S., "Major Steps in Vertebrate Evolution," *Science* (December 29, 1967), Vol. 158, No. 3809, pp. 1629–1637.

U.S. Geological Survey, *The Mississippian and Pennsylvanian (Carboniferous) Systems in the United States.* Professional Paper 1110-A-L. Washington, D.C.: U.S. Government Printing Office, 1979, various paging.

U.S. Geological Survey, *The Mississippian and Pennsylvanian (Carboniferous) Systems in the United States.* Professional Paper 1110-M-DD. Washington, D.C.: U.S. Government Printing Office, 1979, various paging.

The chevalier by this time had arrived with the lady in front of the glass-cases. "May I ask, my dear madam, what this great animal is in the midst of this blue stuff, looking like a potted crocodile?"

"Ah! that glass-case," continued the lady, with a profound look and an emphatic tone, "contains the great family of the Sawruses. That one comes from Dorsetshire, but they are originally of Irish extraction, as you may know by their name."

"May I inquire," said the chevalier, "what is the peculiarity in their appellation which makes you to suppose them to come from the sister island?"

"Bless me!" cried the lady. "Don't you perceive at once? They have all got an O before their family name, just like the O'Donnels, and the O'Connells, and the O'Moores, and the O'Tooles. These are called the O'Sawruses. There's the Pleasy *O'Sawrus and the* Itchy *O'Sawrus, and a great many more besides."*

G. P. R. James
The Commissioner, or De lunatico inquirendo
1st ed., Dublin, 1843, p. 211

Mesozoic means "middle life," and at the time it was named, it was thought to be the middle part of earth history. We now know the great age of the earth and that the Mesozoic is but a part of the late history of the earth. The life of the Mesozoic differs markedly from both the Palezoic and the Cenozoic and was an obvious subdivision of geologic time to the early workers. In North America it was a very active time, as it was throughout the world.

The Mesozoic systems are distinctive in Europe. The Cretaceous was named in 1822 for the chalk beds that not only form the white cliffs of Dover but also are widespread over Europe. The Triassic was named in 1834 for its threefold subdivision in Germany. Thick limestones in the Jura Mountains were named the Jurassic System. Study of the fossils in the Jurassic rocks has been very important to the development of geologic dating of rocks.

In defining systems, fossils from large thicknesses of rocks were used. For more refined dating, the fossils in each thin layer must be studied. The Jurassic rocks of Europe are ideal for this type of study because the ammonite fossils are abundant and evolved very rapidly, as described later in this chapter. The first attempt at refinement was in 1842 by Alcide Dessalines d'Orbigny, who used assemblages of fossils to define his stages. He believed that each stage began with the creation of new life and ended with a catastrophe that killed the animals. Although d'Orbigny's stages were useful in France, they were not elsewhere. In Germany, Friedrich Quenstedt studied Jurassic rocks, collecting fossils centimeter by centimeter. He soon realized that d'Orbigny's stages could not be extended into Germany. In 1856–1858, Quenstedt's student Carl Albert Oppel, after many years of work, subdivided the Jurassic into zones. His zones were based on the stratigraphic extent of each species. He recognized that some species existed for long time spans, while other species were short lived. It is now recognized that his zones are generally limited geographically because, of course, few organisms are found world wide. Even today, the Jurassic is probably the most precisely subdivided system.

In this chapter the Mesozoic history of western North America is completed. The Triassic history of the eastern geosyncline was included in the last chapter with the discussion of the formation of the Appalachian Mountains with which it is related.

WORLD GEOGRAPHY IN THE MESOZOIC

The Mesozoic was a time of drifting apart of the continents. By the end of the Paleozoic, the continents had all joined to set the scene for the great breakup and drifting apart that formed the present earth. Dating of the present ocean floors reveals when the breakup occurred.

The oldest rocks found so far on the sea floors are Jurassic, suggesting that in Triassic and Jurassic, the present oceans, such as the Atlantic, may have begun to

Petrified Forest, Arizona. Logs of the conifer *Araucarioxylon* of Triassic age. Photo courtesy U.S. National Park Service.

449

form. A probable sequence of drifting is shown in Figs. 22–1, 22–2, 22–3. During the Mesozoic, Africa and South America separated, as did Antarctica from Africa. India moved farthest during the Mesozoic. Geosynclines were along the west coasts of North America, South America, and Antarctica. Another important geosyncline that was active during the Mesozoic was between Africa and Eurasia in the vicinity of the present Mediterranean Sea.

The history of the Atlantic Ocean can be deciphered from study of recovered sediments and paleomagnetic studies of the oceanic crust. North America began to separate from Europe about 200 million years ago, in the Triassic. In the Jurassic Period, a narrow sea existed, in which some evaporites were deposited. The south Atlantic began to form in the Jurassic when South America moved away from Africa. That the Atlantic was still narrow in early Cretaceous time is suggested by some apparently stagnant-water deposits. The Atlantic continued to widen to its present size during the rest of the Cretaceous and the Cenozoic. Sometime during the Cretaceous, Greenland apparently separated from Labrador.

THE TRIASSIC OF WESTERN NORTH AMERICA—CYCLE VI

At the beginning of the Triassic, the seas had retreated from the interior and a large part of the geosyncline as well. (See Fig. 22–4.) In part of the geosyncline in Idaho, Lower Triassic marine rocks overlie the Permian conformably. The geosyncline had changed somewhat from its configuration in the Permian. The island chain in central Nevada that had been the site of Paleozoic orogeny was not present. In the far west, great thicknesses of Triassic marine

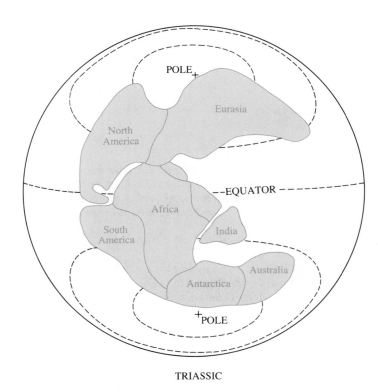

TRIASSIC

FIGURE 22–1   The continents in the Triassic Period.

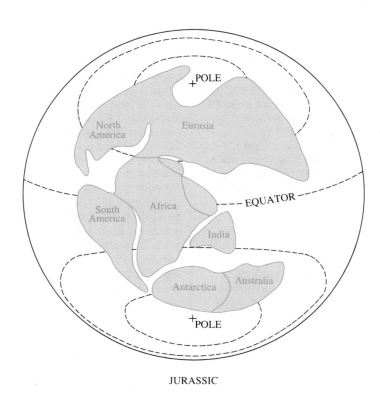

JURASSIC

FIGURE 22–2   The continents in the Jurassic Period.

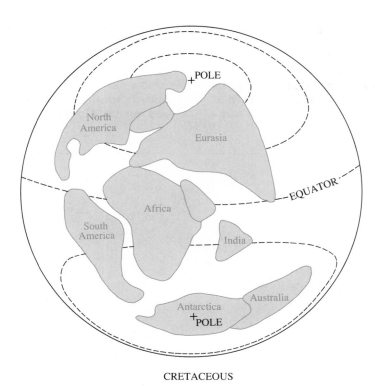

CRETACEOUS

FIGURE 22–3   The continents in the Cretaceous Period.

clastic and volcanic rocks were deposited, suggesting an active subduction zone. At some places these rocks lie unconformably on Permian rocks. Thus the Paleozoic ended with orogeny at places in the active part of the geosyncline. Farther east in the less active part of the geosyncline, marine rocks were deposited. In western Nevada, both clastic and carbonate sections ten thousand meters thick are found. At the edge of the interior, the marine rocks interfinger with nonmarine red clastics.

Lower Triassic marine rocks are found only in the geosyncline but are known from southern Nevada to the Arctic, although in the United States, outcrops are limited. In southern Nevada they are clastic rocks about 2440 meters (8000 feet) thick, and in Idaho they are limestone and shale 1525 meters (5000 feet) thick. These rocks interfinger with red sandstone and siltstone that in the northern Rocky Mountains are rarely over

305 meters (1000 feet) thick. The Triassic sea in that area advanced and retreated a number of times and apparently had an irregular shoreline with large embayments. Highlands in the interior were the source of some clastics. The nonmarine red beds overlie similar Permian red beds, and it is difficult to place the boundary between the two systems because of the paucity of fossils.

In middle Triassic time, the seas retreated farther west so that only the active part of the geosyncline was marine except in Canada where the whole of the geosyncline accumulated marine rocks. Evaporites formed in a restricted basin in Montana-Dakota. With retreat of the sea, some erosion apparently occurred in the areas of earlier deposition as well as elsewhere.

The area of maximum deposition of nonmarine rocks was in Arizona and New Mexico, and here, Lower Triassic rocks are uncon-

formably covered by Upper Triassic rocks with conglomerate at the base. This suggests uplift to the north. In this area, some of the Upper Triassic and Lower Jurassic sandstones appear to be wind deposited. These rocks are well displayed at Zion National Park. Because of the lack of fossils in these rocks, the boundary between Triassic and Jurassic is difficult to place. These wind-deposited rocks and the many red beds suggest that arid conditions prevailed over much of the United States.

In Middle and Late Triassic time, batholiths were emplaced in the Sierra Nevada mountains according to radioactive dates. This activity was the forerunner of later intrusions and metamorphism.

The Jurassic began as a continuation of the Triassic. In the far west thick volcanic and clastic rocks were deposited, especially in a deep trough in the geosyncline in western Nevada. In Oregon, Jurassic rocks lie with an angular unconformity on Upper Triassic rocks, indicating orogeny in that area. The pattern, however, was soon to change, and the trough in western Nevada was the site of thrust faulting during deposition while still in Early Jurassic time. (See Fig. 22–5.) To the east of here, an uplifted area developed as the seas withdrew.

MIDDLE JURASSIC THROUGH CRETACEOUS— CYCLE VII

WEST

When the seas returned in Middle Jurassic time, the shape of the continent had changed. Orogenic events that formed an upland area had occurred in the eastern part of the geosyncline. (See Fig. 22–6.) This upland underwent erosion during most of this cycle and shed clastics into seaways on both sides. Seas occupied the area east of the upland and covered the region of the present Rocky Mountains and Great Plains. This sea finally withdrew almost completely from the continent near the end of the Cretaceous Period, ending the cycle. This was the last time that seas occupied the interior of North America.

The seas returned to the northern Rocky Mountain area in Middle Jurassic time. The seas came from the north, and at the margin of the Canadian Rockies, Early as well as Middle Jurassic sedimentary rocks were deposited. The shallow sea at first came as far south as southern Utah and covered most of Montana and North

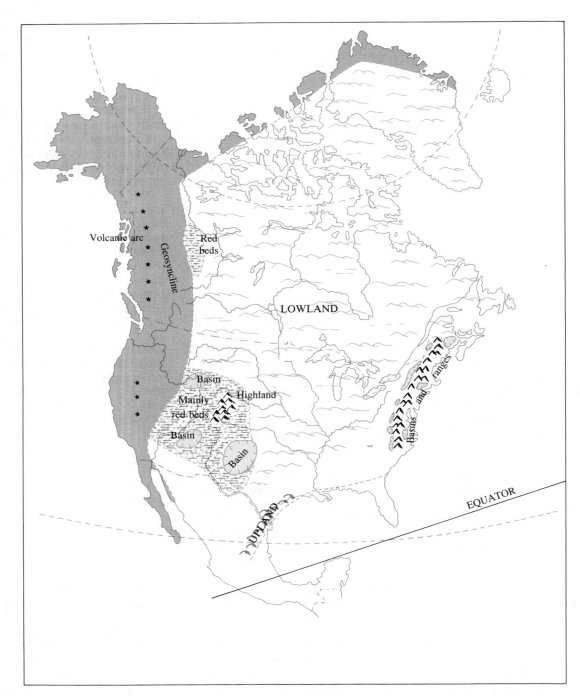

FIGURE 22–4 North America in the Triassic Period. Except in the far west, the sediments are nonmarine.

Dakota. Sand, silt, and clay, with some red beds and evaporites, were laid down; the greatest thick- nesses are found in Wyoming. At the maximum, the seas extended to northern Arizona and New Mex- ico, and possibly to the Gulf of Mexico. In Late Jurassic time the sea retreated to the north. As the

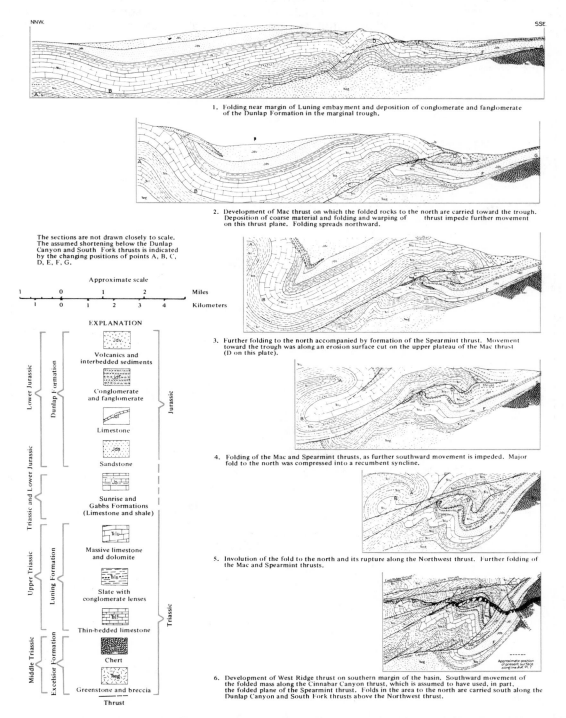

The sections are not drawn closely to scale. The assumed shortening below the Dunlap Canyon and South Fork thrusts is indicated by the changing positions of points A, B, C, D, E, F, G.

Approximate scale

EXPLANATION

Lower Jurassic — Dunlap Formation
- Jdv — Volcanics and interbedded sediments
- Jdf — Conglomerate and fanglomerate
- Jdl — Limestone
- Jds — Sandstone

Triassic and Lower Jurassic
- Js — Sunrise and Gabbs Formations (Limestone and shale)

Upper Triassic — Luning Formation
- Tlu — Massive limestone and dolomite
- Tls — Slate with conglomerate lenses
- Tll — Thin-bedded limestone

Middle Triassic — Excelsior Formation
- Chert
- Teg — Greenstone and breccia

Thrust

1. Folding near margin of Luning embayment and deposition of conglomerate and fanglomerate of the Dunlap Formation in the marginal trough.

2. Development of Mac thrust on which the folded rocks to the north are carried toward the trough. Deposition of coarse material and folding and warping of thrust impede further movement on this thrust plane. Folding spreads northward.

3. Further folding to the north accompanied by formation of the Spearmint thrust. Movement toward the trough was along an erosion surface cut on the upper plateau of the Mac thrust (D on this plate).

4. Folding of the Mac and Spearmint thrusts, as further southward movement is impeded. Major fold to the north was compressed into a recumbent syncline.

5. Involution of the fold to the north and its rupture along the Northwest thrust. Further folding of the Mac and Spearmint thrusts.

6. Development of West Ridge thrust on southern margin of the basin. Southward movement of the folded mass along the Cinnabar Canyon thrust, which is assumed to have used, in part, the folded plane of the Spearmint thrust. Folds in the area to the north are carried south along the Dunlap Canyon and South Fork thrusts above the Northwest thrust.

FIGURE 22-5 Diagrammatic sections showing progressive development in Early Jurassic time of complex structure in the northwestern part of the Pilot Mountains in southwestern Nevada. From H. G. Ferguson and S. W. Muller, U.S. Geological Survey Professional Paper 216, 1949.

sea withdrew, lake- and river-de-posited rocks accumulated. These rocks are only about a hundred meters thick but are of great interest because they contain dinosaur fossils and some rich uranium deposits.

On the western side of the upland, and probably on the upland itself, much activity took place during this cycle. Jurassic rocks are found at many scattered localities in Oregon, Nevada, northern Washington, and especially in California. These rocks are clastics and volcanics for the most part, and some have been metamorphosed. They probably accumulated at a volcanic arc and trench. At some places, such as north-central California, the California-

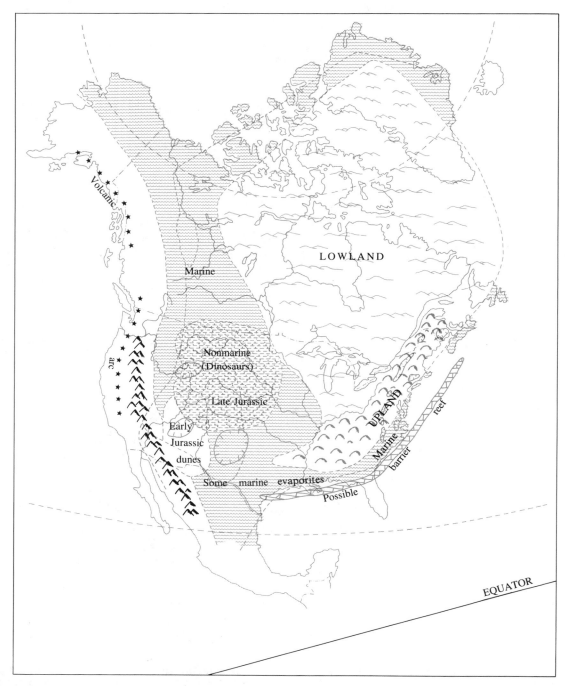

FIGURE 22–6 North America in the Jurassic Period.

Oregon border, central Oregon, and southern British Columbia, the thicknesses measure ten thousand meters or more. In northern California, at least 4880 meters (16,000 feet) of sandstone and shale were deposited in Late Jurassic time, and these rocks are covered by up to 7625 meters (25,000 feet) of similar Cretaceous sediments. Farther west is another sequence up to 15,250 meters (50,000 feet) thick that may be of the same age.

It is difficult to date orogenic events, and this is well illustrated in the Mesozoic of the western United States. As we go westward from the sea in the Rocky Mountain area, it becomes more difficult to date events, because few fossiliferous rocks of Mesozoic age are found, especially in critical areas. Events in the eastern part of this area are, however, reflected in sediments deposited in the sea, enabling some dating. From western Utah all across Nevada is the Basin and Range country of isolated mountain ranges. The rocks in these mountains have been deformed, and the time of orogeny can be told at a few places from scraps of Mesozoic rocks. Detailed field mapping, in some cases of several ranges, may be necessary to date a single event. In California, in the Sierra Nevada, mainly igneous and metamorphic rocks are found, making study more difficult. Dating the granitic batholiths by radioactive methods has helped; this method, together with detailed field mapping, has enabled the recognition of several belts of Mesozoic orogeny in the Great Basin, although it is not yet possible to reconstruct all the events in the Pacific Coast states.

The mountain building so far recognized in the Great Basin begins in the late Paleozoic in central Nevada, as has already been described. In Triassic or Early Jurassic, thrusting occurred in south-central Nevada and southeastern California and extended to west-central Nevada. The evidence is

radioactive dating in the south and stratigraphic dating to the north. In eastern Nevada and western Utah, uplift and deformation occurred mainly in Cretaceous time. The evidence is the Cretaceous rocks deposited in the sea to the east. In Late Cretaceous and Early Cenozoic time, orogeny occurred, both along the Pacific Coast and in the Rocky Mountains; this will be described in the next chapter.

In California some of the events are well dated, but the details remain obscure. The Sierra Nevada mountains are a vast complex of batholiths. The individual plutons have been closely studied, and hundreds of rocks have been dated by radioactive methods. These plutons range in age from

Middle Triassic to Late Cretaceous. Late Jurassic rocks are folded and intruded in the Sierra Nevada. As noted earlier, in central California almost on the flanks of the Sierra Nevada, sedimentation was continuous from Late Jurassic through the Cretaceous. (See Fig. 22–7.) These rocks are largely sandstone, and their source was probably an exposed batholith. Thus it appears that some batholiths were being emplaced as nearby batholiths underwent erosion. (See Fig. 22–8.)

The Mesozoic was a time of batholith emplacement in western North America. Similarly dated batholiths extend from Baja California to Alaska. The Coast Ranges of British Columbia are

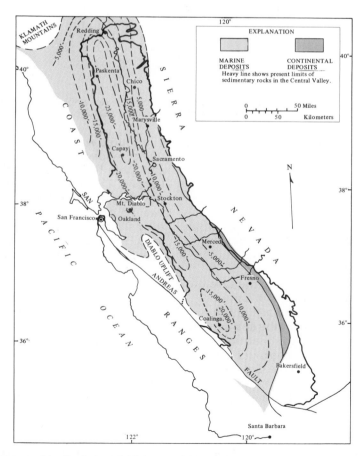

FIGURE 22–7 Distribution and thickness of Cretaceous sedimentary rocks in the Great Valley of California. The lines indicate the thickness that is more than 7620 meters (25,000 feet) in places. From Otto Hackel, in California Divison of Mines and Geology Bulletin 190, 1966.

455

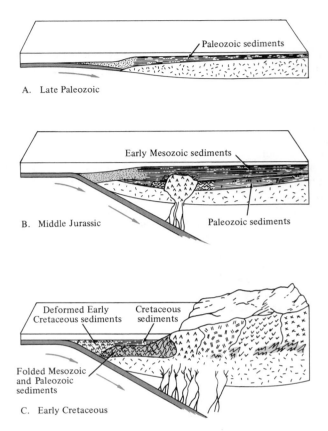

A. Late Paleozoic

Paleozoic sediments

Early Mesozoic sediments

B. Middle Jurassic

Paleozoic sediments

Deformed Early Cretaceous sediments

Cretaceous sediments

Folded Mesozoic and Paleozoic sediments

C. Early Cretaceous

FIGURE 22-8 Hypothetical block diagrams showing the development of the Sierra Nevada batholith and the Mesozoic sedimentary rocks in California.

largely composed of granitic rocks. (See Fig. 22–9.)

Along the Pacific coast, a number of marine embayments existed, especially during the Late Cretaceous. The pattern was probably much like the Cenozoic history of that area. Cenozoic rocks and deformation obscure much of the Cretaceous story.

The seas returned to the area of the Rocky Mountains in Early Cretaceous time. This was the last great inundation of the sea on the North American continent. (See Fig. 22–10.) The seas extended from the Arctic to the Gulf of Mexico, and even along the Atlantic coast, sediments were deposited. In the Rocky Mountains and Great Plains areas, great thicknesses of clastic sediments were deposited. Up to 6100 meters (20,000 feet) of clastics are found at some places.

The source was the orogenic area that extended from western Utah and eastern Nevada to northern Canada. The geosyncline sank as these sediments were rapidly deposited, and at times the sediments accumulated above sea level. The source area was repeatedly uplifted, causing floods of conglomerate and sandstone to be deposited near the western margin of the basin. These rocks gave way to shale farther east, and near the eastern margin, to thin limestone layers. (See Fig. 22–11.) In Late Cretaceous time, the seas withdrew and nonmarine clastic rocks were deposited. These rocks grade into the Cenozoic deposits that will be described in the next chapter. A small, short-lived embayment of the sea in North Dakota lingered into the early Cenozoic. The with-

drawal of the sea marked the end of the Cretaceous and the last great inundation of the continent. Orogeny occurred in part of the Rocky Mountain area near the end of the Cretaceous. This orogeny is more closely related to the Cenozoic history and will be described in the next chapter.

GULF AND ATLANTIC COAST

Deposition in the coastal plain areas of the Gulf of Mexico and the Atlantic coast began near the Gulf in Jurassic time. The Jurassic rocks do not crop out, but they are known from deep drilling for oil. The rocks encountered include evaporites, carbonates, and shale. The salt domes of the Gulf Coast are believed to have come from Jurassic salt deposits.

The Cretaceous deposits are much more extensive, and because of their overlap, the Jurassic rocks are not exposed. In the Gulf area, the Cretaceous sediments are very thick and consist mainly of shale and sandstone. Along the Atlantic coast, thin shale and sandstone beds were the main deposits. The deposition on the Gulf and Atlantic coasts continued into the Cenozoic, and thus the discussion will be continued in the next chapter.

FIGURE 22–9 Mesozoic batholiths of North America.

FAR NORTH

In the Arctic islands, the Jurassic is represented by a hundred to a few hundred meters of marine shale overlain by nonmarine shale and sandstone. This is overlain by a shale unit of Late Jurassic-Early Cretaceous age that is up to 765 meters (2500 feet) thick. In Cretaceous time a clastic unit up to 1375 meters (4500 feet) thick was deposited. (See Fig. 22–12). It is largely nonmarine and contains some coal. Volcanic activity also occurred in the Cretaceous.

In eastern Greenland, Late Jurassic orogeny is recorded by thick

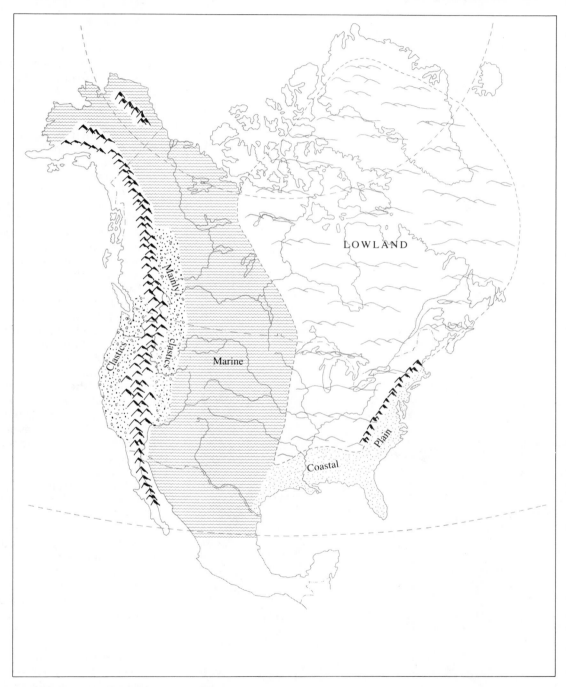

FIGURE 22–10   North America in the Cretaceous Period.

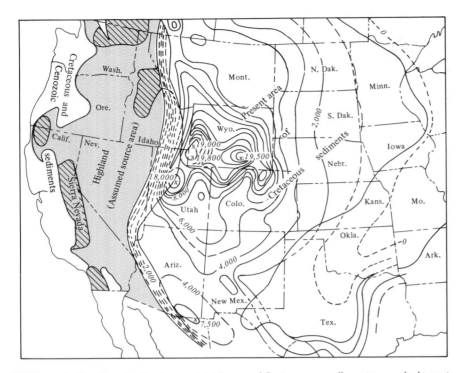

FIGURE 22–11   The distribution and thickness of Cretaceous sedimentary rocks in part of western United States. The lines show the thickness in feet and are dashed where the rocks have been removed by erosion. The rocks are thicker and coarser near the source area. The great volume requires that about 8 kilometers (5 miles) were eroded from the source area, and this does not include the Cretaceous rocks to the west of the highland. The batholiths in the source area are indicated by the diagonal lines. From P. C. Bateman and Clyde Wahrhaftig, in California Division of Mines and Geology Bulletin 190, 1966.

FIGURE 22–12   Oblique aerial view of Axel Heiberg Island in the Canadian Arctic. The folded rocks in the foreground are mainly Cretaceous, and the snow-covered mountains in the background are composed of Triassic sedimentary rocks. The folding occurred in the mid-Cenozoic. Photo from Geological Survey of Canada.

458

clastic rocks of Early Cretaceous age. These are overlain by Cretaceous carbonates and red beds, suggesting a fluctuating shoreline.

LIFE OF THE MESOZOIC

Perhaps the joining together of the continents into one huge continent near the end of the Paleozoic can explain the differences between the life of the late Paleozoic and the early Mesozoic. A single large continent would probably have greater climatic extremes, which would affect plants and terrestrial animals. The total shoreline would be less, creating more competition among the marine animals. A single continent should have more widespread species than a number of isolated continents. All of these effects can be seen in the life of the Mesozoic.

The marine invertebrates of the Mesozoic are very different from those of the late Paleozoic. The extinctions, both in the Permian and at the end of the Paleozoic, were probably caused by the withdrawal of the seas. When the seas returned in the Triassic, the main invertebrates were cephalopods and clams. Interestingly, no corals or foraminifera have ever been found in lower Triassic rocks, but they are present in Middle Triassic rocks, so they did live somewhere. The Triassic corals and foraminifera are very different from those of the Paleozoic. (See Fig. 22–13.)

In the late Paleozoic and the Mesozoic, the cephalopods are almost ideal index fossils. They were abundant, are found in many environments, and they evolved rapidly. The stratigraphy of this interval is based on these animals. They were similar to the present-day nautiloids. (See Fig. 22–14.) They were swimming animals with coiled, chambered shells. They evolved rapidly, with differences between species being the ornamentation of the shell and the shape of the partitions between the chambers. (See Figs. 22–15

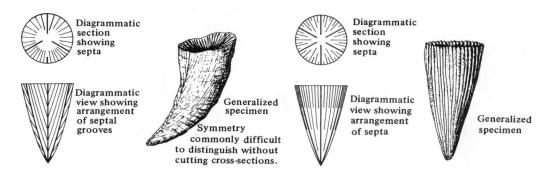

A. Four-fold solitary coral—Paleozoic.   B. Six-fold solitary coral—Mesozoic.

FIGURE 22–13   A. Paleozoic solitary or cup coral, showing fourfold symmetry. B. Mesozoic or Cenozoic solitary coral, showing sixfold symmetry.

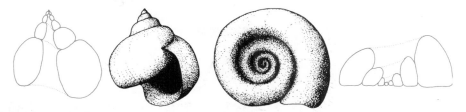

A. Shells and diagrammatic sections through coils of two generalized snails.

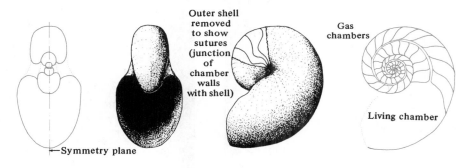

B. Shell and diagrammatic sections of generalized cephalopod.

FIGURE 22–14   Differences between snail and cephalopod. A. Snail has no chambers and no symmetry. B. Cephalopod has chambers, some of complex pattern (see Fig. 22–16), and is coiled in a plane, giving a symmetry plane. Some cephalopods are not coiled. The differences shown here are obvious morphology, and there are other, more fundamental, differences.

and 22–16.) Because they were swimmers, the fossils are found in all environments. It seems likely that they floated after death as well. Some were very large. (See Fig. 22–17.)

Figure 22–18 shows the stratigraphic range of each of the cephalopod suture types. Perhaps one reason for their rapid evolution was that they nearly became extinct twice before finally dying out. At the end of the Permian, only

one group survived; and again near the end of the Triassic, the same thing happened. At the end of the Cretaceous, they did die out. At each of these times, there was a withdrawal of the seas from the interior, but it seems likely that swimmers like these could have lived on in the oceans.

Why they developed such extremely complex partitions is not known, although they probably strengthened the shell much as

corrugations strengthen cardboard. Another possibility is that the chambers were used by the animal to adjust buoyancy and for some reason the complex partitions were an advantage. In any case, the cephalopods died out at the end of the Cretaceous except the nautilus, which has simple, smooth partitions. Perhaps the more specialized forms could not adjust to the new conditions at the end of the Cretaceous, but this

A   B

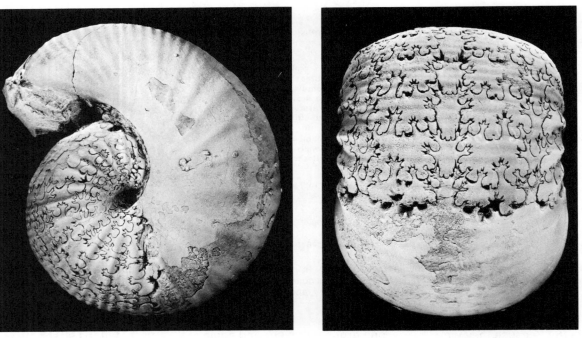

C

FIGURE 22–15   Cephalopods. A. *Baculites compressus.* A 75-millimeter (3-inch) fragment of a straight-shelled ammonite. B. Two views of *Ophiceras commune,* a Lower Triassic ceratite from Axel Heiberg Island, Arctic Canada. About natural size. C. Two views of *Scaphites depressus,* Upper Cretaceous with ammonite suture from Alberta. About natural size. Photo A from Ward's Natural Science Establishment, Inc., Rochester, N.Y.; B and C from Geologic Survey of Canada (109804 and 109804-A, 113737-K and -H).

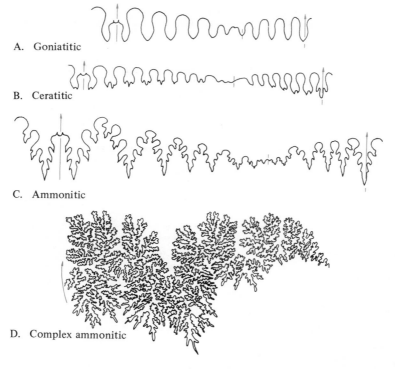

A. Goniatitic

B. Ceratitic

C. Ammonitic

D. Complex ammonitic

FIGURE 22–16 Cephalopod sutures. Arrows point toward the aperture. From *Treatise on Invertebrate Paleontology,* Courtesy of the Geological Society of America and the University of Kansas Press.

FIGURE 22–17 A very large Jurassic ammonite from British Columbia, *Titanites occidentalis,* courtesy Geological Survey of Canada (203171).

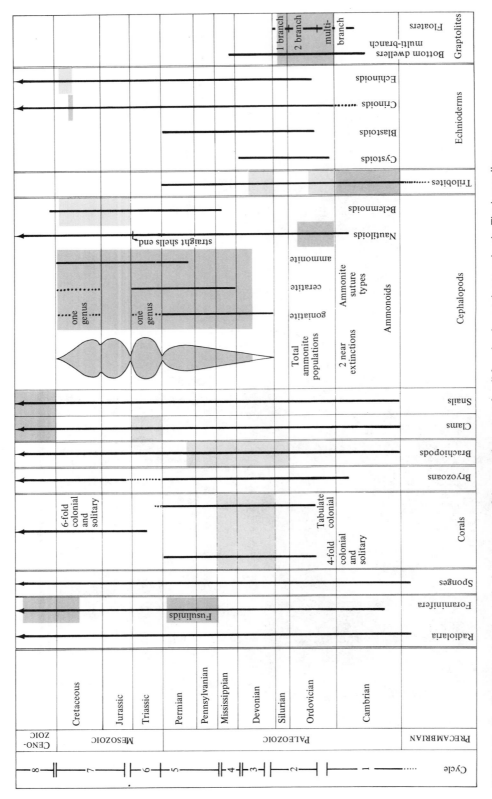

FIGURE 22–18 Chart showing the geologic occurrence of the more important fossil-forming invertebrate animals. The heavy lines show the time span of each animal group. Within some groups, some simple distinctions are indicated that are of value in age determination. Also shown are the parts of the geologic column for which these fossils are useful for worldwide correlation (darker color) and for correlation of more restricted regions (lighter color). Correlation data after Curt Teichert, Geological Society of America, *Bulletin*, Vol. 69, 1958.

seems unlikely for such a successful group.

In the Jurassic, the seas again expanded, and many new forms appeared. One of these was the belemnoids which first appeared in the late Paleozoic. They were a type of cephalopod with an internal cigar-shaped shell. (See Fig. 22–19.) They were much like modern squids. (See Fig. 22–20.) Figure 22–21 shows a somewhat more balanced sea floor in the Late Cretaceous.

In both of the reconstructions, clams, including oysters, are prominent. In the Paleozoic seas, brachiopods were the main bivalves; but in the Mesozoic and Cenozoic, they are largely replaced by clams. The biologic differences between clams and brachiopods are great, and they can be distinguished easily from the symmetry of the shells. (See Fig. 22–22.) Clams live in many habitats, just as the brachiopods

do. Some burrow into the bottom mud, and some are fastened to the bottom. One suggestion for why clams replaced brachiopods is that they burrowed deeper into the bottom mud and so developed a whole new habitat. They took

over many habitats and expanded into many environments. They are very useful in determination of the paleoenvironment. The bottom dwellers attached themselves to the bottom by one shell, and some changed in appearance to resem-

FIGURE 22–20   Jurassic (Cycle VII) sea floor. Belemnoids (squidlike cephalopods) and oysters are the invertebrates shown. Note the marine reptile ichthyosaur in the shadowy background. From Field Museum of Natural History.

FIGURE 22–19   Cretaceous belemnoid, *Belemnitella americana,* from New Jersey. Length, 7.6 centimeters (3 inches). Photo from Ward's Natural Science Establishment, Inc., Rochester, N.Y.

FIGURE 22–21   Late Cretaceous (Cycle VII) sea bottom near Coon Creek, Tennessee. Coiled ammonite and straight-shelled ammonites *(Baculites)* and a number of snails and clams. From Field Museum of Natural History.

463

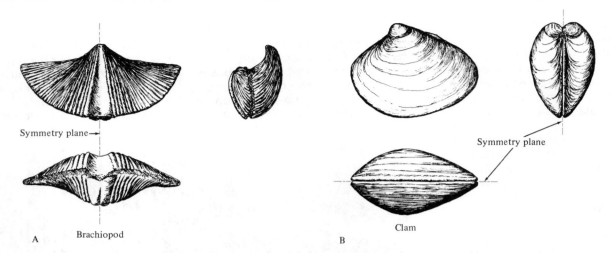

Symmetry plane→

Brachiopod

A

Clam

Symmetry plane

B

FIGURE 22–22 Differences between clam and brachiopod. A brachiopod has two dissimilar shells. A clam has two shells that are mirror images of each other because the clam has a symmetry plane between the two shells. The symmetry plane in a brachiopod cuts through both shells. Some clams (see Fig. 22–23) do not have this symmetry.

ble horn coral. Reefs of these corallike clams formed at some places. Others evolved into oysters, also losing their symmetry. (See Fig. 22–23.)

The Mesozoic plants, too, were different from those of the Paleozoic, although the change is not apparent until Late Triassic. (See Fig. 22–24.) The lush swamps that produced the Pennsylvanian coal beds disappeared in the Permian. This was the result of the climatic change to arid or semiarid conditions that is recorded by the redbed and evaporite deposits and the wind-deposited sandstones. These desert conditions that persisted into Triassic time resulted in the preservation of very few members of the early Mesozoic flora. The few fossil plants found in early Mesozoic rocks, such as horsetails, are holdovers from the Paleozoic. In Late Triassic, the Mesozoic flora of ginkgoes, conifers, ferns, and especially cycads, all of which differed from Paleozoic forms of these plants, was well established. They appear in the background of Figs. 22–28, 22–30 through 22–32, 22–34, and 22–35. A few angiosperms, or flowering plants, also appear in Triassic rocks. The angiosperms, which are important elements of present floras, became abundant

and spread to all continents in the Middle Cretaceous. They will be described in the next chapter.

The Mesozoic was the age of the reptiles, and, of course, the dinosaurs come to mind first. The dinosaurs were not the only reptiles. As noted in the last chapter, all of the reptiles have a *Seymouria*-like reptile of early Permian age as a common ancestor. (See Fig. 22–25.) In the Permian and most of the Triassic periods, the dominant group was the mammallike reptiles. They became greatly reduced in Late Triassic time as the dinosaurs took over. The mammals evolved from the mammallike reptiles, and the first mammal fossils are found in Upper Triassic rocks. The mammals were relatively unimportant until the Cenozoic, and so they will be discussed in the next chapter.

Seymouria's group lasted until near the end of the Triassic. The turtles evolved from them and are found from Early Triassic on. Snakes and lizards also had the same origin; the first fossil lizards are in Upper Triassic rocks, and the first snakes are Cretaceous.

The marine reptiles illustrate how a successful group like the reptiles moved into all environments. They returned to the sea, but they were better adapted than

their amphibian ancestors were. (See Fig. 22–26.) They developed a system whereby the unhatched eggs were kept in the mother's body. Some Triassic ichthyosaurs have been found with unborn young in the body cavity.

The thecodonts (Fig. 22–27) also evolved from *Seymouria*'s group, and they gave rise to the dinosaurs, the crocodiles and alligators, the birds, and the flying reptiles. The thecodonts were a very successful group. They developed legs under the body instead of at the sides like the mammallike reptiles and so improved their locomotion. Some developed a two-legged walk. In spite of this, they were generally small animals and were overshadowed by the mammallike reptiles.

Birds are similar to reptiles in many ways. They differ in having feathers and being warm-blooded. The oldest bird is Jurassic and is known to be a bird rather than a reptile only because three almost complete specimens with feathers have been found in fine-grained limestone. This bird, *Archaeopteryx*, was about the size of a pheasant and had teeth in its beak. (See Figs. 22–28 and 22–29.) Birds are among the rarest of fossils. The toothed flying reptiles are also found in Jurassic and Cretaceous

A

B

C

D

E

FIGURE 22–23 Mesozoic clams. A. *Buchia piochii,* Jurassic from Prince Patrick Island, Canada. Natural size. B. *Inoceramus labiatus,* Cretaceous from Alberta. Natural size. C. *Trigonia thoracica,* a thick-shelled Cretaceous clam. D. *Exogyra arietina,* Cretaceous, 3.8 centimeters (1.5 inches). A thick-shelled clam without the symmetry of most clams. E. Cretaceous clams from north-central Texas. Photos A and B from Geological Survey of Canada (112167-F, 113729-N); C from Smithsonian Institution; D from Ward's Natural Science Establishment, Inc., Rochester, N. Y.; E from U.S. Geological Survey.

465

WERE THE DINOSAURS WARM-BLOODED?

In spite of all that has been written about dinosaurs, they remain enigmatic creatures. They have always been considered to have been reptiles because their skeletons are like those of reptiles. Reptiles are cold-blooded animals that get much of their body heat from their environment; thus, dinosaurs have been considered cold-blooded animals. The argument against this long-standing concept is that the dinosaurs must have been very active animals to rule the lands throughout the Mesozoic Era. Reptiles are generally sluggish animals that spend a lot of time basking in the sun, and this is not our image of dinosaurs.

Dinosaur skeletons show that they had long legs and that they had the erect posture of birds and mammals, not the typical sprawling posture of lizards. The body is above the legs in the erect posture and beside the legs in a sprawler. All modern warm-bloods have erect posture, and all modern cold-bloods have sprawling posture. The erect posture enables much faster locomotion. Of course, erect posture and even high activity level does not prove that dinosaurs were warm-blooded, and many paleontologists are highly skeptical.

Dinosaur size could have been an adaptation to maintain a high body temperature. The mass of a large animal stores more heat, and so the animal is less affected by changing environment. It also seems possible that a large animal could maintain a warm body temperature, even with the slow metabolism of a reptile.

A warm-blooded animal must eat more food than a cold-blooded one. Modern warm-blooded predators require about ten times more prey than do cold-blooded ones. At a number of places, it is possible to count the number of prey fossils and the number of predatory dinosaurs. In each case, the ratio favors warm-blooded dinosaurs, but, of course, such counts are subject to many errors, and the conclusions are not accepted by all.

Dinosaur bones have many microscopic canals for blood circulation, also suggesting that dinosaurs may have been warm-blooded. Such canals are common in the bones of warm-blooded animals and are much fewer in the bones of most, but not all, cold-bloods.

Most paleontologists believe that the birds evolved from the dinosaurs. Birds are warm-blooded and have a four-chambered, two-pump heart. Thus, it is possible that at least one line of dinosaurs had a warm-blooded heart. However, the crocodiles, which are clearly cold-blooded, also have four-chambered hearts. Some argue that because the dinosaurs' heads were so much higher than their lungs, they, too, required two-pump hearts. The reasoning is that the blood pressure necessary to pump blood to a dinosaur's brain would be so great that its lungs would fill with fluid if it had only one blood pump.

The evidence is certainly suggestive, but not conclusive. For more information you might like to read: Adrian J. Desmond, *The Hot-Blooded Dinosaurs: A Revolution in Paleontology.* New York: The Dial Press/James Wade, 1976, 238 pp.

rocks. The largest flying reptile so far found is from Texas and has a wing spread of 15.5 meters (50 feet). It has been proposed that from the dinosaurs evolved the flying reptiles and the birds instead of the conventional evolutionary scheme shown in Fig. 22–25.

The first dinosaurs are found in Upper Triassic rocks. They were small animals, not at all like the huge beasts of Late Jurassic and Cretaceous that were soon to rule the lands. Many of the dinosaurs were small, but it is the big ones that get attention. The term *dinosaur* means "terrible lizard" and has been applied to two different groups of reptiles. It is an informal but useful term. Dinosaurs were very successful and evolved rapidly. Figure 22–25 shows the relationships among the various dinosaurs.

The difference between the two types of dinosaurs is in their hip bones, one type having typical reptile hips (saurischians) and the other having birdlike hips (ornithischians). The saurischians were of two types, carnivores and herbivores. The carnivores were

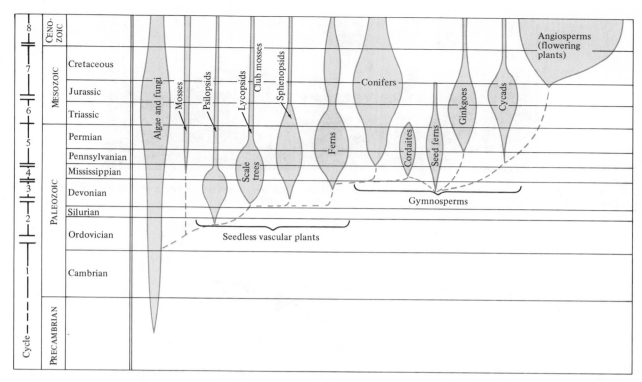

FIGURE 22–24 The development of plants.

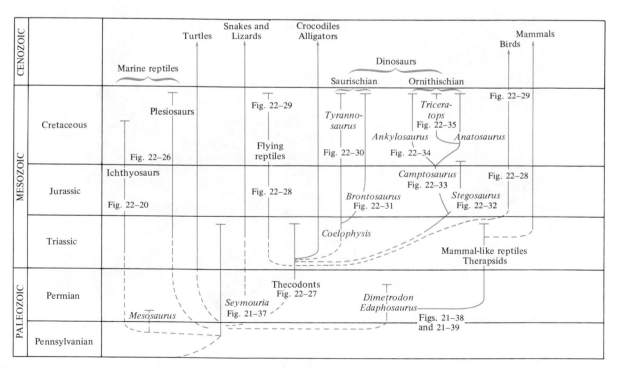

FIGURE 22–25 Evolution of the Mesozoic vertebrates. This figure is part of Fig. 21–29.

FIGURE 22–26 Jurassic marine reptiles. Plesiosaurs are long necked. Ichthyosaurs are porpoiselike and closely resemble fish. About 2 meters (6 feet) long. Painting by C. R. Knight, Field Museum of Natural History.

FIGURE 22–27 Thecodont. This group evolved from *Seymouria,* and the thecodonts gave rise to the dinosaurs, flying reptiles, birds, and crocodiles. Thecodonts had bodies similar to dinosaurs, but were all small, up to 1.2 meters (4 feet), and were confined to the Triassic.

FIGURE 22–28 Jurassic scene. *Archaeopteryx,* the bird, has feathers. The flying reptile, *Rhamphorhynchus,* has no feathers and a wingspread of about 1 meter (3 feet). Also shown are tiny dinosaurs and palmlike cycads. Painting by C. R. Knight, Field Museum of Natural History.

FIGURE 22–29 A. Flying reptile, *Pteranodon*. B. Toothed bird, *Hesperonis*. Both from Upper Cretaceous of Kansas. Photos from Smithsonian Institution.

mainly two-legged, with only small front legs; the largest flesh eater of all times, *Tyrannosaurus* [6.1 meters (20 feet) tall, 15.25 meters (50 feet) long, weighing 7275 to 9090 kilograms (8 to 10 tons)], belongs to this group. (See Fig. 22–30.) The saurischian plant eaters were mainly four-legged; some of the largest animals of all belong to this group (*Brontosaurus*). (See Fig. 22–31.) The ornithischians were apparently all plant eaters, and the bizarre dinosaurs, such as the duckbills, the horned dinosaurs, and the armored dinosaurs, belong to this group. (See Figs. 22–32, 22–33, 22–34, and 22–35.)

Most of the carnivores were two-legged, fast-running predators. The plant eaters developed many defenses. Some could run fast, some had armor, some had horns, some were very large, and some lived in water. Much has been written about the small brains of dinosaurs, but compared with other reptiles, their brains were near average. Compared with mammals, their brains were small.

It is not clear why some of the dinosaurs became so large. It may be because reptiles are cold-blooded and so it is an advantage to be large because a large animal needs a much smaller percentage of its body weight of food each day to maintain its energy. Most of the big dinosaurs were plant eaters and so may have had a problem getting food. A disadvantage of being large is the difficulty of finding shelter in a storm. Size may also have been a defense against the predators. On the other hand, it has been suggested that some of the dinosaurs may have been warm-blooded, but the evidence is not clear.

The reptiles were very similar on all continents during most of the Mesozoic, suggesting that until the Cretaceous, all continents were close enough to allow migration. In the Cenozoic, there was less migration, and, interestingly, 30 orders of mammals evolved during the 65 million years of the Cenozoic, but only 20 orders of reptiles are found in the 200-million-year Mesozoic. This suggests that in the relative isolation of the Cenozoic, more orders formed because each isolated group evolved and radiated.

The orogenic activity that marked the Late Cretaceous and the Early Cenozoic produced a diverse topography that resulted in many climatic variations. These new climatic zones were soon occupied by plants, thus setting the stage for animals to evolve to take over these ecologic niches. The angiosperms, or flowering plants, that appeared in Early Cretaceous include the grains and grasses, and it is the rapid spreading of these to the drier climates that created the new ecologic niches. The plant life of the Cenozoic was essentially modern, consisting mainly of angiosperms and conifers. The many Cenozoic coal beds attest their development.

At the end of the Mesozoic, the dinosaurs, almost all of which had specialized to occupy the many different environments, died out. Some of the dinosaurs were big

ASTEROIDS AND DINOSAURS

In mid-1980, a research team from the University of California at Berkeley published a novel theory to explain the late-Mesozoic extinctions. Their discovery is an example of serendipity. Walter Alvarez, a geologist working in Italy, was attempting to measure sedimentation rates in marine sedimentary rocks at the boundary between the Cretaceous and the Cenozoic. His father, Luis Alvarez, a Nobel Prize winning physicist, suggested measuring the amount of iridium in the rocks. His reasoning was that the small amounts of the element iridium in sedimentary rocks come from the infall of micrometeorites, and the rate of micrometeorite infall is constant. The other two members of the team are physical chemists who did much of the analytical work. They discovered that the amount of iridium in the very thin (2-centimeter) clay layer separating the Mesozoic from the Cenozoic was about 30 times more than average.

This high iridium content was interpreted as having an extraterrestrial origin for two reasons. The other elements enriched in the clay were typical of extraterrestrial sources and very different from the elements associated with iridium from the earth. Similar iridium enrichment was also found at this same stratigraphic horizon everywhere that measurements were made. There seem to be only two possible extraterrestrial sources for the iridium—an asteroid impact or a supernova explosion. The supernova was eliminated on theoretical grounds, and calculations showed that an asteroid impact is a reasonable hypothesis. Calculations suggest that the impact of an asteroid about 10 kilometers in diameter, about the size of Manhattan, could have been the source of the iridium enrichment.

The Berkeley team, feeling that they had established the reality of the asteroid impact, went on to suggest how such an event could explain the extinctions at the close of the Mesozoic. They suggested that the impact threw so much dust into the air that the earth was dark for about three years. Blocking out the sun would kill many plants and so interrupt the food chain of most organisms, causing widespread extinctions. In short, they proposed a modern-day catastrophism.

Such revolutionary ideas are not quickly accepted. At first, the arguments were about the distribution of the asteroid material. These arguments died down after iridium enrichment was found at many places, both in marine and terrestrial sediments. Thus the reality of the impact event was generally accepted. Then the discussion shifted to the extinctions. The paleontological evidence is strong that the extinctions in the oceans did not coincide with those on the lands. The small organisms that float near the ocean surface do appear to have died suddenly, and the micropaleontologists have long suggested that these extinctions were catastrophic. The land organisms showed a different pattern, with the much hardier northern plants dying while the more sensitive tropical plants show few changes. Thus, the fossil record suggests both catastrophic extinctions and the traditional gradual extinctions.

A compromise soon developed. Restudy of distribution of the impact dust suggested that the "darkness at noon" probably only lasted about three months instead of the originally proposed three years. This is long enough to kill the plankton (marine floaters) and so disrupt the marine food chains and account for the marine extinctions. On the lands the darkness only accelerated the extinctions that, judging from the plant extinctions, were being caused by a rapid climatic cooling. So an asteroid probably did not kill the dinosaurs, but a cooling trend, perhaps like the one that caused the Pleistocene glaciers, did.

The iridium-rich layer is apparently a worldwide time line and may eventually enable detailed understanding of the timing of the extinctions at the end of the Mesozoic. At Hell Creek in northeastern Montana, the last dinosaur bone, a femur of *Tyrannosaurus rex,* is found about three meters below the iridium layer. Between that level and the iridium, both Cretaceous and Paleocene fossils are found. Two meters below the iridium, Paleocene vertebrates occur; and 1.5 meters below the iridium, untypical but clearly Cretaceous pollen has been recovered. Thus there is evidence that at least at that place, the dinosaurs died out before the asteroid impact and the marine extinctions.

The development, modification, and acceptance of the impact theory happened quite rapidly, and the details have been published; they make good reading. The original paper is: L. W. Alvarez, Walter Alvarez, Frank Asaro, and Helen V. Michel, "Extraterrestrial Cause for the Cretaceous-Tertiary Extinction," *Science* (June 6, 1980), Vol. 208, No. 4448, pp. 1095–1108. The modifications and acceptance are easier reading: Richard A. Kerr, "Asteroid Theory of Extinctions Strengthened," *Science* (October 31, 1980), Vol. 210, No. 4469, pp. 514–517; and Richard A. Kerr, "Impact Looks Real, the Catastrophe Smaller," *Science* (November 20, 1981) Vol. 214, No. 4523, pp. 896–898. Another summary is: Dale A. Russell, "The Mass Extinctions of the Late Mesozoic," *Scientific American* (January 1982), Vol. 246, No. 1, pp. 58–65.

A

B

FIGURE 22–30 Carnivorous dinosaurs. All carnivorous dinosaurs are saurischian. They could move rapidly on two feet and had only small front feet. A. *Ceratosaurus* from Upper Jurassic Morrison Formation near Canyon City, Colorado. About 7.3 meters (24 feet) long. Note the small horn on the top of the nose. Photo from Smithsonian Institution. B. *Tyrannosaurus,* the largest of the flesh-eating dinosaurs. About 15.25 meters (50 feet) long. Cretaceous. Painting by C. R. Knight, Field Museum of Natural History.

animals, making food gathering in times of changing climate and plant life a difficult task. At this time, the mammals were small and few in number. They apparently were better able to adapt to the new conditions and so replaced the dinosaurs. The mammals were a more successful group than the reptiles for several reasons. They had bigger brains. They had four-chambered hearts instead of the two-chambered hearts of reptiles, and had body hair rather than scales. Both of these features allowed them to maintain their body heat, rendering them less vulnerable to climatic changes. They were better able to preserve their young than the egg-laying reptiles. They also developed jaws and teeth better suited to their food. Their feet also underwent a series of changes that better equipped them to their ecology. In spite of all these reasons why mammals were better equipped than dinosaurs, the dinosaurs were not driven out by the mammals. Rather, for unknown reasons, the dinosaurs died out and the mammals took over.

Many reasons have been suggested for the extinction of the dinosaurs. Most of these reasons could account for the demise of single groups or species but not all of the diverse dinosaurs. Examples are climatic change, loss of food supply because of development of the angiosperms, disease, and predation of their eggs, perhaps by mammals. Because none of these or the many other sug-

A

B

C

FIGURE 22–31 Plant-eating saurischian dinosaurs. Mainly very large four-footed dinosaurs that lived in swampy water. A. *Bronto-saurus,* Late Jurassic. About 21.4 meters (70 feet) long, weighing 31,820 kilograms (35 tons). Eating enough to maintain such a size may have been a problem. Painting by C. R. Knight, Field Museum of Natural History. B. Skeleton of *Brontosaurus.* Photo from Field Museum of Natural History. C. *Diplodocus,* Upper Jurassic, from Dinosaur National Monument, Utah. About 26 meters (85 feet) long, 4 meters (13 feet) high, weighing 11,820 kilograms (13 tons). *Diplodocus* is found in Jurassic and Cretaceous rocks. Photo from Smithsonian Institution.

A

B

FIGURE 22–32 *Stegosaurus.* These plated dinosuars lived in the Jurassic and Cretaceous. About 6.1 meters (20 feet) long, weighing 2 tons. A. Painting by C. R. Knight, Field Museum of Natural History. B. Skeleton. Photo from Smithsonian Institution.

FIGURE 22–33 Duckbilled dinosaur, *Camptosaurus,* Upper Jurassic, Wyoming. *Camptosaurus* was one of the main stems of the ornithischian dinosaurs, and the ceratopsian (horned) dinosaurs evolved from them. *Camptosaurus* lived in the Jurassic and Cretaceous and was 4.6 meters (15 feet) long. Photo from Smithsonian Institution.

FIGURE 22–34   Cretaceous dinosaurs. On the right, *Trachodon,* a duckbill. Other duckbills are the web-footed *Corythosaurus* in the water, and crested *Parasaurolophus* in the left background. In the foreground is *Ankylosaurus,* an armored dinosaur. In the center background is the ostrichlike *Struthiomimus,* a saurischian dinosaur. Painting by C. R. Knight, Field Museum of Natural History.

gestions can explain the extinction of all of the dinosaurs, some have looked to extraterrestrial causes. Cosmic rays and solar wind, especially at times of magnetic reversals, was suggested, but most extinctions here and elsewhere in the geologic record do not generally coincide with times of magnetic reversals. Another possibility is that the impact of a large meteorite or asteroid could have thrown enough dust into the atmosphere to reduce greatly the amount of the sun's energy reaching the earth's surface. Such a loss of sunlight would destroy some organisms in the food chain and so cause extinctions. In support of this theory is the concentration of elements expected in meteorite dust at the Mesozoic-Cenozoic boundary.

A

B

FIGURE 22–35   Ceratopsian (horned) dinosaurs. Cretaceous. They evolved from *Camptosaurus.* A. *Protoceratops.* Eggs of this dinosaur were found in Mongolia. Painting by C. R. Knight, Field Museum of Natural History. B. *Protoceratops* skeleton. About 2 meters (6 feet) long. Photo from Field Museum of Natural History.

474

C

FIGURE 22-35 (cont.) C. *Triceratops* skeleton. About 7.3 meters (24 feet) long. Photo from Smithsonian Institution.

SUMMARY

During the Mesozoic the single supercontinent broke up, forming present oceans such as the Atlantic.

A geosyncline formed between Europe and Africa.

Cycle VI—Triassic of the West

In the far west, thick marine clastics and volcanics overlie Permian rocks unconformably, and to the east are clastics and carbonates.

The marine rocks interfinger with thin red beds in the northern Rocky Mountains.

Nonmarine rocks in Arizona and New Mexico are further evidence of aridity.

Batholiths were emplaced in the Sierra Nevada.

Cycle VII—Middle Jurassic through Cretaceous

In the west, Cycle VII begins with an upland area centered in Nevada

that shed clastics on both sides.

In the Jurassic, marine deposition occurred in the area of the Rocky Mountains and Great Plains.

The sea retreated north, and nonmarine rocks were deposited.

The seas returned in the Cretaceous, and thick clastics were deposited. This was the last great inundation of North America.

The area west of the upland was active throughout the cycle. Batholiths were emplaced throughout the far west.

Marine deposition began in the Gulf Coast in Jurassic and spread to the Atlantic Coast in the Cretaceous.

In the far north are Jurassic marine clastics and Cretaceous nonmarine rocks.

Cephalopods and clams are the most abundant marine invertebrates.

Cephalopods are almost ideal in-

dex fossils because they evolved rapidly and are widespread.

The Mesozoic flora is cycads, ginkgoes, conifers, and ferns.

Angiosperms became abundant in the Middle Cretaceous.

During Permian and Triassic, mammallike reptiles were the dominant group. They evolved into the mammals that are first found in Upper Triassic rocks.

Turtles, lizards, and snakes all evolved during the Mesozoic.

Birds and flying reptiles first appeared in the Jurassic.

The first dinosaurs were small and are found in Upper Triassic rocks.

The reptiles are similar on all continents, suggesting that until the Cretaceous, the continents were close enough to allow migration.

The dinosaurs died out at the end of the Cretaceous and were replaced by the mammals.

QUESTIONS

1 Describe the Triassic rocks of the western United States.
2 What was the extent of the Jurassic seas in North America?
3 Why is it difficult to date Cretaceous events between the Rocky Mountains and the West Coast?
4 Where were batholiths emplaced in North America in the Mesozoic?
5 Describe the Cretaceous rocks of the Rocky Mountains.
6 When did deposition start in the Gulf Coast area?
7 Which continents drifted apart during the Mesozoic?
8 Is there a relationship between continental drift and the locations of Mesozoic geosynclines?
9 Describe the fauna of the Triassic.
10 When did the first flying reptiles appear? The first birds?
11 How do Mesozoic corals differ from Paleozoic corals?
12 How are clams distinguished from brachiopods? Does this distinction hold true for all Mesozoic clams?
13 How are snails distinguished from cephalopods?
14 Outline the history of cephalopods.
15 How are Mesozoic plants different from Paleozoic plants? From Cenozoic plants?
16 What was the time of the dinosaurs?
17 Describe the various types of dinosaurs and other reptiles of the Mesozoic. Which ecologic areas or niches did they occupy?
18 What advantages do mammals have over reptiles?

SUPPLEMENTARY READINGS

Alvarez, L. W., and others, "Extraterrestrial Cause for the Cretaceous-Tertiary Extinction," *Science* (June 6, 1980), Vol. 208, No. 4448, pp.1095–1108.

Bakker, R. T., "Dinosaur Renaissance," *Scientific American* (April 1975), Vol. 232, No. 4, pp. 58–78.

Buffetaut, Eric, "The Evolution of the Crocodilians," *Scientific American* (October 1979), Vol. 241, No. 4, pp. 130–144.

Desmond, A. J., *The Hot-blooded Dinosaurs: A Revolution in Paleontology.* New York: Dial Press, 1975, 238 pp.

Gartner, Stefan, and John Keany, "The Terminal Cretaceous Event: A Geologic Problem with an Oceanographic Solution," *Geology* (December 1978), Vol. 6, No. 12, pp. 708–712.

Hallam, A., "Continental Drift and the Fossil Record," *Scientific American* (November 1972), Vol. 227, No. 5, pp. 56–66.

Imlay, R. W., *Jurassic Paleobiogeography of the Conterminous United States in Its Continental Setting.* U.S. Geological Survey Professional Paper 1062. Washington, D.C.: U.S. Government Printing Office. 1980, 134 pp.

Kielan-Jaworowska, Zofia, "Late Cretaceous Mammals and Dinosaurs from the Gobi Desert," *American Scientist* (March-April 1975), Vol. 63, No. 2, pp. 150–159.

Kurtén, Björn, "Continental Drift and Evolution," *Scientific American* (March 1969), Vol. 220, No. 3, pp. 54–64.

Langston, Wann, Jr., "Pterosaurs," *Scientific American* (February 1981), Vol. 244, No. 2, pp. 122–136.

McLean, D. M., "A Terminal Mesozoic "Greenhouse": Lessons from the Past," *Science* (August 4, 1978), Vol. 210, No. 4354, pp. 401–406.

Mulcahy, D. L., "The Rise of the Angiosperms: A Genecological Factor," *Science* (October 5, 1979), Vol. 206, No. 4414, pp. 20–23.

Ostrum, J. H., "Bird Flight: How Did It Begin?" *American Scientist* (January-February 1979), Vol. 67, No. 1, pp. 46–56.

Valentine, J. W., and E. M. Moores, "Plate Tectonics and the History of Life in the Oceans," *Scientific American* (April 1974), Vol. 230, No. 4, pp. 80–89.

Ward, Peter, Lewis Greenwald, and O. E. Greenwald, "The Buoyancy of the Chambered Nautilus," *Scientific American* (October 1980), Vol. 243, No. 4, pp. 190–203.

23

THE CENOZOIC

Human paleontology shares a peculiar trait with such disparate subjects as theology and extraterrestrial biology: it contains more practitioners than objects for study.

David Pilbeam and Stephen Jay Gould
"Size and Scaling in Human Evolution"
Science, *Vol. 186, No. 4167, p. 892, 1974.*

The Cenozoic was subdivided in a much different way than the older strata were. Charles Lyell, whose books popularizing uniformitarianism were mentioned earlier, subdivided the Cenozoic somewhat earlier than many of the older systems were defined. His method was to use the percentage of living organisms present as fossils in the beds. Gerard Deshayes, a Frenchman, studied the fossils that were mainly snails and clams. In 1833 Lyell recognized four subdivisions:

Newer Pliocene with 90 percent living organisms
Older Pliocene with one-third to one-half living organisms
Miocene with less than 18 percent living organisms
Eocene with about 3.5 percent living organisms.

He also described localities where each of his subdivisions is exposed.

In 1839 he changed the name of the Newer Pliocene to Pleistocene. Soon after, Edward Forbes used the term *Pleistocene* to mean beds deposited during the last glacial age. This usage is still widely followed, making the definition of the Pleistocene somewhat different from other systems.

Other changes occurred as well. In 1854 Heinrich Ernst von Beyrich defined the Oligocene between the Eocene and the Miocene. In 1874 W. P. Schimper described the Paleocene, basing his definition on fossil plants in northern Germany and Belgium. All of the other subdivisions of the Cenozoic are based on marine fossils, so acceptance of the Paleocene was somewhat slower than for the other subdivisions.

WORLD GEOGRAPHY IN THE CENOZOIC

The present distribution of continents on the earth formed during the Cenozoic. Figure 23–1 shows the drift that occurred. North America separated from Europe, and the Atlantic Ocean formed. Greenland separated from North America, and Australia left Antarctica. The south Atlantic widened. India collided with Eurasia.

Orogeny occurred from Alaska to Antarctica. Africa and Europe drifted closer together causing orogeny in the geosyncline between them. This mountain-building event produced the Alps in Europe and a mountain chain across Asia. Parts of Turkey and Iran were transferred from Africa to Asia in this collision. India collided with Asia, forming the Himalaya Mountains.

North America moved somewhat northward, largely by rotation. This may have caused a cooling of the climate reflected in some of the late Cenozoic life. Alaska and Siberia may have collided, causing some deformation.

The Gulf of California and the Red Sea probably formed late in the Cenozoic. The Bay of Biscay opened by the rotation of the Iberian Peninsula. Iceland formed by volcanic activity on the Mid-Atlantic Ridge in late Cenozoic time.

THE CENOZOIC— CYCLE VIII

Following the withdrawal of the seas near the end of the Cretaceous, the seas never again occupied large areas of the interior, but they did cover most coastal areas during parts of the Cenozoic. Although, as shown in Fig. 20–1, some of these marine advances were worldwide, they were not extensive in North America. The geologic history of the Cenozoic in most of North America is concerned with nonmarine rocks, and so this cycle differs greatly from all of the previous cycles. During the Cenozoic, most of the present topographic features of the earth were formed, making its history of great interest to the traveler. It will be possible to describe only a few of the more important areas.

ATLANTIC AND GULF COASTAL PLAINS

Marine sedimentation began in the Atlantic and Gulf coastal plains areas in Mesozoic time and continued throughout most of the Cenozoic. The history is one of repeated advances and retreats of the sea. The Gulf area is one of the most closely studied areas in the world because of the great value of the oil recovered from these rocks. The rocks are penetrated by thousands of oil wells, and some of these are very deep. On the south Atlantic coastal plain, four major advances are recognized. They are mid-Cretaceous, Late Cretaceous to early Cenozoic, and two in the late Cenozoic—middle Miocene to early Pliocene, and Pliocene to Pleistocene. During much of the Ceno-

Fossil fish from the Green River Formation of southwestern Wyoming. Photo by W. T. Lee, U.S. Geological Survey.

FIGURE 23–1 Continental drift during the Cenozoic. Cenozoic mountain building is shown in color. Dotted positions are the end of the Cretaceous; solid line is the present. The movement of Australia and South America is exaggerated by the map projection used here. Based on R. S. Dietz and J. C. Holden, 1970.

zoic, Florida was an area of limestone deposition, separating the Atlantic from the Gulf coastal plain. On the Gulf, eight major advances and retreats are recognized, as well as many more minor ones. The maximum advance occurred in the Eocene when the Mississippi Valley was invaded as far as southern Illinois. This was the first advance of the Cenozoic, and the early Cenozoic rocks overlie the Cretaceous unconformably, indicating that the Cretaceous seas had retreated in late Cretaceous time. A typical advance of the sea is recorded by clastic rocks at the bottom of the sequence. The rocks grade from nonmarine sand or clay to similar brackish water sediments to marine sandstone or shale that thickens greatly toward the Gulf. The advances and retreats of the sea are probably related to uplifts in the Appalachian Mountains, but it has not been possible to correlate these events. Much of the area drained by the Mississippi River was undergoing erosion throughout the Cenozoic and provided debris to the Gulf area. The sedimentary rocks on the Gulf coast are very thick, up to 15,250 meters (50,000 feet) with 4575 meters (15,000 feet) of Pleistocene at some places. Figure 23–2 shows a typical cross-section and shows the rise in the basement rocks offshore.

Through the Cenozoic, limestone was deposited in Florida. It is believed that this area was covered by shallow, warm water. Under such conditions, limestone is being deposited now in the Bahama Banks just east of Florida. Florida was not awash during all of Cenozoic, but at times islands were present.

APPALACHIAN MOUNTAINS

The Appalachian Mountains were above sea level and undergoing erosion during the Cenozoic. Their history cannot be learned from sedimentary rocks except in a very general way because the uplifts of the mountains cannot be related to the coastal plain sediments. Thus we turn to study of the erosion surfaces to determine the history of the Appalachians. This was one of the first places that such studies were made. The culmination of these studies was the classical interpretation shown in Fig. 23–3. The history shown in the cross-section starts with high mountains in post-Triassic time (A). These mountains were eroded to a smooth peneplain before Cretaceous time (B); and in the Cretaceous, coastal plain sediments were deposited on the peneplain (C). This was followed by a broad up-arching (D). Erosion of the up-warped range formed a new peneplain (E). This peneplain was up-arched (F), and erosion developed a partial peneplain on the softer rocks (G). Another uplift caused erosional development of another partial peneplain on the softer rocks (H). Another uplift caused further dissection of the partial peneplains and produced the present Appalachian Mountains.

More recent studies suggest that a simpler history could produce the same topography. In this case, the history would be the same up to block E. A single uplift followed by erosion could result in the ridge tops preserving remnants of the old peneplain at elevations of about 1220 meters (4000 feet), and flat valleys cut on weak rocks.

ROCKY MOUNTAINS AND HIGH PLAINS

Orogeny began in the Rocky Mountains near the end of the Cretaceous and extended to the Oligocene. The age and intensity of this folding vary from place to place. (See Fig. 23–4). At some places Paleocene rocks unconformably overlie folded Late Cretaceous rocks, and at other places rocks as young as Oligocene are involved in the folding. In general, the more intense orogeny occurred near the west shore of the Cretaceous sea where thrust faults are found. The compressive phase was over by middle Eocene except in southern Colorado where late

Eocene thrusts are found. Farther east in Montana, Wyoming, and Colorado, most of the present mountain ranges were uplifted, with folding and faulting that was locally intense. Nearby on the high plains of eastern Montana, undeformed Late Cretaceous nonmarine sediments are conformably overlain by similar Paleocene

rocks, indicating that the orogeny was localized. This sequence of rocks is of great interest because it records the last of the dinosaurs and the rise of mammals. Farther east in North Dakota and adjacent areas, a remnant of the Cretaceous sea remained until the Paleocene.

With the uplift of the mountain ranges in the Rockies, basins were formed between the ranges. (See Fig. 23–5.) These basins were filled with nonmarine sediments during the early Cenozoic. Some volcanic sediments were deposited in some of the basins, and in the Eocene and Oligocene, great thicknesses of volcanic rocks were emplaced in northwestern Wyoming in the Absaroka Range. Volcanic rocks were deposited throughout most of the Cenozoic in the San Juan Mountains of Colorado. Erosion of the ranges filled the basins, mainly in Paleocene to Oligocene time.

From Oligocene to Pliocene time, the streams from the area of the present Rocky Mountains spread fine clastics over the Great Plains. Then, either due to a change in climate or to uplift of the Rocky Mountains, the streams began to erode. It is uncertain whether the uplift occurred at this time or earlier. This erosion removed much of the earlier deposits, forming the present topographic mountains. Before this downcutting, the rivers had been flowing on a relatively smooth surface, in part eroded and in part constructed by the rivers. The courses of the rivers on this surface bore no relationship to the rock types at depth, for they had developed their courses over tens of millions of years. Thus the downcutting rivers at most places removed the soft basin-filling sediments, but at some places cut narrow, deep canyons through the mountain blocks composed of Precambrian metamorphic rocks. Examples of this include Big Horn River Canyon in the Big Horn

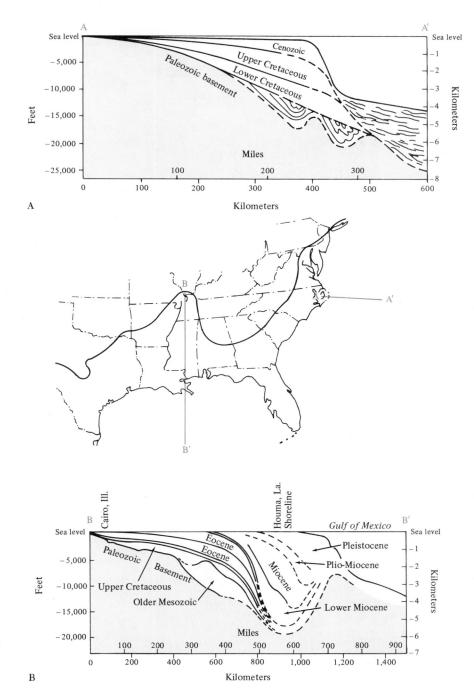

FIGURE 23–2 Cenozoic deposition in the Atlantic and Gulf areas. Note the very thick Cenozoic deposits in the Gulf Coast area. A after K. O. Emery and others, 1970; B after H. A. Bernard and R. J. Leblanc, 1965.

Mountains, Laramie River in the Laramie Mountains, North Platte River, Sweetwater River, and many others. (See Fig. 23–6.)

COLORADO PLATEAU

The events on the Colorado Plateau are related to those in the

Rocky Mountains. In early Cenozoic time, the Colorado Plateau was at a low elevation, and rivers carrying sediments from the

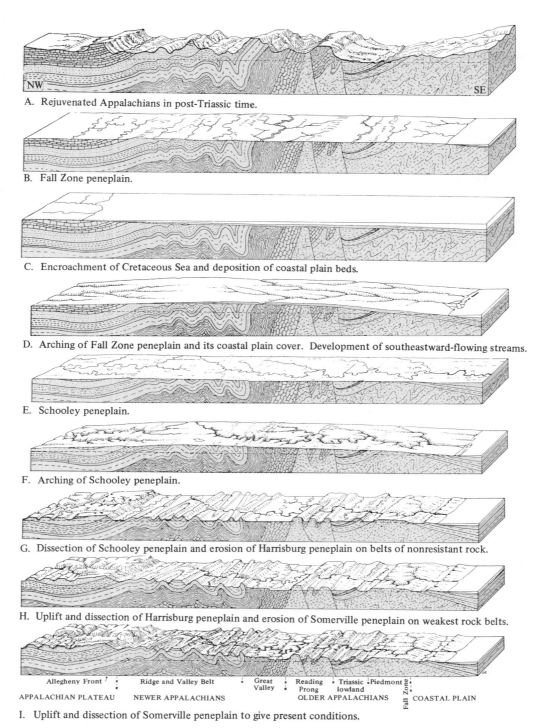

A. Rejuvenated Appalachians in post-Triassic time.

B. Fall Zone peneplain.

C. Encroachment of Cretaceous Sea and deposition of coastal plain beds.

D. Arching of Fall Zone peneplain and its coastal plain cover. Development of southeastward-flowing streams.

E. Schooley peneplain.

F. Arching of Schooley peneplain.

G. Dissection of Schooley peneplain and erosion of Harrisburg peneplain on belts of nonresistant rock.

H. Uplift and dissection of Harrisburg peneplain and erosion of Somerville peneplain on weakest rock belts.

Allegheny Front | Ridge and Valley Belt | Great Valley | Reading Prong | Triassic lowland | Piedmont | Fall Zone

APPALACHIAN PLATEAU   NEWER APPALACHIANS   OLDER APPALACHIANS   COASTAL PLAIN

I. Uplift and dissection of Somerville peneplain to give present conditions.

FIGURE 23–3   The development of the Appalachian Mountains. The history shown in the diagrams is the classical interpretation. New concepts suggest that the present form of the Appalachian Mountains could have developed from a single uplift. From Douglas Johnson, *Stream Sculpture on the Atlantic Slope,* Columbia University Press, N.Y., 1931.

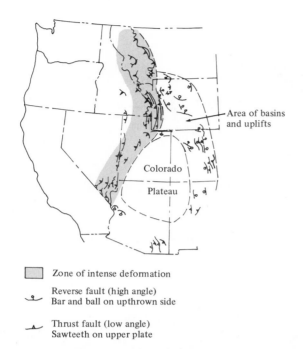

Zone of intense deformation

Reverse fault (high angle)
Bar and ball on upthrown side

Thrust fault (low angle)
Sawteeth on upper plate

FIGURE 23–4 Late Cretaceous and Early Cenozoic deformation in the Rocky Mountain area. In the western zone of intense deformation, folding and thrust faulting are common. In the eastern area of basins and uplifts, a number of basins sank and were filled by detritus from rising highlands. Many of these basins were folded with local thrust faulting during this interval. Very little deformation took place in the Colorado Plateau, an anomalous situation that has not been satisfactorily explained. The west coast was an area of active basins. Between the west coast and the intensely deformed zone, few Cretaceous rocks are present, so deformation during this interval cannot be recognized. Structural details after Gilluly, 1963.

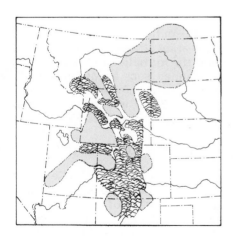

FIGURE 23–5 Early Cenozoic basins and highlands in the Rocky Mountain area.

uplifted areas to the north and east deposited them on the plateau. By Eocene time, the plateau had downwarped, and a vast fresh-water lake covered much of Utah. Organic-rich shale was de-posited in this lake, and the resulting oil shales may someday be an important source of petroleum. Uplift and local volcanic activity occurred during the rest of the Cenozoic. During this time, the Colorado River eroded the Grand Canyon. The rocks on the Colorado Plateau are, for the most part, horizontal, with some faulting and bending. All around this area, similar rocks are folded and deformed. Why the Colorado Plateau escaped this deformation is not known.

BASIN AND RANGE

This area is known variously as the Great Basin, the Basin-Range, or the Basin and Range area because, with the exception of the Colorado River, it is a basin with no external drainage to the ocean, and because the area is character-ized by many fault-block mountain ranges with intervening basins. Like the Colorado Plateau, this is an unusual area. The deformation of the rocks that make up the ranges has been described in the preceding chapters. These de-formed rocks have been uplifted in fault blocks to form the Basin and Range area. The faults appear to be normal faults, and so differ sharply from the compressional features displayed in the ranges. (See Fig. 23–7.) The faulting and uplifting began in the Miocene and are probably still going on, judging from the recent fault scarps and the number of earth-quakes. The best example is prob-ably the westernmost range, the Sierra Nevada in California. (See Figs. 23–8 and 23–9.)

The basins obviously have been sites of deposition since their for-mation. Prior to the faulting, Ceno-zoic rocks were deposited in part of the area, and these rocks are exposed in some of the ranges. Volcanic activity occurred at var-ious times in different parts of the Basin and Range area. In some places, thick, widespread welded tuffs can be traced for many miles from range to range.

NORTHWEST UNITED STATES

The northwest was an area domi-nated by volcanic rocks through-out much of the Cenozoic. Marine sediments and some volcanic rocks accumulated near the Pacific in a number of active embay-ments. A pre-late Miocene uncon-formity is widespread. Inland, the Cenozoic rocks are continental and volcanic. They were deposited in many basins, and the number of unconformities shows the crustal unrest. Farther east, in much of eastern Washington and Oregon, these rocks are covered by the spectacular Columbia River Basalt to a depth of at least a thousand meters. (See Fig. 23–10). Farther east in the Snake River Valley are thick deposits of younger volcanic rocks. In Pliocene time the present Cascade Range was uplifted on a

483

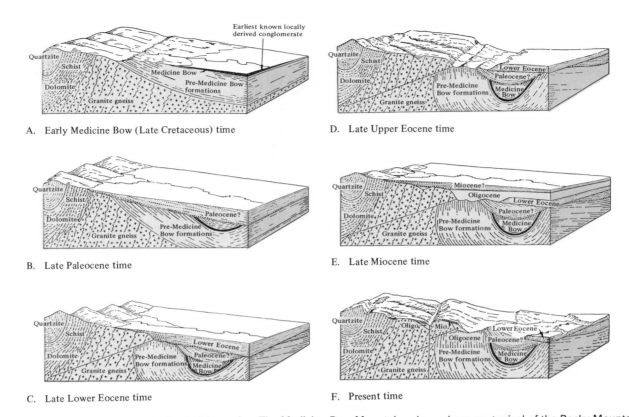

A. Early Medicine Bow (Late Cretaceous) time

B. Late Paleocene time

C. Late Lower Eocene time

D. Late Upper Eocene time

E. Late Miocene time

F. Present time

FIGURE 23–6 Development of the Rocky Mountains. The Medicine Bow Mountains shown here are typical of the Rocky Mountains in Wyoming. In early Cenozoic time deposition and deformation occurred at the same time (Parts A, B, and C). In late Cenozoic times the basins were filled to the level of the now-eroded highlands, and the rivers spread sediments in both the area of the present Rocky Mountains and the Great Plains (E). In very late Cenozoic time uplift or climatic change caused the rivers to erode and downcut (F). The rivers removed much of the soft basin-filling sediments and in many cases cut deep canyons across previously buried ranges, producing the present mountain topography. From S. H. Knight, Wyoming Geological Association, 1953.

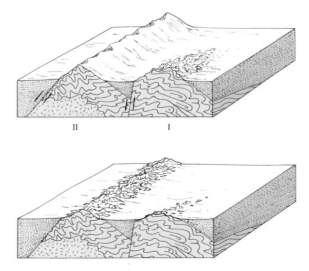

FIGURE 23–7 Typical Basin and Range structure. Both ranges I and II are uplifted fault blocks with intervening basins filled by detritus eroded from the ranges. Range II is younger than range I. The rocks in both ranges were deformed prior to the present uplift. The lower view shows the ranges at a later time.

north-south axis that cuts across the earlier structures. The last step was the formation of the present volcanoes, such as Mt. Shasta, Mt. Hood, Mt. Rainier, Mt. St. Helens and Mt. Baker, in late Pliocene and Pleistocene time. (See Fig. 23–11.)

CALIFORNIA COAST

A brief description of the Cenozoic history of California will, together with the outline of the northwest, give an indication of the types of events that occurred along the Pacific coast. This was an extremely active area throughout the Cenozoic, and a detailed history of the Pacific coast would be very long. In Paleocene and Eocene time, a sea occupied the central and southern part of western California. Eastern California was an upland, and its erosion shed sediments into the sea. Nonmarine clastics were deposited between the sea and the upland. In the area of the present Coast Ranges, several large islands existed. At places, up to 3050 meters (10,000 feet) of clastics were deposited in this sea.

During the Oligocene, nonmarine clastics and red beds were deposited inland with some marine rocks nearer to the coast. The seas returned in the Miocene, and there was much volcanic activity. Orogeny also occurred at this time. In Pliocene many basins developed, and some of these were nonmarine lakes. In the Los Angeles Basin, 4575 meters (15,000 feet) of sediments were deposited. Deposition continued into the Pleistocene. The San Andreas fault, which appears to be a transform fault, was probably active during much of the Cenozoic. In Middle Pleistocene, a major deformation occurred. At this time, the Coast Range mountains were uplifted. Similar uplifts occurred all the way to Washington. As a result of this activity, the Pacific coast has mountains along the coast (the Coast Ranges and Olympic Mountains), an inland valley (the Great Valley in California, Willamette Valley-Puget Sound in Oregon and Washington), and a mountain range (Sierra Nevada and Cascade Range). The only exceptions are southern Washington where the coastal mountains are low hills, the Oregon-California border where there is no valley, and southern California.

FIGURE 23–8   The eastern fault scarp of the Sierra Nevada. The fault is covered by the alluvial fans. In the foreground is Owens River. Owens Valley was the site of a lake that is now dry, and some of the shoreline features can be seen. Photo from U.S. Geological Survey.

FIGURE 23–9   Aerial view of the west slope of the Sierra Nevada, showing Yosemite Valley. Note the relatively smooth surface that was uplifted by the fault whose scarp is shown in Fig. 23–8. The uplifted surface is incised by rivers whose valleys have been modified by glaciation, producing features such as Yosemite Valley. At other places, especially on the skyline, erosion has produced rugged mountain topography. Photo from U.S. Geological Survey.

ARCTIC

The Cenozoic rocks of the Arctic Islands are mainly nonmarine sandstone and shale with some coal. These rocks were apparently deposited in basins, and although most occurrences are only a few hundred feet thick, at one place they are 2135 meters (7000 feet) thick. These rocks were folded sometime in the Cenozoic.

Until the Cenozoic, the Arctic Islands and Greenland were part of the North American continent. The area apparently sunk to form the islands, as the waterways among the islands may be, at least in part, a drowned river system.

PLEISTOCENE

The Pleistocene, although generally thought of as the glacial age, is defined by a type section like other periods. Glaciers occupied only a small part of the earth during the ice age, and they formed and melted at different times at different places. The Pleistocene is generally considered to have begun about one million years ago. Recent radioactive age determinations, however, indicate that glaciers began to form in Iceland about three million years ago.

Much of the Pleistocene history of North America was discussed in a general way in the chapter on glaciation, and this should be referred to at this time. The extent of the glacial advances is shown in Fig. 10–24.

The Pleistocene history of North America is a study worthy of a much longer book than this. Only an outline of events can be presented here because almost all of

the present landforms were made or greatly modified during the Pleistocene. In Chapter 10 the geologic work of glaciers was described and the possible causes of glaciation were discussed. Here, only the effects of glaciation will be mentioned.

The principles involved in Pleistocene geology as outlined here are:

1. Erosion and deposition by alpine and continental glaciers.
2. Drainage changes caused by glaciers, which include damming of rivers by the advance of the glaciers.
3. The effects of the generally rapidly released meltwater from retreating glaciers.
4. The effects of the weight of glacial ice lying on the continents.
5. The effects of the glacial climate in areas away from the ice.
6. The normal nonglacial geological processes, such as erosion, deposition, and orogeny, that have been discussed in the history of the earlier geologic periods.

The most obvious effects of Pleistocene glaciation are seen in the areas actually glaciated. The continental glaciers covered much of the northern part of the United States. The glaciers made four major advances, and many of these advances had one or more smaller advances and retreats. The details of these oscillations have been deciphered by study of the moraines, as shown in Chapter 10. In north-central United States, the long, low morainal hills marking the farthest advance of the last glacier form conspicuous features. The soils of this area and of the midwest are largely glacial deposits scraped from the now almost bare areas to the north in Canada. In New England and New York, the low mountains were covered by the continental ice. The main effects were a rounding and smoothing of the mountains.

FIGURE 23–10   The Columbia River Basalt flows in eastern Washington. Photo by F. O. Jones, U.S. Geological Survey.

FIGURE 23–11   Oblique aerial view of Mount Hood in northern Oregon. From northern California through Washington, the high Cascade volcanoes formed on top of a mountain range composed of deformed Cenozoic rocks. Photo from U.S. Geological Survey.

486

Much of the material eroded by the glaciers was deposited nearby, producing the rocky soils and fields of this region. As in other places, the farthest advance of the glacier is marked by terminal moraines, of which Long Island is the most prominent.

At other places, rivers as well as moraines mark the limit of glaciation. In the midwest, the Ohio River follows the approximate margin of the glacial ice. This river probably developed to carry the meltwater from the margin of the receding ice sheets to the Mississippi River. In the northern Great Plains in Montana, the Missouri River was diverted by the glacial ice. This river probably flowed northeastward toward Hudson Bay in preglacial time. The presence of the glacier forced it into a westerly and southwesterly course to join the Mississippi River. Much of the present course of the Missouri River is within a few tens of miles of the limit of the glacial ice. (See Fig. 10–24.)

Mountain glaciers developed in the high mountains of western United States. These mountain glaciers sculptured the superb scenery of many of our ranges, such as at Yosemite National Park. (See Fig. 10–14.) At some places, so much snow accumulated on the mountains that only the highest peaks were not covered. At places, too, glacial tongues extended some distance out from the mountains. Generally this occurred where several coalescing glaciers extended into the lowland.

The area between the Cascade Mountains and the Northern Rocky Mountains was the scene of a remarkable event in glacial times. A wide path across the Columbia Plateau has been eroded by much rapidly flowing water. This is shown by the stripping off of most of the loose surficial material, eroded stream channels, and large sand bars. (See Fig. 23–12.) This erosion is believed to have been accomplished by glacial meltwater, but obviously

some remarkable events must have occurred to channel enough water through the area to do this work. As the glacier retreated in the northern Rocky Mountains in Montana, meltwater filled the northward-sloping valleys, forming a large, irregularly shaped lake. The glacier to the north dammed this lake. When the water level of the lake overtopped the ice dam, the flowing water eroded the ice dam very rapidly. This is believed to be the source of the rapidly flowing, large amount of water. This water flowed across Idaho and met the Columbia Plateau near Spokane. The Channeled Scablands, as the area eroded in this unusual way is called, extend south from Spokane to Grand Coulee Dam. Apparently many floods of water reached the Columbia River in this way as the ice dam was reformed by forward-moving glacial ice and

destroyed by the overtopping meltwaters again and again. Remember that even in a retreating glacier the ice moves forward. These floodwaters could not follow the Columbia River Valley because the Columbia was also dammed by glacial lobes extending south from the mountains north of the Columbia Plateau. These latter glacial lobes apparently blocked the Columbia River at different places as the various floods reached the Columbia, and they diverted the floods southwesterly across the Columbia Plateau. The diverted floodwaters eroded the scabland and rejoined the Columbia River in southern Washington. (See Fig. 23–13.)

The climatic changes that caused glaciation in high latitudes produced aridity near the equator. The evidence for this is both the type of sediment deposited in the oceans in low latitudes, and sand

FIGURE 23–12 Looking east across the Channeled Scablands in southeastern Washington toward farmlands where the soil was not removed by the glacial flood waters. Compare the many large-scale features of river erosion in the aptly named Channeled Scablands with the smooth farmlands. Photo by Bureau of Reclamation.

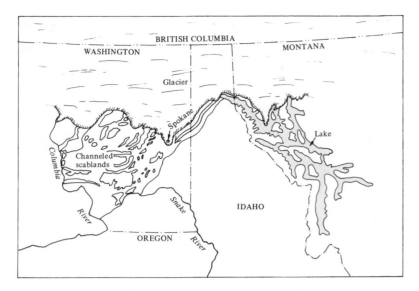

FIGURE 23-13   The development of the Channeled Scablands in eastern Washington. The lake formed from glacial meltwater in the north-sloping valleys in Montana as the glacier retreated. When the glacial ice that dammed the lake was removed by melting, or erosion by water that overtopped it, the lake drained rapidly along a broad valley south of the continental glacier. The water crossed the Columbia Plateau, eroding the Channeled Scablands, and then joined the Columbia River. Based on maps by J. H. Bretz, in Geological Society of America, *Bulletin,* Vol. 67, 1956.

dunes in both northwestern Australia and west Africa that pass under the sea on parts of the shelf that were above sea level when glaciers occupied the northern lands. Nearer the glaciers, the climate was wet.

In southwestern United States during glacial times, many large lakes, such as Great Salt Lake, formed in this region of interior drainage. These lakes were mentioned in Chapter 10.

The Great Lakes of northern United States were also formed during the Pleistocene, but their origin was very different from that of Great Salt Lake. As long as the front of the continental glacier was south of the divide that separates the drainage of the Mississippi River from the St. Lawrence River, the meltwater all flowed out the Mississippi River. When the glacier retreated north of this divide, lakes were impounded between the glacier and the divide. As the water level rose and the glacier retreated, a number of different outlets were uncovered. If nothing else had happened, the lakes

would all have been drained when the glacier melted away. However, the melting of the glacier reduced the weight pressing down on the crust, and the crust began to rise, perhaps isostatically. This uplift acted as the glacier had to impound the Great Lakes. The development of these lakes is known in great detail as a result of much study. When the ice first melted from this area, the crust of the earth had been depressed by the weight of the ice so that it was below sea level, and marine condi-

tions prevailed in part of the St. Lawrence Valley until the crustal rise forced the seas out. (See Fig. 23-14.) Other lakes also formed in this region but were drained when the ice retreated and the crust rose. Lake Winnepeg is a small remnant of one of these other glacial lakes.

LIFE OF THE CENOZOIC

In the Cenozoic all types of life develop into the modern forms, making their study especially interesting. It was an active time, with much mountain making and climatic change. As a result, the changes in life, especially on the land, are great.

In the ocean, the biggest change is the absence of some groups of cephalopods that were so abundant in the Mesozoic. Clams (see Fig. 23-15), oysters, and snails (see Fig. 23-16) were the main elements of Cenozoic seas, along with echinoids (see Fig. 23-17), bryozoans, and foraminifera. The foraminifera underwent a new development and were very widespread. Because they are small enough to be recovered intact in drill cuttings, they have been used extensively to date the rocks encountered when drilling for oil. Limestones composed of large foraminifera were deposited in Eurasia.

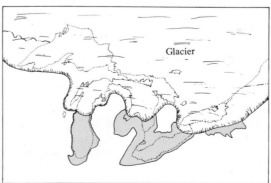

FIGURE 23-14   An early stage in the development of the Great Lakes. The glacier has melted back, exposing large valleys that slope northward. Meltwaters have filled the valleys that are dammed by the glacial ice.

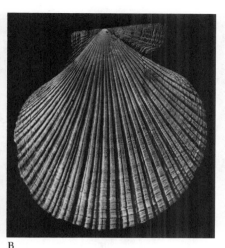

FIGURE 23–16   A Pleistocene snail from Quebec, *Natica clausa.* Photo courtesy Geological Survey of Canada (113093-B).

FIGURE 23–15   Clams. A. *Clinocardium nuttalli,* Pleistocene from British Columbia. B. *Chlamys hindsi,* Pleistocene from British Columbia. Photos courtesy Geological Survey of Canada (113556-N and -L).

The plant life of the Cenozoic was essentially modern. (See Fig. 23–18.) The first fossil angiosperms are found in Jurassic rocks, but such fossils are relatively rare. In Early Cretaceous time, well-differentiated angiosperms became abundant, showing that they had developed somewhere. By the end of the Cretaceous, they replaced the Mesozoic plants. Their radiation continued throughout the Cenozoic.

The Cenozoic is the age of mammals. Mammals have many advantages over reptiles. They have a more efficient heart, and hair, making them warm-blooded. Their brain is larger, and the senses, such as smell and especially hearing, are better. Specialized teeth and jaws improve feeding, and the digestive system does not require inactive periods after eating as reptiles require. Teeth are easily preserved and are excellent mammal fossils. Growth of the young within the mother and a long period of nursing make the young more likely to survive. Just as the reptiles expanded to all habitats in the Mesozoic, so did the mammals in the Cenozoic. Bats are flying mammals; seals, whales, and porpoises are marine mammals.

There are two main types of mammals: the marsupials, such as the kangaroo and opossum, that carry their young in an external pouch; and the placental mammals that carry their young internally. There is one other type of mammal, monotremes, living in Australia. Only two of these are known: the duck-billed platypus and the spiny anteater. They have hair and nurse their young but they lay eggs. There is almost no fossil record of these animals, so they may be "living fossils," almost unchanged from the Cretaceous.

The oldest mammal fossils are small shrewlike skulls found in Upper Triassic rocks. Very few fossils are found until the Late Cretaceous. The largest Mesozoic mammals were the size of house cats. The Cenozoic placental mammals evolved from small, Late Cretaceous insectivores.

The marsupials were much less successful than the placentals. It is

FIGURE 23–17   Echinoid *Eupatagus floridanus,* Eocene, Ocala limestone, Florida. Photo from Ward's Natural Science Establishment, Inc., Rochester, N.Y.

489

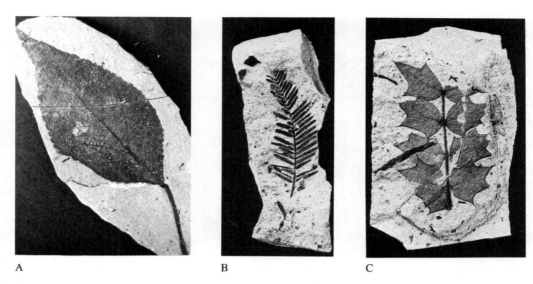

FIGURE 23-18   Cretaceous and later leaf fossils have a modern aspect. A. *Alnus larseni,* Creede, Colorado. B. *Metasequoia,* Collawash River, Oregon. C. *Mahonia marginata,* Creede, Colorado. All are Miocene. Photos from Smithsonian Institution.

estimated that only about five percent of all Cenozoic mammals were marsupials. The oldest marsupial fossils are middle Cretaceous in age. The marsupials appear to have migrated to Australia and South America in the Late Cretaceous and then became isolated. As a result, the marsupials there did not have to compete with the placentals. They evolved into many types and occupied many habitats. Australia is noted for its marsupial fauna. The same thing happened in South America, but here many of the marsupials became extinct at the end of the Pliocene. This occurred because about this time, the Isthmus of Panama formed, and the placentals from North America migrated to South America. The marsupials could not compete with the placentals. The Cenozoic barriers and migrations are shown in Fig. 23-19. During most of the Cenozoic, there was some connection between Eurasia and North America.

In North America, the Paleocene was warm and temperate. The Paleocene mammals were archaic, and most were small. They lived in forests and near streams. Pri-

mates, rodents, insectivores, carnivores, and browsers were present. (See Fig. 23-20.) The browsers had hoofs and five toes. European and North American forms were similar, but the South American types were different. The first horses appeared in the Paleocene.

As the Eocene opened, the climate was subtropical and the animals were largely forest dwellers. All of the modern orders of mammals have been found, although the species are different from those of the present. Rhinoceroses were present, and in the late Eocene, the deer, pig, and camel appeared. (See Fig. 23-21.) The largest land animal was the rhinoceroslike *Uintatherium.* Toothed whales appeared in the sea, indicating that this habitat was soon occupied after the extinction of the marine reptiles. (See Fig. 23-22.) In late Eocene time, more grasslands appeared as the climate became somewhat cooler and drier.

The Oligocene had a mild, temperate climate in North America. The Great Plains was a large floodplain with many rivers and some forests. Figure 23-23 shows

the differences in the climate and vegetation. The archaic forms disappeared, and the fauna took on a more modern aspect. The drier climate favored the faster, long-legged, hoofed grazers and browsers. Cats and dogs appeared. The titanotheres, which had begun in early Eocene as small rhinoceroslike animals, reached the size of elephants in the Oligocene and then died out. *Brontotherium* was the largest mammal ever to live in North America.

The drying and cooling trend continued into the Miocene in North America, and the grasslands expanded. This favored the rapid evolution of grazers such as horses, pigs, rhinoceroses, camels, antelopes, deer, and mastodons. (See Fig. 23-24.) The large cats, bears, and weasels also appeared.

In the Pliocene, the climate cooled, remaining relatively dry. As a result of these changes, evolution continued rapidly. Many forms were weeded out, and those remaining became more specialized. The trend continued into the Pleistocene and produced

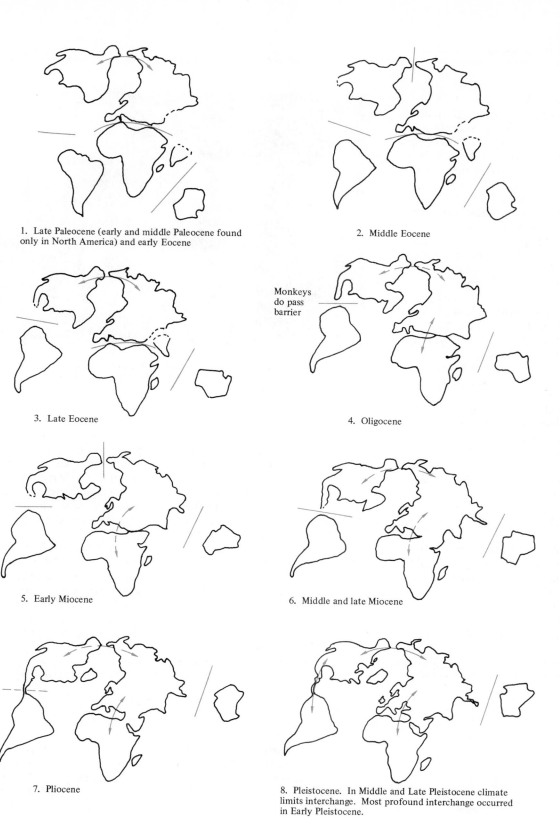

1. Late Paleocene (early and middle Paleocene found only in North America) and early Eocene

2. Middle Eocene

3. Late Eocene

Monkeys do pass barrier

4. Oligocene

5. Early Miocene

6. Middle and late Miocene

7. Pliocene

8. Pleistocene. In Middle and Late Pleistocene climate limits interchange. Most profound interchange occurred in Early Pleistocene.

FIGURE 23–19 Cenozoic vertebrate migrations. Barriers and migration routes are indicated diagrammatically. Continental drift and changing sea level apparently caused many of the barriers and connections.

our present life. Figure 23–25 shows a scene on the high plains. Many of the animals resemble those in the area today. The differences are conspicuous, such as the shovel-tusked mastodon and the giant camel.

The Pleistocene life of North America was dominated by large animals such as mastodons, mammoths, ground sloths, saber-toothed cats, bears, and giant beavers, as well as horses and camels. (See Figs. 23–26 and 23–27.) This was a time of great climatic change everywhere. Most of the animals just listed died out in North America about 8000 years ago, when the last glacial stage was retreating. No one knows why this great extinction occurred, but hunting by early humans has been suggested. It was not the climatic change because they lived through four glacial advances. The present-day fauna of the African plains is somewhat similar to the late Cenozoic of North America.

HUMAN FOSSILS

The development of humans, the most successful of the mammals, is relatively poorly known. The humanlike primates apparently evolved from the apes, but lack of fossils obscures the details. This lack of fossils in part reflects the

FIGURE 23–20 Paleocene scene. *Barylambda,* a primitive hoofed animal with five toes. Western Colorado. About 2.5 meters (8 feet) long. Painting by J. C. Hansen, Field Museum of Natural History.

cleverness of these animals in avoiding accidental deaths, and in part results from the fact that the apes evolved in the narrow transitional environment between the grasslands and the jungle or forest. This environment probably accounts for the development of two-legged locomotion, which freed the hands for other activities.

Fossil humans are about the rarest fossils. In the past, each new discovery of a bone (complete skeletons are extremely rare)

was given a separate name because of disagreement as to where each fossil fits into the evolution of modern humans. Some of the more important early discoveries and their more recent classification are shown in Table 23–1.

The primates, like other mammals, evolved from small insectivores in the Late Cretaceous. They adapted to life in the trees. The most important adaptations were the development of an opposable thumb, enabling grasping and

TABLE 23–1 Some of the more important early discoveries of fossil humans and their classification.

Date	Place	Fossils	Name	Recent Classification
1848	Gibraltar	Skull	Neanderthal	*Homo sapiens*
1856	Germany	Skull and other bones	Neanderthal	*Homo sapiens*
1868	France	Five partial skeletons	Crô-Magnon	*Homo sapiens*
1894	Java	Skull cap, tooth, femur	Pithecanthropus	*Homo erectus*
1907	Heidelberg	Two lower jaws with teeth	*Homo heidelbergensis*	*Homo erectus*
1924	South Africa	Skull of child	*Australopithecus*	*Australopithecus*
1921	Rhodesia	Skull	*Homo rhodesiensis*	*Homo erectus?*
1929	Peking	Skull and molars	Pithecanthropus	*Homo erectus*
1931–1932	Solo River, Java	Several skulls	*Homo soloensis*	*Homo erectus?*

Part 1

Part 2

FIGURE 23–21 Middle Eocene Bridger Formation of Wyoming. Although the plants and animals are reconstructed as accurately as possible, all of these creatures were probably never so close together. Painting by J. H. Matternes, Smithsonian Institution.

493

FIGURE 23-22   Eocene whale, *Basilosaurus cetoides,* from Alabama. These toothed mammals quickly took the place of the Mesozoic marine reptiles. The jaws are 1 meter (3 feet) long. Painting by C. R. Knight, Field Museum of Natural History.

swinging from tree limbs, and the shortening of the face, moving the eyes forward for stereoscopic vision. The latter enabled better judgment of distance—a necessity for a climbing animal. The size of the brain also increased. The higher apes came from the trees to the ground, perhaps because the cooling trend caused the forests to be replaced by grasslands. The gorilla and the chimpanzee live largely on the ground.

The oldest apes are early Oligocene, although other primates are found throughout the Cenozoic. Apes appear to have evolved rather rapidly in the Miocene. One, *Ramapithecus,* the oldest tool user, appears to show the jaw distinctions leading to humans rather than to the other apes; it has been dated as Miocene. (See Fig. 23–28.)

Australopithecus is found with crude stone and bone tools and the remains of increasingly larger animals that it apparently hunted. *Australopithecus* first appears in rocks as old as 3.5 million years. "Lucy," a 40-percent-complete skeleton, was found in 1974, and

in 1975, 13 individuals, "the first family," were also found near Hadar, Ethiopia. Early *Australopithecus* was small but evolved into a much larger, more human-like creature; even the earliest forms walked erect—the first known to do so. The early discoveries led to great confusion, for *Australopithecus* existed together with *Paranthropus,* a larger, more primitive-appearing, erect-walking vegetarian, who apparently evolved little over the same span of time—nearly a million years—until it disappeared.

Olduvai Gorge, Tanzania, is a unique site in which a stratigraphic record going back 2 million years is exposed; its beds have been dated by the potassium-argon method. The lower and middle beds have revealed evolving *Australopithecus,* and the middle and upper beds its contemporary, the primitive *Paranthropus. Homo erectus* has been found in beds dated about a million years ago. (See Fig. 23–29.)

Homo erectus fossils are generally associated with stone tools, and some finds show evidence of

the use of fire. The general characteristics used to distinguish *Homo erectus* are a thick, flat, narrow skull, with an average brain size of 800 to 1000 cubic centimeters (compared with averages of 1400 cubic centimeters in modern humans and 500 cubic centimeters in *Australopithecus* and a modern gorilla); protruding brow ridges; a marked angle for muscle attachment at the rear of the skull; and larger and more primitive teeth and jaw than a modern human's. The legs are close to those of modern humans, unlike the more primitive upright, *Australopithecus.* (See Fig. 23–30.)

Homo erectus has been found in Java, China, and South and East Africa. Tools similar to those associated with those finds have been found in Europe, but so far no fossils, with the exception of a jaw found near Heidelberg. This jaw is dated in the generally accepted mid-range of *Homo erectus,* but it is less primitive than the usual *Homo erectus* fossils, and skull portions which might give more positive assignment as to species are absent; its true position is greatly in doubt. (See Fig. 23–31.)

Although we are now approaching modern humans and the record should begin to clear, it actually does not; there are too many anomalies in the idealized picture. *Homo sapiens* is the species to which modern humans belong. When and where *Homo sapiens* first appeared is now the question, and the answer is ambiguous. Some fossils indicate that *Homo erectus* and *Homo sapiens* may have overlapped considerably in time, or possibly that the progress of evolution of *Homo erectus* varied greatly over its tremendous geographical range, affected locally by climatic conditions and migratory limitations.

Neanderthal fossils occur near the boundary between middle and

Part 1

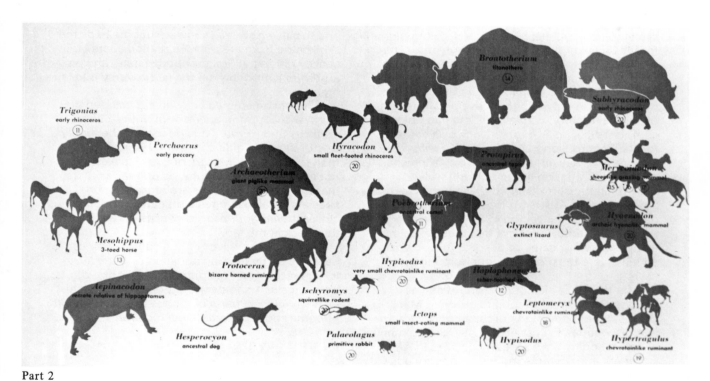

Part 2

FIGURE 23–23 Oligocene of the White River Formation of South Dakota and Nebraska. Painting by J. H. Matternes, Smithsonian Institution.

495

LUCY IN THE SKY WITH DIAMONDS

The only fossils to receive more attention than dinosaurs are fossil hominids, probably because we are fascinated with our distant roots as well as our more recent ones. Each new discovery of a few bones or bone fragments results in new theories or a new hominid family tree. The search for human fossils is very expensive because expeditions to remote areas in Africa are required. Money for such research is always limited, so there is great competition for the funds. The discoverers of new fossils generally become famous, and they are the ones who get the money for expeditions. This situation has produced great competition among the paleoanthropologists, as they call themselves. The legitimate disagreements in the interpretation of the few fossils, together with the competition, have led to widely publicized controversies. One such controversy has developed around a remarkable discovery.

The discovery, made in 1974, was "Lucy," a skeleton about 40 percent complete. Hominid fossils this complete are exceedingly rare. The following year, a huge collection of bones was made in the same area. The discoveries were made by Donald C. Johanson of Case Western Reserve University and the Cleveland Museum of Natural History, who at the time had not yet completed his Ph.D. degree. Lucy was named for the Beatles' song *Lucy in the Sky with Diamonds.* She was about 20 years old when she died, was about one meter tall, and walked upright. Johanson assigned these fossils to a new species, *Australopithecus afarensis,* named for the Afar region of Ethiopia where they were found. The rocks that these fossils came from were dated at over 3.5 million years, so they were the oldest upright hominids yet found.

On the other side of the controversy is Richard E. Leakey, the son of Louis and Mary Leakey, who made many of the important hominid discoveries in East Africa. As a young man he left school to start a safari business in Africa, and then, after completing his secondary education, became an anthropologist like his father. Although Leakey has no college degrees, he is one of the foremost paleoanthropologists in the world. He and Johanson are the same age, but Leakey is considered the elder statesman of the field. However, this does not mean that his opinions are universally accepted.

The controversy concerns when the earliest humans *(Homo)* developed. Louis Leakey, the father, was a very strong personality, and he dominated the field for many years. He believed that humans emerged very long ago. Richard Leakey, the son, has the same opinion, and he feels that *Homo* diverged from *Australopithecus* 5 to 8 million years ago. Both, of course, coexisted. Johanson believes the separation came 2 to 3 million years ago.

The controversy heated up in the middle 1970s as the result of the dating of a skull discovered by one of Leakey's associates. The find was made in 1972, and Leakey assigned it to *Homo habilis* (*H. habilis* means "handyman" and was so named because of tools found in association with the fossils). Leakey thought that the skull was about 3 million years old and so fit his theory. Later dating using radioactive and fossil vertebrates showed that the beds the skull came from are only 1.8 million years old. Leakey did not at first accept this date, and his arguments probably set the controversy into high gear.

Johanson studied Lucy and concluded that she was a common ancestor of both *Australopithecus* and *Homo.* Thus he redrew the family tree of the hominids and made *Homo* younger than the 3.5 million years of Lucy. Leakey could not accept this and argues, as do many others, that the Ethiopian fossils are more than one species. He also suggests that Lucy is much closer to *Ramapithecus,* an ancestor much further back in the family tree.

Some of the flavor of this controversy can be sampled by reading Johanson's book *Lucy.* It is a fascinating but controversial book, with both excellent and poor features. In a book review in *Science,* parts of it were compared to *60 Minutes* and an *International Inquirer.*

REFERENCES

Hay, Richard L., and Mary D. Leakey, "The Fossil Footprints of Laetoli," *Scientific American* (February 1982), Vol. 246, No. 2, pp. 50–57.

Holden, Constance, "The Politics of Paleoanthropology," *Science* (August 14, 1981), Vol. 213, No. 4509, pp. 737–740.

Johansen, Donald C., and Maitland A. Edey, *Lucy: The Beginnings of Humankind.* New York: Simon and Schuster, 1981, 410 pp.

Johanson, Donald C., and Maitland A. Edey, "Lucy," *Science '81* (March 1981), Vol. 2, No. 2, pp. 44–55. (Excerpted from their book *Lucy.*)

Johanson, Donald C., and Maitland A. Edey, "How Ape Became Man: Is It a Matter of Sex?" *Science '81* (April 1981), Vol. 2, No. 3, pp. 45–49. (Adapted from their book *Lucy.*)

Tuttle, Russell H., "Paleoanthropology Without Inhibitions," *Science* (May 15, 1981), Vol. 212, No. 4496, p. 798. (A review of Johanson and Edey's book *Lucy.*)

Part 1

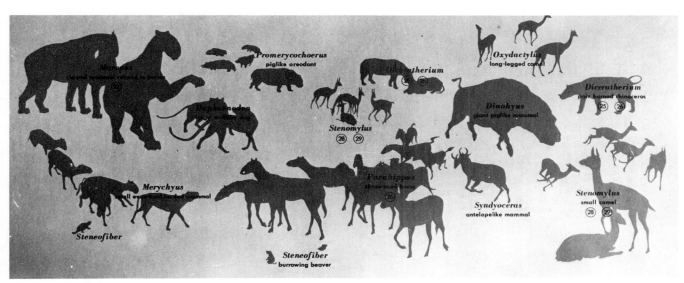

Part 2

FIGURE 23–24 Lower Miocene of the Harrison Formation of western Nebraska. Painting by J. H. Matternes, Smithsonian Institution.

Part 1

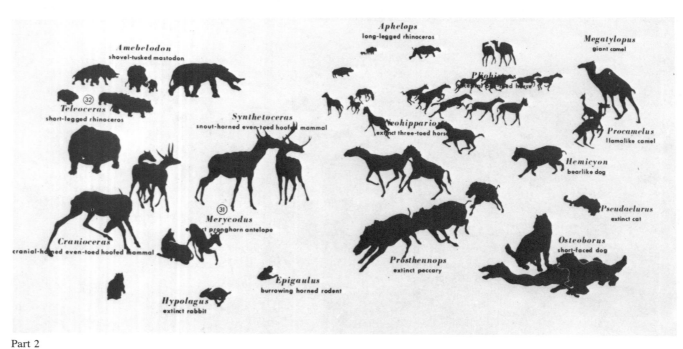

Amebelodon
shovel-tusked mastodon

Teleoceras
short-legged rhinoceros

Synthetoceras
snout-horned even-toed hoofed mammal

Merycodus
extinct pronghorn antelope

Cranioceras
cranial-horned even-toed hoofed mammal

Hypolagus
extinct rabbit

Epigaulus
burrowing horned rodent

Aphelops
long-legged rhinoceros

Pliohippus
ancestor of one-toed horse

Neohipparion
extinct three-toed horse

Prosthennops
extinct peccary

Megatylopus
giant camel

Procamelus
llamalike camel

Hemicyon
bearlike dog

Pseudaelurus
extinct cat

Osteoborus
short-faced dog

Part 2

FIGURE 23–25   Early Pliocene of the high plains region. Painting by J. H. Matternes, Smithsonian Institution.

498

FIGURE 23–26   Pleistocene scene at La Brea tar seeps near Los Angeles, California. Saber-toothed cat, wolf, and vultures have gathered to feed on animals caught in the tar. Painting by C. R. Knight, Field Museum of Natural History.

FIGURE 23–27   Pleistocene scene in cool climate. Wooly mammoth and rhinoceros. Painting by C. R. Knight, Field Museum of Natural History.

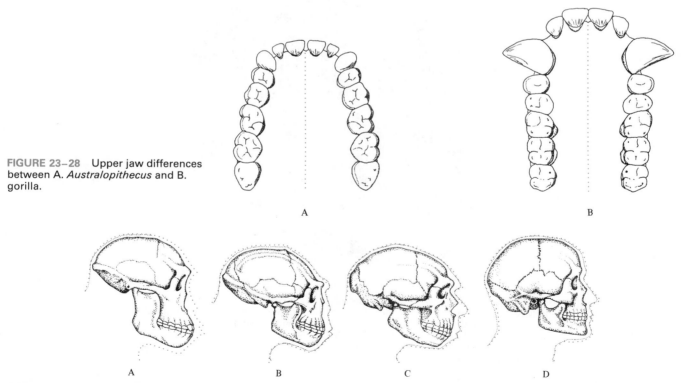

FIGURE 23–28   Upper jaw differences between A. *Australopithecus* and B. gorilla.

A   B

A   B   C   D

FIGURE 23–29   Comparison of skulls of A. *Australopithecus,* B. *Homo erectus,* C. *Homo sapiens neanderthalensis,* and D. *Homo sapiens.*

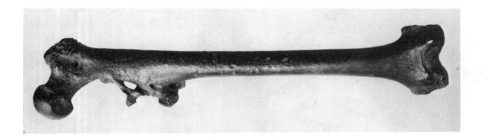

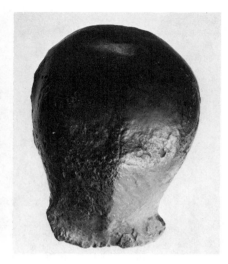

A

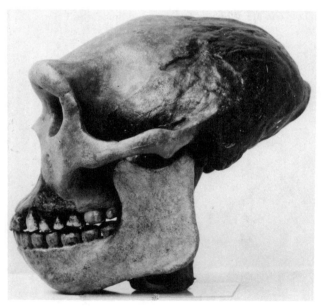

B

FIGURE 23–30   *Homo erectus.* A. Typical fossils—teeth, femur, and skull cap.
B. Restoration of skull by Weidenreich. Note the thick, flat skull with protruding brow
ridges and a ridge at the back. The jaw has a receding chin, and the teeth are
somewhat primitive. Compare with your skull by running your fingers above and
beside your eyes and over the back of your skull and over your chin. Photos from
Smithsonian Institution.

500

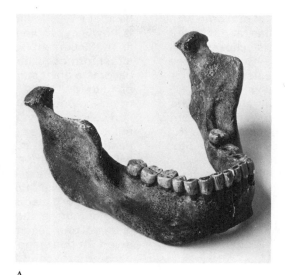

A

B

FIGURE 23–31   The Heidelberg jaw (*Homo erectus*). A. A cast of the fossil. B. The restoration by McGregor, based on the jaw. Photos from Smithsonian Institution.

upper Pleistocene, and they are now classified as *Homo sapiens.* Extensive remains have been found in Europe and around the Mediterranean, dated from nearly 100,000 years ago until about 35,000 years ago. Neanderthal appears to have developed considerable variation. Those inhabiting western Europe, the best and earliest known, are the most different from modern humans in limb and skull shape; and as time progressed they seem to have developed further from, instead of closer to, modern humans until their disappearance. This seeming "regression" may simply represent an isolated race's adaptation to difficult climatic conditions. Finds in the Middle East indicate great variety and possible transitional forms between those with the extreme western-European characteristics and Crô-Magnon, the oldest modern human in the strict sense. This variety exists even in contemporaneous fossils in single locations.

Modern humans emerged approximately 35,000 years ago, and the older forms disappeared. The most studied and best known is Crô-Magnon, whose remains are nearly indistinguishable from modern humans'. Crô-Magnon appears suddenly in Europe in beds of about the same age as late Neanderthal, and then Neanderthal disappears. Apparently these people migrated into Europe, probably from the Middle East, considering the possibly transitional forms recently found there. An important problem is whether Crô-Magnon evolved from Neanderthal or simply first intermingled with them in the Middle East and evolved from some other stock. Crô-Magnon is presumed to be the ancestor of the modern European races. Nothing is known of the recent ancestors of the modern races of humans outside Europe.

The oldest human fossils found in the New World are 11,000 to 13,000 years old on the Palouse River in Washington, and 11,000 to 12,000 years old at Tepexpan, in Mexico. Sites as old as 15,000 years in Alaska and South America, and up to 20,000 years in Mexico, are known from artifacts; but these sites contain no human bones. Many anthropologists believe that human evolution can be described better from artifacts than from bones. (See Fig. 23–32.) Humans must have reached the New World over the Bering land bridge between Siberia and Alaska. They must have crossed the bridge about 35,000 years ago because the bridge was submerged between 35,000 years and 25,000 years ago, and at 25,000 years ago the corridor between the glacial ice sheets was closed by the meeting of the glaciers. Thus the search will continue for early humans in the New World.

THE GEOLOGIC FUTURE

One attribute of science is its ability to predict. A physicist uses laws to predict the outcome when two masses collide, and the astronomer predicts eclipses many years in advance. Geologists can predict future geological events only in a general way.

To find the ultimate fate of the earth, we must look to astronomy, for it is the sun that gives life to the earth. The sun is believed to be about 5500 million years old. It is slowly expanding and is now about 20 percent larger than when it first formed. Ultimately it will become a giant star and will expand to Mercury's orbit. This will occur in about 3000 to 4000 million years. However, in about 2000 million years, the sun will have expanded so much that its radiation will increase temperatures on the earth so that the oceans will boil. It seems unlikely that life, at least as we know it, can survive. After the giant phase, the sun will shrink and cool and become a white dwarf. At this stage, the earth will become very cold, again seemingly precluding any life. These predictions, of course, assume that the sun will not collide with another star, an event with a very small probability. Another possible interruption would be the collision of an asteroid with the earth, but, again, this is an unlikely event.

The moon will continue to recede, making both the day and the month longer. Much more must be known about the earth-moon system to make any more specific predictions. At present, astronomers differ greatly on this question.

On a shorter time span, but still very long by human life spans, other predictions can be made based on principles described earlier. The continents will probably be larger because they have apparently grown during geologic time. The configuration of the continents and their positions on the earth will change. As more is learned about sea-floor spreading, these changes will be predicted by future geologists. If radioactivity is an important source of energy for geologic processes, these processes will slow as more and more of our radioactive elements decay. In a much shorter time, erosion and sedimentation will modify the earth's surface in ways fairly easy to predict. This is about as far as the geologist can predict the future.

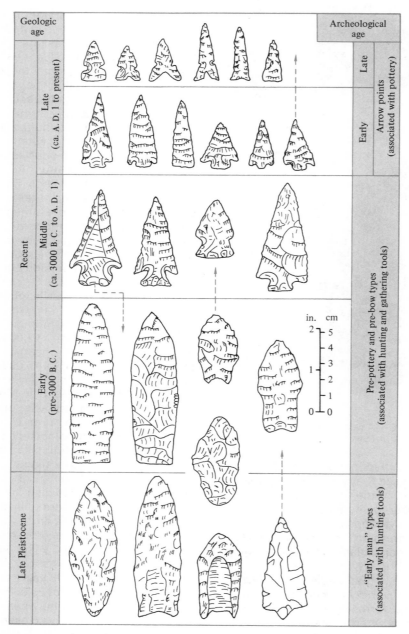

FIGURE 23–32 Ages of projectile points of western United States. Such artifacts are human fossils. From Hunt and Hunt, 1962 (after Krieger, 1950), U.S. Geological Survey Professional Paper 424B.

In the near future, we must learn to use our earth more judiciously. Natural geologic hazards must be avoided or their effects minimized, and human-made hazards eliminated. Better use must be made of our mineral resources, reducing the economic and social changes that will accompany depletion of cheap metal and energy deposits. The lessons of geologic history must be understood. The earth is always changing, and uniformitarianism shows that slow processes can cause great changes given geologic lengths of time. The biologic world is also in constant change, as shown by evolution and extinction, and these processes seem to have occurred in response to physical and climatic changes caused by natural geologic processes. Humans have modified the earth greatly during their short span, suggesting that we must learn to use the earth wisely.

In the West, our desire to conquer nature often means simply that we diminish the probability of small inconveniences at the cost of increasing the probability of very large disasters.

Kenneth E. Boulding
Human Values on the Spaceship Earth *1966*

SUMMARY

During the Cenozoic, the continents continued their movements to the present locations. India collided with Asia, forming the Himalaya Mountains.

Marine deposition occurred on the Atlantic and Gulf Coasts, with repeated advances and retreats of the seas. Up to 15,250 meters (50,000 feet) of clastics were deposited on the Gulf Coast.

Limestone was deposited in Florida.

Erosion surfaces reveal the events in the Appalachian Mountains. Coastal plain deposition was on a pre-Cretaceous peneplain. Uplift caused erosion to produce another peneplain. Later uplifts caused dissection.

Orogeny in the Rocky Mountains in late Cretaceous and early Cenozoic time produced mountains and basins. Erosion reduced the ranges and filled the basins by the Oligocene. In Oligocene and Pliocene time, rivers spread thin clastics over the plains area. Uplift or climatic change caused the rivers to downcut, producing the present topography.

During the Eocene, a fresh-water lake in which oil shale was deposited covered much of Utah.

The Colorado Plateau was uplifted but otherwise was not greatly deformed during the Cenozoic, even though the surrounding areas were severely deformed.

Faulting in the middle and late Cenozoic produced the Basin and Range topography of the Great Basin.

Northwestern United States is dominated by volcanic rocks. Marine clastics near the coast grade inland with volcanics and nonmarine rocks in the Cascade Mountains. East of the Cascades, about a thousand meters of Columbia River Basalt cover thousands of square kilometers. Pleistocene volcanoes cap the Cascade Range.

The California coast was the site of marine clastic deposition in early Cenozoic that gave way to red beds in the Oligocene. Marine deposition, volcanism, and orogeny all occurred in the Miocene. In the Pleistocene, basins developed.

In the Arctic, some nonmarine clastics were deposited and folded. Subsidence produced the present islands.

Much of the present topography of northern North America was produced by glaciation.

The extent of continental glaciation is approximately marked by the Missouri River, the Ohio River, and Long Island.

In east-central Washington, the sudden release of an ice-dammed lake of glacial meltwater eroded the Channeled Scablands.

The Great Lakes were also formed by continental glaciation.

Clams, oysters, and snails were the main elements in Cenozoic seas.

The plant life of the Cenozoic was essentially modern, with angiosperms abundant.

Cenozoic is the age of mammals.

Marsupials migrated to Australia and South America in the Late Cretaceous and became isolated. In the Pliocene, the more successful placentals migrated to South America and replaced most of the marsupials.

The Paleocene was warm and temperate in North America, and mammals were archaic and mainly small.

The Eocene was subtropical. Rhinoceroses, deer, pigs, camels, and toothed whales appeared.

The Oligocene was mild, and the dry climate favored faster, long-legged animals. Cats and dogs appeared. *Brontotherium* was the largest mammal ever in North America.

The Miocene was cool and dry, favoring horses, pigs, rhinoceroses, camels, antelopes, deer, and mastodons.

The drying trend continued through the Pliocene.

The Pleistocene was dominated by large animals such as mastodons, mammoths, ground sloths, saber-toothed cats, bears, and giant beavers. Why these animals died out in North America is not known.

The oldest apes are early Oligocene, although other primates are older.

Australopithecus, about 3.5 million years old, used crude stone and bone tools.

Homo erectus, the first human, has been found in beds as old as 1 million years.

Homo sapiens has been found in beds as old as 100,000 years and appears to overlap *Homo erectus.*

Modern humans emerged about 35,000 years ago.

Humans must have come to the New World about 35,000 years ago, but the oldest artifacts are only 20,000 years old.

In about 2000 million years, the sun will expand and its energy will heat the earth, killing all life as we know it.

QUESTIONS

1 When did the four major advances of the sea occur on the Gulf Coast?
2 Describe the erosional history of the Appalachian Mountains.
3 Describe the topography of the Pacific coast from Washington to California.
4 Outline the Cenozoic history of arctic Canada.
5 Outline the development of the Rocky Mountains.
6 When and how did the Basin and Range structure of the Great Basin develop?
7 Where is the thickest section of marine Cenozoic rocks found in North America?
8 How much of North America was covered by ice during the Pleistocene? How many advances occurred?
9 Describe as many of the features of North America caused by glaciers as you can.
10 Describe the Pleistocene events in northwestern United States.
11 How does the marine life of the Cenozoic differ from that of the Mesozoic?
12 What are the two main types of mammals?
13 Outline the development of marsupials and account for their present distribution.
14 Describe the changing climates of the Cenozoic and how these affected the development of mammals.
15 What problems will future geologists have in correlating the present African "big game" fauna with North American fossils?
16 Why is it difficult to determine the development (i.e., describe the family tree) of modern humans?
17 When and where did modern humans first appear?

SUPPLEMENTARY READINGS

Dorf, Earling, "The Petrified Forests of Yellowstone Park," *Scientific American* (April 1964), Vol. 210, No. 4, pp. 106–114.

Eckhardt, R. B., "Population Genetics and Human Origins," *Scientific American* (January 1972), Vol. 226, No. 1, pp. 94–103.

Holloway, R. L., "The Casts of Fossil Hominid Brains," *Scientific American* (July 1974), Vol. 231, No. 1, pp. 106–115.

Hsü, K. J., "When the Black Sea Was Drained," *Scientific American* (May 1978), Vol. 238, No. 5, pp. 52–63.

Johanson, D. C., and T. D. White, "A Systematic Assessment of Early African Hominids," *Science* (January 26, 1979), Vol. 203, No. 4378, pp. 321–330.

Krantz, G. S., "Human Activities and Megafaunal Extinctions," *American Scientist* (March-April 1970), Vol. 58, No. 2, pp. 164–170.

Raven, P. H., and D. I. Axelrod, "History of the Flora and Fauna of Latin America," *American Scientist* (July-August 1975), Vol. 63, No. 4, pp. 420–429.

Simons, E. L., "Ramapithecus," *Scientific American* (May 1977), Vol. 236, No. 5, pp. 28–35.

Trimble, D. E., *The Geologic Story of the Great Plains.* U.S. Geological Survey Bulletin 1493, Washington, D.C.: U.S. Government Printing Office, 1980, 55 pp.

Trinkaus, Erik, and W. W. Howells, "The Neanderthals," *Scientific American* (December 1979), Vol. 241, No. 6, pp. 118–133.

Walker, Alan, and R. E. F. Leakey, "The Hominids of East Turkana," *Scientific American* (August 1978), Vol. 239, No. 2, pp. 54–66.

Wolfe, J. A., "A Paleobotanical Interpretation of Tertiary Climates in the Northern Hemisphere," *American Scientist* (November-December 1978), Vol. 66, No. 6, pp. 694–703.

APPENDICES

A MINERALS

TABLE A–1 Mineral identification key.

This table helps to identify the more common minerals. To use the chart, first decide the sort of luster (metallic or nonmetallic) the mineral has; next, its color (light or dark); third, its hardness (compared to a knife blade); fourth, whether it has cleavage; and finally, read the brief descriptions to identify it.

Dark-colored nonmetallic luster	**Hard—not scratched by knife**	**Shows cleavage**	Black to dark green; cleavage 2 planes at nearly right angles; hardness, 5–6.	AUGITE
			Black to dark green; cleavage 2 planes about 60°; hardness, 5–6.	HORNBLENDE
		No cleavage	Red to red-brown; fracture resembles poor cleavage; brittle equidimensional crystals; hardness, 6.5–7.5.	GARNET
			Various shades of green and yellow; glassy luster; granular masses and crystals in rocks; hardness, 6.5–7 (apparent hardness may be much less).	OLIVINE
			White, clear, or any color; glassy luster; transparent to translucent; hexagonal (6-sided) crystals; hardness, 7; conchoidal fracture.	QUARTZ
			Any color or variegated; glassy luster; hardness, 5–6; conchoidal fracture.	OPAL
			Any color or variegated; waxy luster; hardness, 7; conchoidal fracture.	CHALCEDONY (AGATE)
			Red to brown; red streak; earthy appearance; hardness, 5.5–6.5 (apparent hardness may be less).	HEMATITE
			Yellow-brown to dark brown, may be almost black, streak yellow-brown; earthy; hardness, 5–5.5 (may have lower apparent hardness).	LIMONITE (GOETHITE)
	Soft—scratched by knife	**Shows cleavage**	Brown to black; cleavage, 1 direction; hardness, 2.5–3 (black mica).	BIOTITE
			Various shades of green; cleavage, 1 direction; hardness, 2–2.5 (green "mica").	CHLORITE
			Yellow-brown, dark brown, or black; streak white to pale yellow; resinous luster; cleavage, 6 directions; hardness, 3.5–4.	SPHALERITE
		No cleavage	Red to brown; red streak; earthy appearance; hardness, 5.5–6.5 (apparent hardness may be less).	HEMATITE
			Scarlet to red-brown; scarlet streak; hardness, 2–2.5; high specific gravity.	CINNABAR
			Lead-pencil black, smudges fingers; hardness, 1; (one cleavage that is apparent only in large crystals).	GRAPHITE
			Yellow-brown to dark brown, may be almost black; streak yellow-brown; earthy; hardness, 5–5.5 (may have lower apparent hardness).	LIMONITE (GOETHITE)
			Dark to light green; greasy or waxy luster; some varieties are fibrous; hardness, 2–5, generally 4.	SERPENTINE

TABLE A–1 Mineral identification key. (cont.)

Light-colored nonmetallic luster	Hard—not scratched by knife	Shows cleavage	White or flesh-colored; 2 cleavage planes at nearly right angles; hardness, 6. Large crystals which show irregular veining are PERTHITE. — ORTHOCLASE (POTASSIUM FELDSPAR)
			White or green-gray; 2 cleavage planes at nearly right angles; hardness, 6; striations on one cleavage. — PLAGIOCLASE
		No cleavage	White, clear, or any color; glassy luster; transparent to translucent; hexagonal (6-sided) crystals; hardness, 7; conchoidal fracture. — QUARTZ
			Various shades of green and yellow; glassy luster; granular masses and crystals in rocks; hardness, 6.5–7 (apparent hardness may be much less). — OLIVINE
			Any color or variegated; glassy luster; hardness, 5–6; conchoidal fracture. — OPAL
			Any color or variegated; waxy luster; hardness, 7; conchoidal fracture. — CHALCEDONY (AGATE)
	Soft—scratched by knife	Shows cleavage	Colorless to white; salty taste; cubic cleavage; hardness, 2.5. — HALITE
			White, yellow to colorless; rhombohedral cleavage; hardness, 3; effervesces with dilute hydrochloric acid. — CALCITE
			Pink, colorless, white, or dark; rhombohedral cleavage; hardness, 3.5–4; effervesces with dilute hydrochloric acid only if powdered. — DOLOMITE
			White to transparent; 3 unequal cleavages; hardness, 2. — GYPSUM
			Green to white; feels soapy; 1 cleavage; hardness, 1. — TALC
			Colorless to light yellow or green; transparent in thin sheets which are very elastic; 1 cleavage; hardness, 2–2.5 (white mica). — MUSCOVITE
			Green to white; fibrous cleavage; may form veins. — ASBESTOS
		No cleavage	Green to white; feels soapy; hardness, 1. — TALC
			White to transparent; hardness, 2. — GYPSUM
			Yellow to greenish; resinous luster; hardness, 1.5–2.5. — SULFUR
Metallic luster			Black; strongly magnetic; hardness, 6. — MAGNETITE
			Lead-pencil black; smudges fingers; hardness, 1; one cleavage that is apparent only in large crystals. — GRAPHITE
			Brass yellow; black streak; cubic crystals, commonly with striations; hardness, 6–6.5. — PYRITE
			Brass-yellow; may be tarnished; black streak; hardness, 3.5–4; massive. — CHALCOPYRITE
			Shiny gray; black streak; very heavy; cubic cleavage; hardness, 2.5 — GALENA

TABLE A–2 Summary of common minerals.

Mineral	Composition	Color	Luster	Streak	Hardness	Cleavage	Other Properties–Uses
Amphibole—*see hornblende*							
Andalusite	Al_2SiO_5 Aluminum silicate	Reddish-brown to olive green	Vitreous to dull	White	7.5	2 poor	In metamorphic rocks
Asbestos	Hydrous magnesium silicate	Shades of green	Silky	None	Low apparent hardness	Fibrous	Fireproofing & insulation against heat & electricity
Augite	$(Ca,Na)(Mg,Fe^{II},Fe^{III},Al)(Si,Al)_2O_6$ Calcium, sodium, magnesium, iron, aluminum silicate	Black–dark green	Vitreous	None	5–6	2 at nearly right angles	Rock-forming mineral
Biotite	$K(Mg,Fe)_3(AlSi_3O_{10})(OH)_2$ Hydrous potassium, magnesium, iron, aluminum silicate	Black–dark green	Vitreous	None	2.5–3	1 perfect	Rock-forming mineral
Calcite	$CaCO_3$ Calcium carbonate	Generally white or colorless	Vitreous to earthy	None	2.5–3	Rhombohedral, 3 not at right angles	Effervesces in cold dilute hydrochloric acid. Manufacture of cement
Chalcedony	SiO_2 Silicon dioxide	Any	Waxy	None	7	None	Conchoidal fracture
Chalcopyrite	$CuFeS_2$ Copper–iron sulfide	Brass-yellow—may be tarnished to bronze or iridescent	Metallic-dull	Greenish-black	3.5–4	None	Brittle
Chlorite	$(Mg,Fe,Al)_6(Al,Si)_4O_{10}(OH)_8$ Hydrous magnesium, iron, aluminum silicate	Green	Vitreous	None	2–2.5	1 perfect	Rock-forming mineral
Cinnabar	HgS Mercury sulfide	Vermillion-red to brownish-red	Adamantine to dull earthy	Scarlet	2.5	1	High specific gravity. Only important ore of mercury.

510

TABLE A-2 Summary of common minerals. (cont.)

Mineral	Composition	Color	Luster	Streak	Hardness	Cleavage	Other Properties–Uses
Clay	A family of hydrous aluminum silicates which may contain potassium, sodium, iron, magnesium, etc.	Generally light	Earthy	None	Very low apparent hardness	None	Rock-forming mineral
Cordierite	$(Mg,Fe)_2Al_4Si_5O_{18}$ Magnesium aluminum silicate	Blue	Vitreous	White	7–7.5	1 poor	In metamorphic rocks
Dolomite	$CaMg(CO_3)_2$ Calcium-magnesium carbonate	Generally some shade of pink, flesh color, or white	Vitreous to pearly	None	3.5–4	Rhombohedral, 3 not at right angles	Powder effervesces in cold dilute hydrochloric acid
Epidote	$Ca_2(Al,Fe)_3Si_3O_{12}(OH)$ Hydrous calcium, aluminum, iron silicate	Green to dark green	Vitreous	White to gray-white	6–7	1	In metamorphic rocks
Feldspar–see plagioclase and orthoclase (potassium feldspar)							
Galena	PbS Lead sulfide	Lead-gray	Bright metallic	Lead-gray	2.5	Cubic, 3 at right angles	High specific gravity. Ore of lead
Garnet	A family of silicate minerals containing aluminum, iron, magnesium, calcium, etc.	Commonly red	Vitreous	None	6.5–7.5	None	Characteristic isometric crystals
Graphite	C Carbon	Black to steel-gray	Metallic or dull earthy	Black	1–2	1	Greasy feel—marks paper
Gypsum	$CaSO_4 \cdot 2H_2O$ Hydrous calcium sulfate	Colorless, white, gray	Vitreous, pearly, silky to earthy	None	2	1 perfect. Others poorer	Plaster of Paris
Halite	$NaCl$ Sodium chloride	Colorless to white	Vitreous	None	2.5	Cubic, 3 at right angles	Salty taste
Hematite	Fe_2O_3 Iron oxide	Reddish-brown to black or gray	Earthy to bright metallic	Light to dark Indian-red	5.5–6.5	None	Most important ore of iron
Hornblende	$Ca_2Na(Mg,Fe^{II})_4(Al,Fe^{III},Ti)(Al,Si)_8O_{22}(O,OH)_2$ Hydrous calcium, sodium, magnesium, iron, aluminum silicate	Black–dark green	Vitreous	None	5–6	2 at 60°	Rock-forming mineral

TABLE A-2 Summary of common minerals. (cont.)

Mineral	Composition	Color	Luster	Streak	Hardness	Cleavage	Other Properties–Uses
Kyanite	Al_2SiO_5 Aluminum silicate	Blue	Vitreous to pearly	White	4–5 along crystal 7 across	1	In metamorphic rocks
Laumontite	$CaAl_2Si_4O_{12} \cdot 4H_2O$ Hydrous calcium aluminum silicate	White	Vitreous	White	4	3	A zeolite
Lawsonite	$CaAl_2(Si_2O_7)(OH)_2 \cdot H_2O$ Calcium aluminum silicate	Pale blue to blue-gray	Vitreous to greasy	White	8	2	In metamorphic rocks
Limonite (Goethite)	$FeO(OH) \cdot nH_2O$ Hydrous iron oxide	Yellow-brown to black	Generally earthy	Yellowish-brown	5–5.5	None	Low apparent hardness
Magnetite	Fe_3O_4 Iron oxide	Black	Metallic to submetallic	Black	6	None	Strongly magnetic. Important iron ore
Mica–see biotite and muscovite							
Muscovite	$KAl_2(AlSi_3O_{10})(OH)_2$ Hydrous potassium-aluminum silicate	Clear–light green	Vitreous	None	2–2.5	1 perfect	Rock-forming mineral
Olivine	$(Mg,Fe)_2SiO_4$ Magnesium-iron silicate	Green	Vitreous	None	6.5–7	None	Rock-forming mineral
Opal	$SiO_2 \cdot nH_2O$ Hydrous silicon dioxide	Any	Vitreous or resinous	None	5–6	None	Conchoidal fracture
Orthoclase (Potassium feldspar)	$KAlSi_3O_8$ Potassium-aluminum silicate	White to pink	Vitreous	None	6	2 at right angles	Rock-forming mineral
Plagioclase	$CaAl_2Si_2O_8$ & $NaAlSi_3O_8$ Calcium & sodium aluminum silicate	White to green	Vitreous	None	6	2 at right angles—striations on one cleavage direction	Rock-forming mineral
Prehnite	$Ca_2Al_2(Si_3O_{10})(OH)_2$ Hydrous calcium aluminum silicate	Light green to white	Vitreous	White	6–6.5	1	In metamorphic rocks
Pyrite	FeS_2 Iron sulfide	Pale brass-yellow	Metallic bright	Greenish or brownish-black	6–6.5	None	Brittle. Cubic crystals
Pyroxene–see augite							

TABLE A–2 Summary of common minerals. (cont.)

Mineral	Composition	Color	Luster	Streak	Hardness	Cleavage	Other Properties–Uses
Quartz	SiO_2 Silicon dioxide	Any	Vitreous	None	7	None	Rock-forming mineral
Serpentine	$Mg_6Si_4O_{10}(OH)_8$ Hydrous magnesium silicate	Dark to light green	Greasy to waxy	None	2–5	None	Metamorphic rocks. Fibrous variety is asbestos
Sillimanite	Al_2SiO_5 Aluminum silicate	Brown, pale green, white	Vitreous	White	6–7	1	In metamorphic rocks
Sphalerite	ZnS Zinc sulfide	Commonly yellow-brown to black	Resinous to adamantine	White to yellow and brown	3.5–4	6	Zinc ore
Staurolite	$(Fe,Mg)_2Al_9Si_4O_{23}(OH)$ Ferrous iron aluminum silicate	Reddish-brown to brown-black	Vitreous to resinous	Gray	7–7.5	1 poor	In metamorphic rocks
Sulfur	S Sulfur	Yellow	Resinous	None	1.5–2.5	None	Fracture conchoidal to uneven. Brittle
Talc	$Mg_3(Si_4O_{10})(OH)_2$ Hydrous magnesium silicate	Apple-green, gray, white, or silver-white	Pearly to greasy	None	1	1	Cosmetics

B TOPOGRAPHIC MAPS

He had bought a large map representing the sea,
  Without the least vestige of land:
And the crew were much pleased when they found it to be
  A map they could all understand.

"What's the good of Mercator's North Poles and Equators,
  Tropics, Zones and Meridian Lines?"
So the Bellman would cry: and the crew would reply
  "They are merely conventional signs!"

                              Lewis Carroll
                              The Hunting of the Snark
                              Fit the Second, Stanzas 2 and 3

TOPOGRAPHIC MAPS

Maps are simply scale drawings of a part of the earth's surface. Thus they are similar to a blueprint of an object or to a dress pattern. Most maps are drawn on sheets of paper and so show only the two horizontal dimensions. Geologists, together with civil engineers and many other users of maps, require that the third dimension, elevation, be shown on maps. Maps that indicate the shape of the land are called *topographic maps.* A number of different ways of showing topography are in current use, such as the familiar color tint that uses shades of green for low elevations and brown for higher elevations. The most accurate method, which will be described here, involves the use of contour lines.

MAP SCALES

Map scales are designated in several ways. (1) A scale can be stated as a specified unit of length on the map corresponding to a specified unit of length on the ground—for example, one inch equals one mile, meaning that one inch on the map represents one mile on the ground. (2) A scale can also be stated as a scale ratio or representative fraction, meaning that one unit (any unit) on the map equals a specified number of the same units on the ground. In the example just used, the scale would be given as 1:63,360, and 1 inch on the map would represent 63,360 inches

515

on the ground (1 mile = 12 × 5,280 = 63,360 inches). (On that same map, 1 centimeter would also represent 63,360 centimeters on the ground.) (3) Most maps contain a graphic scale, generally in the lower margin, for measuring distances.

CONTOUR LINES

A contour line is an imaginary line, every part of which is at the same elevation. Thus a shore line is an example of a contour. Because a single point can only have one elevation, two contour lines can never cross (except on an overhanging cliff).

By convention, contour lines on each map are a fixed interval apart. This is called the contour interval. Typical contour intervals are 5, 10, 20, 40, 80, or 100 feet or meters. The contour lines are always at an elevation that is a whole number times the contour interval. That is, the contour lines are drawn at 20, 40, 60, 80, and so forth, feet or meters, never at 23, 43, or 63.

A little imagination will reveal that if the contour lines are close together, the surface is steeper than if they are far apart. (See Fig. B-1.)

If we think of an island and remember that the shore line is the zero contour, we can then imagine the successive contours, or shore lines, if the water level rises. From this we learn that a hill is represented on a contour map by a series of concentric closed lines, and, of course, a ridge would be similar, but elongate.

If we think of a valley, we see that the contour lines form V's pointing upstream. Remember a stream flows downhill at the lowest point in a valley; hence, if we start on one of the valley sides and walk in the upstream direction at a given elevation, we will eventually reach the stream.

A few minutes study of Fig. B-2 will reveal much information about reading contour maps.

Contour lines have the inherent limitation that the elevations of points between contours are not designated, although the approximate eleva-

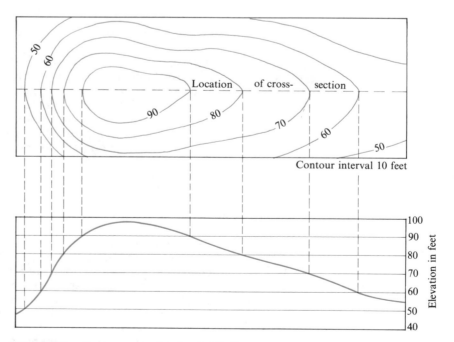

FIGURE B-1. Contour map of a simple hill. The cross-section shows that where the contour lines are closer together the hill is steeper.

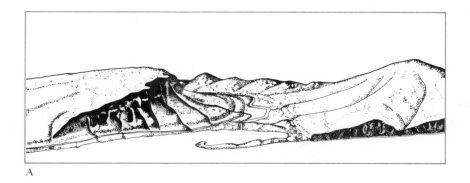

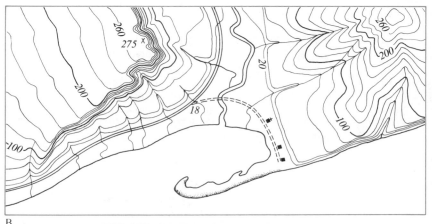

THE USE OF SYMBOLS IN MAPPING

These illustrations show how various features are depicted on a topographic map. The upper illustration is a perspective view of a river valley and the adjoining hills. The river flows into a bay which is partly enclosed by a hooked sandbar. On either side of the valley are terraces through which streams have cut gullies. The hill on the right has a smoothly eroded form and gradual slopes, whereas the one on the left rises abruptly in a sharp precipice from which it slopes gently, and forms an inclined tableland traversed by a few shallow gullies. A road provides access to a church and two houses situated across the river from a highway which follows the seacoast and curves up the river valley.

The lower illustration shows the same features represented by symbols on a topographic map. The contour interval (the vertical distance between adjacent contours) is 20 feet.

FIGURE B-2. Perspective view of an area and a contour map of the same area. After U.S. Geological Survey.

tion can be determined by interpolation. In an effort to overcome this limitation, the elevations of hills, road junctions, and other features are commonly given on the maps.

CONVENTIONS USED ON TOPOGRAPHIC MAPS

Because the United States Geological Survey produces most of the topographic maps in use in the United States, these are described here. Until recently these maps were made by actual field surveying, but are now made from aerial photographs with some field surveying to establish horizontal and vertical control points. The maps made since 1942 meet high accuracy standards. Ninety percent of all well-defined features are

517

within one-fiftieth of an inch of their true location on the published map, and the elevations of ninety percent of the features are correct within one-half of the contour interval.

The maps, which are called quadrangles, are bounded by latitude and longitude lines. They are named for features within the map area, and because the names are not duplicated within a state, some are named for less important features. The sizes and scales used are shown in Table B-1. The number of square miles in a map of a given scale varies because of the convergence of longitude lines at the north pole.

TABLE B-1. Sizes and scales of U.S.G.S. topographic maps.

Map Designation	Scale		Quadrangle Size (Lat.-Long.)	Quadrangle Area (Sq. Miles)	Paper Size (Inches)
	Ratio	One inch equals			
7½ minute	1:24,000	exactly 2,000 feet	7½' × 7½'	49-70	22 × 27
					23 × 27
	1:31,680	exactly ½ mile	7½' × 7½'	49-68	17 × 21
15 minute	1:62,500	approximately 1 mile	15' × 15'	197-282	17 × 21
					19 × 21
30 minute	1:125,000	approximately 2 miles	30' × 30'	789-1,082	17 × 21
1 degree	1:250,000	approximately 4 miles	1° × 1°	3,173-4,335	17 × 21
1:250,000	1:250,000	approximately 4 miles	1° × 2°	6,349-8,669	24 × 34

COLORS

Water features are in blue.
Human structures—roads, houses, and so forth—are in black.
Contour lines are in brown.
In addition, some maps show:
　　Important roads, urban areas, and public land subdivision lines in red.
　　Woodlands, orchards, and so forth, in green.

If you would like to obtain topographic maps of any area, write to the addresses below for an index to topographic maps for the state desired. They are free on request and show all of the maps available.

For states east of the Mississippi River:

Eastern Distribution Branch
U.S. Geological Survey
1200 South Eads Street
Arlington, VA 22202.

For states west of the Mississippi River:

Western Distribution Branch
U.S. Geological Survey
Box 25286, Federal Center
Denver, CO 80225.

C GEOLOGICAL MAPS

Geological mapping is one of the most important methods used in the study and interpretation of the earth and earth history. A geologic map is simply a map on which the various rock types that make up the earth's surface are plotted. Generally, the rock types are plotted on topographic (contour) maps because the topography is commonly related to the rock types, and knowing the topography helps another geologist to interpret the geologic map. To further aid in the interpretation of geologic maps, other symbols are used to indicate such things as the attitude or slope of planar features such as the bedding planes of sedimentary rocks. From the information on the geologic map, which is the surface geology, the underlying structure of the area is interpreted. The main difficulty in geologic mapping is that outcrops of bedrock are relatively rare in most regions because weathered material, soil, and alluvium cover much of the earth's surface.

The preparation of a geologic map begins with the field geologist locating the outcrops of the various rock types on a map or an aerial photograph of the area (Fig. C–1B). Generally, the units that are mapped are *formations.* Formations are rock units that are generally easy to recognize and so reasonably easy to map. The geologic map should show the boundaries between formations, and these boundaries are called *contacts* (Fig. C–1C). If the units shown on the map have directional properties, such as bedding in sedimentary rocks or foliation in metamorphic rocks, the *attitude* of such features should also be noted. The attitude is shown by the "T"-shaped *strike and dip* symbol shown on the maps. The longer line shows the strike, which is the direction of the intersection of the bed and a horizontal plane. The shorter line shows the direction of the dip, the down slope perpendicular to the strike, and the number near it shows the dip angle (Fig. C–2). In addition, a geologic map should show the contacts of such features as intrusive rock bodies. Faults should also be shown.

Typically, different colors are used to show the various rock units on a geologic map. The ages and relationships among the rock units, and the color used for each, are shown in the legend of the map. The legend is generally at the side or bottom of a geologic map. Any other symbols used on the map will also be explained in the legend. Cross-sections may also be included to show the structures (Fig. C–3).

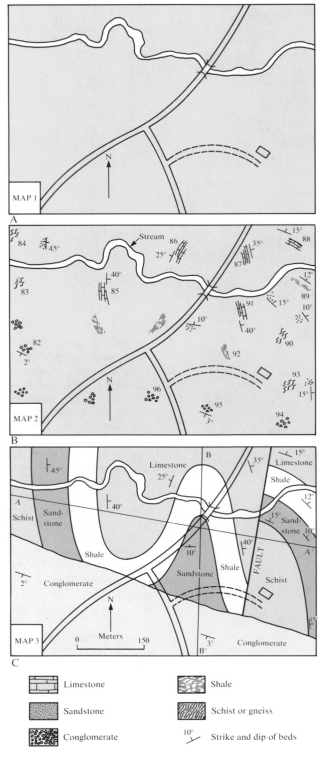

A

B

C

FIGURE C–1 The stages in making a geologic map.

Legend:

Limestone

Sandstone

Conglomerate

Shale

Schist or gneiss

10° — Strike and dip of beds

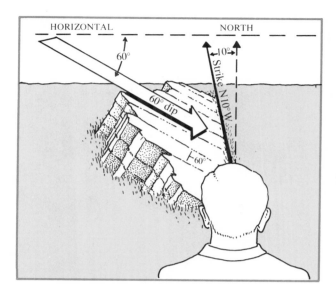

FIGURE C–2 Strike and dip.

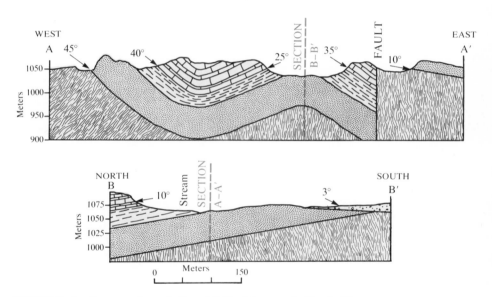

FIGURE C–3 Cross sections *A-A'* and *B-B'* of the map in Fig. C–1C.

D ROCK TEXTURES

Size in mm	This Book (all rocks)	Igneous and Metamorphic	Sedimentary		Pyroclastic	
			Sediment	Rock	Sediment	Rock
256		Very coarse *	Boulder	Conglomerate (fragment size is generally indicated as pebble con-glomerate, etc.)	Block **	Breccia
128		Pegmatitic	Cobble		Bomb **	
64						
32						
16		Coarse *	Pebble			
8	Coarse-grained	Phaneritic			Lapilli	Lapilli
4					Cinder	
2		Medium *	Granule			
1				Coarse sandstone	Coarse Ash	Coarse Tuff
1/2						
1/4		Fine *	Sand			
1/8				Fine sandstone		
1/16						
1/32	Fine-grained	Aphanitic				
1/64		Dense	Silt	Siltstone	Fine Ash	Fine Tuff
1/128						
1/256						
			Clay	Shale (most shales contain some silt)		

*These terms are generally used as modifiers such as medium-grained granite.
**Blocks are fragments broken from solidified rocks; bombs are molten when ejected.

FIGURE D–1 Summary of texture terms in common use and those used in this book. Terms used in this book are in italics.

NATURAL RADIOACTIVITY

The naturally radioactive isotopes are shown in Table E–1 and Fig. E–1. Note that there are three radioactive series that go through many steps before a stable isotope is formed, as well as a number of isotopes that decay to a stable product in a single step. Also some isotopes, such as potassium 40, decay by two processes to two different daughter products. In addition to the three series shown in the figures is the neptunium series; however, the half-life of this series is much shorter than that of the others, so only very small amounts of some of the members of this series are found in nature.

TYPES OF RADIOACTIVE DECAY

Every element has a certain equal number of protons and electrons, and a variable number of neutrons that determines the isotope of that element. If the element decays by alpha emission, it loses the two protons and two neutrons that make up the alpha particle. Thus, the atomic number is reduced by two, and the atomic mass is reduced by four as shown in the following equation:

$$^{147}_{62}\text{Sm} \rightarrow {}^{143}_{60}\text{Nd} + {}^{4}_{2}\alpha$$

Alpha particles can be seen on a luminous dial watch. In a dark room the flash of light that is produced each time an alpha particle hits the zinc sulfide on the dial of the watch can be seen with a magnifying lens.

In beta decay an electron is lost. This can be thought of as occurring because a neutron separates into a proton and an electron; the electron is emitted at high velocity, and the proton is retained by the nucleus. Thus the atomic number is increased by one, and the atomic mass remains the same. The additional proton in the nucleus is balanced by gaining an electron for the outer shell. Free electrons are generally available and can be captured by the atom. In equation form

$$^{87}_{37}\text{Rb} \rightarrow {}^{87}_{38}\text{Sr} + \beta^{-}$$

The third geologically important type of radioactivity occurs by electron capture. In this case, the nucleus takes an electron from one of the inner shells. This electron unites with a proton to form a neutron. Thus

TABLE E–1 Summary of naturally occurring radioactive isotopes. A few isotopes with very long half-lives and of very small abundances are not included.

A. SINGLE-STEP DECAY

Isotope	Decay	Half-life (in years unless noted)	Percent of Total Amount of Element in Crust
Neutron	$^1n \rightarrow {}^1H^+ + \beta^-$	12 minutes	—
Tritium (^3_1H)	$^3_1H \rightarrow {}^3He + \beta^-$	12.26	0.0001—0.00001
Carbon 14	$^{14}_6C \rightarrow {}^{14}_7N + \beta^-$	5660 ± 30	(in biologic material)
Potassium 40	$^{40}_{19}K \xrightarrow{89\%} {}^{40}_{20}Ca + \beta^-$ $\xrightarrow{11\%} {}^{40}_{18}A$ (electron capture)	1.3×10^9	0.0118
Rubidium 87	$^{87}_{37}Rb \rightarrow {}^{87}_{38}Sr + \beta^-$	4.7×10^{10}	27.85
Indium 115	$^{115}_{49}In \rightarrow {}^{115}_{50}Sn + \beta^-$	6×10^{14}	95.72
Lanthanum 138	$^{138}_{57}La \xrightarrow{85\%} {}^{138}_{58}Ce + \beta^-$ $\xrightarrow{15\%} {}^{138}_{56}Ba$ (electron capture)	1.1×10^{11}	0.089
Samarium 147	$^{147}_{62}Sm \rightarrow {}^{143}_{60}Nd + {}^4_2\alpha$	1.06×10^{11}	15.1
Lutetium 176	$^{176}_{71}Lu \xrightarrow{33\%} {}^{176}_{72}Hf + \beta^-$ $\xrightarrow{67\%} {}^{176}_{70}Yb$ (electron capture)	2.1×10^{10}	2.59
Rhenium 187	$^{187}_{75}Re \rightarrow {}^{187}_{76}Os + \beta^-$	4×10^{10}	62.93

B. MULTIPLE-STEP DECAY

All isotopes with atomic number greater than 83 are radioactive, and all of these that occur naturally are members of the three radioactive families listed below. Some isotopes with atomic number less than 83 also occur in these families. The complete decay of these families is shown in Fig. E–1.

Isotope	Decay	Half-Life (in years)	Percent of Total Amount of Element in Crust
Uranium 235	$^{235}_{92}U \rightarrow {}^{207}_{82}Pb + 7{}^4_2\alpha + 4\beta$	7.1×10^8	0.72
Uranium 238	$^{238}_{92}U \rightarrow {}^{206}_{82}Pb + 8{}^4_2\alpha + 6\beta$	4.5×10^9	99.28
Thorium 232	$^{232}_{90}Th \rightarrow {}^{208}_{82}Pb + 6{}^4_2\alpha + 4\beta$	1.4×10^{10}	100

the atomic number is reduced by one, and the mass remains the same, as in

$$^{40}_{19}K + e^- \rightarrow {}^{40}_{18}A$$

The only changes in rate of radioactive decay occur in decay by electron capture, in which the rate changes slightly as the abundance of nearby electrons changes.

In most radioactivity, gamma rays are also emitted. Emission of a particle may leave the nucleus in an excited state, and emission of gamma rays can allow the atom to settle down. In any case, mass must be con-

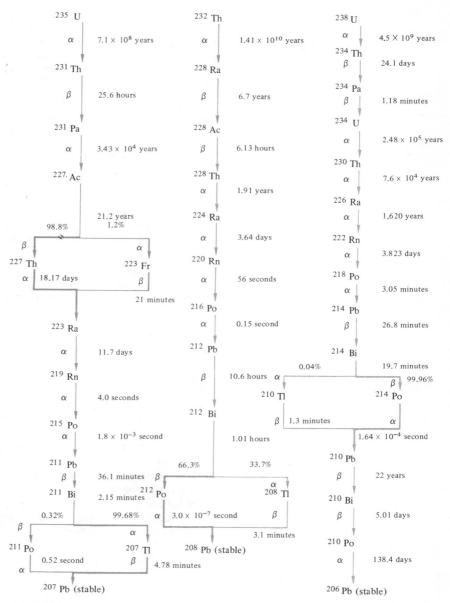

FIGURE E–1 Decay schemes for uranium 235, uranium 238, and thorium 232. Type of decay, branching ratios, and half-lives are indicated. This figure is an expansion of Table E–1B.

served in a radioactive event, and this can explain why gamma rays are emitted. Note in Table E–2 that the mass of a neutron is slightly greater than the combined masses of a proton and an electron. The question then, using beta decay as an example, is: Where did the excess mass go when the neutron changed into a proton and an electron? The answer is that the mass was changed into energy in the form of gamma rays. The equivalence of mass and energy was postulated by Einstein, and experiments have so far shown this to be correct. His famous equation is

$$E = mc^2$$

or, in words, energy equals mass times the speed of light squared. However, in beta decay the energy of the gamma rays is less than the amount

TABLE E–2 Mass and charge of some particles and atoms. Masses are in atomic mass units (amu) based on oxygen atom = 16.0000 amu (old unit).* Note that the mass of an electron plus a proton is less than the mass of a neutron.

	Charge	Mass		Charge	Mass
Electron	−1	0.000549	$^{16}_{8}O$	0	16.000000 (standard)
Proton	+1	1.007593	$^{18}_{8}O$	0	18.004855
Neutron	0	1.008982	$^{204}_{82}Pb$	0	204.0361
α-particle	+2	4.002675	$^{206}_{82}Pb$	0	206.0386
$^{1}_{1}H$	0	1.008142	$^{207}_{82}Pb$	0	207.0403
$^{2}_{1}H$	0	2.014735	$^{208}_{82}Pb$	0	208.0414
$^{3}_{1}H$	0	3.016997	$^{232}_{90}Th$	0	232.11034
$^{3}_{2}He$	0	3.016977	$^{235}_{92}U$	0	235.11704
$^{4}_{2}He$	0	4.003873	$^{238}_{92}U$	0	238.12493

*1 amu = 1/16 of the weight of an oxygen atom. A molecular weight of oxygen is 16 grams and there are 6.02×10^{23} atoms in a molecular weight (Avogadro's number). Therefore,

$$1 \text{ amu} = 1/16 \times \frac{16 \text{ grams}}{6.02 \times 10^{23}} = 1.66 \times 10^{-24} \text{ gram}$$

10^{23} means 1 followed by 23 zeros
10^{-24} means 1/1 followed by 24 zeros

of mass lost. This led Fermi to postulate the emission of a small neutral particle he called the *neutrino*. Many years later the existence of neutrinos was proved. The neutrino is only one of a few dozen tiny particles found or postulated by atomic scientists. How they are formed and whether they exist in the nucleus as particles or in the form of energy is not known. With this in mind, it is clear that the descriptions above of how radioactive decay takes place are only convenient fictions to account for the main results.

Calculation of the amount of energy released by radioactivity is simple. One needs only to know the masses of the parent and daughter elements and the masses of the particles released. (See Table E–2.) These data and Einstein's famous equation, $E = mc^2$ (energy equals loss of mass multiplied by the speed of light squared), tell the amount of energy released by the decay of one atom of the parent isotope. The rate of decay (half-life) is then used to determine how often an atom of the parent decays.

F PHYSICAL METHODS OF AGE DETERMINATION

Some of the methods are described in Chapter 16. Table F–1 on the following four pages after E. J. Zeller, "Modern Methods for Measurement of Geologic Time," Mineral Information Service, California Division of Mines and Geology, January, 1965, Vol. 18, No. 1, pp. 9–15.

TABLE F-1 Physical methods of age determination.

Method	Kind of Sample Used	Parameter Measured	Range and Accuracy	Remarks
		Isotopic or radioactive yield methods		
Uranium-Lead	Zircon, xenotime, and other igneous and metamorphic minerals with original uranium, but with no original lead.	Uranium and lead by analyses and α-counts per unit of mass per unit of time.	10 million to 3-4 thousand million years. Not as accurate as with isotopic analysis.	There should be no leaching of uranium during life of mineral and no original lead in mineral at time of crystallization.
Uranium 238-Lead 206	Any igneous or metamorphic mineral or whole rock with original uranium 238 (especially uraninite, pitchblende, samarskite, cyrtolite, thorianite, zircon, and thorite.)	Quantities of uranium 238, lead 206, and lead 204 by mass spectrometer analysis. Lead 204 gives indication of amount of original lead, of all isotopes, that is present.	10 million to approximately 10,000 million years. More accurate than the chemical uranium-lead method.	Uranium 238 is more abundant than uranium 235 so less chance of analytical error. Radon 222 half-life is long enough (3.82 days) to suggest that there may be error through radon escape.
Uranium 235-Lead 207	Any igneous or metamorphic mineral or whole rock with original uranium 235. (Same minerals as above.)	Quantities of uranium 235, lead 207, and lead 204 by mass spectrometer analysis. Lead 204 gives indication of amount of original lead, of all isotopes, that is present.	10 million to several thousand million years. More accurate than the chemical uranium-lead method.	Radon 219 in series has such short half-life (3.92 sec.) that escape is much less likely than in uranium 238-lead 206 series.
Thorium 232-Lead 208	Accessory minerals in igneous rocks.	Thorium 232, lead 208 by mass spectrometer analysis.	10 million to several thousand million years. Accuracy 10 percent.	Lead 208 may be removed selectively from thorium minerals resulting in ages which appear too low.
Lead-Lead	Galena or uranium minerals.	Lead 206, lead 207, lead 208, lead 204 by mass spectrometer analysis.	10 million to several thousand million years. 1 percent over 100 million years. (Younger ages may be calculated with accuracy of 1 million years.)	Has the advantage that it involves an isotopic ratio instead of absolute measurement.

TABLE F–1 Physical methods of age determination. (cont.)

Method	Kind of Sample Used	Parameter Measured	Range and Accuracy	Remarks
		Isotopic or radioactive yield methods (continued)		
Lead 210	Fresh and marine water and glacial ice. Recently deposited sediments.	Count rate of bismuth 210 derived by radioactive decay of separated lead 210.	Up to about 100 years. Sequence of analyses gives fairly accurate estimate of rates of accumulation.	Part of chain of uranium 238-lead 206. Radon 222, to atmosphere from rocks, yields lead 210. Lead 210 washed from atmosphere by rain. This disintegrates to bismuth 210-polonium 210-lead 206. Lead 210 is rapidly removed from water by precipitation or absorption on sedimentary particles. Rate of removal from water and original amount of lead 210 in sediments are necessary for data for age determination.
Potassium-Argon	Micas and feldspars.	Potassium 40, argon 40. Potassium measured by flame photometer, argon by mass spectrometer.	30,000 to several thousand million years. Error in micas less than 10 percent.	Feldspars lose argon (up to 60 percent). Metamorphism affects the results if rocks are heated.
Rubidium-Strontium	Mica, muscovite, biotite. Feldspar. Whole rock. Glauconite (in sediments).	Amount of strontium 86. Amount of strontium 87. Amount of rubidium 87. Ratios determined by mass spectrometry.	Not suited for ages below 50 million years due to long rubidium 87 half-life. Accuracy of method dependent on rubidium 87 half-life which is not accurately determined.	Decay constant of rubidium 87 uncertain due to low energy of β-particles emitted. Metamorphism may cause rehomogenation of rubidium and strontium isotopes so method may be used to date metamorphic events.
Thorium 230	Oceanic sediments, cores.	Count rate of thorium 230	Up to 300,000 years.	Part of chain of uranium 238-lead 206. Used to date marine sediments. Uranium 238 in seawater decays to thorium 230 which is precipitated and deposited on sea floor.
Thorium 230-Protactinium 231	Oceanic sediments, cores.	Count rate of thorium 230 and protactinium 231.	Up to 300,000 years.	Protactinium 231 is part of chain of uranium 235-lead 207. Accumulates similar to thorium 230 described above.

TABLE F–1 Physical methods of age determination. (cont.)

Method	Kind of Sample Used	Parameter Measured	Range and Accuracy	Remarks
Isotopic or radioactive yield methods (continued)				
Helium	Uranium and thorium bearing minerals. (The helium is the α particles emitted.)	Abundance of helium. Uranium 238, uranium 235, and thorium 232 abundance by their α activity. Uranium/thorium ratio.	Several million to several thousand million years. Accuracy in dispute.	Method is not extensively used at present due to discovery of the ease of diffusion by helium.
Carbon 14	Charcoal, wood, shell.	β-activity of carbon in sample by β-counting techniques.	0 to 50,000 years. Accuracy good for last 20,000 years.	Assumes a steady state for the production of carbon 14 in the atmosphere by cosmic radiation. Assumes uniform distribution of carbon 14 throughout the earth's atmosphere.
Hydrogen 3 (Tritium)	Water samples, 2-6 liters.	β-activity of tritium.	Last 30 years.	Recent bomb testing has greatly affected the usefulness of this method.
Radiation damage methods				
Metamict	Zircon (best); euxenite; sphene; uraninite; davidite.	Amount of radiation damage (i.e. metamict condition) in mineral usually by X-ray diffraction. Amount of radioactive material by α-count.	Now considered a comparatively inaccurate method.	Heating will cause crystal to become crystalline again; therefore, method may be used to date metamorphic or igneous events.
Pleochroic halo (Radiation halo)	Biotite with an inclusion of zircon, apatite, or monazite.	Intensity of the coloration of the biotite. Width of the coloration. α-activity of the inclusion.	Late Precambrian to approximately Eocene. Useful for relative age determination.	Coloration is temperature sensitive and may reverse at a high level of radiation. Assumes identical irradiation sensitivity of all biotites and that the coloration, once formed, does not fade through time.
Thermo-luminescence	Carbonate rocks, fluorite, quartz, feldspars.	Luminescence of sample when heated to temperatures up to 400°C. Glow-curve is measured by a recording microphotometer.	One thousand million year maximum. Accuracy ± 15 percent.	May be used to determine periods of metamorphism. Has specific uses for paleoclimatological measurements.
Radiation pitting (Fission particle tracks)	Mica. Other minerals may also be usable.	Particle tracks counted by optical microscopy. Uranium determined by α activity.	One thousand million years. Accuracy has not yet been tested fully.	Very promising new method. May be sensitive to heating in metamorphic processes.

TABLE F–1 Physical methods of age determination. (cont.)

Method	Kind of Sample Used	Parameter Measured	Range and Accuracy	Remarks
		Chemical methods		
Devitrification of glass	Glass, either natural obsidian or manufactured.	Crystallinity of the glass, especially on the surface.	Useful for relative age determination in past 5 million years for obsidian.	Rate of crystallization dependent upon chemical composition, temperature, and water present in proximity of the glass.
Degradation of amino acids	Shell or bone fragments.	Amino acids and peptide bonding.	Accuracy best in last 1 million years but amino acids are detectable in rocks over 500 million years old.	Involves a chemical process and rate is temperature dependent.
Protein-nitrogen	Shells of mature forms of mollusks.	Organic nitrogen content of water-soluble residues (by micro Kjeldahl method).	Probably limited to Pleistocene and Late Tertiary.	Only relative ages may be determined.

G UNIT MEASUREMENT SYSTEMS

METRIC AND ENGLISH UNITS

LENGTH

1 millimeter (mm) = 0.001 meter (m) = 0.0394 inch
1 centimeter (cm) = 0.01 meter = 0.394 inch
1 kilometer (km) = 1000.0 meters = 0.62 mile
1 meter = 39.4 inches

MASS

1 gram = 0.0022 pound = 0.0352 ounce
1 milligram (mg) = 0.001 gram (g)
1 kilogram (kg) = 1000.0 grams = 2.2 pounds

VOLUME

1 milliliter (ml) = 0.001 liter (1) = (approx) 1 cubic centimeter (cc) =
  0.00106 quart = 0.0339 ounce

APPROXIMATE EQUIVALENTS

1 meter = 39.4 inches
1 mile = 1.6 kilometers
1 kilometer = 0.62 mile
1 inch = 2.5 centimeters
1 avoir. ounce = 28 grams
1 kilogram = 2.2 pounds
1 pound = 454 grams
1 fluid ounce = 30 milliliters
1 quart = 946 milliliters
1 liter = 1.06 quarts

TEMPERATURE SCALES

Fahrenheit (F)	Celsius (C)	
3,632°	2,000°	
1,832°	1,000°	
1,000°	538.6°	
212°	100°	(boiling point of water)
32°	0°	(freezing point of water)

$$\frac{F - 32}{180} = \frac{C}{100}$$

$$C = \frac{5}{9}(F - 32)$$

$$F = \frac{9}{5}C + 32$$

FOSSIL-FORMING ORGANISMS

TABLE H–1 Classification of the more important fossil-forming organisms.

A. Classification scheme and an example.

Kingdom	Animalia
Phylum	Chordata
Class	Mammalia
Order	Primate
Family	Hominidae
Genus	*Homo*
Species	*sapiens*
Individual	John Doe

Most organisms identified by genus and species (in italics).

B. Classification of more important fossil-forming organisms.

Kingdom Protista
  Phylum Schizomycetes—bacteria
  Phylum Sarcodina—foraminifers, radiolarians
Kingdom Animalia
  Phylum Porifera—sponges
  Phylum Coelenterata
    Class Anthozoa—corals
  Phylum Bryozoa
  Phylum Brachiopoda
  Phylum Mollusca
    Class Pelecypoda—clams, oysters
    Class Gastropoda—snails
    Class Cephalopoda
      Subclass Tetrabranchiata
        Order Nautiloidea—nautiloids
        Order Ammonoidea—ammonites, goniatites, ceratites
      Subclass Dibranchiata
        Order Belemnoidea
  Phylum Arthropoda
    Class Arachnida—scorpions, spiders, eurypterids
    Class Trilobita
    Class Insecta
  Phylum Echinodermata
    Subphylum Pelmatozoa (attached forms)

TABLE H–1 (cont.)

Class Cystoidea
Class Blastoidea
Class Crinoidea
Subphylum Eleutherozoa (free moving)
Class Stelleroidea—starfish
Class Echinoidea—echinoids
Phylum Chordata
Subphylum Hemichordata
Class Graptolithina
Order Dendroidea—dendroids
Order Graptoloidea—graptolites
Subphylum Vertebrata
Superclass Pisces—fish
Class Agnatha
Class Placodermi
Class Chondricthyes—cartilaginous (sharks, rays)
Class Osteicthyes—bony
Superclass Tetrapoda
Class Amphibia
Class Reptilia
Class Aves—birds
Class Mammalia
Kingdom Plantae
Phylum Thallophyta—algae, fungi, lichens
Phylum Bryophyta—liverworts and mosses
Phylum Tracheophyta—vascular plants
Subphylum Psilopsida
Subphylum Lycopsida
Subphylum Sphenopsida
Subphylum Pteropsida
Class Filicineae—ferns
Class Gymnospermae—plants with naked seeds
Order Coniferales
Order Ginkgoales
Order Cordaitales
Order Cycadales
Order Cycadofilicales
Class Angiospermae—flowering plants
Subclass Monocotyledonae
Subclass Dicotyledonae

In the following pages, the organisms in the B portion of Table H–1 are described.

Kingdom Protista
Phylum Schizomycetes

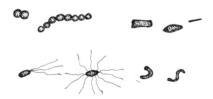

FIGURE H–1   Bacteria are rarely sought as fossils except in ancient Precambrian rocks. Bacteria are shown in Fig. 19–7.

Phylum Sarcodina

Many of the one-celled organisms in this phylum are important rock formers, and many are used widely to date rocks. Because these small organisms are recovered in drill cores, they are especially useful in correlating strata in oil wells.

FIGURE H–2   Foraminifera are mainly less than 1 millimeter (0.04 inch) in diameter, although a few types are much larger. Only a few of the many diverse forms are shown.

FIGURE H–3   Fusulinids are large foraminifers and were abundant in the late Paleozoic. Their size and shape are similar to wheat grains. See Fig. 21–26.

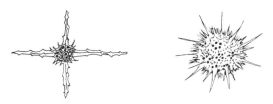

FIGURE H–4   Radiolaria are very small (less than 1 millimeter) and have shells composed of silica. They are seldom used in geologic dating in spite of their great diversity and numbers.

Kingdom Animalia
Phylum Porifera

FIGURE H–5   Sponges are simple aquatic animals that feed by drawing water into their bodies through pores and expelling it through a central opening. The most common fossil remains are the spicules that help to maintain the body shape. See Figs. 21–23 and 21–24.

Phylum Coelenterata

This phylum includes jellyfish, hydra, and corals. Of these, only the corals are important fossils. Coelenterates have a simple sacklike body with a mouth surrounded by tentacles at one end.

Class Anthozoa

Corals are both colonial and solitary. All are marine. Corals are widespread in Paleozoic rocks and are less important in Mesozoic and younger rocks. They are shown in many figures, and the differences between some Paleozoic and Mesozoic corals are shown in Fig. 22–13.

Phylum Bryzoa

FIGURE H-6 Bryozoans are small colonial animals. Their fossil remains are generally lacy or stony fragments with many holes, or the framework that holds such fronds. See Figs. 20–18, 21–19, and 21–20.

Phylum Brachiopoda

Brachiopods are bivalved animals. They were important in the Paleozoic and have declined in numbers since then. They are illustrated in many figures, especially Figs. 20–15, 21–17, 21–24, and 21–25. Figure 22–22 shows the differences in symmetry between clams and brachiopods.

Phylum Mollusca

This diverse phylum includes clams, oysters, snails, and cephalopods, as well as other forms that are rarely fossilized.

Class Pelecypoda

Clam   Pecten   Oyster   Rudistid

FIGURE H-7 Clams and oysters. Clams have two shells that are mirror images. The forms are diverse. Most are bottom dwellers, but *Pecten,* a scallop, is a swimming type. The shells of oysters and rudistids do not have the bilateral symmetry of clams. See Figs. 22–22 and 22–23.

Class Gastropoda

FIGURE H-8 Snails have a single shell, generally without symmetry. They are the only terrestrial molluscs, although most are aquatic. See Figs. 17–5A and B, and 22–14.

Class Cephalopoda
Subclass Tetrabranchiata
Order Nautiloidea
Order Ammonoidea

Cephalopods are both straight and coiled. See Figs. 20–16, 20–19, 21–20, 21–23 and 22–15. They can be distinguished from snails by their symmetry as shown in Fig. 22–14. Nautiloids have simple sutures separating the chambers, and ammonites have more complex sutures. The distinctions among the ammonoids are shown in Fig. 22–16.

Subclass Dibranchiata
Order Belemnoidea

Belemnoids had cigar-shaped shells that were internal. See Figs. 22–19 and 22–20.

Phylum Arthropoda—the joint-footed animals
Class Arachnida
The arachnids include spiders, scorpions, and eurypterids. Eurypterids are shown in Figs. 20–22 and 20–23.
Class Trilobita
Trilobites were important in the Paleozoic, especially in the Cambrian Period. They are illustrated in Figs. 20–13, 20–14, 20–19, and 21–20.
Class Insecta
The insects are the largest group of animals. They are relatively uncommon fossils and so are of little stratigraphic value. See Fig. 17–5D.
Phylum Echinodermata
The echinoderms are all marine and are "five-sided."
Subphylum Pelmatozoa
The pelmatozoans are all attached to the bottom by a flexible stem.
Class Cystoidea (See Fig. 20–19.)
Class Blastoidea (See Figs. 21–21 and 21–22.)
Class Crinoidea (See Figs. 21–20, 21–21, and 21–23.)

Cystoid   Blastoid   Crinoid

FIGURE H–9   Cystoids are less regular than other pelmatozoans. Blastoids are budlike. Crinoids are sea lilies.

Subphylum Eleutherozoa
These echinoids are free moving.
Class Stelleroidea

FIGURE H–10   Starfish. See also Fig. 21–21.

Class Echinoidea

FIGURE H–11   Echinoids differ from starfish in their lack of arms. See Fig. 23–17.

Phylum Chordata
Subphylum Hemichordata
Class Graptolithina
Order Dendroidea
Order Graptoloidea

541

The graptolites are extinct marine, colonial animals. They are too poorly known to classify accurately. The Graptoloidea are important Ordovician and Silurian fossils.

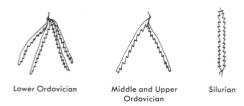

Lower Ordovician   Middle and Upper   Silurian
                              Ordovician

FIGURE H–12 Graptolites. See also Fig. 20–21.

Subphylum Vertebrata
    Superclass Pisces
        Class Agnatha

FIGURE H–13 The agnathids were primitive jawless fish with bony armor. See also Fig. 21–27.

Class Placodermi

FIGURE H–14 The placoderms were early jawed fish. Some were very large. This fish *Dinichthys* reached 9 meters (30 feet) long. See also Fig. 21–28.

Class Chondrichthyes

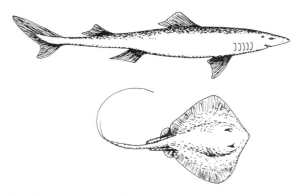

FIGURE H–15 Sharks and rays have been predators since the Devonian.

Class Osteichthyes
This class includes all of the bony fish.
    Subclass Actinopterygii
    This subclass includes most of the modern fish.
    Subclass Sarcopterygii—lobe-finned fish
        Order Dipnoi—lungfish
        Order Crossopterygii—Crossopterigians
        These fish are shown in Figs. 21–35 and 21–36.

Superclass Tetrapoda
Class Amphibia
All amphibians must return to water to lay their eggs. Most live on land most of their lives. Familiar amphibians include toads, frogs, and salamanders. See Figs. 21–34 and 21–38.
Class Reptilia
Order Cotylosauria
The stem reptiles include *Seymouria,* shown in Fig. 21–37.
Orders Pelycosauria and Therapsida
These orders include the mammallike reptiles.
Order Chelonia—turtles. See Fig. 23–21.
Orders Sauropterygia and Ichthyosauria
The plesiosaurs and ichthyosaurs were marine reptiles. See Fig. 22–26.
Order Squamata—lizards and snakes. See Fig. 23–23.
Order Thecodonta
A thecodont is shown in Fig. 22–27.
Order Crocodilia—crocodiles. See Fig. 23–21.
Orders Saurischia and Ornithischia
These orders include the dinosaurs. See Figs. 22–30 through 22–35.
Class Aves—birds. See Figs. 22–28, 22–29, and 23–26.
Class Mammalia
Order Monotremata
The monotremes are egg-laying mammals. The duck-billed platypus and the spiny anteater are examples. Monotremes are unimportant as fossils.
Order Marsupialia

FIGURE H–16   Marsupials carry their young in pouches.

Subclass Eutheria—placental mammals
Order Insectivora

FIGURE H–17   The insectivores include shrews, moles, and hedgehogs. See Fig. 23–23.

Order Chiroptera—bats
Order Creodonta
The creodonts were primitive carnivores from which the present carnivores evolved.

543

Order Carnivora

FIGURE H–18 The carnivores include dogs, cats, bears, weasels, hyenas, and civets. They are swift, smart hunters. See Figs. 23–21, 23–23, 23–25, and 23–26.

Order Perissodactyla

These are the "odd-toed" hoofed animals. They are characterized by having the axis of the foot go through the middle toe and by the loss of the some of the other toes by straightening out of the foot.

FIGURE H–19 The "odd-toed" animals include horses, rhinoceroses, and tapirs. See Figs. 23–20, 23–21, 23–23, 23–24, and 23–25.

Order Artiodactyla

These are the "even-toed" hoofed animals. They are more abundant than the "odd-toed" mammals. See Figs. 23–21, 23–23, 23–24, and 23–25.

FIGURE H–20 The "even-toed" mammals include camels, deer, and cattle.

Order Proboscidea

This order includes the elephants and mammoths that have trunks. See Figs. 17–5E, 23–25, and 23–27.

Order Edentata

FIGURE H–21 Sloths and armadillos are modern edentates.

Order Cetacea—whales, porpoises. See Fig. 23–22.
Order Rodentia—mice, rats. See Figs. 23–21, 23–23, and 23–25.
Order Lagomorpha—hares, rabbits. See Figs. 23–23 and 23–25.
Order Primates
  Suborder Prosimii

FIGURE H–22   The prosimians include lemurs, tarsiers, and lorises.

Suborder Anthropoidea
  Superfamily Ceboidea—New World monkeys
  Superfamily Cercopithecoidea—Old World monkeys
  Superfamily Hominoidea—apes, humans. See Figs. 23–28 through 23–31.

FIGURE H–23   Monkeys, apes, and humans.

Kingdom Plantae
    Phylum Thallophyta

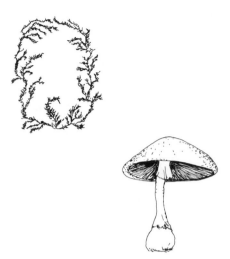

FIGURE H–24 The thallophytes include algae, fungi, and lichens. See also Fig. 19–12.

Phylum Bryophyta

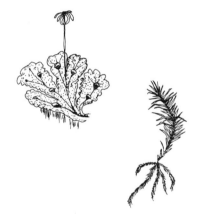

FIGURE H–25 The bryophytes include liverworts and mosses.

Phylum Tracheophyta—vascular plants
    Subphylum Psilopsida
These are the earliest vascular plants, and they did not have true roots.

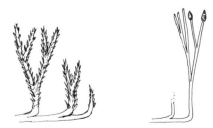

FIGURE H–26 Psilopsids. See Fig. 21–31.

Subphylum Lycopsida

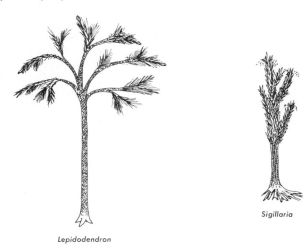

Sigillaria

Lepidodendron

FIGURE H–27 Lycopsids such as *Lepidodendron* and *Sigillaria* were spore-bearing plants with true roots. See Figs. 21–32 and 21–33.

Subphylum Sphenopsida

FIGURE H–28 *Equisetum,* a "scouring" rush, or horsetail. See Fig. 21–31, and *Calamites,* Fig. 21–32.

Subphylum Pteropsida—ferns and seed-bearing plants
Class Filicineae

FIGURE H–29 Ferns. See Fig. 21–33.

Class Gymnospermae—plants with naked seeds
Order Cycadofilicales
Glossopteris, shown in Fig. 14–18A, is typical of this order.
Order Cordaitales
Cordaites was abundant in late Paleozoic.
Order Coniferales
Pines, firs, cedars, junipers, and redwoods are common members of this order. See Fig. 23–18B.
Order Ginkgoales

FIGURE H–30   Ginkgo leaf.

Order Cycadales
Cycads are shown in Fig. 22–28.
Class Angiospermae—flowering plants
Subclass Monocotyledonae
Subclass Dicotyledonae. See Figs. 17–5C and 23–18A and C.
The angiosperms are predominant in modern floras.

GLOSSARY

"When I use a word," Humpty Dumpty said, in rather a scornful tone, "it means just what I choose it to mean—neither more nor less."

"The question is," said Alice, "whether you can make words mean so many different things."

"The question is," said Humpty Dumpty, "which is to be master—that's all."

Lewis Carroll
Through the Looking Glass
Chapter 6

Aa lava   Hawaiian term for blocky, rough-surfaced lava.

Abrasion   Erosion by friction and impact—generally caused by rock particles carried by running water, ice, or wind.

Adamantine luster   Diamondlike luster.

Adaptation   The adjustment of organisms to the environment by improvements such as the development of new features.

Aftershock   An earthquake following a larger earthquake, originating at or near the focus of the larger earthquake.

Agate   Chalcedony with colors in bands, irregular clouds, or mosslike forms.

Agglomerate   Pyroclastic rock composed of rounded pyroclasts such as bombs.

A horizon, soil   Uppermost layer of the soil. Topsoil.

Algae   A group of organisms that includes single-celled plants and seaweeds.

Alkali lake, flat   Lake or level area in desert region where water or soil contains soluble alkalis. The alkalis, potassium, sodium, and calcium salts, come from weathering and are concentrated by evaporation and poor drainage.

Alluvial fan   Low, cone-shaped deposit built by a river where it issues from a mountain into a lowland.

Alpha decay   Radioactive decay in which an alpha particle is emitted.

Alpha particle   Helium nucleus. Consists of two protons and two neutrons, and is positively charged.

Alpine glacier   Mountain glacier. A glacier in a mountain valley.

Amino acid   One of the nitrogenous organic compounds that are the building blocks of proteins.

Amniote egg   An egg in which the embryo is surrounded by membranes, one of which is the amnion, filled with watery amniotic fluid. This egg is an adaptation to protect the developing embryo against drying and physical shock.

Amphibian   A "cold-blooded" vertebrate that lives in water and breathes through gills in the early stages of development and breathes air through lungs in later stages.

Amphibole   A ferromagnesian mineral group. The most important rock-forming amphibole is hornblende.

Amphibolite   A metamorphic rock composed mainly of amphibole, such as hornblende, and plagioclase feldspar.

Anaerobic   An adjective describing an organism that is able to live in the absence of oxygen.

549

Andesite   Fine-grained igneous rock. Intermediate in color and composition between rhyolite and basalt.

Andesite line   A line drawn on a map separating the oceanic areas of basaltic volcanism from the continental and island arc areas of andesitic volcanism.

Angiosperm   A plant with true flowers that bears seeds enclosed in ovaries.

Angle of repose   Maximum angle of slope at which loose material will come to rest.

Angstrom   A unit of length = 0.00000001 centimeter (approximately 0.000000004 inch).

Angular unconformity   An unconformity in which the older strata dip or slope at a different angle (generally steeper) from the younger strata.

Anorthosite   A coarse-grained rock composed almost entirely of plagioclase.

Anthracite   A type of coal formed by mild metamorphism of lower-grade coal.

Anticline   A type of fold in which the strata slope downward in opposite directions away from a central axis, like the roof of a house. The oldest rocks are in the core.

Apatite   A mineral, Number 5 on the hardness scale. $Ca_5(PO_4)_3F$

Aphanitic   A rock texture in which the minerals are too small to see with the naked eye.

Arête   A narrow, jagged ridge typical of glacially eroded mountains.

Arkose   A sandstone containing much feldspar.

Artesian well   A well in which the water is under pressure so that it rises up the well bore. The water may or may not flow out onto the surface.

Arthropod   A member of the phylum Arthropoda. They have segmented bodies and jointed legs.

Asbestos   A name given to fibrous mineral material widely used as insulation. It may be composed of either an amphibole or a variety of serpentine.

Ash grain   Pyroclasts smaller than two millimeters.

Asteroid   One of a group of small planetlike bodies whose general orbit around the sun is between Mars and Jupiter.

Asthenosphere   The layer of the mantle below the low-velocity zone in which the rocks appear to flow slowly. See *lithosphere.*

Atoll   A ringlike coral island or islands with a central lagoon.

Atom   The smallest unit of an element. Composed of protons, electrons, and neutrons.

Atomic fission   A reaction in which an atom is split. The atomic bomb reaction is an example.

Atomic fusion   A reaction in which two or more atoms are combined to form a different element. The hydrogen bomb reaction is an example.

Atomic mass number   The total number of protons and neutrons in the nucleus of an atom.

Atomic mass unit   The unit used by physicists to express the weight of a nucleus. 1.66×10^{-24} gram (based on oxygen).

Atomic number   The number of protons, or positive charges, in the nucleus of an atom.

Augite   The common rock-forming pyroxene mineral. Generally black with two cleavages at right angles. (Ca,Na) $(Mg,Fe^{II}, Fe^{III}, Al)$ $(Si, Al)_2O_6$

Aurora   Luminous bands or areas of colored light seen in the night sky near the poles. Northern or southern lights.

Autoradiograph   A "photograph" of radioactive material taken by placing film directly on a radioactive specimen.

Avalanche   The fall or slide of a snow mass. The term is sometimes used for *landslide.*

Avogadro's number   The number of atoms in a molecular weight of a gas. 6.02×10^{23}

Backarc   Area between a volcanic arc and a continent or another volcanic arc.

Banded iron formation   A sedimentary rock consisting of alternate layers of iron-rich and iron-poor silica.

Barchan dune   A crescent-shaped dune in which the cusps face downwind.

Barrier island   A low, long, sandy, wave-built island between the ocean and the continent.

Barrier reef   A reef between the main ocean and an island or continent.

Basalt   A fine-grained volcanic rock. It is dark in color and composed mainly of plagioclase feldspar and pyroxene. It is one of the most common rock types.

Base level of river   The lowest level to which a river can erode. It is sea level for a river flowing into the ocean. The elevation of the mouth of a river.

Basement rocks   The rocks that underlie the sedimentary cover. The basement is generally a complex of igneous and metamorphic rocks.

Batholith   A very large discordant intrusive rock body, generally composed of granitic rocks. Some batholiths may be of metamorphic origin.

Bauxite   Aluminum ore. Bauxite is the term given to a weathering product composed of several aluminum-rich minerals.

Bedding   The layers or laminae of sedimentary rocks.

Bed load   The material not in solution or suspension that is moved along the bottom of a stream by the moving water.

Belemnoid   A squidlike animal with an internal shell. A member of the phylum Mollusca, which includes clams, snails, oysters, and ammonites.

Bench, surf-cut   A flat surface cut or eroded by wave action.

Beryl   A mineral, an ore of beryllium (a metal). $Be_3Al_2Si_6O_{18}$

Beta decay   Radioactive decay in which a beta particle is released.

Beta particle   In radioactivity, an electron moving at high velocity.

B horizon, soil   Second or middle layer of the soil. See Fig. 4–13.

Biotite   A common mineral, a member of the mica family. Recognized by its shiny black or dark color and perfect cleavage. $K(Mg,Fe)_3(AlSi_3O_{10})(OH)_2$

Bituminous coal   Soft coal.

Blastoid   A stalked echinoderm. Also called "sea buds."

Block fault   A fault that bounds an uplifted or down-dropped segment of the crust.

Block, volcanic   Angular pyroclastic fragment greater than 64 millimeters in size.

Blocky lava   Lava with a blocky surface. See *aa lava*.

Blowout   Hollow caused by wind erosion.

Blueschist   A foliated metamorphic rock containing blue minerals such as glaucophane. Blueschist forms under high pressure and low temperature. See *greenschist*.

Bomb, volcanic   Pyroclastic material ejected from a volcano in a melted or plastic state. Its shape forms either in flight or upon impact.

Bornite   A mineral, an ore of copper. Cu_5FeS_4

Boulder   A rock fragment greater than 256 millimeters in size.

Boulder train   A number of boulders from the same source deposited in a line by a glacier.

Bowen's reaction series   The sequence and reactions that take place in the crystallization of a basalt melt.

Brachiopod   A member of the phylum Brachiopoda. Marine animals with two unequal shells, each of which is bilaterally symmetrical. Also called "lamp shells."

Braided river   A river with many channels that divide and join.

Breccia   A clastic sedimentary rock with angular fragments larger than 2 millimeters in diameter, a sharpstone conglomerate. A volcanic breccia is composed of pyroclastic fragments greater than 2 millimeters in size.

Breeder reactor   An atomic reactor that produces more fissionable material than it uses.

Brown clay   A deep-sea deposit formed by fine material, mainly volcanic dust. (The color is probably caused by oxidation.) Also *red clay*.

Bryozoa   A phylum of colonial animals that build lacy, branching, or other types of structures with many small holes. Also called "moss animals."

Calcic plagioclase   Plagioclase feldspar with a high calcium content. Plagioclase is a continuous series between $CaAl_2Si_2O_8$ and $NaAlSi_3O_8$.

Calcite   A common rock-forming mineral, composes limestone and marble. Recognized by three directions of cleavage not at right angles, and reaction with dilute acid. $CaCO_3$

Caldera   A large volcanic depression containing volcanic vents.

Caliche   A whitish accumulation of calcium carbonate in soil in dry areas.

Capillary action   The rising of water in narrow passages, caused by surface tension. See Fig. 12–2.

Carbon 14   Radioactive isotope of carbon having an atomic mass number of 14. Used in radioactive dating.

Carbonate   A rock or mineral containing carbonate (CO_3). The main minerals are calcite and dolomite, and the main rock types are limestone and dolomite.

Carbonation   A reaction in chemical weathering in which carbon dioxide (CO_2) combines with the rock to form a carbonate (CO_3).

Carbonic acid   An acid important in weathering that is formed by the solution of carbon dioxide in water. H_2CO_3

Carboniferous Period   A part of the geologic time scale. It is so called because many coal deposits were formed at that time. It lasted from about 345 million years ago to about 280 million years ago, the Mississippian and Pennsylvanian Periods combined.

Carbonization   A fossilization process in which the organism becomes a film of carbon, generally on a bedding surface of a sedimentary rock. Many leaf fossils form in this way.

Carnotite   An ore of uranium and vanadium. $K_2(UO_2)_2(VO_4)_2 \cdot 3H_2O$

Cassiterite   An ore of tin. SnO_2

Cast, fossil   A fossil formed by the infilling of a cavity produced by the decay of an organism.

Cataclastic   A metamorphic texture in which at least some minerals have been broken and flattened during dynamic metamorphism.

Catastrophism   A doctrine that held that most geologic changes were caused by sudden violent events.

Cementation   The process whereby binding material is precipitated or deposited between clastic grains, forming a rock.

Cenozoic   An era of the geologic time scale. It lasted from about 65 million years ago to the present.

Cephalopod   A marine invertebrate characterized by tentacles and a straight or coiled calcareous shell with chambers.

Chalcedony   An extremely fine-grained variety of quartz. Agate and chert are made of chalcedony.

Chalcocite   An ore of copper. Cu_2S

Chalcopyrite   An ore of copper. Fool's gold. $CuFeS_2$

Chalk   A soft, usually white variety of limestone. Composed of shells of microorganisms and very fine calcite.

Chemical weathering   The reactions that occur between a rock and its surroundings. The reactions tend to bring the minerals into equilibrium with the environment at or near the surface of the earth. Typically water, oxygen, and carbon dioxide are involved.

Chert   An extremely fine-grained variety of quartz. See *chalcedony.*

Chlorite   A group of platy, mica-like, usually green minerals. Also, a compound containing chlorine.

C horizon, soil   The third layer of a soil, partially decomposed bedrock.

Chromite   An ore of chromium. $(Mg,Fe^{II})(Cr,Al,Fe^{III})_2O_4$

Cinder cone   A cone-shaped small volcano composed of pyroclastic material.

Cinder, volcanic   Pyroclastic material between 4 and 32 millimeters in size.

Cinnabar   An ore of mercury. HgS

Cirque   A glacial erosion feature. The steep amphitheater-like head of a glacial valley.

Clam   A bivalved animal, which has two shells that are mirror images. A member of the phylum Mollusca. Also called pelecypod.

Clastic   A textural term for rocks or sediments composed of fragments broken from some other pre-existing rock.

Clay   A mineral family. Most clays are very fine grained. Much of the material in soil and shales is clay.

Claystone   A rock composed largely of fine clay.

Cleavage   The tendency to split along planes of weakness.

Coal   A sedimentary rock formed of altered plant material.

Cobaltite   An ore of cobalt. $CoAsS$

Cobble   A rock fragment between 64 and 256 millimeters in size.

Col   A mountain pass, especially the high, sharp pass formed by the meeting of two headward-eroding cirques.

Colonial animal   Most of the colonial animals mentioned in this book live in groups or colonies that are organized so that there is a division of labor among the individuals. This term can also mean a type of animal that lives in close association with other similar individuals.

Columnar jointing   Jointing that breaks the rock into typically rough five- or six-sided columns. Common in basalt, and probably caused by shrinking due to cooling.

Compaction   Reduction in volume. Generally caused by pressure in sedimentary rocks.

Composite volcano   A volcano composed of both lava and pyroclastic material.

Conchoidal fracture   The rounded shell-like fracture typical of brittle materials such as glass.

Concordant intrusive rock   An intrusive igneous body whose boundaries are parallel to the enclosing sedimentary rocks.

Cone of depression   The cone-shaped depression of the water table caused by withdrawing water from a well faster than it can move through the rocks to the well.

Conglomerate   A coarse-grained clastic sedimentary rock. The fragments are larger than 2 millimeters.

Contact metamorphic rocks   Thermally metamorphosed rocks near the boundary or contact of an intrusive body such as a batholith.

Continental drift   The movement of continents or plates on the earth's surface.

Continental glacier   A large, thick glacier caused by the buildup of an ice cap on a continent. Greenland has an example.

Continental rise   The transition between the continental slope and the deep ocean floor.

Continental shelf   The shallow ocean area between the shoreline and the continental slope that extends to about a 180-meter (600-foot) depth surrounding a continent. The underlying crust is generally continental.

Continental slope   The relatively steep slope between the continental shelf and the deep ocean basin.

Continuous mineral series   A mineral group such as plagioclase feldspar in which complete or continuous change in composition between end members is possible.

Contour line   A line on a map connecting points of equal elevation.

Convection current   Movement in a fluid caused by differences in density. The density differences are generally caused by heating.

Convergence   In evolution, the formation of similar structures in different types of organisms.

Convergent plate boundary   A place where two plates move toward each other, as in a collision or at a volcanic arc.

Coprolites   Fecal pellets or castings. Fossilized excrement or feces.

Coquina   A type of limestone composed largely of shells and shell fragments.

Coral   Bottom-dwelling marine animals. They are both solitary and colonial, and secrete external calcium carbonate skeletons.

Core of earth   The innermost layer of the earth. Consists of an inner core believed to be solid and an outer liquid core.

Correlation   The determination of the equivalence of two exposures of a rock unit.

Corundum   A mineral. Number 9 on the hardness scale. Al_2O_3

Cosmic ray   Very high-speed nuclei coming from space.

Covalent bonding   A bond formed by the sharing of electrons.

Crater, impact   A depression caused by impact of a falling object. Meteorites cause most large impact craters.

Crater, volcanic   A volcanic depression typically developed at the top of a volcano.

Creep   Slow downslope movement of surface material under the influence of gravity.

Cretaceous   A period in the geologic time scale. It lasted from about 136 million years ago to about 65 million years ago.

Crevasse   A deep crack in the surface of a glacier.

Crinoid   A type of echinoderm attached to the sea bottom by a stalk.

Cross-bedding, cross-strata   Beds or laminations at an oblique angle to the main bedding.

Crust of the earth   The outermost layer of the earth. Above the Moho.

Crustal plate   A relatively thin plate of the outer part of the earth. The depth is much less than the other two dimensions.

Crystal   A geometrical solid bounded by smooth planes that result from an orderly internal atomic arrangement.

Crystalline   A material with a crystal structure; that is, with an orderly internal atomic arrangement.

Crystallization   The processes of crystal formation, of becoming crystalline.

Crystallography   The study of crystals.

Crystal settling   The sinking of early-formed crystals in a melt.

Crystal structure   An orderly internal (atomic) arrangement in a crystal.

Crystal system   All crystals can be classified into one of the six crystal systems.

Curie point   The temperature above which a substance loses its magnetism.

Cycle of erosion   The sequence of changes that occur during erosion of a landscape.

Darcy's law   A formula relating the flow of fluids to the pressure gradient and the properties of the rocks.

Daughter element   An element formed from another element in radioactive decay.

Debris slide   Landslide involving mainly unconsolidated material.

Decay, radioactive   The spontaneous change of one element into another.

Deep-focus earthquake   Earthquake with a focus below 290 kilometers (180 miles).

Deflation   Sorting and removal of material by wind erosion.

Delta   The material deposited by a river at its mouth.

Density   Denseness or compactness. The mass of a material in grams per cubic centimeter.

Deuterium   An isotope of hydrogen.

Devonian   A period in the geologic time scale. It lasted from about 395 to 345 million years ago.

Diamond   A mineral composed of carbon. Number 10 on the hardness scale.

Diatomite   A rock composed largely of the siliceous shells of diatoms.

Differentiation   The process by which different rock types are derived from a single magma.

Dike   A tabular discordant intrusive body.

Dike swarm   A number of dikes, either more or less parallel, or radial from a single source.

Dinosaur   A group of extinct Mesozoic reptiles. They are characterized by the structures of their skulls and hips.

Diorite   A coarse-grained igneous rock intermediate in composition between granite and gabbro.

Dipole magnetic field   The type of magnetic field produced by a bar magnet. See Fig. 14–23.

553

Discharge The quantity of water flowing past a point in a river; volume of water per unit of time.

Disconformity A type of unconformity in which the beds above and below the unconformity are parallel.

Discontinuity, Mohorovičić (M-discontinuity, Moho) The level at which a distinct change in seismic velocity occurs. It separates the crust from the mantle.

Discontinuous mineral series That part of the Bowen reaction series in which abrupt changes in mineral occur.

Discordant intrusive An intrusive rock body that cuts across the bedding of the enclosing sedimentary rocks.

Disseminated magmatic mineral deposit A mineral deposit in which the ore is scattered throughout an igneous body.

Dissolved load The material carried in solution by a river.

Divergent plate boundary A mid-ocean ridge.

Dolomite A mineral $CaMg(CO_3)_2$, or a rock composed largely of dolomite.

Dominant character A character inherited from one parent that appears in the offspring to the exclusion of the corresponding recessive character from the other parent.

Dreikanter A pebble faceted by sandblasting by windblown sand. Same as *ventifact.* Also, specifically having three wind-cut facets.

Drift See *continental drift.*

Drift, glacial A general term for glacier-transported and -deposited material.

Drumlin A smooth hill of glacial drift with its long axis parallel to the direction of glacial movement.

Dune An accumulation of windblown sand.

Dunite A coarse igneous rock composed almost entirely of olivine.

Dynamic metamorphism Metamorphism caused by deformation of the rock.

Earthflow A slow flow of wet overburden.

Earthquake Vibrations of the earth caused by sudden internal movements in the earth.

Echinoderm A member of the phylum Echinodermata. Marine invertebrate animals, most of which have five-sided symmetry, and many of which have spines.

Eclogite A rock that forms under very high pressure, composed of garnet and pyroxene. Its chemical composition is similar to that of basalt.

Electron A fundamental subatomic particle. It has a very small mass and a negative charge.

Element A material that cannot be changed into another element by ordinary chemical means. Just over 100 elements are known. All atoms of an element have the same atomic number (number of protons).

Emergent coastline A coastline formed by either uplift of the continent or lowering of sea level in which areas formerly under water are now dry land.

Enargite An ore of copper. Cu_3AsS_4

End moraine The moraine or deposit formed at the front or snout of a glacier at any time. See also *terminal moraine.*

Eocene A subdivision of the most recent era in the geologic time scale. It lasted from about 53 to 38 million years ago.

Eolian A term applied to wind erosion or deposition. (*aeolian,* obsolete)

Epicenter The point on the earth's surface directly above the focus of an earthquake.

Epiclastic volcanic rock A rock formed of material eroded from pre-existing volcanic rocks.

Epidote A metamorphic mineral. $Ca_2(Al,Fe^{III})_3(SiO_4)_3OH$

Epoch A subdivision of a period in the geologic time scale.

Equilibrium assemblage In metamorphic rocks, a group of minerals that indicates the metamorphic facies.

Era A major subdivision of the geologic time scale.

Erosion The wearing away and removal of material on the earth's surface.

Erratic A boulder transported, commonly far, from its source and deposited on a different bedrock. Also, such a boulder, specifically transported by a glacier.

Esker Long curving ridges of glacially deposited material.

Estuary Tidal inlet along a sea coast. Many form at river mouths.

Eurypterid An extinct arthropod.

Evaporite A sedimentary rock formed by the evaporation of water, leaving the dissolved material behind.

Extrusive igneous rock An igneous rock erupted on the earth's surface. A volcanic rock.

Faceted spur A ridge that has been cut off by glacial, stream, or wave action, or by faulting.

Facies, metamorphic A group or assemblage of minerals that formed in response to a physical environment (metamorphic conditions).

Fault   A break in the earth's crust along which movement has occurred.
Fauna   All of the animals of a given area.
Feldspar   The most abundant mineral family. Consists of plagioclase $CaAl_2Si_2O_8$ and $NaAlSi_3O_8$, and potassium feldspar $KAlSi_3O_8$ (orthoclase and sanidine). Mixtures of potassium and sodium feldspar are termed perthite.
Felsite   A light-colored, fine-grained igneous rock. Most have chemical compositions similar to rhyolite.
Ferromagnesian mineral   A mineral containing iron and magnesium. Most are dark colored.
Fissility   The ability of a rock to split easily along parallel planes.
Fission, atomic   A reaction in which an atom is split. The atomic bomb reaction is an example.
Flint   A variety of chalcedony generally dark in color. See *chalcedony*.
Flood basalt   Basalt that forms thick extensive flows that appear to have flowed rapidly (flooded).
Floodplain   The flat part of a river valley that is subject to floods at times.
Flora   All of the plants of a given area.
Fluorescence   The emission of light when exposed to ultraviolet light.
Fluorite   A mineral. Number 4 on the hardness scale. CaF_2
Fluvial   Pertaining to rivers.
Focus, earthquake   The point where an earthquake originates.
Folding   The bending of strata.
Foliation   A directional property of metamorphic rocks caused by the layering of minerals. Generally caused by crystallization under pressure. Also the flat arrangement of features in any rock type.
Foraminifera   One-celled animals. Their generally microscopic shells are useful in dating some rocks.
Forearc   Area between a submarine trench and a volcanic arc.
Formation   A rock unit that can be mapped.
Fossil   Any evidence of a once-living organism.
Fracture zone   Long linear areas of apparent faulting and irregular topography on the deep-sea floor.
Frost heaving   The lifting of the surface by the expansion of freezing water.
Frost polygons   Polygonal patterns of stones caused by repeated freezing.
Frost wedging   Mechanical weathering caused by expansion of freezing water.
Fusion, atomic   A reaction in which two or more atoms are combined to form a different element. The hydrogen bomb reaction is an example.
Fusulinid   A type of foraminifer shaped like a wheat grain. They are important fossils of the Pennsylvanian and Permian Periods.

Gabbro   A coarse-grained igneous rock composed mainly of plagioclase feldspar and augite.
Galena   An ore of lead. PbS
Garnet   A mineral family.
Genes   The units of inheritance, transmitted in the sex cells of the parents, that control the characters of the offspring.
Genus   A group of species believed to have descended from a single ancestor.
Geode   A hollow space in a rock, generally with a lining of another mineral such as chalcedony and an inner lining of crystals.
Geosyncline   A large elongate downsinking area in which many thousand feet of sedimentary rocks are deposited. It is later deformed to form a mountain range.
Geyser   An intermittent hot spring or fountain from which hot water and/or steam issues.
Glacier   A mass of moving ice.
Glass   A supercooled liquid. A glass does not have an orderly internal structure like a crystal.
***Glossopteris* flora**   A fossil assemblage characterized by the distinctive *Glossopteris* leaves. The flora occurs in rocks in the southern hemisphere and is an important line of evidence for continental drift.
Gneiss   A coarse-grained foliated metamorphic rock. This rock contains feldspar and is generally banded.
Goethite   A mineral. FeO(OH)
Gouge, fault   The fine-grained ground rock in a fault zone.
Grade, metamorphic   A measure of the intensity of metamorphism.
Graded bedding   Bedding that has a gradation from coarse at the bottom to fine nearer the top.
Graded river   An equilibrium river that has just the slope and velocity to carry all the debris carried into it.
Gradient of a river   The slope of a river.
Granite   A coarse-grained rock containing quartz and feldspar. See Fig. 3–43.
Granule   A fragment between 2 and 4 millimeters in size.

Graphite   A soft mineral composed of carbon.

Graptolite   An extinct colonial animal. Found in early Paleozoic rocks.

Gravel   Unconsolidated fragments larger than 2 millimeters.

Gravity anomaly   An area where the attraction of gravity is larger or smaller than expected.

Gravity folding   Folding caused by beds sliding down a slope.

Gravity meter   An instrument for measuring the attraction of gravity.

Graywacke   A type of sandstone, generally poorly sorted with a clay or chloritic matrix.

Greenschist   A schist with much chlorite and other green minerals. Generally formed by the metamorphism of basaltic rocks.

Guyot   A flat-topped seamount. Generally basalt volcanoes bevelled by wave erosion.

Gypsum   A mineral. Number 2 on the hardness scale. $CaSO_4 \cdot 2H_2O$

Half-life   The time required for one-half of the atoms of a radioactive element to decay.

Halite   A mineral. Common salt. $NaCl$

Hanging valley   A tributary valley whose floor is higher than that of the main stream valley at their junction. Commonly caused by glaciation.

Hardness   A mineral's resistance to scratching.

Hematite   An ore of iron. Fe_2O_3

Holocene   The youngest subdivision in the geologic time scale. The *Recent.*

Horn   A sharp, pyramidal peak shaped by headward erosion of glacial cirques.

Hornblende   A common rock-forming mineral. An amphibole. $Ca_2Na(Mg,Fe^{II})_4(Al,Fe^{III},Ti)(Al,Si)_8O_{22}(O,OH)_2$

Hornfels   A metamorphic rock formed by thermal or contact metamorphism of any rock type.

Humus   Dark, decomposed organic material in soil.

Hydration   A process in chemical weathering that involves the addition of water.

Hydrothermal deposit   A mineral deposit emplaced by hot water solutions.

Icecap glacier   See *continental glacier.*

Ichthyosaur   An extinct fishlike reptile.

Igneous rock   A rock that formed by the solidification of a magma. A rock that was once melted or partially melted.

Ilmenite   An ore of titanium. $FeTiO_3$

Index fossil   A fossil that can be used to date the enclosing rocks.

Index mineral   A mineral, the presence of which indicates the metamorphic grade of rocks of a given chemical composition.

Infiltration   The seeping of water into rocks or soil.

Insect   An arthropod of the class Insecta, which has a three-part body, three pairs of legs attached to the middle body part, one pair of antennae, and usually wings.

Intensity, earthquake   A measure of the surface effects of an earthquake.

Interlocking texture   A texture in igneous and metamorphic rocks in which the mineral grains grew together, filling all of the available space.

Intrusive igneous rock   A rock that solidified below the surface.

Invertebrate   An animal without a backbone.

Ion   An atom that has gained or lost electrons and so has an electrical charge.

Ionic bonding   The atoms are attracted to each other by electrical charges caused by concurrent gain and loss of electrons.

Island arc   A curving group of volcanic islands associated with deep-focus earthquakes, deep ocean trenches, and negative gravity anomalies.

Isograd   On a map, a line drawn to show the first appearance of the index minerals that indicate metamorphic grade.

Isostasy   The theory that the crust floats in equilibrium on a substratum in the mantle.

Isotope   A variety of an element having a different number of neutrons in its nucleus.

Jointing   A fracture in a rock along which no appreciable movement has occurred.

Jurassic   A period in the geologic time scale. It lasted from about 195 to 136 million years ago.

Kettle   A depression in glacial deposits probably caused by the melting of a buried block of ice.

Laccolith   A concordant intrusive igneous rock body that domed up the overlying sedimentary rocks to some extent.

Lacustrine   Pertaining to lakes.

Lagoon   A shallow body of water with restricted connection to the ocean.

Landslide   Rapid downslope movement of bedrock and/or overburden. Also used as a general term for all downslope movements.

Lapilli   Pyroclastic fragments in the size range of 2 to 64 millimeters.
Lapilli-tuff   A pyroclastic rock made of consolidated lapilli.
Lateral fault   A fault with horizontal movement. Also called strike-slip fault, and wrench fault.
Lateral moraine   The material deposited by a glacier along its sides.
Laterite   The red-brown soil formed by chemical weathering in the tropics.
Lava   Molten volcanic flows on the earth's surface; also, the rock solidified from it.
Leaching   The dissolving of minerals by water passing through rocks.
Levee   The raised bank along a river.
Lignite   A brown-black coal that forms from the alteration of peat.
Limestone   A sedimentary rock composed largely of calcite.
Limonite   A general term for brown hydrous iron oxides.
Lithification   The process whereby a sediment is turned into a sedimentary rock.
Lithosphere   The crust and upper part of the mantle above the low-velocity zone in which the rocks behave as solids. See *asthenosphere.*
Loess   Wind-deposited silt.
Longitudinal dune   An elongate sand dune parallel to the wind direction. *Seif dune* is a type of longitudinal dune.

Magma   A natural hot melt composed of a mutual solution of rock-forming materials (mainly silicates) and some volatiles (mainly steam) that are held in solution by pressure. It may or may not contain suspended solids.
Magmatic (volcanic) arc   The area of volcanic and intrusive igneous rocks associated with a descending tectonic plate at a subduction zone.
Magnetite   A mineral. An ore of iron. Fe_3O_4
Magnitude, earthquake   A measure of the amount of energy released by an earthquake.
Mammal   A warm-blooded vertebrate that bears live young and produces milk to feed them.
Mantle   The middle layer of the earth, between the crust and the core.
Marble   A metamorphic rock most commonly composed largely of calcite. Formed by the metamorphism of limestone. There is also dolomite marble.
Mare   A large, low, dark, smooth area on the moon. (Plural: *maria*)
Marsupial   One of a subclass of mammals in which the mother's nipples are located inside a pouch, or marsupium. The relatively undeveloped young are carried in the pouch for several months after birth.
Mascon   From "mass concentration." Areas on the moon with higher gravity, believed to be caused by concentrations of mass.
Mass number, mass unit   See *atomic mass number, atomic mass unit.*
Matrix   Fine-grained interstitial material.
Maturity   The stage between youth and old age in the cycle of erosion.
Meanders   A series of more or less regular loop-like bends on a river.
Mechanical weathering   The disintegration of a rock by physical processes.
Mesozoic   An era in the geologic time scale. It lasted from about 225 to 65 million years ago.
Metallic bonding   The atoms are closely packed and the electrons move freely among the atoms.
Metallic luster   The luster characteristic of metals.
Metamorphic facies   A group or assemblage of minerals that formed in response to a physical environment (metamorphic conditions).
Metamorphic grade   A measure of the intensity of metamorphism.
Metamorphic rock   A rock that has undergone change in texture, mineralogy, or composition as a result of conditions below the depth affected by weathering processes.
Metamorphic zone   Essentially equivalent to metamorphic grade. Zones are generally thought of as related to depth at which metamorphism occurs. Generally recognized by index minerals.
Metaquartzite   Quartzite.
Meteor   A shooting star. A meteorite burning due to friction in the atmosphere.
Meteorite   Matter that has fallen on the earth from outer space.
Methane   Marsh gas. CH_4
Mica   A mineral family. Biotite and muscovite are common rock-forming micas.
Microfossil   Small fossils seen only by using a microscope.
Micrometeorite   A tiny meteorite that must be viewed under the microscope. (See *meteorite.*)
Microtektite   A glassy, usually spherical object, less than a millimeter in diameter, which may be related to tektites.
Millibar   A unit of pressure, one one-thousandth of a bar. One thousand dynes per square centimeter.
Mineral   A naturally occurring, crystalline, inorganic substance with a definite small range in chemical composition and physical properties.

Miocene A subdivision of the Cenozoic Era of the geologic time scale. It lasted from about 26 to 7 million years ago.

Mississippian A period in the geologic time scale. It lasted from about 345 to 310 million years ago.

Mohorovičić discontinuity (M-discontinuity, Moho) The level at which a distinct change in seismic velocity occurs. It separates the crust from the mantle.

Mold, fossil The impression left in the surrounding rock by the decay of organic material.

Molybdenite An ore of molybdenum. MoS_2

Monocline A flexure changing the level of any given bed. See Fig. 15–11.

Monotreme One of a primitive subclass of mammals which lay reptilelike eggs, and which secrete milk through a number of nonunited glands rather than through true nipples. Duck-billed platypus and spiny anteaters are monotremes.

Moraine Unsorted material deposited by a glacier. Also, the distinctive landforms that characterize these deposits.

Mountain An area elevated above the surrounding area.

Mountain glacier *Alpine glacier*. A glacier in a mountain valley.

Mountain root The low-density rocks below a mountain range. The thickening of the crust below mountain ranges.

Mud cracks The irregular polygonal cracks that form due to the shrinkage of mud when it dries.

Mudflow The downslope movement of water-"lubricated" debris.

Mudstone A fine-grained sedimentary rock, without fissility, that formed from the consolidation of mud.

Muscovite White mica. Recognized from its color and perfect cleavage. An important rock-forming mineral. $KAl_2(AlSi_3O_{10})(OH)_2$

Mutation A spontaneous, inheritable change in an organism.

Mylonite A fine-grained metamorphic rock formed by the milling of rocks on fault surfaces.

Nappe A large plate of rocks moved from its place of origin by thrust faulting or recumbent folding.

Nautiloid A cephalopod in which the septa separating the chambers are either straight or have simple curves.

Nebula A cloud of gas or dust found in space.

Neutron A fundamental particle with no electrical charge found in the nuclei of atoms.

Nonclastic sedimentary rock A chemically or biologically precipitated sedimentary rock.

Nonconformity A type of unconformity in which younger layered rocks overlie older intrusive or metamorphic rocks.

Normal fault A fault caused by tension or uplift in which the hanging wall is depressed. See Fig. 15–15.

Nuées ardentes A French term meaning "glowing clouds." A volcanic pyroclastic eruption of hot ash and gas.

Obsidian Volcanic glass.

Oceanic crust That part of the earth above the Moho and under the oceans. Composed largely of basaltic rocks.

Old age A stage in the cycle of erosion near its end.

Oligocene A subdivision of the geologic time scale. It lasted from about 37 to 26 million years ago.

Olivine A common rock-forming mineral. $(Mg,Fe)_2SiO_4$

Omphacite A dense, green pyroxene found in the rock eclogite.

Oölite A small spherical or ellipsoidal body, generally of calcite. Also a rock composed of oölites cemented together.

Ooze A fine-grained sediment found on deep ocean floors composed of more than 30 percent organic material.

Opal An amorphous material with composition $SiO_2 \cdot nH_2O$. Loosely considered a mineral by most people.

Ordovician A period in the geologic time scale. It lasted from about 500 to 435 million years ago.

Ore Material that can be mined at a profit.

Ornithiscian A dinosaur with pelvic bones similar in construction to those of birds.

Orogeny The process of making the internal structures of mountains, especially folding, faulting, igneous and metamorphic processes.

Orthoclase A common rock-forming mineral. Potassium feldspar. $KAlSi_3O_8$

Overburden The surface material overlying bedrock.

Overturned fold A fold at least one limb of which has been rotated more than 90 degrees so that the beds are overturned.

Oxbow lake   A crescent-shaped lake formed in an abandoned river curve.
Oxidation   The process of combining with oxygen.
Oxide   A mineral formed by the combination of oxygen and another element.

Pahoehoe lava   Hawaiian term for lava with a smooth ropy surface.
Paleocene   A subdivision of the geologic time scale. It lasted from about 65 to 54 million years ago.
Paleoenvironment   The environment of some time in the past.
Paleogeography   The geography of some time in the past.
Parabolic dune   A sand dune with the general shape of a parabola. The cusps point in the opposite direction to the movement of the wind.
Peat   Dark brown material formed by the partial decomposition of plant material.
Pebble   A fragment between 4 and 64 millimeters in size.
Pedalfer   A type of soil found extensively in eastern United tates. It is enriched in iron and aluminum.
Pediment   An erosion surface cut in bedrock and superficially resembling an alluvial fan.
Pedocal   A type of soil found extensively in western United States. It is enriched in calcium.
Pegmatite   A very-coarse-grained igneous rock. The crystals are larger than 1 centimeter.
Peneplain   The smooth, rolling erosion surface that develops very late in the cycle of erosion.
Pennsylvanian   A period in the geologic time scale. It lasted from about 310 to 280 million years ago.
Percolation   Movement of water through rocks.
Peridotite   A coarse-grained igneous rock composed mainly of olivine and pyroxene with little or no feldspar.
Period, geologic time   A subdivision of the geologic time scale.
Period, ocean wave   The time it takes a wave to travel one wavelength.
Permafrost   Permanently frozen subsoil.
Permeability   The interconnection of pore space allowing fluids to move through a rock.
Permian   A period in the geologic time scale. It lasted from about 280 to 225 million years ago.
Permineralization   A process of fossilization in which material is deposited in pore spaces.
Perthite   A type of feldspar formed by the intergrowth of potassium feldspar and sodic plagioclase.
Petrifaction   The process of changing organic material into stone.
Phaneritic   A textural term meaning that the crystals or grains can be seen with the naked eye.
Phosphate rock   A sedimentary rock containing calcium phosphate.
Photosynthesis   Production of carbohydrates from water, carbon dioxide, and solar energy by chlorophyll in plants.
Phylum   One of the major divisions of the plant and animal kingdoms. A group of closely related classes.
Pillow lava   Lava extruded underwater or in wet mud develops a structure that resembles a pile of pillows.
Piracy, stream   The diversion of the headwaters of a stream by the headward erosion of another stream. Also *stream capture.*
Pitchblende   An ore of uranium.
Placental mammal   A mammal whose young develop within the mother's body. The embryo's nourishment and waste disposal take place through means of an organ called the placenta, which is a highly selective filter between the mother's bloodstream and that of the offspring.
Placer deposit   Accumulation of valuable heavy, durable minerals. Made by agents of transportation and deposition such as running water, wind, or waves.
Placoderm   Extinct primitive jawed fish.
Plagioclase   One of the feldspar minerals. A continuous series from $CaAl_2Si_2O_8$ to $NaAlSi_3O_8$. Recognized by its hardness, two cleavages, and striations on one of the cleavages.
Plastic   Can change shape permanently without rupture.
Plateau   An elevated fairly flat surface.
Pleistocene   A subdivision of the geologic time scale. The last glacial age.
Pleochroic halo   A colored zone or halo surrounding a tiny radioactive mineral grain. The color of the halo changes when the grain is rotated under a polarizing microscope. Also *radiation halo.*
Pliocene   A subdivision of the geologic time scale.
Plucking   Glacial erosion in which blocks of rock are torn out by the action of freezing water.

Plunging fold A fold that is not horizontal. See Fig. 15–9.

Pluton Any body of igneous rock that formed below the surface.

Polar wandering Movement of the geographic poles (as opposed to continental drift).

Polymorph Two or more minerals with the same composition but different crystal form, such as graphite and diamond.

Porosity The open space in a rock.

Porphyritic texture An igneous rock texture in which some of the crystals are much larger than the matrix.

Porphyry A rock with a porphyritic texture.

Pothole A hole worn into bedrock by moving water.

Precambrian The oldest subdivision of the geologic time scale. It includes all the time from the origin of the earth to about 570 million years ago.

Precession of the equinoxes A slow change from year to year of the position of the stars at the time of equinox.

Precipitate Separate from a solution.

Precipitation Rain or snowfall, *OR* being separated from a solution.

Primate One of an order of mammals that includes apes, monkeys, lemurs, and humans.

Progressive metamorphism Metamorphism of increasing grade.

Proton A fundamental subatomic particle with a positive charge. Found in the nucleus of an atom.

Pumice Very light, cellular, glassy lava.

P wave (push wave, primary wave) One of the waves caused by earthquakes. The fastest traveling wave. A compression wave.

Pyrite A common mineral. Fool's gold. FeS_2

Pyroclast A fragment ejected by volcanic action.

Pyroclastic rock A rock formed by volcanic ejecta.

Pyroxene A mineral family. Augite is the common rock-forming member.

Quadrangle Topographic map produced by the United States Geological Survey.

Quartz A common rock-forming mineral. SiO_2 Recognized by its hardness, 7, and lack of cleavage.

Quartz diorite A coarse-grained igneous rock. Composed largely of quartz, plagioclase, and dark minerals.

Quartz (quartzose) sandstone A sandstone composed largely of quartz.

Quartzite A metamorphic rock composed mainly of quartz. It forms from the metamorphism of sandstone or chert.

Quick clay A rock type that becomes fluid if vibrated or disturbed.

Radial stream pattern A stream pattern like the spokes of a wheel.

Radiation, biologic Spreading into new habitats, resulting in the branching into separate lines from a single ancestor.

Radiation halo A colored zone or halo surrounding a tiny radioactive mineral grain. The color of the halo changes when the grain is rotated under a polarizing microscope. Also *pleochroic halo.*

Radioactivity The property of some elements to change spontaneously into other elements by the emission of particles from the nucleus.

Rain shadow Area of diminished rainfall on the lee side of a mountain range.

Reaction series See *Bowen's reaction series.*

Recent Most recent subdivision of the geologic time scale. The *Holocene.*

Recessional moraine A glacial moraine deposited by a retreating glacier, either during a temporary pause in the retreat, or during a minor readvance in a general retreat.

Recessive character A character inherited from one parent that is not developed in offspring when the corresponding dominant character is provided by the other parent.

Red beds Red-colored sedimentary rocks due to the presence of iron oxide.

Red clay A deep-sea deposit formed by fine material, mainly volcanic dust. (The color is probably caused by oxidation.) Also *brown clay.*

Reef A rock built of organic remains, especially of corals.

Regional metamorphic rocks Rocks such as schist and gneiss formed by metamorphism affecting a large region, rather than a restricted location.

Remanent magnetism Permanent magnetism induced in a rock by a magnetic field.

Reptile A vertebrate which has a dry, hardened, usually scaled skin, breathes with lungs, lays shelled eggs, and whose temperature is dependent on environment.

Retrogressive metamorphism Metamorphism at conditions of lower grade than those under which the original rock formed.

Reversed magnetism A magnetic field with polarity reversed from that of the present field.

Reverse fault A fault caused by compression. The hanging wall has been raised. See Fig. 15–14. See also *thrust fault.*

Rhyolite A fine-grained igneous rock similar in composition to granite.
Rift fault See *lateral fault.*
Rift valley The valley formed along a rift fault or lateral fault.
Rille A valley-like feature found on the moon.
Rip current A short-lived, rapid current caused by the return to the ocean of water carried landward by waves.
Ripple-mark An undulating surface on a sediment caused by water movement.
Roches moutonées Rounded knobs of bedrock that have been overrun by a glacier. French for *sheep's backs* which they resemble in the distance.
Rock A natural aggregate of one or more minerals *OR* any essential and appreciable part of the solid portion of the earth (or any other part of the solar system).
Rock cycle The sequence through which rocks may pass when subjected to geological processes.
Rock flour Very finely ground material from glacial abrasion.
Rockslide Rapid downslope movement of bedrock.
Root of mountain The low-density rocks below a mountain range. The thickening of the crust below mountain ranges.
Ropy lava Lava with a ropy surface. See *pahoehoe lava.*
Roscoelite An ore of vanadium. $K(V,Al)_2(OH)_2AlSi_3O_{10}$
Runoff The water that flows on the earth's surface.
Rutile An ore of titanium. TiO_2

Salt dome A structure caused by the upward movement of rock salt through overlying sedimentary rocks.
Sand Fragments that range in size from 1/16 to 2 millimeters.
Sand bar An accumulation of sand built up by waves or running water.
Sandblasting Erosion by impact of windblown sand.
Sandstone A clastic sedimentary rock with fragments between 1/16 and 2 millimeters in size.
Sandstorm Strong winds carrying sand. Most of the sand moves close to ground level.
Sanidine A potassium feldspar found in volcanic rocks. $KAlSi_3O_8$
Saurischian A dinosaur with reptilelike pelvic bones.
Scarp A steep slope, cliff.
Scheelite An ore of tungsten (wolfram). $CaWO_4$
Schist A foliated metamorphic rock generally containing conspicuous mica.
Scoria Volcanic rock with coarse bubble holes.
Seamount A submarine mountain. Generally basalt volcanoes.
Secondary wave (shake wave, S wave) One of the waves generated by an earthquake. This transverse wave is second to arrive.
Sedimentary rock Rocks formed near the earth's surface, generally in layers.
Seif dune An elongate sand dune parallel to the wind direction. A type of *longitudinal dune.*
Seismic sea wave (tsunami) A commonly destructive wave generated by submarine earthquakes. Commonly mistermed *tidal wave.*
Seismograph A device that records earthquakes.
Serpentine A mineral family. $Mg_3Si_2O_5(OH)_4$
Serpentinite A rock composed largely of serpentine.
Shake wave, shear wave, secondary wave See *S wave.*
Shale A fine-grained sedimentary rock composed largely of clay and silt and characterized by its fissility.
Sharpstone conglomerate A conglomerate composed of angular pebbles. (See also *breccia.*)
Shear To force two adjacent parts of a body to slide past one another.
Shelf, continental See *continental shelf.*
Shield A continental area that has been relatively stable for a long time. Shields are mainly composed of ancient Precambrian rocks.
Shield volcano A broad, gently sloping volcano built by flows of very fluid lava.
Silicate mineral A mineral containing SiO_4 tetrahedra.
Silicon-oxygen tetrahedron (SiO_4 tetrahedron) The basic unit of the silicate minerals. Composed of four oxygen atoms in the shape of a tetrahedron surrounding a silicon atom.
Sill A concordant igneous intrusive body.
Sillimanite A mineral found in metamorphic rocks. Al_2SiO_5
Silt A clastic sediment ranging in size between 1/256 and 1/16 millimeter.
Siltstone A clastic sedimentary rock composed largely of silt.
Silurian A period in the geologic time scale that lasted from about 435 to 395 million years ago.
Sink A depression with no outlet.
Slate A fine-grained metamorphic rock characterized by well-developed foliation.

Slipface   The steep lee side of a sand dune. The sand grains move up the wind-struck side and fall off the steeper slipface.

Slope, continental   See *continental slope.*

Slump   A landslide that develops where strong, resistant rocks overlie weak rocks. A block slips down a curving plane.

Smaltite   An ore of cobalt. $(Co,Ni)As_{2-3}$

Soapstone   A metamorphic rock composed of talc.

Sodic plagioclase   The sodium-rich end member of the plagioclase feldspar series. $NaAlSi_3O_8$

Soil   The mixture of altered mineral and organic material at the earth's surface that supports plant life.

Solifluction   Downslope movement of overburden saturated with water.

Sorting   A measure of the range in size of the fragments in a clastic sediment or sedimentary rock. If the fragments that compose a sediment are all similar in size, the rock is well sorted.

Species   A group of similar organisms that can interbreed to produce fertile offspring.

Specific gravity   Ratio of the mass of a material to the mass of an equal volume of water.

Sphalerite   An ore of zinc. $(Zn,Fe)S$

Spit   A sand bar connected on one end to the shore.

Spodumene   An ore of lithium. $LiAlSi_2O_6$

Spring   A place where water issues naturally from the ground.

Spur, truncated   A minor ridge that has been cut off by a moving glacier.

Stalactite   A deposit formed on the roof of a cave.

Stalagmite   A deposit formed on the floor of a cave.

Staurolite   A metamorphic mineral. $FeAl_4Si_2O_{10}(OH)_2$

Stibnite   An antimony ore. Sb_2S_3

Stock   An igneous intrusive body smaller than a batholith.

Stoping   The sinking of blocks of heavier pre-existing rock through molten magma.

Strain   Deformation caused by an applied force.

Stratigraphic correlation   See *correlation.*

Streak   The color of a powdered mineral.

Stress   Force per unit area. Force that causes strain.

Striation, glacial   Scratch on bedrock made by rocks carried by a glacier.

Striations, mineral   Parallel lines that result from intergrown crystals (twins). Commonly seen on one cleavage of plagioclase feldspar.

Strike-slip fault   A fault with horizontal movement. Also called *transcurrent fault.*

Subduction   Descent of a plate, generally of oceanic plate, into the asthenosphere.

Submergent coastline   A coastline along which either the land has sunk or the ocean level has risen.

Subsoil   The middle layer or B horizon of a soil profile.

Supernova   A very luminous exploding star.

Surf   Nearshore wave activity.

Surface wave   One of the waves caused by an earthquake. They move slowly with large amplitude and cause much damage.

Suspended load   The material actually carried by wind or water, as opposed to bed load which is rolled along.

S wave (secondary, shake, or **shear wave)**   One of the waves generated by an earthquake. This transverse wave is second to arrive.

Syncline   A fold in which the beds slope inward toward a common axis. The youngest rocks are in the core.

System, geologic   The rocks deposited during a geologic period.

Tachylite   Volcanic glass of basaltic composition.

Talc   A metamorphic mineral. Number 1 on the hardness scale. $Mg_3(Si_4O_{10})(OH)_2$

Talus   The slope at the foot of a cliff composed of fragments weathered from the cliff.

Tarn   A small lake in a glacially gouged basin in a cirque. Also, any landlocked small lake.

Tectonic   Pertaining to deformation of the earth's crust, especially the rock structure and surface forms that result.

Tektite   A piece of glass commonly of unusual shape. Such pieces are believed to result from either meteorite impact or to have come from space.

Telluride   A mineral containing tellurium.

Tephra   Unconsolidated pyroclastic material.

Terminal moraine   The moraine formed by a glacier at its farthest advance down-valley.

Terrace   A flat-topped surface.

Tertiary   A period in the geologic time scale.

Tetrahedron   A solid body with four faces, each of which is an equilateral triangle.

Texture   The size and arrangement of grains in a rock.

Thecodont   An extinct group of reptiles. They were the ancestors of the dinosaurs, as well as several other groups of reptiles.

Thermal metamorphic rocks   Rocks whose metamorphism was controlled primarily by temperature near the boundary or contact of an intrusive body such as a batholith. Also *contact metamorphic rocks*.

Thrust fault   A reverse fault with a fairly flat fault plane.

Tidal current   A current caused by tidal movement of water.

Tidal wave   See *seismic sea wave*.

Tide   The periodic rise and fall of the ocean, caused by the gravitational attraction of the moon and to a lesser extent the sun.

Till   Material deposited directly by glacial ice.

Tillite   A sedimentary rock made of till (or a till-like sediment).

Titanothere   An extinct group of hoofed Cenozoic animals. Some of the later titanotheres were large.

Topaz   A mineral. Number 8 on the hardness scale. $Al_2(SiO_4)(F,OH)_2$

Topsoil   The upper or A horizon of a soil profile.

Tourmaline   A mineral group.

Transcurrent fault   A fault with horizontal movement. Also called *strike-slip fault*.

Transform fault   A fault associated with mid-ocean ridges in which the actual motion is opposite to the apparent offset of the mid-ocean ridge.

Transverse sand dune   An elongate dune with its long axis at a right angle to the wind direction.

Trench   A very deep, elongate area of the ocean, associated with volcanic island arcs.

Triassic   A period of the geologic time scale. It lasted from about 225 to 190 million years ago.

Trilobite   An extinct arthropod that lived in the Paleozoic and was most abundant in the Early Paleozoic. Characterized by segmented bodies with longitudinal grooves that divide the body into three segments.

Tritium   An isotope of hydrogen with two neutrons and a proton in its nucleus.

Troilite   A mineral. FeS

Truncated spur   A minor ridge that has been cut off by a moving glacier.

Tsunami   See *seismic sea wave*.

Tuff   A pyroclastic volcanic rock with fragments smaller than 2 millimeters.

Turbidity current   A current or flow caused by the movement downslope of water heavy because of its suspended material (muddy water).

Turbulence   A type of flow in which the fluid moves in a disorderly, irregular way.

Type area   The place at which a formation or system is well exposed and is defined.

Unconformity   A surface of erosion and/or nondeposition. A gap in the geologic record.

Undertow   The bottom, nearshore water returning to the ocean after waves moved the water onshore.

Uniformitarianism   The concept that ancient rocks can be understood in terms of the processes presently occurring on the earth.

Uraninite   An ore of uranium.

Van Allen radiation belt   A more or less doughnut-shaped zone of radiation in space high above the equator.

Varved clay   Fine, thinly bedded glacial sediments deposited in still water. Each bed is assumed to be a yearly deposit.

Ventifact   A pebble faceted by sandblasting by windblown sand. See *dreikanter*.

Vertebrate   An animal with a backbone.

Viscosity   The resistance to flowage of a liquid. Caused by internal friction.

Vitreous luster   Glassy luster. The shiny luster of broken glass.

Volatile   A material easily vaporized.

Volcanic (magmatic) arc   The area of volcanic and intrusive igneous rocks associated with a descending tectonic plate at a subduction zone.

Volcanic island arc   A volcanic arc at which the volcanoes form islands.

Volcanic sedimentary rocks (volcanic sandstone and **volcanic conglomerate)**   Sandstone and conglomerate composed of volcanic fragments.

Water table   The surface below which the pores are saturated with water.

Weathering   The alteration that occurs to a rock at the earth's surface.

Weight   The mutual attractive force between a mass and the earth.

Welded tuff   A tuff that was erupted when hot and whose fragments have fused together as a result of the action of heat and the gases present.

Wolframite   An ore of tungsten (wolfram). $(Fe,Mn)WO_4$

Youth   The first stage in the cycle of erosion.

Zircon   A mineral. $ZrSiO_4$

INDEX